Results and Problems in Cell Differentiation

Volume 60

More information about this series at http://www.springer.com/series/400

Rachel K. Miller
Editor

Kidney Development and Disease

Springer

Editor
Rachel K. Miller
Department of Pediatrics
The University of Texas Health Science
Center at Houston
Houston, Texas
USA

ISSN 0080-1844 ISSN 1861-0412 (electronic)
Results and Problems in Cell Differentiation
ISBN 978-3-319-51435-2 ISBN 978-3-319-51436-9 (eBook)
DOI 10.1007/978-3-319-51436-9

Library of Congress Control Number: 2017934474

Printed on acid-free paper

This Springer imprint is published by Springer Nature
The registered company is Springer International Publishing AG
The registered company address is: Gewerbestrasse 11, 6330 Cham, Switzerland

Preface

The study of kidney development and maintenance is critical in understanding the progression of human disease. Disruption of renal development can give rise to congenital pathologies including cystic diseases and pediatric kidney cancer. Maintenance of the kidney is critical to prevent renal damage later in life, which may lead to fibrosis and chronic kidney disease (CKD). Current research aimed at understanding kidney development and homeostasis is enabling the field to progress toward regenerative and stem cell therapies to repair damaged nephrons. Our improved imaging capabilities as well as our growing understanding of the molecular mechanisms behind nephrogenesis are leading to novel strategies to generate nephrons de novo as potential treatment for CKD and other renal pathologies.

Research in animal models has made substantial contributions to our fundamental understanding of nephrogenesis. Models including the fly, fish, and frog have simplified nephron organization and structure, enabling their use to assess the formation and function of the kidney. Although their kidneys are structurally less complex, they perform similar functions as those of mammals, such as maintaining physiological water and electrolyte balance and removing waste and toxins. Additionally, the molecular processes involved in nephrogenesis are highly conserved among these models, allowing researchers to uncover the fundamental molecular events required for nephron formation and function through their study. These models are also being used in studies of human kidney diseases. The first section of this volume focuses on the use of animal models to study kidney development, function, disease, and repair.

During kidney formation, molecular signaling leads to coordination of cell renewal and differentiation, leading to the development of specific cell fates. The intricate signaling networks among different cell types within the kidney give rise to the development of functional nephrons. Researchers have invested significant effort in understanding the molecular signaling involved in kidney development within the nephrogenic zone, collecting duct system, and stromal compartment. By understanding the processes involved in the formation of renal tissues, some of the mechanisms underlying renal disease are becoming apparent. The second section of this volume examines the developmental and disease processes that occur within the nephrogenic zone, collecting duct system, and the stromal components of the kidney.

In order for the developing kidney to function properly, its epithelial tissue must undergo morphogenesis to generate functioning nephric tubes. Recent imaging advances have accelerated our study of morphogenetic processes within the kidney. Additionally, significant advances have been made in understanding developmental anomalies leading to morphogenetic defects within the urinary tract. The most prevalent among these abnormalities are cystic kidney diseases. Maintenance of the kidney is also important in order to prevent the establishment of renal diseases, including fibrosis. The final section of this volume describes morphogenesis and maintenance of the kidney and the diseases resulting from the disruption of these processes.

Our current understanding of developmental and disease processes in the kidney continues to expand with the establishment of novel techniques and approaches along with imaging technologies to study nephrogenesis. Additionally, the classical approaches to explore nephron development and maintenance have provided a strong foundation for our understanding of the molecular processes involved. Current studies in kidney development and disease are taking advantage of our molecular understanding along with novel strategies to generate nephrons de novo for potential therapeutic use in the long run.

Department of Pediatrics
Pediatric Research Center
University of Texas McGovern Medical School
Houston, TX, USA

Rachel K. Miller

Program in Genes and Development
The University of Texas MD Anderson Cancer Center
UTHealth Graduate School of Biomedical Sciences
Houston, TX, USA

Program in Cell and Regulatory Biology
The University of Texas MD Anderson Cancer Center
UTHealth Graduate School of Biomedical Sciences
Houston, TX, USA

Department of Genetics
University of Texas MD Anderson Cancer Center
Houston, TX, USA

Acknowledgments

Significant contributions from the kidney research community went into the preparation of this volume. I am grateful for their valuable insights and thoughtful suggestions in critically reviewing and evaluating each chapter. Additionally, I would like to thank the series editors, Dr. Malgorzata Kloc and Dr. Jacek Kubiak, for the invitation to serve as a volume editor. This work was also supported by a National Institutes of Health NIDDK grant (K01DK092320) and startup funding from the Department of Pediatrics at the University of Texas McGovern Medical School.

Contents

Book Abstract

Kidney form and function have been extensively studied for centuries, leading to discoveries related to their development and disease. Recent scientific advances in molecular and imaging techniques have broadened our understanding of nephron development and maintenance as well as the diseases related to these processes. *Kidney Development and Disease* brings together established and new investigators who are leading authorities in nephrology to describe recent advances in three primary areas of research. The first section describes the use of animal models as powerful tools for the discovery of numerous molecular mechanisms regulating kidney development. The second section focuses on nephric cell renewal and differentiation, which lead to diverse cell fates within the developing kidney, and discusses diseases resulting from the aberrant regulation of the balance between cell fate decisions. The final section concentrates on morphogenesis of the developing kidney and its maintenance after formation as well as the diseases resulting from failures in these processes.

Part I
Model Systems of Kidney Development and Disease

Chapter 1
Drosophila Malpighian Tubules: A Model for Understanding Kidney Development, Function, and Disease

Naveen Kumar Gautam, Puja Verma, and Madhu G. Tapadia

Abstract The Malpighian tubules of insects are structurally simple but functionally important organs, and their integrity is important for the normal excretory process. They are functional analogs of human kidneys which are important physiological organs as they maintain water and electrolyte balance in the blood and simultaneously help the body to get rid of waste and toxic products after various metabolic activities. In addition, it receives early indications of insults to the body such as immune challenge and other toxic components and is essential for sustaining life. According to National Vital Statistics Reports 2016, renal dysfunction has been ranked as the ninth most abundant cause of death in the USA. This chapter provides detailed descriptions of *Drosophila* Malpighian tubule development, physiology, immune function and also presents evidences that Malpighian tubules can be used as a model organ system to address the fundamental questions in developmental and functional disorders of the kidney.

1.1 Introduction

Kidneys play a vital role as it filters blood, maintains homeostasis, and eliminates toxicants from the body. They are made up of millions of nephrons, which are the functional units, and each nephron consists of glomerulus and a tubular duct known as the renal tubule. The glomerulus is a ball of capillaries where the filtering of

N.K. Gautam
Embryotoxicology Laboratory, Environmental Toxicology Division, CSIR-Indian Institute of Toxicology Research (CSIR-IITR), Vishvigyan Bhavan, 31, Mahatma Gandhi Marg, Lucknow 226001, Uttar Pradesh, India
e-mail: naveen.gautam@iitr.res.in; gautam.naveen6@gmail.com

P. Verma • M.G. Tapadia (✉)
Department of Zoology, Cytogenetics Laboratory, Banaras Hindu University, Varanasi 221005, Uttar Pradesh, India
e-mail: pujaaiu@gmail.com; madhu@bhu.ac.in

R.K. Miller (ed.), *Kidney Development and Disease*, Results and Problems in Cell Differentiation 60, DOI 10.1007/978-3-319-51436-9_1

blood takes place, and the selective reabsorption of useful substances and water takes place in the renal tubule. A series of interactions between the epithelial mesonephric duct and the surrounding metanephric mesenchyme leads to the formation of nephrons (Saxena 1987). Sustaining life becomes impossible when filtrations from glomerulus and absorption as well as reabsorption from renal tubules get affected.

Renal dysfunction can result from varieties of pathologies and mortality resulting from renal diseases ranked as the ninth most abundant cause of death in the USA (Heron 2016). Acute kidney injury (AKI) is commonly associated with bacterial infection, sepsis, or ischemia–reperfusion injury (I/R that can transform to chronic renal disease). Inflammation and immune system activation are important causal factors in the development of both acute and chronic renal disease, and their contribution to renal function and disease is now more widely recognized (Imig and Ryan 2013). Several key components of innate immunity have been implicated in the progression of renal disease including the complement system, Toll-like receptors (TLRs), dendritic cells, natural killer (NK) cells, and inflammatory cytokines.

To enhance our understanding of the pathogenesis and treatment of kidney diseases, understanding molecular basis of kidney development and function will be supportive and selection of appropriate model is important for addressing the questions raised. Flies have been used to model several human diseases (Alzheimer disease, Angelman syndrome, Amyotrophic lateral sclerosis, muscular dystrophies, Parkinson disease, cardiovascular diseases, metabolic disorders, etc.) based on the conservation of human disease genes in *Drosophila* genome (http://flystocks.bio.indiana.edu/Browse/HD/HDintro.htm). There are many human renal disease loci that are also conserved between *Drosophila* and human (Dow and Davies 2003; Dow 2009; Dow and Romero 2010). Developmental and functional similarities between vertebrate kidney and *Drosophila* excretory system offers a good model system to study the genetic basis of several human kidney diseases.

The excretion in *Drosophila* is a coordinated effort of two functionally and anatomically distinct organs, the nephrocytes and the Malpighian tubules (MTs). The nephrocytes are present around the heart and esophagus and remove the waste products from the hemolymph by filtration in a manner similar to the endocytic processes in podocytes in human glomerulus (Weavers et al. 2009). On the other hand, MTs have similar function (absorption, reabsorption, and secretion) to tubular part of nephrons (Fig. 1.1B) that generates urine via active transport of ions, water and organic solutes from the hemolymph into the MTs lumen (Maddrell and O'Donnell 1992; Dow and Romero 2010). This chapter is mainly focused on kidney diseases that can be studied with Malpighian tubules.

There are four MTs which make the anterior and the posterior pair, anterior being longer than the posterior. The tubules in each pair joins to form a common ureter, which connects at the junction between the midgut and hindgut. Anterior MTs are structurally and genetically divided into different regions: initial, transition, main, lower segments and ureter. Initial and lower segments are also known as distal and proximal segments respectively. MT segments are enriched with different transporters that express in their specialized cell i.e. principal and stellate cells

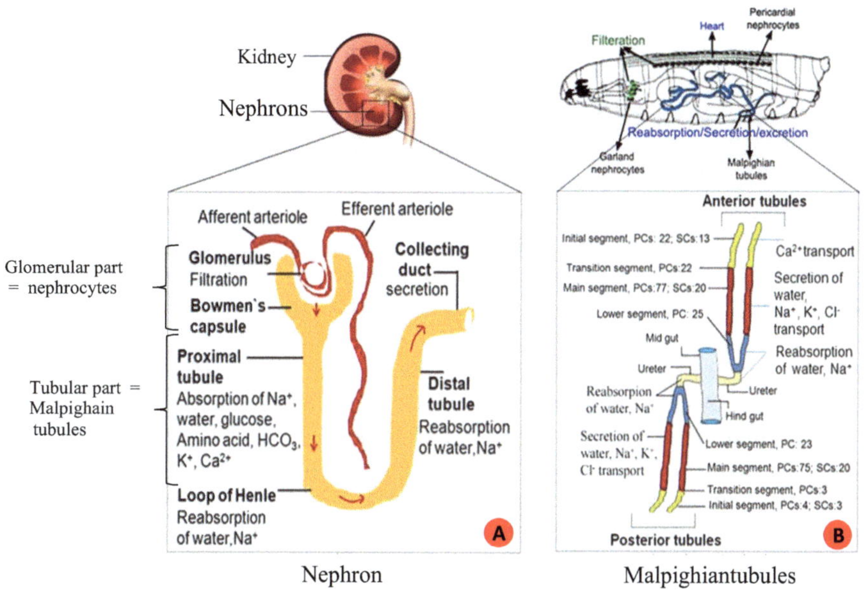

Fig. 1.1 Nephron versus *Drosophila* renal system: *Drosophila* renal system is made up of nephrocytes and Malpighian tubules (MTs). Nephrocytes are present around the heart and esophagus and remove the waste products from the hemolymph by filtration in a manner similar to podocytes in glomerulus in human kidney, while MTs have similar function (absorption, reabsorption, and secretion) to tubular part of nephrons (**B**) that generates urine via active transport of ions, water, and organic solutes from the hemolymph into the Malpighian tubules lumen. Two pairs of Malpighian tubules (MTs) are freely floating in body in anterior and posterior direction and are known as anterior and posterior MTs. Structurally and functionally, MTs are divided into initial, transition, main and lower segments and attached at the junction between midgut and hindgut through ureter. MTs are made up of mainly principal cells (PCs) and stellate cells (SCs), and their number remains fixed in each segment throughout life. Different segments of MTs perform different functions as different segments of nephron do in kidney (**A**). In anterior MTs, initial segment works as storage segment and transports Ca^+ ion into lumen. Main segment is secretory segment enriched with several transporters like Na^+/K^+-ATPase, V-ATPase, Organic solute transporters, Potassium channel, Chloride channel and aquaporins, while lower segment and ureter play reabsorptive role. Posterior MTs have only few cells in initial and transition segments; thus, posterior tubules mainly play secretory function through main segment and reabsorptive function through lower segment and ureter

which facilitate different functions. The initial segment is used as a storage segment, storing calcium, magnesium and other solutes in its lumen (Dube et al. 2000; Wessing et al. 1992). This segment does not play role in fluid secretion (Dow et al. 1994; Rheault and O'Donnell 2001), however, main segment plays major secretary function and produces primary urine. It secretes a near-isosmotic fluid containing ~120 mmol l^{-1} K^+, ~30 mmol l^{-1} Na^+, and ~150 mmol l^{-1} Cl^-. The lower segment secretes Ca^{2+} into the lumen, acidifies the luminal fluids, and reabsorbs K^+, Cl^-, and water to make concentrated urine (O'Donnell and Maddrell 1995). Posterior MTs are relatively smaller having only few cells in initial and transition segments, so there is lack of storage segment (Sözen et al. 1997). These tubules play major role in secretion and reabsorption through main segment and lower segment, respectively (Fig. 1.1B).

Similar to MTs, nephron is a tubular structure composed of single layer of epithelial cells lining a series of segments: proximal tubule, thin descending tubule of the loop of Henle, thin and thick ascending tubule of the loop of Henle, the distal tubule and the collecting duct. Each segment of the nephron has specific function in terms of absorption, reabsorption and secretion, and the specificity of function is dependent upon the transport proteins that reside in the apical (lumen facing) and the basolateral membranes (serosal facing). The proximal segment (tubule) of nephron reabsorbs 2/3 of filtered Na^+ and water, all the glucose and amino acids, fraction of bicarbonate, potassium, phosphate, and calcium (Curthoys and Moe 2014). Loop of Henle again reabsorbs water and allows urine to get concentrated. Na^+, K^+, and Cl^- ions are also actively reabsorbed through this region (Palmer and Schnermann 2015). Distal segment (tubule) is relatively impermeable to water, but in the presence of antidiuretic hormone (ADH), its permeability to water increases making urine concentrated (Sands and Layton 2009). Collecting duct is the final segment of the renal tubule. It has two types of cells, principal and intercalated cells. Principal cells reabsorb sodium as well as water and secrete potassium, while α-intercalated cells secrete hydrogen ions and reabsorb K^+ through H^+/K^+-ATPase. Similarly, intercalated beta cells secrete HCO_3^- by having a basolateral H^+-ATPase and apical pendrin channel protiens (Roy et al. 2015) (Fig. 1.1A).

1.2 Development of Malpighian Tubules

Similar to nephrons in human kidney, MTs develop from ectodermal and mesodermal cells during embryonic development. Ectodermal tubule buds evert from the gut and increase in size by cell division and later mesodermal cells integrate into ectodermal tubular bud that undergoes cell rearrangements and grow to make physiologically active tubules (Skaer 1993; Denholm et al. 2003). Entire process of MTs development can be divided into different developmental processes: (1) primordium specification, (2) bud evagination, (3) bud extension, and (4) tubule elongation (Fig. 1.1).

1. *Primordium specification*: Initially, primordium is established by overlapping expression of several genes like *tailless* (*tll*), *huckebein* (*hkb*), *forkhead* (*fkh*), and *wingless* (*wg*) at the posterior region of the embryo from stages 5 to 10 of embryogenesis (Fig. 1.2.II A). Later, tubule buds evaginate from this proctodeum (Weigel et al. 1989; Gaul and Weigel 1990; Skaer 1993; Harbeck and Lengyel 1995; Wu and Langyel 1998).
2. *Bud evagination*: Two pairs of MTs evaginate as buds at the junction of hindgut and posterior midgut under the influence of *krüppel (Kr)* and *wg* at stage 11 of embryonic development (Harbecke and Janning 1989;Skaer 1993). Hedgehog

(Hh) is also required for the completion of bud evagination (Fig. 1.2.II B) (Hoch and Pankratz 1996).

3. *Bud extension*: After bud evagination, *kr* regulates the expression of transcription factor *cut* (*ct*) in a ring of cells in bud, as a result of cell shape changes. At the same time, *kr*, *cut*, and *wg* play an important role in directing the bud to extend cylindrically and become narrower both proximally and distally resulting in a crescent-shaped morphology during stages 12–13 (Fig. 1.2.II C).
4. *Tubule elongation*: Followed by bud extension, tip cell is specified within the tubule primordial via notch signaling which then starts expressing *kr* and epidermal growth factor (*Egfr*) which stimulates mitosis in neighboring cells and growth of tubules by addition of new cells. At the same time, *wg* is also required for cell division and morphogenesis of tubules (Skaer and Martinez 1992; Hoch et al. 1994, Harbecke and Lengyel 1995; Singh et al. 2007). Tip cell works as dynamic cellular anchor in the morphogenesis of looped tubules (Weavers and Skaer 2013).

At the end of stage 13 of embryonic development, there is no further cell division, and the circumference of each primitive tubule comprises about eight cells (Janning et al. 1986; Skaer and Martinez 1992). Once the cells stop dividing, the tubular circumference is reduced to only two cells and the tubule enters the vicinity of the caudal mesoderm and interacts with a subpopulation of the caudal mesodermal cells (Fig. 1.2.II D). At this stage, the surrounding mesodermal cells undergo a mesenchymal-to-epithelial transition, and by the process of cell rearrangements, these mesodermal cells get recruited into the ectodermal bud cells at regular intervals which later form the SCs (Fig. 1.2.II E) and the ectodermal cells from the PCs (Denholm et al. 2003). Later, by tubule elongation the MTs are concomitantly extended further during stages 14 and 15 of embryonic development (Skaer 1993). The PCs and SCs are also known as primary and secondary cells or type I and type II cells, respectively (Fig. 1.1). Thus, *Drosophila* MTs develop from two sources: hindgut primordium (ectodermal epithelia) and visceral mesoderm (Denholm et al. 2003; Jung et al. 2005) similar to human kidney.

Similar to MTs, in human kidneys, the metanephric mesenchyme cells induce branching of mesonephric duct to form the ureteric bud initially, and then they respond to signals derived from them (Saxena 1987). Resulting mesenchymal cells are recruited to the buds, where they undergo a mesenchymal-to-epithelial transition as they condense to form nephrons (Cho and Dressler 2003). Failure in the process of recruitment leads to polycystic kidneys and, in the severest cases, to renal agenesis (Khoshnoodi and Tryggvason 2001; Knier et al. 2016). Therefore, MT offers opportunity to understand the molecular events underlying recruitment of mesenchymal cells and transition to epithelial cells to make mature functional tubular nephrons.

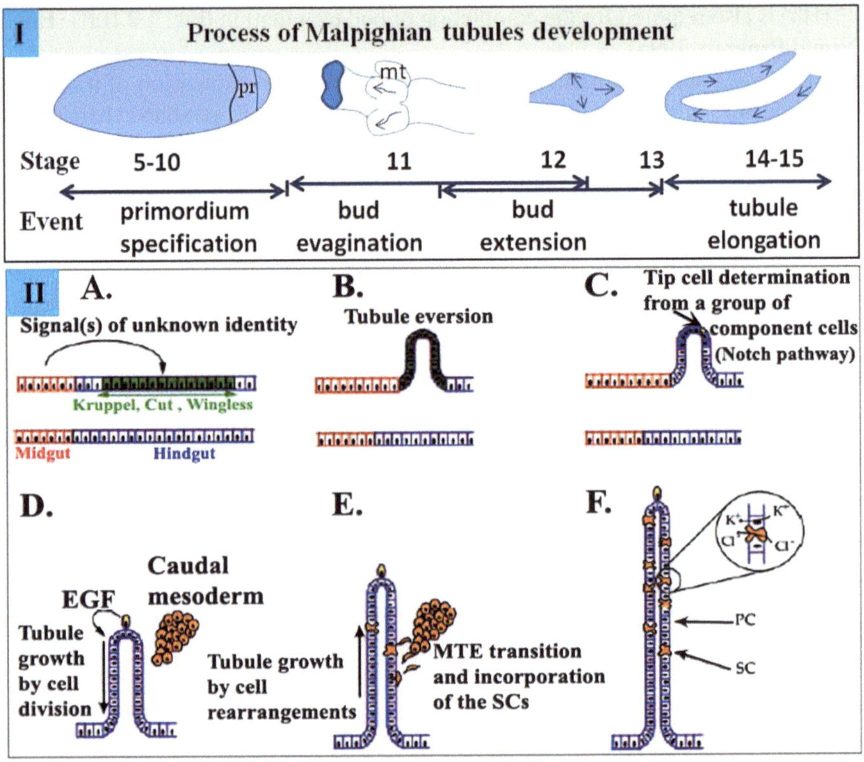

Fig. 1.2 Process of Malpighian tubule development: Malpighian tubules development begins at embryonic stages through different developmental processes: primordium specification, bud evagination, bud extension, and bud elongation. At early stages of embryonic development when midgut and hindgut anlagen interacts, signals of unknown identity activate the expression of *kruppel* (*kr*) in the hindgut to give rise to future principal cells in MTs. *kr* triggers the expression of *cut* in the hindgut. Transcription factors, *kr* and *cut*, along with *wingless* play role in tubule eversion. Furthermore, a new cell lineage called tip cell is produced from these cell population by the activation of *Notch*. The specialized tip cell produces *epidermal growth factor* (*Egf*) and stimulates mitosis in neighboring cells at the distal ends and tubule grows by the addition of new cells. During development, tubule comes in the vicinity of the caudal mesoderm. At this stage, the Malpighian tubule cells stop dividing, and tubule growth proceeds through extensive cell rearrangement and intercalation. The caudal mesoderm cells interact with the tubule; individual cells undergo a mesenchymal-to-epithelial transition (MET) and progressively incorporate into tubule epithelium as stellate cells. By the end of embryogenesis, in mature tubule principal cells and stellate cells specialized for different functions. Image modified from Jung et al. (2005) and Liu et al. (1999)

1.3 Cells in Malpighian Tubules

The mature tubules are made up of an average of 484 principal cells (PCs) and 110 stellate cells (SCs) divided into four tubules (Janning et al. 1986; Sözen et al. 1997). The number of PCs and SCs is decided during embryonic development and remains fixed throughout life in different segments of tubules. In anterior tubules, initial segment contains 22 PCs and 13 SCs, transition segment contains 22 PCs, main segment contains 77 PCs and 20 SCs, and lower segment contains 25 PCs, while in posterior tubules, initial segment contains 4 PCs and 3 SCs, transition segment contains 3 PCs, main segment contains 75 PCs and 20 SCs, and lower segment contains 23 PCs. The total number of PCs and SCs differs in anterior and posterior tubules. In anterior tubules, the number of PCs and SCs is 145 and 33, respectively, while in posterior, the PCs and SCs number are comparatively less which are 111 and 22 (Sözen et al. 1997).

The major physiological activities of Malpighian tubules are carried out by these two cell types. PCs regulate the ionic balance through channel proteins, (Na^+/K^+)-ATPase (Lebovitz et al. 1989), Vacuolar-type H^+-ATPase (V-ATPase) (Day et al. 2008), potassium pump (Wu et al. 2015), organic cations, and anion channels (Rheault and O'Donnell 2004). Inwardly rectifying potassium channels, Irk1 and Irk2, are responsible for transepithelial K^+ flux in the PCs (Wu et al. 2015). Several transcription factors also control the activities of PCs (Wang et al. 2004; Dow 2009) (Fig. 1.3).

SCs maintain fluid secretion through *Drosophila* aquaporin (Drip) and chloride ion gradient through chloride channel (Pannabecke'r 1995; Dow and Davies 2003; Dow 2009, Kaufmann et al. 2005; Gautam and Tapadia 2010). Transcription factor, teashirt, expresses in stellate cells and regulates the expression of water transport channel, *Drosophila* aquaporin (Drip), chloride conductance channel, CLC-a, and leukokinin receptor (LR) that are required for primary urine production (Denholm et al. 2013) (Fig. 1.3).

Capability peptides and diuretic hormone stimulate urine production through cGMP and cAMP pathways in PCs (Coast et al. 2001; Cabrero et al. 2002; Kean et al. 2002; Johnson et al. 2005). Diuretic hormone 44 (DH44) acts through DH receptor in PCs, and kinin neuropeptide acts through Dromekinin (DK) receptor in SCs and plays essential role in the regulation of desiccation and starvation tolerance (Cannell et al. 2016).

Apart from these two cell types, some other cells are also characterized in the Malpighian tubules, such as the "tiny" cells in the proximal part of Malpighian tubules (including lower tubules and ureters) which could be homologues of myoendocrine cells of the ant *Formica* (Garayoa et al. 1994) that collect urine in the renal duct and secrete neurohormones in the hemolymph (Sözen et al. 1997).

The tiny cells in the region of lower tubules and ureters function as multipotent renal stem cells, and an autocrine JAK-STAT signaling regulates the stem cells' self-renewal ability (Singh et al. 2007; Zeng et al. 2010). Sav/Scrib and Ras pathways regulate transformation of these renal stem cells into cancer stem cells in Malpighian tubules (Zeng et al. 2010). EGFR/MAPK signaling regulates the proliferation of renal stem cells (Li et al. 2015). Differential Notch activity is required for maintaining homeostasis of Malpighian tubules in adult *Drosophila* (Li et al. 2014).

Similarly, in human kidney, nephrons perform their functions through different transporters like sodium and hydrogen exchanger 3 (NHE3), Na^+/K^+-ATPase, Na^+/

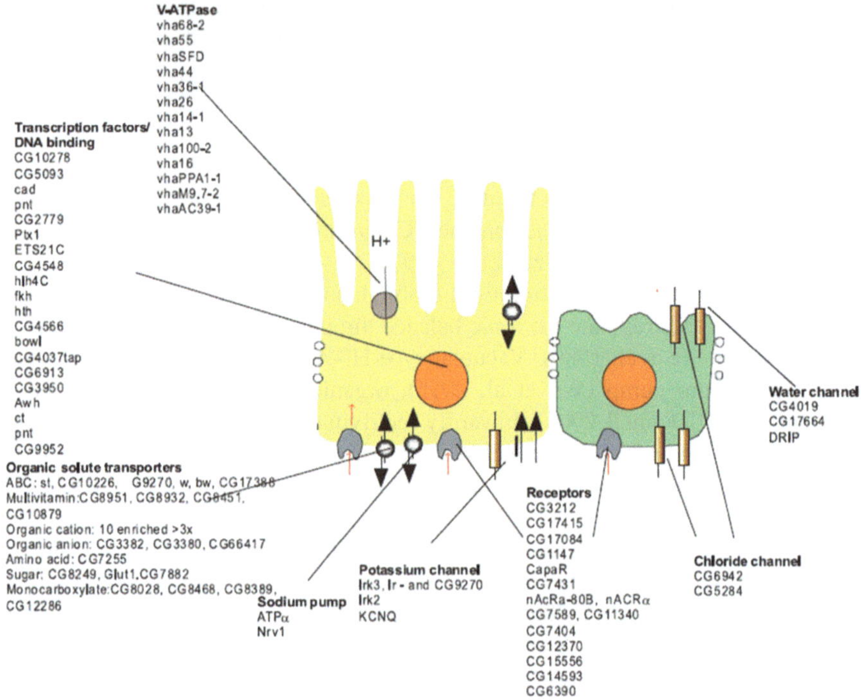

Fig. 1.3 Expression of different classes of genes in Malpighian tubules: Principal cells and stellate cells are specialized for different processes through different functional classes of genes that express in adult MTs (Wang et al. 2004; Dow 2009)

HCO_3 (Curthoys and Moe 2014), V-ATPase (Nakhoul and Hamm 2002), membrane water channel proteins, aquaporins (King et al. 2001; Nielsen et al. 1993; Agarwal and Gupta 2008), glucose transporter (Aronson and Sacktor 1975), and organic ion transporters (Breljak et al. 2016). Reabsorption of Na^+ occurs in all segments; proximal segments mainly reabsorb sodium through sodium and hydrogen exchanger 3 (NHE3), Na^+/K^+-ATPase, and Na^+/HCO_3. Loop of Henle mainly reabsorbs Na^+ through Na/K/2Cl cotransporter and hydrogen exchanger 3 (NHE3). Reabsorption of Na^+ in distal convoluted tubule (DCT) occurs through $NaCl^-$ cotransporter NCC, Na-H exchange process mediated by NHE2 (Wang et al. 2001), and epithelial Na^+ channel (Costanzo 1984) (Fig. 1.1a). Water concentration is regulated by aquaporin that facilitates water transport across cell membranes. At least seven aquaporins (AQP) express in the kidney (Agarwal and Gupta 2008). Therefore, Malpighian tubules provide the opportunity to understand the genetic and molecular basis of function of different transporters in different segments.

1.4 Ecdysone Signaling and Malpighian Tubules

Steroid hormone ecdysone is critical for the development of *Drosophila* at all stages starting from embryogenesis to larval molting, pupation, metamorphosis and in adults. Heterodimer of ecdysone receptor (EcR) and ultraspiracle (USP) is essential for ecdysone-induced signaling and binds to the ecdysone receptor (Yao et al. 1992; Thomas et al. 1993).

USP is the *Drosophila* homologue of vertebrate Retinoid X receptor (RXR) (Oro et al. 1990), while EcR gene encodes three functional isoforms, EcR-A, EcR-B1 and EcR-B2, which have tissue-specific roles (Koelle et al. 1991; Talbot et al. 1993). EcR-A predominantly expresses in imaginal discs that develop into adult specific structures while EcR-B1 predominantly expresses in those larval structures which are destined to die (Schubiger et al. 1998; Cherbas et al. 2003; Davies et al. 2005).

MTs are the specialized tissues as they do not undergo ecdysone-induced histolysis and proliferation during larval to adult transition as other larval tissues do. Therefore, role of ecdysone signaling during MT development has been investigated. All three isoforms, EcR-A, B1, and B2, express in tubules at third instar larval stage; however, EcR-B2 starts expressing in MTs from early embryonic developmental stages and plays important role in its development (Gautam et al. 2015). Ecdysone signaling regulates proper intercalation of mesodermal SCs into ectodermal tubule bud during Malpighian tubules' development (Gautam and Tapadia 2010), which is required for the physiological maturation of tubules (Denholm et al. 2003). Epithelial morphogenesis is dependent on proper cytoskeletal organization which is regulated by ecdysone. Ecdysone signaling plays specific roles in different cell types of MTs, as disruption of ecdysone signaling in SCs leads to disorganization of SCs but not the PCs (Gautam and Tapadia 2010), while disruption of ecdysone signaling in PCs leads to disorganization of both SCs and PCs (Gautam et al. 2015).

Ecdysone also plays important role in fluid secretion of tubules via regulating the transporter function of SCs and PCs. Ecdysone regulates the expression of Drip in SCs and also maintains the ionic balance by regulating the expression of (Na^+/K^+)-ATPase in PCs (Gautam et al. 2015). Crystals of calcium oxalate, Ca-PO4, xanthine, and uric acid are deposited in MTs as excretory products and removed from the body (Dow and Davies 2003; Hirata et al. 2012; Chi et al. 2015; Browne and O'Donnell 2016). Functional disability of SCs and PCs through disruption of ecdysone signaling affects the deposition of uric acid crystal in MTs (Gautam and Tapadia 2010; Gautam et al. 2015). A recent study shows that expression of a stretch of 127 glutamine repeat leads to degeneration of MTs, while in normal condition these tissues do not undergo ecdysone-induced histolysis (Yadav and Tapadia 2016). Therefore, MTs offer a model system to study the role of steroid hormone in the development and physiology of human renal system.

1.5 Malpighian Tubules as an Immune Tissue

Innate immune response is an evolutionarily conserved phenomenon in mammals and *Drosophila*. Similar to humans, *Drosophila* protects against microbes and parasites via epithelial barriers that exist in trachea, gut, genital tract and MTs. Apart from excretion, the excretory epithelium of MTs has an important role in immune defense (McGettigan et al. 2005; Stergiopoulos et al. 2009; Terhzaz et al. 2010; Overend et al. 2012; Verma and Tapadia 2012, 2014, 2015), and by virtue of their position in the hemocoel, they are the immediate sensors of toxic insults. *Drosophila* relies only on innate immune response, which is manifested in different ways: humoral response, cellular response, melanization, and coagulation reaction. In humoral response, the antimicrobial peptides (AMPs) are generated by the immune deficiency (IMD) and Toll pathways (Lemaitre et al. 1995; Lemaitre and Hoffmann 2007) activated by Gram-negative bacterial infection and Gram-positive bacterial and fungal infections, respectively (Silverman and Maniatis 2001; Tanji and Ip 2005). The AMPs are effector molecules of humoral response which act against bacteria, fungi, parasites, and, in some cases, enveloped viruses. AMPs are found in evolutionarily diverse organisms ranging from prokaryotes to invertebrates and vertebrates, and to plants (Tossi et al. 2002; Ganz 2003; Bulet et al. 2004). In *Drosophila*, there are 34 AMPs grouped into eight different families including lysozyme (Hultmark 2003; Imler and Bulet 2005; Lemaitre and Hoffmann 2007). MTs can sense threat and mount effective killing response by secreting AMPs of IMD and Toll pathways, which include diptericin, attacin, cecropin, metchnikowin, defensin, and drosomycin (Tzou et al. 2000; McGettigan et al. 2005). The expression of IMD-mediated AMPs in MTs is constitutive as well as inducible (Verma and Tapadia 2012) and they express all the components of the innate immune pathways like PGRP-LC, IMD, Dfadd, DTak1, Dredd, Diap2, Ird5, Kenny, and Relish (Chintapalli et al. 2007; Robinson et al. 2013).

A known activator of immune response, nitric oxide (NO), is also involved in tubule systemic immune response as targeted overexpression of dNOS, an enzyme in MTs, confers increased survival of adult flies upon septic infection. NO-stimulated cGMP via cGMP-dependent kinases (DG1 and DG2) activates IMD signaling in tubule (McGettigan et al. 2005; Davies and Dow 2009; Davies et al. 2012).

The kidney is the target of immune-mediated injury in a variety of disease situations with glomerulus being particularly vulnerable to damage by the immune system (Mathieson 2003). Systemic inflammation is known to target kidney tubular epithelial cells (TECs) leading to acute kidney injury and chemokine production (Van Kooten et al. 2000). TECs contribute to the inflammatory reaction through generation of chemotactic cytokines such as TNF-α, IL-8, IL-6, IL-1β, TGF-β, MCP-1, RANTES, ENA-78 and also activates T lymphocytes (Van Kooten et al. 2000). Different pathogen-associated molecular patterns, PAMPs, including lipopolysaccharides (LPS) and other microbial products such as lipoteichoic acid and porins, have been reported to directly interact with resident kidney cells. TLRs expressing on TECs are upregulated in response to endogenous ligands and play a critical role in detecting exogenous microbial products (Frei et al. 2010). Harmful effects of LPS after binding to TLR-4 are mainly due to its influence on tubular cells (Frei et al. 2010). TECs directly interact with Toll-like receptor and induce release of several cytokines and inflammatory mediators, including IL-6, IL-10, IL-18, IP-10, KC, MCP-1, and tumor necrosis factor-α (TNF-α), via the TLR adapter protein MyD88 (Myeloid Differentiation Factor 88) (Ho et al. 2008; Allam et al. 2012). Inflammatory mediators derived from TECs may contribute to the development and progression of renal injury and may mediate distant organ dysfunction (Cantaluppi et al. 2015). The podocytes or glomerular epithelial cells that act as active participants of immune response are important in human glomerular physiology and disease (Pavenstadt et al. 2003). TECs operate as professional immune cells, modulating both innate and adaptive immune responses and play a central role in the local inflammatory response via cytokine production (Zhang et al. 2008).

Inflammation and immune system activation are important causal factors in the development of both acute and chronic renal disease; thus, the conservation of the immune system in *Drosophila* and human has made it possible to study immunity and inflammation-related renal dysfunction in humans.

1.6 Malpighian Tubules as Model for Kidney Development and Diseases

MTs resemble the human kidney in terms of their cellular structure and functional activities. They share common principle during development, for example, recruitment of cells, divisions of cells in a regulated fashion, rearrangement of cells to form elongated tubule buds, and intercalation of two distinct cell populations ectodermal and mesodermal cells to make mature functional tubules (Woolf and Bard 2003). Several signaling pathways and signaling molecules involved in MT development are evolutionarily conserved and their role in mammalian kidney is proven. The branched nature of MTs provides a very good opportunity to explore the inborn abnormalities during nephron development.

In early stages of MT development, interaction between the midgut and hindgut anlagen redefines the expression of the transcription factor, *kruppel* (*Kr*), to a small number of precursor cells. *Kr* is homologue of human *kruppel-like family* of transcription factors (KLFs) which further controls the expression of another transcription factor, *Cut*, homologue of human *Clox* (*Cut-like homeo box*). Combined expression of *Kr* and *Cut* signals for the evagination of MTs from the junction of midgut and hindgut (Denholm et al. 2003). Signaling pathways that are conserved in humans also play very important role during MT development, such as Wnt and tll (tlxin human) in primordium specification, Wingless (Wg) in cell allocation, Hedgehog (Hh) in bud evagination, Transforming growth factor beta (TGF-β) in branching of the tubules, Notch defining the single tip cell, and epidermal growth factor (Egf) in elongation and cell division in the tubule. EcR-USP heterodimer (homologue of human RXR) functions in cell intercalation and cell arrangement in tubules (Liu et al. 1999; Denholm et al. 2003; Jung et al. 2005; Gautam and Tapadia 2010; Gautam et al. 2015).

Similar to Malpighian tubules, in human kidneys, the metanephric mesenchyme cells induce branching of mesonephric duct to form the ureteric bud. These buds induce signals (Saxena 1987) resulting in recruitment of mesenchymal cells to the buds, where they undergo a mesenchymal-to-epithelial transition as they condense to form nephrons (Cho and Dressler 2003). *Nephrin*, human homologue of *Drosophila hibris*, plays an essential role in recruitment and fusion of cells (Artero et al. 2001; Dworak et al. 2001; Dworak and Sink 2002). Failure in the process of recruitment leads to polycystic kidneys and, in the severest cases, to renal agenesis (Khoshnoodi and Tryggvason 2001; Knier et al. 2016). *Nephrin* also expresses in podocytes of glomerulus in human kidney and nephrocytes in *Drosophila* and its mutation leads to impaired glomerular and nephrocyte function (Weavers et al. 2009). The interacting partner of *nephrin* is *CMS/CD2AP*, homologue of *Drosophila CG11316* (www.flybase.bio.indiana.edu). Nephrin/CD2AP complex also binds to actin and to p130^{cas} (Kirsch et al. 1999), *Drosophila* homologue *CG1212*.

Several other genes like *rolling pebbles (rols)*, *myoblast city (mbc)*, and *Rac* are involved in cell intercalation and arrangement in tubules (Erickson et al. 1997; Nolan et al. 1998; Galletta et al. 1999; Pütz et al. 2005); *faint sausage (fas)* in cell adhesion during tubule elongation (Lekven et al. 1998); *walrus (wal)* in tubule bud evagination (Liu et al. 1999); and *star, rhomboid, spitz, flb, pointed*, and *sevenup* in tubule elongation (Skaer 1989; Bumann and Skaer 1993; Kerber et al. 1998). Genes like *barr, thr, pimples* (*Pim*), and *string* (*stg*) are essential for providing particular number and size of cells and arrangement in tubules. *Short gastrulation (sog)* does not express in Malpighian tubules but is essential during development as it antagonizes the activity of *Bone Morphogenetic Protein 4 (BMP4)* homologue, *decapentaplegic* (*Dpp*) (Francois et al. 1994; Harbeck and Lengyel 1995). *Numb* plays essential role in determining stereotypic track of MTs through the body cavity (Ainsworth et al. 2000; Wan et al. 2000).

Study from lower organisms has proved that cytoskeletal regulation plays essential role in organogenesis (Kim et al. 2011). Cytoskeletal proteins, F-actin and β-tubulin, control diverse activities like cytokinesis, polarized intracellular trafficking, adhesion, migration, and morphogenesis (Pruyne and Bretscher 2000; Mitchison and Cramer 1996; Vasioukhin et al. 2000; Lanier and Gertler 2000; Lehmann 1995; Sullivan and Theurkauf 1995). In *Drosophila* MTs, altered organization of F-actin and β-tubulin was observed when ecdysone signaling was disrupted in these cells (Gautam et al. 2015).

Malpighian tubules are polarized epithelial structures consisting of tight junctions (TJs) and adherens junctions (AJs) (Tepass and Hartenstein 1994, Tepass 1997). Tight junctions are present at the most apical region and bring adjacent plasma membranes in close proximity to prevent transepithelial diffusion and restrict membrane proteins to their respective domains (Balda and Matter 1998). Adherens junctions form the zonula adherens and participate in cell–cell interactions and intercellular communication (Gumbiner 1996; Yap et al. 1997).

Formation of polarized epithelia in *Drosophila* is controlled by specific transmembrane proteins, Crumbs (CRB). Loss of CRB protein leads to loss of polarity (Wodarz et al. 1993, 1995; Tepass 1996, 1997). The *Drosophila* Disc large (Dlg), a homologue of tight junctional protein ZO-1, and Armadillo, the *Drosophila* homologue of adherence junction protein β-catenin, express in Malpighian tubules and are regulated through EcR-USP heterodimer function (Gautam 2012, Ph.D. Thesis). Dlg and Armadillo are required for the localization of junctional complexes or for the maintenance of apico-basal polarity in epithelial cells and maintaining cell–cell adhesion, cell polarity, and cytoskeletal integrity, respectively.

An important role of ions in the regulation of immune response has also been identified. Intracellular concentrations of Zn^{++} ions regulate immune cell activities (Haase and Lothar 2009). Voltage-gated Ca^{++} channels maintain a constant flux of Ca^{++} ions which is necessary for the survival and function of T-lymphocyte (Jha et al. 2009). Ions also play a role in the regulation of immune response in Malpighian tubules. The expression of AMPs is regulated through ion channels, Na^{+}/K^{+}-ATPase and V-ATPase (Verma and Tapadia 2014). Hence, *Drosophila* renal system provides opportunities to decipher molecular mechanisms of various kidney diseases. Some examples of kidney diseases are listed below that can be modeled with *Drosophila* renal system:

A. *Nephrolithiasis*

The formation of crystal aggregates in kidney results in kidney stones, the clinical condition referred to as nephrolithiasis. Ectopic calcification is the primary reason for nephrolithiasis; however, its initiating factors remain largely unknown. Identifying fundamental principal factors for ectopic calcification may have broad translational significance to deal with nephrolithiasis. Murine and porcine models have been used to study nephrolithiasis but have limitations like relatively longer time for onset of stone formation, and lack of genetic tools has limited genetic screening as a means of exploring mineralization. Insect model *Drosophila* overcomes the above problem, and it allows exploration of the complex interplay among diets, genes, and environmental exposures which are known to influence kidney stone formation (Dow 2009; Hirata et al. 2012; Miller et al. 2013; Bagga et al. 2013 Chi et al. 2015).

Mutation in *xanthine dehydrogenase* (*Xdh*) gene results in recurrent nephrolithiasis in humans and dogs (Arikyants et al. 2007; Jacinto et al. 2013). Inhibition of *Xdh* in *Drosophila melanogaster* leads to ectopic calcification in Malpighian tubules (Bonse 1967; Chi et al. 2015). Selective knockdown of *prestin* gene (CG5485) in MTs results in the decrease of calcium oxalate crystal formation (Hirata et al. 2012). *Prestin* codes for proteins homologous to the human Slc26a5 and Slc26a6 transporters that perform as Cl^{-} exchange systems for HCO^{3-}, oxalate $^{2-}$, SO4 $^{2-}$, and formate. Drosophila epithelial transport of Cl^{-}, oxalate^{2-}, and $SO4^{2-}$ in Malpighian tubule is regulated by a WNK/OSR1 signaling via dPrestin protein activation (Dorwart et al. 2008; Sindić et al. 2007, Hirata et al. 2012) (Fig. 1.2).

Identification of MTs initial segment as major site for Ca^{++} ion uptake will be relevant to understanding the fundamental mechanisms of Ca^{++}containing kidney stone formation (Browne and O'Donnell 2016). Moreover, exogenous exposure of ethylene glycol and oxalate to flies also initiates the formation of calcified particles in tubules (Chen et al. 2011). Furthermore, it was observed that inhibition of Zn transporter genes (CG3994, CG11163, and CG17723) in ZnT family leads to suppression of crystal formation for uric acid, calcium oxalate, calcium citrate, xanthin, and hypoxanthin in *Drosophila* Malpighian tubules (Chi et al. 2015).

Effect of sulfate and thiosulfate on crystal formation was investigated by using *Drosophila* model and observed that both compounds were effective at decreasing CaOx crystallization via dPrestin transporter that specifically expresses in PCs of Malpighian tubules (Landry et al. 2016). Recently, Wu et al. used *Drosophila* models to screen medicinal plants having anti-nephrolithiasis potential (Wu et al. 2014). Therefore, Malpighian tubules offer a good model to understand molecular basis for kidney stone formation which will have major translational impact to treat kidney stone and its prevention.

B. *Nephrotic syndrome*

Nephrotic syndrome (NS) is a kidney disorder that causes excretion of too much protein in the urine due to damage of nephrons in the kidneys that help to filter waste and excess water from blood. In the Bowman's capsule, podocytes are polarized epithelial cells that form glomerular filtration barrier through their interdigitated foot processes and cytoskeletal proteins. The filtration barrier is responsible for ultrafiltration of the blood and ensures that essential plasma proteins are retained. Dysfunction of podocytes leads to removal of proteins and other essential materials in the urine.

C. Nephrotic syndrome also affects the function of the tubular nephrons which modify the filtrate that filters from glomerulus, for example, abnormalities of proximal tubular function such as glycosuria, aminoaciduria, and defects in acidification (Syed et al. 2012) of urine. *Drosophila* renal system, nephrocytes, and Malpighian tubules provide a very good model to reveal the molecular basis of nephrotic syndrome. *Drosophila* nephrocytes are functionally analogous to human glomerular podocytes. In particular, both cell types possess a specialized filtration diaphragm, known as the slit diaphragm (sd) in podocytes and nephrocyte diaphragm (nd) in nephrocytes. Nephrocyte diaphragm expresses genes *stick and stone (sns), dumbfounded (duf), polychateoid (pyd)Mec2*, and *CG31012* which are orthologues of nephrin, neph1, ZO-1, podocin, CD2AP, respectively, and form a complex of interacting proteins that closely mirrors the vertebrate slit diaphragm complex. Nephrocyte diaphragm is completely lost in flies' mutant for sns or duf, a phenotype resembling loss of the slit diaphragm in the absence of either nephrin (as in the human kidney disease NPHS1) or neph1. These changes drastically impair filtration function in the nephrocyte (Weavers et al. 2009). The function of nephrocytes can be easily monitored by assaying the filtration and uptake of secreted fluorescent proteins from the *Drosophila* hemolymph (Gee et al. 2015).

D. Malpighian tubules play similar function as tubular parts of the nephron. Main segment of Malpighian tubules is responsible for fluid secretion, lower segment and ureter for reabsorption, and initial segment for storing ions and other metabolic products. Since, Malpighian tubules are enriched with several transporters, fluid transport and ion transport assay offer opportunity to evaluate the function of transporters. Nitrogenous waste products and uric acid gets deposited in lumen of MTs and finally excretes out through alimentary canal. Capability of transporter and functions of MTs can also be judged through uric acid deposition assay. Junctional protein Armadillo (orthologue of vertebrate beta-catenin), tight junctional protein Dlg (orthologue of vertebrate Zonula occludens ZO-1), cell adhesion protein Fas-2 (orthologue of vertebrate neural cell adhesion molecule N-CAM) expresses in MTs and offers opportunity to judge the effects of nephritic syndrome on junctional and adhesion proteins. It will also be helpful to identify new proteins/molecules interacting with these proteins. MTs help in maintaining homeostasis, intestinal pH, regulation and secretion of calcium (Maddrell et al. 1991), immunological defense, and clearance to toxic substances (Beyenbach et al. 2010). Therefore, MTs act as a quality controller for the hemolymph as nephrons control the quality of blood in humans.

E. *Diabetic nephropathy*

Impairment of kidney function through diabetes is known as diabetic nephropathy. Long-term diabetes leads to chronic kidney diseases and renal failure; 44% cases of renal failures are due to diabetes in the USA (National Diabetes Statistics Report 2014). Since in *Drosophila,* type 1 (Rulifson et al. 2002) as well as type 2 diabetes (Musselman et al. 2011; Morris et al. 2012, Pendse et al. 2013; Trinh and Boulianne 2013) can be induced, and its renal system similar to nephrons, it offers the best opportunity to study molecular basis of diabetic nephropathy (Na and Cagan 2013; Na et al. 2015; Betz and Conway 2016). Very recently, Delanoue et al. identified that insulin release is triggered by stunted (Sun), a ligand of Methuselah (Mth) receptor, in insulin producing cells of Drosophila brain. Therefore, Mth and Sun delineate a new cross-organ circuitry that modulates physiological insulin level in response to nutrients (Delanoue et al. 2016).

F. *Renal Cancer*

Fault in regulation of stem cell behavior results in tissue degeneration, premature aging, and cancer formation. Human kidney has a low rate of cellular turnover but has a great ability for tissue regeneration after an ischemic injury. Multipotent stem cells were identified in MTs of adult *Drosophila.* These stem cells are relatively quiescent and only divide once in one week. However, they can become very active and even develop stem cell tumors upon activating the JAK-STAT signal transduction pathway or expressing the activated form of the Ras oncogene (Singh et al. 2007; Zeng et al. 2010). Therefore, *Drosophila* MTs provide an excellent in vivo system for understanding the molecular mechanisms of stem cell self-renewal and differentiation which may increase our understanding of the mechanisms underlying cancer formation, aging, and degenerative diseases. It will be helpful for using stem cells for future regenerative medicine and gene therapy for acute and chronic kidney diseases in humans.

G. *Xenobiotic nephrotoxicity*
Kidney is the major target for xenobiotics, which includes drugs, industrial chemicals, environmental toxicants, and other compounds. Accurate methods for screening of large number of potential nephrotoxic xenobiotics with diverse chemical structure are currently not available. *Drosophila* can provide in vivo tool for screening of nephrotoxicants affecting glomerulus and tubular part of nephron as nephrocytes play similar function to podocytes (Weavers et al. 2009) and MTs play role similar to tubular part of nephrons. MTs are enriched with several conserved genes that participate in metabolism and detoxification, for example, alcohol dehydrogenase, glutathione transferase, and cytochrome P450. Glutathione transferases play a role in metabolism of xenobiotics, conjugating reduced glutathione to lipophilic substrates, making them more hydrophilic and thus more easily excreted (Dow 2009). Therefore, *Drosophila* renal system offers a good opportunity for dissection of molecular mechanisms of nephrotoxicants.

1.7 Conclusion

As the basic developmental process and genes are conserved in Malpighian tubules and human kidneys, the genetic tools available in *Drosophila* can be utilized to understand the kidney development and associated disorder. Since Malpighian tubules develop from ectodermal and mesodermal cell lineages, it provides opportunity to identify novel genes involved in the integration of cells from different origins that makes mature and functional structure. Different segments of tubules are specialized to perform different functions that provides opportunity to dissect pathways involved in the regulation of absorption, reabsorption and secretion. Additionally Malpighian tubules are immune tissues and express several components of immune pathways hence they can be used to understand innate immune mechanisms pertaining to humoral response. Therefore, *Drosophila* renal system provides an opportunity to decipher molecular basis of many kidney diseases which will be helpful for identifying potential target for the alleviation of these diseases and preventing renal failure.

References

Agarwal SK, Gupta A (2008) Aquaporins: the renal water channels. Indian J Nephrol 18 (3):95–100

Ainsworth C et al (2000) Coordinating cell fate and morphogenesis in *Drosophila* renal tubules. Philos Trans R Soc Lond 355:931–937

Allam R et al (2012) Histones from dying renal cells aggravate kidney injury via TLR2 and TLR4. J Am Soc Nephrol 23(8):1375–1388

Arikyants N et al (2007) Xanthinuria type I: a rare cause of urolithiasis. Pediatr Nephrol 22:310–314

Aronson PS, Sacktor B (1975) The Na^+ gradient-dependent transport of D-glucose in renal brush border membranes. J Biol Chem 250:6032–6039

Artero RD et al (2001) The immunoglobulin-like protein Hibris functions as a dose-dependent regulator of myoblast fusion and is differentially controlled by Ras and Notch signaling. Development 128:4251–4264

Bagga HS et al (2013) New insights into the pathogenesis of renal calculi. Urol Clin North Am 40:1–12

Balda MS, Matter K (1998) Tight junctions. J Cell Sci 111:541–547

Betz B, Conway B (2016) An update on the use of animal models in diabetic nephropathy research. Curr Diab Rep 16:18

Beyenbach KW et al (2010) The developmental, molecular, and transport biology of Malpighian tubules. Annu Rev Entomol 55:351–374

Bonse A (1967) Studies on the chemical nature and formation of the urinary conglomerate in the Malpighian vessels of the rosy mutant of *Drosophila melanogaster*. Z Naturforsch B 22:1027–1029

Breljak D et al (2016) Distribution of organic anion transporters NaDC3 and OAT1-3 along the human nephron. Am J Physiol Renal Physiol 311(1):F227–F238

Browne A, O'Donnell MJ (2016) Segment-specific Ca^{2+} transport by isolated Malpighian tubules of *Drosophila melanogaster*: a comparison of larval and adult stages. J Insect Physiol (in press)

Bulet P, Stocklin R, Menin L (2004) Anti-microbial peptides: from invertebrates to vertebrates. Immunol Rev 198:169–184

Cabrero P et al (2002) The Dh gene of *Drosophila melanogaster* encodes a diuretic peptide that acts through cyclic AMP. J Exp Biol 205:3799–3807

Cannell E et al (2016) The corticotrophin releasing factor like diuretic hormone 44 (DH44) and kinin neuropeptides modulate desiccation and starvation tolerance in *Drosophila melanogaster*. Peptides 9781(16):30020–30021 (S0196)

Cantaluppi V et al (2015) Endothelial progenitor cell-derived extracellular vesicles protect from complement-mediated mesangial injury in experimental anti-Thy1.1 glomerulonephritis. Nephrol Dial Transplant 30(3):410–422

Centers for Disease Control and Prevention (2014) National diabetes statistics report: estimates of diabetes and its Burden in the United States, 2014. U.S. Department of Health and Human Services, Atlanta, GA

Chen et al (2011) Ethylene glycol induces calcium oxalate crystal deposition in Malpighian tubules: a *Drosophila* model for nephrolithiasis/urolithiasis. Kidney Int 80:369–377

Cherbas L et al (2003) EcR isoforms in *Drosophila*: testing tissue-specific requirements by targeted blockade and rescue. Development 130:271–284

Chi T et al (2015) A *Drosophila* model identifies a critical role for zinc in mineralization for kidney stone disease. PLoS One 10(5):e0124150

Chintapalli VR et al (2007) Using FlyAtlas to identify better *Drosophila melanogaster* models of human disease. Nat Genet 39:715–720

Cho E, Dressler G (2003) Formation and development of nephrons. In: Vize P, Woolf A, Bard J (eds) The kidney: from normal development to congenital disease. Academic, San Diego, pp 195–210

Coast GM et al (2001) The *Drosophila melanogaster* homologue of an insect calcitonin-like diuretic peptide stimulates V-ATPase activity in fruit fly Malpighian tubules. J Exp Biol 204:1795–1804

Costanzo LS (1984) Comparison of calcium and sodium transport in early and late rat distal tubules: effect of amiloride. Am J Phys 246:F937–F945

Curthoys NP, Moe OW (2014) Proximal tubule function and response to acidosis. Clin J Am Soc Nephrol 9(9):1627–1638

Davies SA, Dow JA (2009) Modulation of epithelial innate immunity by autocrine production of nitric oxide. Gen Comp Endocrinol 162(1):113–121

Davies MB et al (2005) Phenotypic analysis of EcR-A mutants suggests that EcR isoforms have unique functions during *Drosophila* development. Dev Biol 282:385–396
Davies SA et al (2012) Immune and stress response 'cross-talk' in the *Drosophila* Malpighian tubule. J Insect Physiol 58(4):488–497
Day JP et al (2008) Identification of two partners from the bacterial Kef exchanger family for the apical plasma membrane VATPase of Metazoa. J Cell Sci 121:2612–2619
Delanoue R et al (2016) *Drosophila* insulin release is triggered by adipose Stunted ligand to brain Methuselah receptor. Science 353(6307):1553–1556
Denholm B et al (2003) Dual origin of the renal tubules in *Drosophila*: mesodermal cells integrate and polarize to establish secretory function. Curr Biol 13:1052–1057
Denholm B et al (2013) The tiptop/teashirt genes regulate cell differentiation and renal physiology in *Drosophila*. Development 40(5):1100–1110
Dorwart MR et al (2008) The solute carrier 26 family of proteins in epithelial ion transport. Physiology 23:104
Dow JAT (2009) Insights into the Malpighian tubule from functional genomics. J Exp Biol 212:435–445
Dow JAT, Davies SA (2003) Integrative physiology and functional genomics of epithelial function in a genetic model organism. Physiol Rev 83:687–729
Dow JAT, Romero MF (2010) *Drosophila* provides rapid modeling of renal development, function, and disease. Am J Physiol Renal Physiol 299(6):F1237–F1244
Dow JAT et al (1994) The Malpighian tubules of *Drosophila melanogaster*, a novel phenotype for studies of fluid secretion and its control. J Exp Biol 197:421–428
Dube KA, McDonald DG, O'Donnell MJ (2000) Calcium transport by isolated anterior and posterior Malpighian tubules of *Drosophila* melanogaster, roles of sequestration and secretion. J Insect Physiol 46:1449–1460
Dworak HA, Sink H (2002) Myoblast fusion in *Drosophila*. Bioessays 24:591–601
Dworak HA et al (2001) Characterization of *Drosophila hibris*, a gene related to human nephrin. Development 128:4265–4276
Erickson MRS et al (1997) Drosophila myoblast city encodes a conserved protein that is essential for myoblast fusion, dorsal closure, and cytoskeletal organization. J Cell Biol 138(3):589–603
Francois V et al (1994) Dorsal-ventral patterning of the *Drosophila* embryo depends on a putative negative growth factor encoded by the short gastrulation gene. Genes Dev 8:2602–2616
Frei R et al (2010) MHC class II molecules enhance toll-like receptor mediated innate immune responses. PLoS One 5(1):e8808
Galletta BJ et al (1999) Identification of a *Drosophila* homologue to vertebrate Crk by interaction with MBC. Gene 228(1–2):243–252
Ganz T (2003) Microbiology: Gut defence. Nature 422(6931):478–479
Garayoa M et al (1994) Myoendocrine-like cells in invertebrates: occurrence of noncardiac striated secretory-like myocytes in the gut of the ant Formica polyctena. Gen Comp Endocrinol 95:133–142
Gaul U, Weigel D (1990) Regulation of kruppel expression in the anlage of the Malpighian tubules in *Drosophila* embryo. Mech Dev 33:57–67
Gautam N (2012) Studies on the role of ecdysone signaling in development and function of Malpighian tubules of *Drosophila melanogaster*. PhD Thesis, Banaras Hindu University, Varanasi
Gautam NK, Tapadia MG (2010) Ecdysone signaling is required for proper organization and fluid secretion of stellate cells in the Malpighian tubules of *Drosophila melanogaster*. Int J Dev Biol 54(4):635–642
Gautam NK et al (2015) Ecdysone regulates morphogenesis and function of Malpighian tubules in *Drosophila* melanogaster through EcR-B2isoform. Dev Bio 398:163–176
Gee et al (2015) KANK deficiency leads to podocyte dysfunction and nephrotic syndrome. J Clin Invest 125(6):2375–2384
Gumbiner BM (1996) Cell adhesion: the molecular basis of tissue architecture and morphogenesis. Cell 84:345–357

Harbeck R, Lengyel J (1995) Genes controlling posterior gut development in the *Drosophila* embryo. Roux's Arch Dev Biol 204:308–329
Harbecke R, Janning W (1989) The segmentation gene Kruppel of *Drosophila melanogaster*has homeotic properties. Genes Dev 3:114–122
Heron M (2016) Deaths: leading causes for 2013. Natl Vital Stat Rep 65:11
Hirata T et al (2012) In vivo *Drosophila* genetic model for calcium oxalate nephrolithiasis. Am J Physiol Renal Physiol 303(11):F1555–F1562
Ho AW, Wong CK, Lam CW (2008) Tumor necrosis factor-alpha up-regulates the expression of CCL2 and adhesion molecules of human proximal tubular epithelial cells through MAPK signaling pathways. Immunobiology 213(7):533–544
Hoch M, Pankratz M (1996) Control of gut development by fork head and cell signaling molecules in *Drosophila*. Mech Dev 58:3–14
Hoch M et al (1994) Sequential fates in a single cell are established by the neurogenic cascade in the Malpighian tubules of *Drosophila*. Development 120:3439–3450
Hultmark D (2003) *Drosophila* immunity: paths and patterns. Curr Opin Immunol 15:12–19
Imig JD, Ryan MJ (2013) Immune and inflammatory role in renal disease. Compr Physiol 3 (2):957–976
Imler JL, Bulet P (2005) Antimicrobial peptides in *Drosophila*: structures, activities and gene regulation. Chem Immunol Allergy 86:1–21
Jacinto AML et al (2013) Urine concentrations of xanthine, hypoxanthine and uric acid in UK Cavalier King Charles spaniels. J Small Anim Pract 54:395–398
Janning W et al (1986) Clonal analysis of the blastoderm anlage of the Malpighain tubules in *Drosophila melanogaster*. Roux's Arch Dev Biol 195:22–32
Jha MK et al (2009) Defective survival of naive CD8+ T lymphocytes in the absence of the beta3 regulatory subunit of voltage-gated calcium channels. Nat Immunol 10:1275–1282
Johnson EC et al (2005) A novel diuetic hormone receptor in Drosophila: evidence for conservation of CGRP signaling. J Exp Biol 208:1239–1246
Jung CA et al (2005) Renal tubule development in *Drosophila*: a closer look at cellular level. J Am Soc Nephrol. 16:322–328
Kaufmann N et al (2005) Developmental expression and biophysical characterization of a *Drosophila melanogaster* aquaporin. Am J Physiol Cell Physiol 289:397–407
Kean L et al (2002) Two nitridergic peptides are encoded by the gene capability in *Drosophila melanogaster*. Am J Physiol Regul Integr Comp Physiol 282:1297–1307
Kerber B et al (1998) Seven-up, the *Drosophila* homolog of the COUP-TF orphan receptors, controls cell proliferation in the insect kidney. Genes Dev 12:1781–1786
Khoshnoodi J, Tryggvason K (2001) Congenital nephritic syndromes. Curr Opin Genet Dev 11:322–327
Kim HS et al (2011) VAB-10 spectraplakin acts in cell and nuclear migration in *Caenorhabditis elegans*. Development 138:4013–4023
King LS et al (2001) Defective urinary-concentrating ability due to a complete deficiency of aquaporin-1. N Engl J Med 345:175–179
Kirsch K et al (1999) CMS: an adapter molecule involved in cytoskeletal rearrangements. Proc Natl Acad Sci 96:6211–6216
Knier CG et al (2016) Bicaudal-C in *Drosophila* as a model of polycystic kidney disease (PKD) and intersection of oxalate nephrolithiasis FASEB J 30: Sup 1224.27
Koelle MR et al (1991) The *Drosophila* EcR gene encodes an ecdysone receptor, a new member of the steroid receptor super family. Cell 67:59–77
Landry GM et al (2016) Sulfate and thiosulfate inhibit oxalate transport via a dPrestin (Slc26a6)-dependent mechanism in an insect model of calcium oxalate nephrolithiasis. Am J Physiol Renal Physiol 310(2):152–159
Lanier LM, Gertler FB (2000) From Abl to actin: Abl tyrosine kinase and associated proteins in growth cone motility. Curr Opin Neurobiol 10:80–87
Lebovitz RM et al (1989) Molecular characterization and expression of the (Na^+/K^+)-ATPase-sub unit in *Drosophila melanogaster*. EMBO 8:193–202

Lehmann R (1995) Cell-cell signaling, microtubules, and the loss of symmetry in the *Drosophila* oocyte. Cell 83:353–356

Lekven AC et al (1998) Faint sausage encode a novel extracellular protein of the immunoglobulin superfamily required for cell migration and the establishment of normal axonal pathways in the *Drosophila* nervous system. Development 125:2747–2758

Lemaitre B, Hoffmann J (2007) The host defense of *Drosophila melanogaster*. Annu Rev Immunol 25:697–743

Lemaitre B et al (1995) A recessive mutation, immune deficiency (imd), defines two distinct control pathways in the *Drosophila* host defense. Proc Natl Acad Sci USA 92(21):9465–9469

Li et al (2014) Further, differential Notch activity is required for homeostasis of Malpighian tubules in adult *Drosophila*. J Genet Genomics 41:649–652

Li et al (2015) EGFR/MAPK signaling regulates the proliferation of Drosophila renal and nephric stem cells. J Genet Genomics 42:9e20

Liu X et al (1999) Identification of genes controlling malpighian tubule and other epithelial morphogenesis in *Drosophila melanogaster*. Genetics 151:685–695

Lothar HH (2009) Significance of zinc-related signaling pathways in immune cells. Annu Rev Nutr 29:133–152

Maddrell, O'Donnell (1992) Insect Malpighian tubules: V-ATPase action in ion and fluid transport. J Exp Biol 172(1):417–429

Maddrell SHP et al (1991) 5-hydroxytryptamine: a second diuretic hormone in Rhodniusprolixus. J Exp Biol 156:557–566

Mathieson PW (2003) Immune dysregulation in minimal change nephropathy. Nephrol Dial Transplant 18(Suppl 6):vi26–vi29

McGettigan J et al (2005) Insect renal tubules constitute a cell-autonomous immune system that protects the organism against bacterial infection. Insect Biochem Mol Biol 35(7):741–754

Miller J et al (2013) *Drosophila melanogaster* as an emerging translational model of human nephrolithiasis. J Urol 190(5):1648–1656

Mitchison TJ, Cramer LP (1996) Actin-based cell motility and cell locomotion. Cell 9:371–379

Morris SNS et al (2012) Development of diet-induced insulin resistance in adult *Drosophila melanogaster*. Biochim Biophys Acta 1822:1230–1237

Musselman LP et al (2011) A high-sugar diet produces obesity and insulin resistance in wild-type *Drosophila*. Dis Model Mech 4(6):842–849

Na J, Cagan RL (2013) The *Drosophila* nephrocyte: back on stage. J Am Soc Nephrol 24:161–163

Na et al (2015) Diet-induced podocyte dysfunction in Drosophila and Mammals. Cell Rep 12 (4):636–647

Nakhoul NL, Hamm LL (2002) Vacuolar H(+)-ATPase in the kidney. J Nephrol (Suppl 5):S22–S31

Nielsen S et al (1993) CHIP28 water channels are localized in constitutively water-permeable segments of the nephron. J Cell Biol 120:371–383

Nolan KM et al (1998) Myoblast city, the *Drosophila* homolog of DOCK180/CED-5, is required in a Rac signaling pathway utilized for multiple developmental processes. Genes Dev 12:3337–3342

O'Donnell, Maddrell (1995) Fluid reabsorption and ion transport by the lower Malpighian tubules of adult female *Drosophila*. J Exp Biol 198:1647–1653

Oro AE et al (1990) Relationship between the product of *Dosophila* ultra spiracle locus and vertebrate retinoid X receptor. Nature 347:298–301

Overend G et al (2012) The receptor guanylate cyclase Gyc76C and a peptide ligand, NPLP1-VQQ, modulate the innate immune IMD pathway in response to salt stress. Peptides 34 (1):209–218

Palmer LG, Schnermann J (2015) Integrated control of Na transport along the nephron. Clinical J Am Soc Nephrol 10(4):676–687

Pannabecke'r T (1995) Physiology of the Malpighian tubules. Annu Rev Entomol 40:493–510

Pavenstadt H, Kriz W, Kretzler M (2003) Cell biology of the glomerular podocyte. Physiol Rev 83 (1):253–307

Pendse J et al (2013) A *Drosophila* functional evaluation of candidates from human genome-wide association studies of type 2 diabetes and related metabolic traits identifies tissue-specific roles for dHHEX. BMC Genomics 14:136–147
Pruyne D, Bretscher A (2000) Polarization of cell growth in yeast. I Establishment and maintenance of polarity states. J Cell Sci 113:365–375
Pütz M et al (2005) In *Drosophila melanogaster*, the rolling pebbles isoform 6 (Rols6) is essential for proper Malpighian tubule morphology. Mech Dev 122:1206–1217
Rheault MR, O'Donnell MJ (2001) Analysis of K^+ transport in Malpighian tubules of *Drosophila melanogaster*: evidence for spatial and temporal heterogeneity. J Exp Biol 204:2289–2299
Rheault MR, O'Donnell MJ (2004) Organic cation transport by Malpighian tubules of *Drosophila melanogaster*: application of two novel electrophysiological methods. J Exp Biol 207:2173–2184
Robinson SW et al (2013) FlyAtlas: database of gene expression in the tissues of *Drosophila melanogaster*. Nucleic Acids Res 41(Database issue):D744–D750
Roy A, Al-bataineh MM, Pastor-Soler NM (2015) Collecting duct intercalated cell function and regulation. Clin J Am Soc Nephrol 10(2):305–324
Rulifson EJ et al (2002) Ablation of insulin-producing neurons in flies: growth and diabetic phenotypes. Science 296:1118–1120
Sands JM, Layton HE (2009) The physiology of urinary concentration: an update. Semin Nephrol 29(3):178–195
Saxena L (1987) Organogenesis of the kidney. Cambridge University Press, Cambridge
Schubiger M et al (1998) *Drosophila* EcR-B ecdysone receptor isoforms required for larval molting and for neuron remodling during metamorphosis. Development 125:2053–2062
Silverman N, Maniatis T (2001) NF-kappa B signaling pathways in mammalian and insect innate immunity. Genes Dev 15(18):2321–2342
Sindić A et al (2007) Renal physiology of SLC26 anion exchangers. Curr Opin Nephrol Hypertens 16:484
Singh SR et al (2007) The adult *Drosophila* Malpighian tubule are maintained by pluripotent stem cells. Cell Stem Cell 16:191–203
Skaer H (1989) Cell division in Malpighian tubule development in *D. melanogaster* is regulated by a single tip cell. Nature 342:566–569
Skaer H (1993) The alimentary canal. In: Bate M, Arias AM (eds) Development of *Drosophila melanogaster*. Cold Spring Harbor Laboratory Press, Plain view, pp 941–1012
Skaer H, Martinez A (1992) The wingless product is required for cell proliferation in the Malpighian tubules anlage of *Drosophila melanogaster*. Development 116:745–754
Sözen MA et al (1997) Functional domains are specified to single-cell resolution in a *Drosophila* epithelium. Proc Natl Acad Sci USA 94(10):5207–5212
Stergiopoulos K et al (2009) *Salty* dog, an SLC5 symporter, modulates *Drosophila* response to salt stress. Physiol Genomics 37(1):1–11
Sullivan W, Theurkauf WE (1995) The cytoskeleton and morphogenesis of the early *Drosophila* embryo. Curr Opin Cell Biol 7:18–22
Syed KH et al (2012) Proximal renal tubular acidosis: a not so rare disorder of multiple etiologies. Nephrol Dial Transplant 27(12):4273–4287
Talbot WS et al (1993) *Drosophila* tissues with different metamorphic responses to ecdysone express different ecdysone receptor isoforms. Cell 73(7):1323–1337
Tanji T, Ip YT (2005) Regulators of the Toll and Imd pathways in the *Drosophila* innate immune response. Trends Immunol 26(4):193–198
Tepass U (1996) Crumbs, a component of the apical membrane, is required for zonula adherens formation in primary epithelia of *Drosophila*. Dev Biol 177:217–225
Tepass U (1997) Epithelial differentiation in *Drosophila*. Bioessays 19:673–682
Tepass U, Hartenstein V (1994) The development of cellular junctions in the *Drosophila* embryo. Dev Biol 161:563–596
Terhzaz S et al (2010) Mislocalization of mitochondria and compromised renal function and oxidative stress resistance in *Drosophila* SesB mutants. Physiol Genomics 41(1):33–41

Thomas HE et al (1993) Heterodimerization of *Drosophila* ecdysone receptor with retinoid X receptor and ultra spiracle. Nature 362:471–475
Tossi A, Sandi L et al (2002) Molecular diversity in gene-encoded, cationic antimicrobial polypeptides. Curr Pharm Des 8(9):743–761
Trinh I, Boulianne GL (2013) Modeling obesity and its associated disorders in *Drosophila*. Physiology 28:117–124
Tzou P et al (2000) Tissue-specific inducible expression of antimicrobial peptide genes in *Drosophila* surface epithelia. Immunity 13(5):737–748
Van Kooten C et al (2000) Immunological function of tubular epithelial cells: the functional implications of CD40 expression. Exp Nephrol 8(4-5):203–207
Vasioukhin V et al (2000) Directed actin polymerization is the driving force for epithelial cell-cell adhesion. Cell 21:209–219
Verma P, Tapadia MG (2012) Immune response and anti-microbial peptides expression in Malpighian tubules of *Drosophila melanogaster* is under developmental regulation. PLoS One 7(7):e40714
Verma P, Tapadia MG (2014) Epithelial immune response in *Drosophila* malpighian tubules: interplay between Diap2 and ion channels. J Cell Physiol 229(8):1078–1095
Verma P, Tapadia MG (2015) Early gene broad complex plays a key role in regulating the immune response triggered by ecdysone in the Malpighian tubules of *Drosophila melanogaster*. Mol Immunol 66(2):325–339
Wan S et al (2000) Multiple signalling pathways establish cell fate and cell number in*Drosophila* Malpighian tubules. Dev Biol 217:153–165
Wang T et al (2001) Role of NHE isoforms in mediating bicarbonate reabsorption along the nephron. Am J Physiol 281:F1117–F1122
Wang J et al (2004) Function-informed transcriptome analysis of Drosophila renal tubule. Genome Biol 5:R6
Weavers H, Skaer H (2013) Tip cells act as dynamic cellular anchors in the morphogenesis of looped renal tubules in *Drosophila*. Dev Cell 27(3):331–344
Weavers H et al (2009) The insect nephrocyte is a podocyte-like cell with a filtration slit diaphragm. Nature 15:322–326
Weigel D et al (1989) Primordeum specific requirement of the homeotic gene fork head in the developing gut of the *Drosophila* embryo. Roux's Arch Dev Biol 98:201–210
Wessing A et al (1992) Two types of concretions in *Drosophila* Malpighian tubules as revealed by X-ray microanalysis: a study on urine formation. J Insect Physiol 38:543–554
Wodarz A et al (1993) Crumbs is involved in the control of apical protein targeting during *Drosophila* epithelial development. Mech Dev 44:175–187
Wodarz et al (1995) Expression of crumbs confers apical character on plasma membrane domains of ectodermal epithelia of *Drosophila*. Cell 82:67–76
Woolf A, Bard J (eds) (2003) The kidney: from normal development to congenital disease. Academic Press, San Diego, pp. 165–179
Wu L, Langyel J (1998) Role of caudal in hind gut specification and gastrulation suggests homology between *Drosophila* amnoproctodeal invagination and vertebrate blastopore. Development 125:2433–2442
Wu et al (2014) An emerging translational model to screen potential medicinal plants for nephrolithiasis, an independent risk factor for chronic kidney disease. Evid Based Complement Alternat Med. eID 972958:1–7
Wu Y et al (2015) Two inwardly rectifying potassium channels, Irk1 and Irk2, play redundant roles in *Drosophila* renal tubule function. Am J Physiol Regul Integr Comp Physiol. 309(7):R747–R756
Yadav S, Tapadia MG (2016) Expression of polyQ aggregates in Malpighian tubules leads to degeneration in *Drosophila melanogaster*. Dev Biol 409(1):166–180
Yao TP et al (1992) *Drosophila* ultraspiracle modulates ecdysone receptor function via heterodimer formation. Cell 71:63–72
Yap AS et al (1997) Molecular and functional analysis of cadherin-based adherens junction. Annu Rev Cell Dev Biol 13:9119–9146

Zeng X et al (2010) Tumor suppressors Sav/Scrib and oncogene Ras regulate stem-cell transformation in adult *Drosophila* malpighian tubules. J Cell Physiol 224(3):766–774

Zhang B et al (2008) TLR4 signaling mediates inflammation and tissue injury in nephrotoxicity. J Am Soc Nephrol. 19(5):923–932

Chapter 2
Zebrafish Pronephros Development

Richard W. Naylor, Sarah S. Qubisi, and Alan J. Davidson

Abstract The pronephros is the first kidney type to form in vertebrate embryos. The first step of pronephrogenesis in the zebrafish is the formation of the intermediate mesoderm during gastrulation, which occurs in response to secreted morphogens such as BMPs and Nodals. Patterning of the intermediate mesoderm into proximal and distal cell fates is induced by retinoic acid signaling with downstream transcription factors including wt1a, pax2a, pax8, hnf1b, sim1a, mecom, and irx3b. In the anterior intermediate mesoderm, progenitors of the glomerular blood filter migrate and fuse at the midline and recruit a blood supply. More posteriorly localized tubule progenitors undergo epithelialization and fuse with the cloaca. The Notch signaling pathway regulates the formation of multi-ciliated cells in the tubules and these cells help propel the filtrate to the cloaca. The lumenal sheer stress caused by flow down the tubule activates anterior collective migration of the proximal tubules and induces stretching and proliferation of the more distal segments. Ultimately these processes create a simple two-nephron kidney that is capable of reabsorbing and secreting solutes and expelling excess water—processes that are critical to the homeostasis of the body fluids. The zebrafish pronephric kidney provides a simple, yet powerful, model system to better understand the conserved molecular and cellular progresses that drive nephron formation, structure, and function.

2.1 Introduction

The vertebrate kidney functions to maintain the composition of the blood so as to permit healthy bodily function. In order to achieve this, the kidney has evolved one of the most structurally and physiologically complex organ subunits in the body, the

R.W. Naylor • S.S. Qubisi • A.J. Davidson (✉)
Department of Molecular Medicine and Pathology, School of Medical Sciences, Faculty of Medical and Health Sciences, The University of Auckland, Private Bag 92019, Auckland, 1142, New Zealand
e-mail: a.davidson@auckland.ac.nz

R.K. Miller (ed.), *Kidney Development and Disease*, Results and Problems in Cell Differentiation 60, DOI 10.1007/978-3-319-51436-9_2

nephron. Each nephron consists of a blood filter (the renal corpuscle containing the glomerular tuft) that selectively transports fluid from the blood vasculature into epithelial tubules where passive and active reabsorption and secretion of solutes such as ions, sugars, amino acids, and water occurs. Nitrogenous waste and excess minerals and water are then expelled to the exterior via a duct system.

In amniotic vertebrates, three kidney forms arise during development: the pronephros, the mesonephros, and the metanephros. Each kidney is created sequentially and requires the previous kidney type for its induction. While their topographical complexity increases progressively, each kidney type maintains the nephron as its functional subunit. The pronephros, at least in frogs and fish, is comprised of two nephrons, while the mesonephros contains tens to hundreds of nephrons. The metanephros (found only in mammals, reptiles, and birds) has a much greater nephron number, although this number is highly divergent, even between members of the same species. In humans, healthy metanephric kidneys contain between 200,000 and >2.5 million nephrons, with an average of one million nephrons (Bertram et al. 2011).

Anamniotic vertebrates, such as the teleost fish *Danio rerio* (zebrafish), do not develop a metanephros. Instead, kidney development ceases upon formation of the mesonephros. Despite this, development of the zebrafish kidney utilizes molecular pathways and cell types that are conserved in the mammalian kidney. The amenability of zebrafish as an experimental model means it has become a powerful tool to unlock the secrets of kidney development and disease. In this chapter, we will describe in detail how the pronephros forms during early development. The studies that have characterized zebrafish pronephrogenesis have been important in aiding our understanding of key processes in mammalian kidney development.

2.2 Pronephric Origins and Anatomical Organization

Early development in the zebrafish embryo involves rapid cleavage of blastomeres before activation of genomic transcription at the mid-blastula transition. Soon after this time point, the single-layered blastula embryo initiates extensive cell movements (epiboly) to form the tri-laminar gastrula embryo. The three germ layers (the ectoderm, the mesoderm, and the endoderm) form between 4 and 10 hours post-fertilization (hpf) (Kimmel et al. 1995). The pronephros is derived from the mesoderm germ layer, and progenitors for the pronephros are first detected at the late gastrula stage (Drummond et al. 1998; Pfeffer et al. 1998). This bilateral stripe of cells is anatomically positioned between paraxial mesoderm and lateral plate mesoderm and thus is termed the intermediate mesoderm. The zebrafish intermediate mesoderm gives rise to erythroid cells, endothelium, and the pronephros (Davidson and Zon 2004). The origin of glomerular precursors is not so easily discernable, but a population of cells expressing *Wilms' tumor suppressor 1a* (*wt1a*), an early marker of glomerular fate, can be detected at the anterior end of the intermediate mesoderm stripe from the 3-somite stage (11 hpf) onwards

(Fig. 2.1) (Bollig et al. 2006). By 48 hpf, the pronephros consists of a fused midline glomerulus linked via a neck segment to paired tubules that run the length of the trunk to the cloaca (Fig. 2.1) (Drummond et al. 1998). Excluding the neck segment, the pronephric tubules are segmented along the anterior–posterior axis into four segments: the proximal convoluted tubule (PCT), the proximal straight tubule (PST), the distal early tubule (DE), and the distal late segment (DL). The PCT, PST, and DE segments are identifiable by the expression of multiple different solute carriers that perform the transport functions of the nephron (see Fig. 2.1 for examples) (Wingert and Davidson 2008, 2011; Wingert et al. 2007). The DL segment expresses genes encoding solute carriers (such as *slc12a3*) and genes that are also expressed in the mammalian nephric duct (such as *gata3*) suggesting that this segment is a tubule/duct hybrid (Fig. 2.1) (Wingert et al. 2007). At its most posterior end, the DL segment is fused with the cloaca, which acts as an opening to the exterior where waste products are expelled (Pyati et al. 2006).

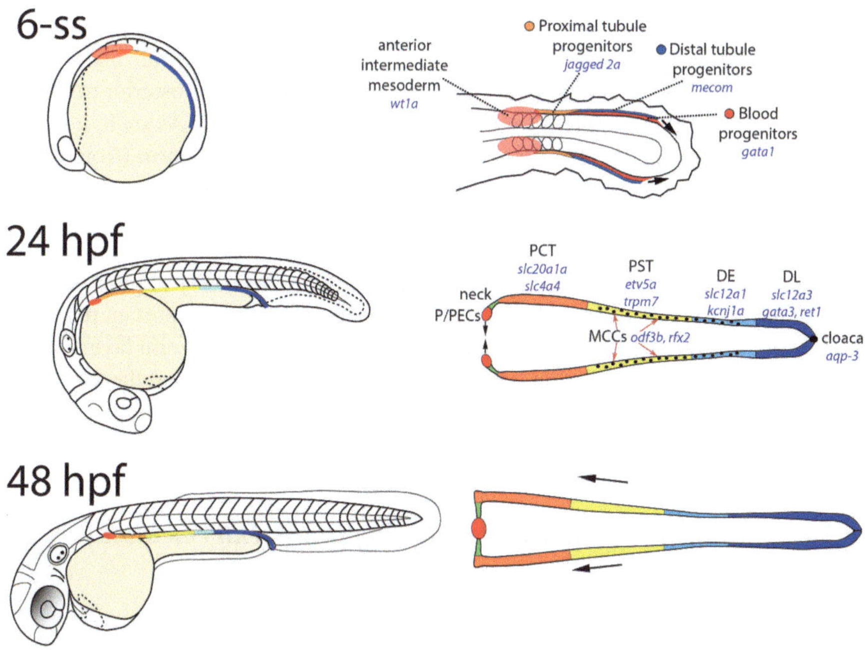

Fig. 2.1 Early pronephros development. Whole embryo views on the *left* are lateral views outlining the spatial arrangement of the pronephros at the stages indicated. Schematic images on the *right* show dorsal views of the complementary stages. *Arrows* in 6-ss stage embryo highlight caudal migration of the distal tubule to the cloaca. *Arrows* in 24 hpf embryo are to show the subsequent migration of the P/PEC lineage towards the *midline*. *Arrows* in 48 hpf embryo indicate anterior collective cell migration that occurs in the pronephros from 29 hpf. Images are not to scale. Abbreviations: *6-ss*, 6-somite stage, *P/PEC* podocytes and parietal epithelial cells, *PCT* proximal convoluted tubule, *PST* proximal straight tubule, *DE* distal early tubule, *DL* distal late tubule/duct hybrid segment, *MCC* multi-ciliated cell

2.3 Early Embryonic Development Establishes Distinct Mesodermal Territories

The formation of the pronephric glomerulus and tubules is primarily dependent upon early embryonic patterning events that allocate the intermediate mesoderm and its different progenitor subtypes. The establishment of different territories of mesoderm along the rostral-caudal (RC) axis, the left–right axis, and the dorsal-ventral (DV) axis initiates at the blastula stage and continues during gastrulation. In order to comprehend the spatial arrangement of these mesoderm territories in the early embryo, we first need to define an accurate layout of these axes. In this chapter, we assign spatial orientation of embryonic axes based on the fate map of *Xenopus laevis* (Kumano and Smith 2002; Lane and Sheets 2002b). This fate map fits lineage-labeling work in zebrafish (Warga and Nusslein-Volhard 1999; Woo and Fraser 1995) and reconciles previous classical experiments that were misinterpreted. For example, UV treatment of *Xenopus* or zebrafish embryos ablates a cortical rotation in the zygote cytoplasm that is required to establish the RC axis (Elinson and Rowning 1988; Jesuthasan and Stahle 1997; Zust and Dixon 1975). The phenotype caused by UV irradiation was termed "ventralization" even though this treatment favors cranial and anterior trunk fates over posterior trunk and tail fates, thus could be more accurately termed "caudalization." As such, it can be generally applied that the classic DV axis of the non-axial mesoderm (from which the pronephros descends) should be redefined as the RC axis [sometimes referred to as the anterior-posterior (AP) axis] (Fig. 2.2) (Lane and Sheets 2006). This orientation of the mesoderm is dissimilar to the neurectoderm, whose RC fates (forebrain/midbrain/hindbrain rhombomeres/spinal cord) are generally aligned with the animal vegetal axis (Kozlowski et al. 1997). This highlights the important point that the embryonic axes are initially oriented differently for different germ layers, and it is only at the end of gastrulation, after extensive cell movements, that the axes of all three germ layers become entrained.

2.3.1 Establishing the RC Positioning of the Intermediate Mesoderm

The earliest definitive marker of intermediate mesoderm is *pax2a*, which is expressed from the late gastrula stage in zebrafish (Fig. 2.2) (Thisse and Thisse 2005). Initially, *pax2a* expression is restricted to a caudolateral domain, but eventually *pax2a*$^{+}$ cells can be observed as a stripe extending around the caudal end of the late gastrula embryo (Fig. 2.2). This expression pattern suggests that the RC axis of the intermediate mesoderm is initially oriented orthogonal to the animal-vegetal axis of the embryo. As such, AP fates in the intermediate mesoderm may be influenced by classic DV patterning signals. Study of patterning across the RC axis (the classic DV axis) was pioneered by Mangold and Spemann who showed

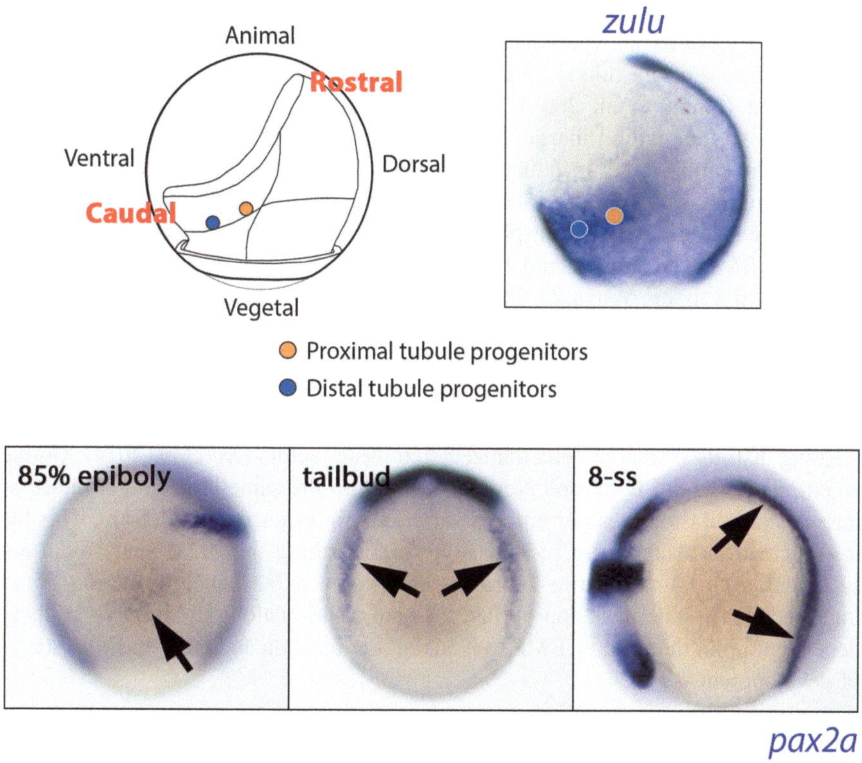

Fig. 2.2 Axis orientation in the zebrafish gastrula stage embryo. *Top left* panel is a schematic representation of the mesendoderm layer (hypoblast) of a zebrafish embryo with the positions of proximal and distal pronephric progenitors shown. The same positions are shown on a late gastrula embryo labeled with *zulu*, a pan mesodermal marker that is highly expressed in the posterior lateral mesoderm from which the pronephros descends. *Bottom* panels are a stage series of zebrafish embryos stained for *pax2a*, the earliest marker of pronephric fate. The *arrows* indicate regions of pronephric *pax2a* expression at the stages indicated. All images are lateral views apart from the tailbud stage *pax2a* stained embryo, which is a sagittal view from the posterior

grafts of the blastopore lip of a salamander embryo transplanted to the opposite side of another embryo induced an entire secondary axis (Spemann and Mangold 1923). Importantly, the difference in pigmentation of the donor and engrafted salamanders demonstrated that dorsal structures induced on the ventral side came from the host. This groundbreaking work highlighted the blastopore lip as an "organizer" region (capable of inducing fates in neighboring tissues). Modern molecular analyses have shown that the organizer largely emits antagonizing signals to prevent caudal structures from forming (as reviewed in De Robertis 2009; Langdon and Mullins 2011; Niehrs 2004). Caudally, bone morphogenetic proteins (BMPs) and Wnts are expressed, whereas the more rostrally positioned organizer emits BMP inhibitors (Chordin, Noggin, Follistatin-like 1b) and Wnt inhibitors (Sfrp3, Frzb, Dickkopf1).

In zebrafish, when rostral, lateral or caudal regions of the embryonic margin (the equatorial ring of tissue where gastrulation movements initiate) are transplanted to the animal pole of blastula stage embryos, tissue induction occurs in, and around, the explant (Fauny et al. 2009). The rostral margin was found to induce axial mesendoderm, the lateral margin induces anterior somites and proximal pronephros, and the caudal margin induces posterior somites, distal pronephros, and tail. This finding showed that the entire embryonic margin in zebrafish has organizer activity. Startlingly, these transplants are recapitulated by injection of mRNAs encoding BMP and Nodal, and the ratio of BMP to Nodal determines the RC identity of the tissue. When 25 times more *bmp2b* mRNA was injected relative to *nodal-related 2* mRNA, caudal fates (mostly tail) were preferred. In 1:1 mRNA injections, posterior head and anterior trunk fates were preferred (Fauny et al. 2009). In an elegant follow-up experiment, BMP and Nodal mRNAs were separately injected into two opposing animal pole blastomeres at the 128-cell stage, creating distinct clones of cells that secreted these factors (Xu et al. 2014). Unlike when injecting both BMP and Nodal mRNAs into the same blastomere, injection into separate blastomeres permitted a variation of concentrations to form. This led to the creation of an entire secondary axis, confirming that BMP and Nodal are sufficient to form the entire zebrafish embryonic axis. Thus, formation of an intermediate mesoderm of appropriate size and position along the RC axis is likely dependent upon the ratio of BMP to Nodal at the blastula and early gastrula stages of development.

2.3.2 Establishing the DV Positioning of the Intermediate Mesoderm

While BMP and Nodal are required for determining the size of rostral versus caudal domains in the early embryo, a similar understanding of pathways regulating dorsal versus ventral fates is lacking. The DV axis specifies the separation of paraxial mesoderm, intermediate mesoderm, and lateral plate mesoderm. In zebrafish and *Xenopus* studies, misexpression of early kidney markers, such as *lhx1* or *pax2*, promotes ectopic kidney formation but only in paraxial mesoderm and not in lateral plate mesoderm (Bedell et al. 2012; Carroll and Vize 1999). Additionally, perturbation of anterior somite formation in *Xenopus* embryos precludes pronephrogenesis (Seufert et al. 1999). These findings suggest that the paraxial mesoderm contains an inductive signal for intermediate mesoderm formation. In chick embryonic development, a gradient of BMP4 across the DV axis has been suggested to separate the different mesodermal subdomains across this axis (James and Schultheiss 2003, 2005; Obara-Ishihara et al. 1999). However, the signal from the paraxial mesoderm to induce intermediate mesoderm in zebrafish has not been discovered. Instead, other mechanisms for DV patterning of the mesoderm have been proposed. The size of the DV extent of mesendoderm versus ectoderm in the blastula embryo is dependent upon the AV gradient of Nodal, which has been directly visualized and quantitated at blastula stages (Harvey and Smith 2009). This study found that nodal signaling is highest at the margin and lowest at the animal pole. In *squint* (*sqt*) and *cyclops* (*cyc*) mutants that are deficient in *nodal-related 1* (*ndr1*) and *nodal-related 2* (*ndr2*), respectively, no mesendoderm forms (Dougan et al. 2003). This fits with the known function of nodals in mesendoderm induction (Rodaway et al. 1999). Recently, a novel LIM-domain binding protein, Ldb2, was shown to negatively modulate ndr1 activity (Gu et al. 2015). In *ldb2* morphants, the size of the mesendoderm increased at the expense of ectoderm. Nodal is therefore required for determining the relative amounts of ectoderm versus mesendoderm that form across the AV axis. One potential model for DV patterning of the mesoderm that arises from these studies is that the concentration of Nodal could also directly induce different DV fates across this axis. This model is challenging to test, as it will be difficult to separate the role of Nodal in mesendoderm induction from a parallel role in DV patterning. The *Progressive Critical Intervals* model proposes that DV and RC fates are assigned simultaneously but progressively along the embryo during gastrula stages (Tucker et al. 2008). This model fits well for the neurectoderm, where the forebrain is specified first, then the midbrain, then finally the hindbrain and spinal cord, but further work is needed to understand if this model applies to the mesoderm.

2.3.3 Specification of Blood Versus Kidney Fates in the Intermediate Mesoderm

In addition to its contribution to the kidney anlage, the intermediate mesoderm gives rise to erythroid and endothelium precursors. The splitting of the intermediate mesoderm into kidney versus blood appears dissimilar in *Xenopus* and zebrafish. For *Xenopus*, kidney and blood fates are assigned along the RC axis, creating a caudolateral *pax8*$^{+}$ pronephric anlage and caudoventral blood anlage that becomes the ventral blood island (Kyuno et al. 2008; Lane and Sheets 2002a). In zebrafish, the intermediate mesoderm is split along the DV axis (Davidson and Zon 2004; de Jong et al. 2010). Early zebrafish red blood progenitors express *pax2a*, but later in development a medial domain of the intermediate mesoderm stripe is specified as blood and migrates towards the midline to form the inner cell mass. In *spadetail* (*spt*) mutants, which lack the T-box transcription factor *tbx16*, erythroid precursors are lost and the pronephros is expanded (Warga et al. 2013). This phenotype is suggested to be a result of elevated *fgf8a* expression in the posterior trunk of *spt* mutants, which Warga et al. (2013) propose may favor kidney fate over blood fate. Zebrafish studies into the *odd-skipped related* (*osr*) class of zinc finger transcription factors showed *osr1* expression in the endoderm during gastrulation is also important in establishing the ratio of kidney versus blood/vascular lineages (Mudumana et al. 2008). Intriguingly, the amount of endoderm in the embryo is negatively regulated by *osr1*, and *osr1*-depleted embryos have ectopic angioblast formation at the expense of anterior kidney fates. This suggests that the endoderm contains an inductive signal that promotes blood/vascular fates over kidney fates in zebrafish.

2.4 Development of the Renal Corpuscle

The next stage in pronephros development involves the expression of genes within the intermediate mesoderm that specify distinct nephron cell types. These genes are largely conserved between zebrafish and mammals. In this section, we will investigate the developmental pathways involved in the formation of the renal corpuscle (the blood filter).

2.4.1 Formation of the Renal Corpuscle Ultrastructure

The rostral-most cells of the intermediate mesoderm contribute two epithelial cell populations that will establish the renal corpuscle: podocytes and parietal epithelial cells (aka Bowman's capsule) that we herein refer to collectively as P/PECs. The *wt1a* gene is the earliest marker of the P/PEC lineage. Transcripts for *wt1a* are initially detected in a broad domain before restricting to being highly expressed in P/PEC progenitors lateral to somite 3 between the 18-somite to 24 hpf stages (Bollig et al. 2006; Drummond et al. 1998; O'Brien et al. 2011; Serluca and Fishman 2001). Other early markers of the P/PEC lineage include the transcription factors *wt1b*, *mafba*, *hey1*, and *lhx1a* (O'Brien et al. 2011) (see Sect. 2.4.2 for a detailed discussion of the roles these factors play in glomerulogenesis). At the 48 hpf stage, *wt1a*, *wt1b*, and *mafba* expression persist in the P/PEC lineage, but between 24 and 48 hpf, *hey1* and *lhx1a* levels reduce and markers of podocyte differentiation begin to be expressed (such as *podocalyxin*, *nephrin*, *podocin*, and *integrinα3*).

This commencement of molecular differentiation can be correlated to the physical formation of the renal corpuscle. Between the 24 and 40 hpf stages, the bilateral P/PEC populations migrate to, and fuse at, the midline beneath the dorsal aorta and notochord (Drummond et al. 1998; Majumdar and Drummond 2000). Between 40 and 48 hpf, capillaries extending out from the dorsal aorta enter the mass of P/PECs and form the glomerular tuft. Small molecular weight fluorescent dextrans (10 kDa) are able to filter through the glomerulus by the 48 hpf stage, indicating the onset of blood filtration function (Drummond et al. 1998). However, the renal corpuscle is leaky at this early stage and does not fully mature until 4 dpf when it has an upper size limit of 70 kDa (Ichimura et al. 2012; Kramer-Zucker et al. 2005b).

The glomerular filtration barrier is established by the cumulative functionality of three components: the fenestrated endothelium, the glomerular basement membrane (GBM), and the podocyte slit diaphragm. The fenestrations in the afferent capillaries of the glomerulus are trans-cellular passages that permit contents of the blood to pass through the surrounding GBM, but are too small to allow large proteins and blood cells to egress from the blood. The GBM consists of a feltwork of proteins such as laminin and collagen and is dually created by the podocytes and the endothelial cells of the capillaries. While the GBM filters only molecules of a certain size and charge, it is thought that the slit diaphragm created by podocytes also determines which components of the blood can pass into the tubules. The podocytes envelope the endothelium with a complex cytoarchitecture of interdigitating foot processes. The slit diaphragm forms as a protein "zipper" between the foot processes. The glomerulus is surrounded by a Bowman's capsule, which is created by PECs. The common ontogeny of the P/PEC lineage in zebrafish indicates these cells must be separated later in development by an as yet unknown mechanism. The final cellular component of the renal corpuscle is the mesangial cell, which is a pericyte-like cell type whose function is to provide contractile support to the glomerulus (Ichimura et al. 2012). When pericytes enter the glomerular tuft in zebrafish is not known, but potential candidate markers to examine in the future include *acta2*, *timp3*, and *adamts1* (Schrimpf et al. 2012; Whitesell et al. 2014).

2.4.2 Molecular Regulation of Renal Corpuscle Formation

2.4.2.1 The Role of the WT1 Transcription Factors in Determining Podocyte Fate and Function

Knockdown of *wt1b* has no effect on P/PEC specification and development, whereas knockdown of *wt1a* reduces *nephrin* and *podocin* expression (Bollig et al. 2006; O'Brien et al. 2011; Perner et al. 2007; Schnerwitzki et al. 2014). *wt1a* has been shown to interact synergistically with *forkhead box c1a* (*foxc1a*) and *notch mediator recombination signal binding protein for immunoglobulin kappa J* (*rbpj*) as double knockdowns of *wt1a/foxc1a* or *wt1a/rbpj* completely prevents expression of early podocyte markers such as *wt1b*, whereas singly depleted *wt1a*, *foxc1a* or *rbpj* embryos only reduce *wt1b* expression (O'Brien et al. 2011). Glutathione S-transferase pull-down assays indicated that *wt1a*, *foxc1a*, and *NotchICD3* can physically interact (O'Brien et al. 2011). While further analysis is required to decipher the importance of these interactions and to determine if wt1a is able to interact with other transcription factors and Notch pathway molecules, it appears that cross talk between wt1a and the Notch signaling pathway is important for the formation of the P/PEC lineage, most likely through fate determination and control of proliferation.

2.4.2.2 Retinoic Acid Patterns the Intermediate Mesoderm to Promote P/PEC Fates

A conserved 299 bp element of the *wt1a* promoter (approximately 4.2 kb upstream from the start codon) was found to be necessary and sufficient to drive transgenic *eGFP* expression in a spatial and temporal manner that recapitulated endogenous *wt1a* expression (Bollig et al. 2009). Within this region of the *wt1a* promoter is a Retinoic Acid Response Element. The Retinoic Acid (RA) signaling pathway is dependent on the concentration of free RA in the cell, which will bind to RA receptors (RAR/RXR). When bound to RA, RARs/RXRs translocate to the nucleus and interact with DNA binding sites in the genome to modulate gene expression. Embryos treated with an inhibitor of RA synthesis (diethylaminobenzaldehyde) fail to express *wt1a, wt1b*, and *mafba* in the intermediate mesoderm (Wingert et al. 2007). Conversely, application of RA to *wt1a::eGFP* transgenic embryos greatly increased the expression of eGFP (Bollig et al. 2009). These results support a model whereby RA induces *wt1a* expression and, together with notch, *wt1a* specifies the P/PEC lineage from the intermediate mesoderm.

2.4.2.3 pax2a Restricts the Size of the Glomerulus

In the *pax2a* mutant *no isthmus* (*noi*), the *wt1b*$^+$ P/PEC lineage appears to expand into the neck segment, which suggests a requirement for *pax2a* in appropriate formation of the renal corpuscle and neck lineages (Majumdar et al. 2000). In mice, Wt1 is a negative regulator of *Pax2* (Ryan et al. 1995), and a similar epistatic relationship between *wt1a* and *pax2a* may be present in zebrafish as the anterior *pax2a* expression domain initially overlaps with *wt1a* in P/PECs, but then restricts to the neck segment at 24 hpf (O'Brien et al. 2011). In addition, *pax2a* overexpression reduces *wt1a* expression and promotes formation of an aglomerular pronephros (Bedell et al. 2012). These data suggest that *pax2a* and *wt1a* are negative regulators of each other. Such genetic interactions may be important for the appropriate formation of the glomerulus/neck/proximal tubule boundaries.

2.4.2.4 Odd-Skipped Related 1 and Lhx1a as Downstream Regulators of Podocyte Fate

Podocyte progenitors co-express *wt1a* and *osr1* (Tomar et al. 2014). In embryos depleted of *osr1*, *wt1a* expression is normal, but *nephrin* and *podocin* are not expressed. However, in *wt1a*-depleted embryos, *osr1* expression is reduced (Tomar et al. 2014). In summary, these results suggest *osr1* acts downstream of *wt1a* to regulate *nephrin* and *podocin* expression. Interestingly, *osr1* morphant embryos fail to express *lhx1a* in P/PECs, and the effects of *osr1* depletion on *nephrin* and *podocin* expression can be rescued by forced expression of an activated form of *lhx1a* (ldb1-lhx1a) (Tomar et al. 2014). Taken together, it appears that the following hierarchical transcriptional pathway operates during P/PEC formation: RA > *wt1a* > *osr1* > *lhx1a* > *podocin/nephrin*.

2.4.2.5 Podocalyxin Is Necessary for Proper Formation of the Slit Diaphragm

Podocalyxin is a highly sialyated glycoprotein that localizes specifically to the apical pole of podocytes by 34 hpf (Ichimura et al. 2013). The exact mechanism for how podocalyxin regulates foot process formation is not fully understood. In *Podocalyxin*-deficient mice, regular foot processes do not form and the cell body attaches directly to the GBM (Doyonnas et al. 2001). A similar phenotype is observed in zebrafish injected with a splicing morpholino targeting the *podocalyxin* mucin-domain encoding exon 2 (Ichimura et al. 2013). This mucin domain is extensively glycosylated and sialyated. The presence of sialic acid in the mucin domain creates a negative charge on the apical surface of the podocyte that is hypothesized to generate and maintain spacing between foot processes by charge repulsion and by acting as an anti-adhesive (Schnabel et al. 1989; Takeda et al. 2000).

2.4.2.6 A Role for von Hippel–Lindau in Glomerular vegfa Signaling

The *von Hippel–Lindau* (*vhl*) tumor suppressor gene encodes an E3-ubiquitin ligase that under normal oxygen levels targets the *hypoxia-inducing factor* (*hif*) for degradation (Haase 2006). The vhl/hif oxygen-sensing pathway enables organisms to sense and adapt to a low-oxygen environment. Depletion of zebrafish genes that regulate the vhl/hif pathway, such as *proly 4-hyrdoxylase* (Hyvärinen et al. 2010) or *vhl* (Chen et al. 2015), results in disrupted glomerular development. Knockdown of *vhl* creates an embryo that is unable to recognize oxygen levels in the body. Consequently, hif protein is not degraded and pathological hypoxia-driven angiogenesis results. Vascular endothelial growth factor-a (Vegf-a) is required for endothelial cell differentiation and angiogenesis during vertebrate development and in *vhl* morphants, *vegfa* expression increases (Chen et al. 2015). It is hypothesized that this increased *vegfa* expression causes overgrowth of the vasculature, which is likely to disrupt vascularization of the glomerulus of *vhl* morphants.

2.5 Genes Involved in Pronephros Tubule Formation

The tubules are essential for reabsorption of solutes, amino acids, glucose, and water. Their formation involves a number of key developmental processes, including molecular patterning, mesenchymal-to-epithelial transitions, tubulogenesis, cell migration, and tissue morphogenesis. In this section, we will concentrate on the genes expressed in the intermediate mesoderm that are involved in the earliest stages of pronephric tubule formation.

2.5.1 Pax2a and Pax8: Critical Regulators of Nephric Specification

In mice and zebrafish, the paired-box transcription factors *Pax2* (*pax2a* in zebrafish) and *Pax8* are the earliest expressed genes that label the intermediate mesoderm (Bouchard et al. 2000; Thisse and Thisse 2004). In mice lacking Pax2 and Pax8, neither the pronephros nor later nephric structures form (Bouchard et al. 2002). This same phenotype is observed in zebrafish depleted of *pax2a* and *pax8* (Naylor et al. 2013). *Pax2* or *Pax8* singly deficient mice and zebrafish embryos do not develop such a dramatic early kidney agenesis phenotype, consistent with these related factors having redundant functions in the specification of the intermediate mesoderm (Ikenaga et al. 2011; Majumdar et al. 2000; Torres et al. 1995). *pax2a* and *pax8* are expressed throughout the intermediate mesoderm up to the ~15-somite stage (16 hpf). Subsequently, levels of *pax8* expression reduce, but *pax2a* remains strongly expressed in the neck region, multi-ciliated cells in the PST and DE tubule segments and throughout the DL segment. Singly deficient *pax2a* zebrafish develop edema and die at around 5 dpf, most likely as a consequence of perturbed glomerulus, neck and tubule differentiation, and variable failure in the fusion of the DL segment to the cloaca (Majumdar et al. 2000). Thus, while *pax2a* and *pax8* share functionality, differing expression patterns result in *pax2a* playing a more dominant role during later events in nephrogenesis.

The biochemical activities of Pax2 in the cell and how these actions affect nephrogenesis remain an ongoing area of investigation. In zebrafish, the *hepatocyte nuclear factor 1b* (*hnf1b*) transcription factor has been reported to suppress *pax2a* expression in the proximal tubule (Naylor et al. 2013). Zebrafish embryos deficient in *hnf1ba* and *hnf1bb* paralogs maintain *pax2a* expression and ectopically express glomerular markers such as *nephrin* and *podocin* in the proximal tubule. A similar phenotype is observed in embryos depleted of *atypical protein kinase C iota* and *zeta* (*prkcι*, *prkcζ*) where *pax2a* expression increases and *wt1a* and *wt1b* are ectopically expressed in the tubule (Gerlach and Wingert 2014). Interestingly, this phenotype can be rescued by *pax2a* depletion, suggesting that *prkcι* and *prkcζ* support epithelial identity by inhibiting *pax2a* expression. Together, it could be deduced from these studies that hnf1ba/b and prkcι/ζ regulate *pax2a* levels in the tubule, and this activity may be important for the suppression of glomerular fate in the proximal tubule.

pax2a has also been shown to participate in a negative feedback loop by inducing the expression of *plac8 onzin related protein 1* (*ponzr1*), a poorly studied gene that acts to antagonize *pax2a* expression. Morpholino-mediated knockdown of *ponzrl* causes persistent expression of *pax2a* in P/PECs, resulting in the downregulation of *wt1a* and the formation of an aglomerular pronephros (Bedell et al. 2012).

2.5.2 The Role of Odd-Skipped Genes in Tubule Formation

The Odd-skipped-related class of transcription factors are zinc finger proteins that have been shown to be important in embryonic patterning and tissue morphogenesis. In *Odd1*$^{-/-}$ mutant mice, heart development is disrupted and nephrogenesis is perturbed as the metanephric mesenchyme fails to form (Wang et al. 2005). In chick, *Odd1* has been shown to be important in tubule differentiation (James et al. 2006). As described in Sect. 2.3.3, knockdown of *osr1* in zebrafish inhibits the formation of the proximal tubule and concomitantly promotes angioblast fate (Mudumana et al. 2008).

While *osr1* is not thought to be expressed in the pronephric tubules, the closely related *osr2* gene is found in the anteriormost portion of the pronephros (Neto et al. 2012; Tena et al. 2007). Double knockdown of *osr1* and *osr2* inhibits formation of the proximal tubule in zebrafish and causes severe edema (Tena et al. 2007). Interestingly, RA signaling activates *osr2* expression in the intermediate mesoderm and *osr1/2* are required for inducing *wnt2ba* expression in this region (Neto et al. 2012). While *wnt2ba* knockdown has no effect on kidney development, it does inhibit pectoral fin formation, suggesting a relay mechanism is in place between the anterior paraxial/intermediate mesoderm and lateral plate mesoderm to regulate fin development. These findings nicely highlight how cross talk between different tissues, in this case kidney and fin precursors, is fundamentally important for organogenesis.

2.5.3 *hnf1b* as a Critical Regulator of Pronephric Tubule Differentiation

The *hnf1ba* and *hnf1bb* transcription factors are expressed in the zebrafish intermediate mesoderm that will form the pronephric tubules from the 5-somite stage of development (12 hpf) onwards (Naylor et al. 2013; Sun and Hopkins 2001). In embryos depleted of both *hnf1ba* and *hnf1bb*, nearly all markers of tubule differentiation fail to be expressed (Naylor et al. 2013). These include the many segment-specific solute carriers (such as *slc4a4*, *slc12a1*, and *slc12a3*), later expressed transcription factors (such as *irx3b*) and cell adhesion molecules (such as *cdh17*). In addition, *hnf1ba/b*-deficient embryos maintain expression of early acting transcription factors such as *pax2a*, *pax8*, *lhx1a*, and *jag1b*. Despite these early markers not being restricted, the pronephros still undergoes a degree of epithelialization and tubulogenesis, though the lumen does not inflate. As such, this study highlights the importance of Hnf1b factors for tubule differentiation, but not tubulogenesis. This integral role for Hnf1b in zebrafish pronephric tubule differentiation is conserved in mice as conditional deletion of *Hnf1b* in metanephric nephrons results in a similar phenotype (Heliot et al. 2013; Massa et al. 2013).

2.5.4 Roles for Notch/Jagged in Pronephric Tubule Differentiation

The Notch signaling pathway is a paracrine-signaling network that is activated by binding of the trans-membrane Notch receptor to its Jagged/Delta ligands (Guruharsha et al. 2012). This binding permits γ-secretase to cleave the Notch receptor, releasing the Notch intra-cellular domain (NICD). NICD then translocates to the nucleus where it forms a core transcriptional complex with Suppressor of Hairless (a DNA-binding protein) and Mastermind (a nuclear effector protein required to stabilize the transcription complex). This complex then modulates the expression of downstream effector genes. Commonly, Notch signaling is utilized in a "lateral inhibition" pathway where a group of cells signal locally to each other in order to control which cells will adopt one of two different fates. In the zebrafish pronephros, multi-ciliated cells (MCCs) are important for promoting fluid flow through the lumen of the tubule and are present in a "salt and pepper" pattern in the intermediate tubule region (comprising the PST and DE tubule segments) (Fig. 2.1). This dispersed positioning of MCCs between solute transporter cells is mediated by a Notch lateral inhibition mechanism. From the 5-somite stage (12 hpf), *jag2a* is expressed throughout the anterior portion of the pronephric tubules (which will give rise to the PCT, PST, and DE segments) (Thisse and Thisse 2005). Similarly, *notch1a* and *notch3* are expressed in the pronephros from as early as the 10-somite stage (14 hpf) (Ma and Jiang 2007). From the 20-somite stage (19 hpf), *jag2a* expression restricts to a "salt and pepper" pattern in the intermediate region of the pronephros (within the PST and DE segments). In *mindbomb*$^{-/-}$ mutants (*mib*$^{-/-}$), *jag2a* expression does not restrict and is uniformly expressed in the intermediate region of the tubule. These embryos do not express markers associated with solute transport function, such as *trpm7*, *Na*$^{+}$*K*$^{+}$*ATPase*, or *slc13a1*, and instead have ectopic expression of MCC markers, such as *odf3b* and *rfx2* (Liu et al. 2007; Ma and Jiang 2007). Injection of *NICD* mRNA induces the opposite phenotype to *mib*$^{-/-}$ mutants, inhibiting *odf3b* expression and promoting transporter fates over MCCs. Thus, a Notch signaling lateral inhibition pathway determines MCC or solute transporter cell fate in the pronephric tubule. Knock-down of ETS transcription factors *etv4* or *etv5a* also reduces the number of MCCs, but not when embryos are treated with the γ-secretase inhibitor DAPT (Marra and Wingert 2016). This result suggests a novel role for *etv4* and *etv5a* downstream of Notch signaling to promote MCC fate in the pronephros.

2.5.5 *Roles for Cilia in Pronephrogenesis*

Cilia are membrane-bounded, centriole-derived, microtubule-containing organelles that project out from the cell surface. The ciliary cytoskeleton (axoneme) is arranged into two major patterns: 9+2, in which microtubule doublets surround a central pair of singlet microtubules or 9+0, where the central microtubules are absent (Satir and Christensen 2007). The two microtubule patterns are indicative of the functional role of a cilium. 9+2 cilia are motile and present in bundles on the apical surface of epithelial cells, whereas 9+0 cilia are nonmotile, sometimes referred to as primary or sensory cilia, which are able to alter intracellular biochemistry based on extracellular cues. Over 600 cilia proteins have been identified (Pazour 2004), and dysfunction of a number of these proteins can cause an array of diseases collectively termed ciliopathies. These broad set of developmental and adult diseases include polycystic kidney disease, nephronophthisis (NPHP), Bardet–Biedl Syndrome, Joubert syndrome, and Meckel Gruber syndrome (Hildebrandt et al. 2011). The zebrafish kidney is particularly vulnerable to aberrant ciliogenesis as both multi-ciliated cells (MCCs) and nonmotile mono-ciliated cells are required for pronephros development.

2.5.5.1 MCCs Drive Fluid Flow in the Zebrafish Pronephros

9+2 cilia are found in MCCs across the animal kingdom, where they act to promote and direct fluid flow (Brooks and Wallingford 2014). For example, in humans, MCCs are important for cerebrospinal fluid flow in the spinal cord and ventricles of the brain (Sawamoto et al. 2006), for ovum transport in the fallopian tubes (Lyons et al. 2006), and for clearance of mucus in the airways (Wanner et al. 1996). In zebrafish, mutants with defective ciliogenesis develop pronephric cysts during early development (Drummond 2005; Obara et al. 2006; Sullivan-Brown et al. 2008). Examples of such mutants include *oval* (containing a single point mutation in *ift88* (Tsujikawa and Malicki 2004)), *double bubble* [unknown mutation (Drummond et al. 1998)], *fleer* [which has a nonsense point mutation that truncates the ift70 protein (Pathak et al. 2007)], and a number of cilia genes [including *pkd2*, *ruvbl1*, *lrcce6l*, and *arl13b*] found through the Hopkins retroviral insertion mutagenesis screen (Sun et al. 2004). Similarly, morpholino-mediated knockdown of intraflagellar transport proteins *ift88*, *ift57* (Kramer-Zucker et al. 2005a), and *ttc26* (Zhang et al. 2012) causes pronephric cysts. The deglutamylase *ccp5*, a critical regulator of microtubule glutamylation (which is essential for ciliogenesis), also causes pronephric cysts when it is depleted (Pathak et al. 2014). Ciliogenesis and cilia function requires the highly conserved family of GTP-binding proteins called Septins (Kim et al. 2010). The pronephric tubules have enriched expression of *sept7b*, and morpholino knockdown results in fewer and shorter cilia as well as cyst formation (Dash et al. 2014). Such "cystic" phenotypes are believed to be due to fluid accumulation and distension of the pronephric tubules/neck region, consistent with the notion that MCCs drive fluid flow in the lumen of the pronephric tubule (Kramer-Zucker et al. 2005a). MCCs are not normally found in adult mammalian nephrons, perhaps because normal blood pressure is sufficient to propel fluid through the glomerulus and tubules of the nephron [normal systolic blood pressure in humans is ~110 mmHg, whereas in zebrafish it is ~0.68 mmHg (Hu et al. 2001)]. As such, mammalian renal ciliopathies are associated with nonmotile cilia dysfunction.

2.5.5.2 Nonmotile Cilia Are Important for Distal Tubule Morphogenesis

Apart from MCCs, all other cells in the pronephric tubules contain a solitary nonmotile cilium that can sense both physical and biochemical extracellular signals. Discerning between the pronephric phenotypes associated with aberrant motile versus nonmotile cilia function remains an ongoing challenge given the shared assembly mechanisms for both cilia types (Ishikawa and Marshall 2011). As such, factors that impede cilia assembly will prevent function of both cilia forms. Despite this, numerous signaling pathways connect nonmotile cilia to proper pronephros morphogenesis (Hossain et al. 2007; Makita et al. 2008; Tian et al. 2007). Downstream co-activators of the Hippo pathway, *yap* and *taz*, are expressed in the distal tubules of the zebrafish pronephros, and their depletion causes pronephric cyst phenotypes (He et al. 2015; Skouloudaki et al. 2009; Zhang et al. 2015). These morphants have reduced cilia number and length in the PST and DE tubule segments where MCCs reside, but also in the DL segment that contains mono-ciliated cells. Such ciliary defects are correlated with aberrant cell migration and apical-basal polarity in the distal tubule. Similarly, morpholino knockdown of *nphp4* reduces cilia number and length in the DL segment and perturbs cell migration and DL fusion with the cloaca (Burckle et al. 2011; Slanchev et al. 2011). These studies found that the perturbed cloacal rearrangements in *nphp4* morphants, which are attributed to the nonmotile cilia defect, result in a failure in non-canonical Wnt signaling, a pathway implicated in the orientation of cells within a single plane (Gao 2012). Taken together, these results favor a model whereby mono-ciliated sensory cilia in the zebrafish distal nephron are required to mediate the cellular rearrangements needed to fuse the pronephric tubules to the cloaca.

2.6 Pronephric Tubule Segmentation

In all vertebrates, the nephron is segmented along its axis in order to enable efficient reabsorption/secretion of solutes. The proximal tubule is considered the "workhorse" of the nephron as it performs the bulk reabsorption of these solutes. More distal segments are involved in fine-tuning the filtrate, in particular by reabsorption of sodium, chloride, and bicarbonate ions. The mammalian nephron contains additional nephron segments that are not found in zebrafish nephrons, such as the loop of Henle and principal and intercalated cells of the distal nephron/collecting duct. Nevertheless, zebrafish tubule segments are well conserved functionally in comparison to mammals, and many transporter genes show similar expression patterns in both phyla (Wingert et al. 2007). In this section, we will provide an overview of our current understanding of how the zebrafish pronephric tubule is patterned in response to RA signaling.

2.6.1 Retinoic Acid and Establishment of the Tubule Segmentation Pattern

In Sect. 2.3.1, we showed that opposing gradients of BMP and Nodal establish the RC axis of the gastrulating zebrafish embryo. Factors acting downstream of BMP/Nodal are likely involved in committing cells to particular fates along this axis. One candidate is RA, which has been shown to be important for the relative sizes of the PCT/PST/DE/DL tubule segments (Wingert et al. 2007). In embryos where RA synthesis is inhibited by treatment with diethylaminobenzaldehyde (DEAB), pronephroi form with larger distal segments at the expense of proximal segments (Wingert et al. 2007). Inhibiting RA synthesis can completely preclude formation of all proximal segments (PCT, PST, and DE) if DEAB treatment is commenced from early gastrula stages. Progressively later treatments have gradually reduced effects and treatment from the 8-somite stage has no effect on pronephric fate. Thus, RA is important in regulating pronephros tubule fate during gastrulation and early somitogenesis stages of development.

RA is synthesized from retinol by retinaldehyde dehydrogenase (also called aldhehyde dehydrogenase). At gastrula stages of development, a major retinaldehyde dehydrogenase expressed in the zebrafish embryo is *aldh1a2*. *aldh1a2* is initially expressed in all cells of the embryonic margin but restricts to the rostral side of the embryo by the ~60% epiboly stage (6 hpf) (Grandel et al. 2002). From mid-to-late gastrulation, the RA catabolic enzyme *cyp26a1* (a member of the cytochrome p450 family of enzymes) begins to be expressed on the caudal side of the embryo (Kudoh et al. 2001; Thisse et al. 2001). These expression patterns support the interpretation that there is an RA "source" on the rostral side of the embryo and an RA "sink"on the caudal side of the embryo. This suggestion is aided by analyses that directly observed RA abundance in the zebrafish embryo using a fluorescent resonance energy transfer-based system called GEPRA (Shimozono et al. 2013). GEPRA zebrafish embryos show a clear accumulation of RA on the rostral side of the late gastrula embryo, which appears to be maintained in the anterior trunk region at early somitogenesis stages of development. Taken together, the timing of RA proximalizing actions (during gastrulation), the high expression of RA synthesis genes rostrally and RA catabolic genes caudally, as well as the accumulation of RA rostrally when observed by GEPRA analysis suggest RA acts to pattern kidney fates across the RC axis of the embryo during gastrula stages. As RA can act as a morphogen, it can further be proposed that PCT cells form in response to high levels of RA, PST cells in response to medium levels of RA, DE cells in response to low RA levels, and the DL segment in the absence of RA (Wingert et al. 2007).

2.6.2 A Role for sim1a in the Formation of the PST Segment

RA is the major determinant of pronephric segment identity; however, the downstream effectors of RA are a focus of ongoing research. One possible mechanism by which RA affects gene expression is via the modulation of transcription factor activities. The *single-minded family bHLH transcription factor 1a* (*sim1a*) gene is dynamically expressed from early stages of zebrafish pronephros development. At the 2-somite stage (11 hpf), it is expressed in the caudal region of the intermediate mesoderm, but later in development [at the 22-somite stage (20 hpf)] it is expressed in a more proximal subdomain before restricting to a distal population of cells that contribute to a kidney-derived endocrine gland called the Corpuscles of Stannius (Cheng and Wingert 2015). Intriguingly, *sim1a* knockdown prevents formation of the PST segment and concomitantly expands the PCT segment. When overexpressed, the opposite phenotype is observed, and PST segment size increases at the expense of PCT fates. The expression pattern of *sim1a* is influenced by RA levels, suggesting RA induces *sim1a* expression, and sim1a acts downstream of RA to mediate the boundary between the PCT and PST segments.

2.6.3 mecom as an RA Inhibitor that Regulates Formation of the DL Segment

Another transcription factor whose expression is regulated by RA is *mecom* (*mds1/evi1 complex*). This gene is expressed in the caudal non-RA responsive region of the intermediate mesoderm that gives rise to the DL segment (Li et al. 2014). In *mecom* morphants, the size of the DL segment reduces and the PCT and PST segments expand caudally (Li et al. 2014). Surprisingly, *Mecom* morphants have increased numbers of MCCs. As discussed in Sect. 2.5.4, Notch signaling via lateral inhibition is required to induce MCC fate in the intermediate region of the pronephros tubule. Li et al. suggest that Mecom interacts with the Notch signaling although it is also possible that the increased numbers of MCCs in *mecom*-deficient embryos is caused indirectly, such as, by increased proliferation of the MCC-bearing PST and DE segments in response to a shortened DL segment.

2.6.4 *irx3b* as a Determinant of DE Segment Formation

The *Iroquois* (*Irx*) gene family encode homeodomain transcription factors that are regulators of tissue patterning and cell fate specification. In zebrafish, *irx3b* is expressed in the PST and DE segments from the 15-somite stage (16 hpf). In *irx3b* zebrafish morphants, *slc12a1* expression is lost in the DE segment and markers of the proximal tubule are expanded distally (Wingert and Davidson 2011). In *Xenopus*, *irx3* knockdown also disrupts the formation of the segment equivalent to the DE (Alarcon et al. 2008; Reggiani et al. 2007). In mice, *Irx1* and *Irx2* are downregulated in *Hnf1b*-deficient nephrons, and this is associated with the proximal tubule and loop of Henle segments failing to form, consistent with *Irx1/2* being potential regulators of segmentation.

In zebrafish, the related *Iroquois* family member, *irx1*, is expressed in a similar domain to *irx3b*. However, knockdown of *irx1* does not cause an overt tubule segmentation phenotype. Expression of *irx1* is negatively regulated by the homeobox transcription factors *mnx1* and *mnx2b* that are expressed in mutually exclusive regions of the intermediate mesoderm to *irx1* (Ott et al. 2015). When *mnx1* and *mnx2b* are knocked down, *irx1* becomes ectopically expressed in the DL segment and these embryos exhibit defects in cilia arrangement and apical microvilli morphology (Ott et al. 2015). Taken together, these observations indicate that the *Iroquois* genes are conserved regulators of tubule segmentation identity and morphogenesis in zebrafish, frogs, and mammals.

2.7 Mechanobiological Regulation of Pronephrogenesis

It is increasingly being recognized that mechanical forces are required for the development of nephric structures. The process of mechanotransduction involves the interpretation by a cell of mechanical inputs that lead to changes in its biochemistry. During early development, the intermediate mesoderm undergoes extensive remodeling, during which cells experience a number of mechanical stresses. Between the 12-somite and 20-somite stage (15–19 hpf), the intermediate mesoderm undergoes mesenchymal-to-epithelial transitions (MET) and forms a tubule (Gerlach and Wingert 2014). In addition, there is a degree of caudal migration of the DL segment towards the cloaca in an analogous manner to nephric duct migration observed in *Xenopus*, chick, and mouse (Slanchev et al. 2011). Such caudal migration means cells will be stretched, a mechanical process that can alter rates of proliferation in the pronephros (Vasilyev et al. 2012). Also, at the onset of lumenal flow, the apical surfaces of the pronephric tubules experience sheer stress. How cells in the pronephros respond to these multiple mechanical inputs has become a new area of interest in the field.

2.7.1 Shear Stress Induction of Collective Cell Migration in the Pronephric Tubule

Collective cell migration is the movement of multiple cells in a specific direction with the major feature being that if these cells were individually isolated, they would not migrate as efficiently (Mayor and Etienne-Manneville 2016). In the zebrafish pronephros, collective cell migration plays a key role in tissue morphogenesis. From the 29 hpf stage of development, the cells of the pronephros collectively move rostrally (Vasilyev et al. 2009). This migration compacts and convolutes the proximal tubule. The initiation of collective cell migration in the zebrafish pronephros correlates with the onset of fluid flow in the lumen of the tubule. When fluid flow is halted by simple obstruction, collective migration halts and the proximal tubule fails to convolute (Vasilyev et al. 2009). In follow-up experiments, it was shown that the capacity of the pronephros to accommodate such a dramatic compaction in the proximal segments is afforded by the corresponding proliferation of the DE and DL segments. This can be clearly seen when older embryos are compared to young larvae: at 24 hpf the size of the DL segment is roughly three somite widths (Wingert et al. 2007) but by the 8 mm stage (~21 days post fertilization), it is much longer, spanning between the posterior end of the swim bladder to the region just proximal to the cloaca (Diep et al. 2015). Cell division is observed in the DE and DL tubule segments between the 48 and 72 hpf stages of development, and this can be inhibited by anterior obstruction of the tubule (to halt collective cell migration) or by treating embryos with the Phosphoinositide-3 Kinase (Pi3K) inhibitor LY294002. Treatment with LY294002 produced pronephric tubules that still underwent collective cell migration towards the glomerulus but distally the cells became severely stretched (Vasilyev et al. 2012). These data have led to a model in which shear stress induced by fluid flow promotes rostral cell migration, and this compaction is accommodated by mechanical stretch-induced cell proliferation in the more distal segments.

2.7.2 Vascular Shear Stress Is Required for Capillary Formation in the Glomerulus

As discussed in Sect. 2.4.1, the process of glomerular capillary tuft formation involves the invasion of dorsal aorta endothelial cells into the fused mass of P/PEC progenitors at the midline (Carmeliet et al. 1996; Pham et al. 2001). This process requires vascular sheer stress as zebrafish mutants that have reduced or no blood flow fail to form a glomerular capillary tuft (Bedell et al. 2012; Drummond and Davidson 2010; Majumdar and Drummond 1999; Rottbauer et al. 2001; Sehnert et al. 2002). Studies have implicated matrix-metalloproteinase 2 (MMP2) in this process. Zebrafish embryos treated with an MMP2 inhibitor do not form a glomerulus but have otherwise normal blood circulation (Serluca et al. 2002). As *mmp-2* is expressed by smooth muscle cells and endothelium in response to stretching it suggests that blood flow-induced sheer stress activates *mmp-2*, which then regulates vascularization of the renal corpuscle (Bassiouny et al. 1998; Singhal et al. 1996; Yasuda et al. 1996).

2.8 Summary

The pronephros descends from the intermediate mesoderm, and its formation involves the concerted actions of secreted morphogens, transcription factors, and cellular rearrangements including migration, stretch, and epithelialization. Patterning of the intermediate mesoderm into different proximal and distal cell fates occurs in response to RA signaling and downstream acting transcription factors such as *wt1a*, *pax2a*, *pax8*, *hnf1b*, *sim1a*, *mecom*, and *irx3b*. Glomerular progenitors migrate and fuse at the midline and recruit a blood supply while tubule progenitors undergo epithelialization and fuse with the cloaca. The Notch signaling pathway regulates the formation of multi-ciliated cells in the tubules that help propel the urine down the tubule. The lumenal sheer stress caused by this flow activates anterior collective migration of the proximal tubules and induces stretching and proliferation of the more distal segments. Ultimately these processes create a simple two-nephron kidney capable of reclaiming and balancing vital metabolites and expelling excess water and waste products to the exterior. This simple, yet dynamic, kidney provides a powerful model system to better understand the conserved molecular and cellular progresses that drive nephron formation, structure, and function.

References

Alarcon P, Rodriguez-Seguel E, Fernandez-Gonzalez A, Rubio R, Gomez-Skarmeta JL (2008) A dual requirement for Iroquois genes during Xenopus kidney development. Development 135:3197–3207

Bassiouny HS, Song RH, Hong XF, Singh A, Kocharyan H, Glagov S (1998) Flow regulation of 72-kD collagenase IV (MMP-2) after experimental arterial injury. Circulation 98:157–163

Bedell VM, Person AD, Larson JD, McLoon A, Balciunas D, Clark KJ, Neff KI, Nelson KE, Bill BR, Schimmenti LA (2012) The lineage-specific gene ponzr1 is essential for zebrafish pronephric and pharyngeal arch development. Development 139:793–804

Bertram JF, Douglas-Denton RN, Diouf B, Hughson MD, Hoy WE (2011) Human nephron number: implications for health and disease. Pediatr Nephrol 26:1529–1533

Bollig F, Mehringer R, Perner B, Hartung C, Schafer M, Schartl M, Volff JN, Winkler C, Englert C (2006) Identification and comparative expression analysis of a second wt1 gene in zebrafish. Dev Dyn 235:554–561

Bollig F, Perner B, Besenbeck B, Kothe S, Ebert C, Taudien S, Englert C (2009) A highly conserved retinoic acid responsive element controls wt1a expression in the zebrafish pronephros. Development 136:2883–2892

Bouchard M, Pfeffer P, Busslinger M (2000) Functional equivalence of the transcription factors Pax2 and Pax5 in mouse development. Development 127:3703–3713
Bouchard M, Souabni A, Mandler M, Neubuser A, Busslinger M (2002) Nephric lineage specification by Pax2 and Pax8. Genes Dev 16:2958–2970
Brooks ER, Wallingford JB (2014) Multiciliated cells. Curr Biol 24:R973–R982
Burckle C, Gaude HM, Vesque C, Silbermann F, Salomon R, Jeanpierre C, Antignac C, Saunier S, Schneider-Maunoury S (2011) Control of the Wnt pathways by nephrocystin-4 is required for morphogenesis of the zebrafish pronephros. Hum Mol Genet 20:2611–2627
Carmeliet P, Ferreira V, Breier G, Pollefeyt S, Kieckens L, Gertsenstein M, Fahrig M, Vandenhoeck A, Harpal K, Eberhardt C, Declercq C, Pawling J, Moons L, Collen D, Risau W, Nagy A (1996) Abnormal blood vessel development and lethality in embryos lacking a single VEGF allele. Nature 380:435–439
Carroll TJ, Vize PD (1999) Synergism between Pax-8 and lim-1 in embryonic kidney development. Dev Biol 214:46–59
Chen YH, Chang CF, Lai YY, Sun CY, Ding YJ, Tsai JN (2015) von Hippel-Lindau gene plays a role during zebrafish pronephros development. In vitro cellular and developmental biology. Animal 51:1023–1032
Cheng CN, Wingert RA (2015) Nephron proximal tubule patterning and corpuscles of Stannius formation are regulated by the sim1a transcription factor and retinoic acid in zebrafish. Dev Biol 399:100–116
Dash SN, Lehtonen E, Wasik AA, Schepis A, Paavola J, Panula P, Nelson WJ, Lehtonen S (2014) Sept7b is essential for pronephric function and development of left-right asymmetry in zebrafish embryogenesis. J Cell Sci 127:1476–1486
Davidson AJ, Zon LI (2004) The 'definitive' (and 'primitive') guide to zebrafish hematopoiesis. Oncogene 23:7233–7246
de Jong JL, Davidson AJ, Wang Y, Palis J, Opara P, Pugach E, Daley GQ, Zon LI (2010) Interaction of retinoic acid and scl controls primitive blood development. Blood 116:201–209
De Robertis EM (2009) Spemann's organizer and the self-regulation of embryonic fields. Mech Dev 126:925–941
Diep CQ, Peng Z, Ukah TK, Kelly PM, Daigle RV, Davidson AJ (2015) Development of the zebrafish mesonephros. Genesis 53:257–269
Dougan ST, Warga RM, Kane DA, Schier AF, Talbot WS (2003) The role of the zebrafish nodal-related genes squint and cyclops in patterning of mesendoderm. Development 130:1837–1851
Doyonnas R, Kershaw DB, Duhme C, Merkens H, Chelliah S, Graf T, McNagny KM (2001) Anuria, omphalocele, and perinatal lethality in mice lacking the CD34-related protein podocalyxin. J Exp Med 194:13–27
Drummond IA (2005) Kidney development and disease in the zebrafish. J Am Soc Nephrol 16:299–304
Drummond IA, Davidson AJ (2010) Zebrafish kidney development. Methods Cell Biol 100:233–260
Drummond IA, Majumdar A, Hentschel H, Elger M, Solnica-Krezel L, Schier AF, Neuhauss SC, Stemple DL, Zwartkruis F, Rangini Z, Driever W, Fishman MC (1998) Early development of the zebrafish pronephros and analysis of mutations affecting pronephric function. Development 125:4655–4667
Elinson RP, Rowning B (1988) A transient array of parallel microtubules in frog eggs: potential tracks for a cytoplasmic rotation that specifies the dorso-ventral axis. Dev Biol 128:185–197
Fauny JD, Thisse B, Thisse C (2009) The entire zebrafish blastula-gastrula margin acts as an organizer dependent on the ratio of Nodal to BMP activity. Development 136:3811–3819
Gao B (2012) Wnt regulation of planar cell polarity (PCP). Curr Top Dev Biol 101:263–295
Gerlach GF, Wingert RA (2014) Zebrafish pronephros tubulogenesis and epithelial identity maintenance are reliant on the polarity proteins Prkc iota and zeta. Dev Biol 396:183–200
Grandel H, Lun K, Rauch GJ, Rhinn M, Piotrowski T, Houart C, Sordino P, Kuchler AM, Schulte-Merker S, Geisler R, Holder N, Wilson SW, Brand M (2002) Retinoic acid signalling in the

zebrafish embryo is necessary during pre-segmentation stages to pattern the anterior-posterior axis of the CNS and to induce a pectoral fin bud. Development 129:2851–2865
Gu W, Monteiro R, Zuo J, Simoes FC, Martella A, Andrieu-Soler C, Grosveld F, Sauka-Spengler-T, Patient R (2015) A novel TGFbeta modulator that uncouples R-Smad/I-Smad-mediated negative feedback from R-Smad/ligand-driven positive feedback. PLoS Biol 13:e1002051
Guruharsha KG, Kankel MW, Artavanis-Tsakonas S (2012) The Notch signalling system: recent insights into the complexity of a conserved pathway. Nat Rev Genet 13:654–666
Haase VH (2006) The VHL/HIF oxygen-sensing pathway and its relevance to kidney disease. Kidney Int 69:1302–1307
Harvey SA, Smith JC (2009) Visualisation and quantification of morphogen gradient formation in the zebrafish. PLoS Biol 7:e1000101
He L, Xu W, Jing Y, Wu M, Song S, Cao Y, Mei C (2015) Yes-associated protein (Yap) is necessary for ciliogenesis and morphogenesis during pronephros development in zebrafish (Danio Rerio). Int J Biol Sci 11:935–947
Heliot C, Desgrange A, Buisson I, Prunskaite-Hyyrylainen R, Shan J, Vainio S, Umbhauer M, Cereghini S (2013) HNF1B controls proximal-intermediate nephron segment identity in vertebrates by regulating Notch signalling components and Irx1/2. Development 140:873–885
Hildebrandt F, Benzing T, Katsanis N (2011) Ciliopathies. N Engl J Med 364:1533–1543
Hossain Z, Ali SM, Ko HL, Xu J, Ng CP, Guo K, Qi Z, Ponniah S, Hong W, Hunziker W (2007) Glomerulocystic kidney disease in mice with a targeted inactivation of Wwtr1. Proc Natl Acad Sci USA 104:1631–1636
Hu N, Yost HJ, Clark EB (2001) Cardiac morphology and blood pressure in the adult zebrafish. Anat Rec 264:1–12
Hyvärinen J, Parikka M, Sormunen R, Rämet M, Tryggvason K, Kivirikko KI, Myllyharju J, Koivunen P (2010) Deficiency of a transmembrane prolyl 4-hydroxylase in the zebrafish leads to basement membrane defects and compromised kidney function. J Biol Chem 285:42023–42032
Ichimura K, Bubenshchikova E, Powell R, Fukuyo Y, Nakamura T, Tran U, Oda S, Tanaka M, Wessely O, Kurihara H (2012) A comparative analysis of glomerulus development in the pronephros of medaka and zebrafish. PLoS One 7(9):e45286
Ichimura K, Powell R, Nakamura T, Kurihara H, Sakai T, Obara T (2013) Podocalyxin regulates pronephric glomerular development in zebrafish. Phys Rep 1
Ikenaga T, Urban JM, Gebhart N, Hatta K, Kawakami K, Ono F (2011) Formation of the spinal network in zebrafish determined by domain-specific pax genes. J Comp Neurol 519:1562–1579
Ishikawa H, Marshall WF (2011) Ciliogenesis: building the cell's antenna. Nat Rev Mol Cell Biol 12:222–234
James RG, Schultheiss TM (2003) Patterning of the avian intermediate mesoderm by lateral plate and axial tissues. Dev Biol 253:109–124
James RG, Schultheiss TM (2005) Bmp signaling promotes intermediate mesoderm gene expression in a dose-dependent, cell-autonomous and translation-dependent manner. Dev Biol 288:113–125
James RG, Kamei CN, Wang Q, Jiang R, Schultheiss TM (2006) Odd-skipped related 1 is required for development of the metanephric kidney and regulates formation and differentiation of kidney precursor cells. Development 133:2995–3004
Jesuthasan S, Stahle U (1997) Dynamic microtubules and specification of the zebrafish embryonic axis. Curr Biol 7:31–42
Kim SK, Shindo A, Park TJ, Oh EC, Ghosh S, Gray RS, Lewis RA, Johnson CA, Attie-Bittach T, Katsanis N, Wallingford JB (2010) Planar cell polarity acts through septins to control collective cell movement and ciliogenesis. Science (New York, NY) 329:1337–1340
Kimmel CB, Ballard WW, Kimmel SR, Ullmann B, Schilling TF (1995) Stages of embryonic development of the zebrafish. Dev Dyn 203:253–310

Kozlowski DJ, Murakami T, Ho RK, Weinberg ES (1997) Regional cell movement and tissue patterning in the zebrafish embryo revealed by fate mapping with caged fluorescein. Biochem Cell Biol 75:551–562

Kramer-Zucker AG, Olale F, Haycraft CJ, Yoder BK, Schier AF, Drummond IA (2005a) Cilia-driven fluid flow in the zebrafish pronephros, brain and Kupffer's vesicle is required for normal organogenesis. Development 132:1907–1921

Kramer-Zucker AG, Wiessner S, Jensen AM, Drummond IA (2005b) Organization of the pronephric filtration apparatus in zebrafish requires Nephrin, Podocin and the FERM domain protein Mosaic eyes. Dev Biol 285:316–329

Kudoh T, Tsang M, Hukriede NA, Chen X, Dedekian M, Clarke CJ, Kiang A, Schultz S, Epstein JA, Toyama R, Dawid IB (2001) A gene expression screen in zebrafish embryogenesis. zfin.org

Kumano G, Smith WC (2002) Revisions to the Xenopus gastrula fate map: implications for mesoderm induction and patterning. Dev Dyn 225:409–421

Kyuno J, Masse K, Jones EA (2008) A functional screen for genes involved in Xenopus pronephros development. Mech Dev 125:571–586

Lane MC, Sheets MD (2002a) Primitive and definitive blood share a common origin in Xenopus: a comparison of lineage techniques used to construct fate maps. Dev Biol 248:52–67

Lane MC, Sheets MD (2002b) Rethinking axial patterning in amphibians. Dev Dyn 225:434–447

Lane MC, Sheets MD (2006) Heading in a new direction: implications of the revised fate map for understanding Xenopus laevis development. Dev Biol 296:12–28

Langdon YG, Mullins MC (2011) Maternal and zygotic control of zebrafish dorsoventral axial patterning. Annu Rev Genet 45:357–377

Li Y, Cheng CN, Verdun VA, Wingert RA (2014) Zebrafish nephrogenesis is regulated by interactions between retinoic acid, mecom, and Notch signaling. Dev Biol 386:111–122

Liu Y, Pathak N, Kramer-Zucker A, Drummond IA (2007) Notch signaling controls the differentiation of transporting epithelia and multiciliated cells in the zebrafish pronephros. Development 134:1111–1122

Lyons RA, Saridogan E, Djahanbakhch O (2006) The reproductive significance of human Fallopian tube cilia. Hum Reprod Update 12:363–372

Ma M, Jiang YJ (2007) Jagged2a-notch signaling mediates cell fate choice in the zebrafish pronephric duct. PLoS Genet 3:e18

Majumdar A, Drummond IA (1999) Podocyte differentiation in the absence of endothelial cells as revealed in the zebrafish avascular mutant, cloche. Dev Genet 24:220–229

Majumdar A, Drummond IA (2000) The zebrafish floating head mutant demonstrates podocytes play an important role in directing glomerular differentiation. Dev Biol 222:147–157

Majumdar A, Lun K, Brand M, Drummond IA (2000) Zebrafish no isthmus reveals a role for pax2.1 in tubule differentiation and patterning events in the pronephric primordia. Development 127:2089–2098

Makita R, Uchijima Y, Nishiyama K, Amano T, Chen Q, Takeuchi T, Mitani A, Nagase T, Yatomi Y, Aburatani H, Nakagawa O, Small EV, Cobo-Stark P, Igarashi P, Murakami M, Tominaga J, Sato T, Asano T, Kurihara Y, Kurihara H (2008) Multiple renal cysts, urinary concentration defects, and pulmonary emphysematous changes in mice lacking TAZ. Am J Physiol Ren Physiol 294:F542–F553

Marra AN, Wingert RA (2016) Epithelial cell fate in the nephron tubule is mediated by the ETS transcription factors etv5a and etv4 during zebrafish kidney development. Dev Biol 411:231–245

Massa F, Garbay S, Bouvier R, Sugitani Y, Noda T, Gubler MC, Heidet L, Pontoglio M, Fischer E (2013) Hepatocyte nuclear factor 1beta controls nephron tubular development. Development 140:886–896

Mayor R, Etienne-Manneville S (2016) The front and rear of collective cell migration. Nat Rev Mol Cell Biol 17:97–109

Mudumana SP, Hentschel D, Liu Y, Vasilyev A, Drummond IA (2008) odd skipped related1 reveals a novel role for endoderm in regulating kidney versus vascular cell fate. Development 135:3355–3367

Naylor RW, Przepiorski A, Ren Q, Yu J, Davidson AJ (2013) HNF1beta is essential for nephron segmentation during nephrogenesis. J Am Soc Nephrol 24:77–87

Neto A, Mercader N, Gomez-Skarmeta JL (2012) The Osr1 and Osr2 genes act in the pronephric anlage downstream of retinoic acid signaling and upstream of Wnt2b to maintain pectoral fin development. Development 139:301–311

Niehrs C (2004) Regionally specific induction by the Spemann-Mangold organizer. Nat Rev Genet 5:425–434

Obara T, Mangos S, Liu Y, Zhao J, Wiessner S, Kramer-Zucker AG, Olale F, Schier AF, Drummond IA (2006) Polycystin-2 immunolocalization and function in zebrafish. J Am Soc Nephrol 17:2706–2718

Obara-Ishihara T, Kuhlman J, Niswander L, Herzlinger D (1999) The surface ectoderm is essential for nephric duct formation in intermediate mesoderm. Development 126:1103–1108

O'Brien LL, Grimaldi M, Kostun Z, Wingert RA, Selleck R, Davidson AJ (2011) Wt1a, Foxc1a, and the Notch mediator Rbpj physically interact and regulate the formation of podocytes in zebrafish. Dev Biol 358:318–330

Ott E, Wendik B, Srivastava M, Pacho F, Tochterle S, Salvenmoser W, Meyer D (2015) Pronephric tubule morphogenesis in zebrafish depends on Mnx mediated repression of irx1b within the intermediate mesoderm. Dev Biol 411:101–114

Pathak N, Obara T, Mangos S, Liu Y, Drummond IA (2007) The zebrafish fleer gene encodes an essential regulator of cilia tubulin polyglutamylation. Mol Biol Cell 18:4353–4364

Pathak N, Austin-Tse CA, Liu Y, Vasilyev A, Drummond IA (2014) Cytoplasmic carboxypeptidase 5 regulates tubulin glutamylation and zebrafish cilia formation and function. Mol Biol Cell 25:1836–1844

Pazour GJ (2004) Comparative genomics: prediction of the ciliary and basal body proteome. Curr Biol 14:R575–R577

Perner B, Englert C, Bollig F (2007) The Wilms tumor genes wt1a and wt1b control different steps during formation of the zebrafish pronephros. Dev Biol 309:87–96

Pfeffer PL, Gerster T, Lun K, Brand M, Busslinger M (1998) Characterization of three novel members of the zebrafish Pax2/5/8 family: dependency of Pax5 and Pax8 expression on the Pax2.1 (noi) function. Development 125:3063–3074

Pham VN, Roman BL, Weinstein BM (2001) Isolation and expression analysis of three zebrafish angiopoietin genes. Dev Dyn 221:470–474

Pyati UJ, Cooper MS, Davidson AJ, Nechiporuk A, Kimelman D (2006) Sustained Bmp signaling is essential for cloaca development in zebrafish. Development 133:2275–2284

Reggiani L, Raciti D, Airik R, Kispert A, Brandli AW (2007) The prepattern transcription factor Irx3 directs nephron segment identity. Genes Dev 21:2358–2370

Rodaway A, Takeda H, Koshida S, Broadbent J, Price B, Smith JC, Patient R, Holder N (1999) Induction of the mesendoderm in the zebrafish germ ring by yolk cell-derived TGF-beta family signals and discrimination of mesoderm and endoderm by FGF. Development 126:3067–3078

Rottbauer W, Baker K, Wo ZG, Mohideen MA, Cantiello HF, Fishman MC (2001) Growth and function of the embryonic heart depend upon the cardiac-specific L-type calcium channel alpha1 subunit. Dev Cell 1:265–275

Ryan G, Steele-Perkins V, Morris JF, Rauscher FJ 3rd, Dressler GR (1995) Repression of Pax-2 by WT1 during normal kidney development. Development 121:867–875

Satir P, Christensen ST (2007) Overview of structure and function of mammalian cilia. Annu Rev Physiol 69:377–400

Sawamoto K, Wichterle H, Gonzalez-Perez O, Cholfin JA, Yamada M, Spassky N, Murcia NS, Garcia-Verdugo JM, Marin O, Rubenstein JL, Tessier-Lavigne M, Okano H, Alvarez-Buylla A (2006) New neurons follow the flow of cerebrospinal fluid in the adult brain. Science (New York, NY) 311:629–632

Schnabel E, Dekan G, Miettinen A, Farquhar MG (1989) Biogenesis of podocalyxin – the major glomerular sialoglycoprotein – in the newborn rat kidney. Eur J Cell Biol 48:313–326
Schnerwitzki D, Perner B, Hoppe B, Pietsch S, Mehringer R, Hänel F, Englert C (2014) Alternative splicing of Wilms tumor suppressor 1 (Wt1) exon 4 results in protein isoforms with different functions. Dev Biol 393:24–32
Schrimpf C, Xin C, Campanholle G, Gill SE, Stallcup W, Lin SL, Davis GE, Gharib SA, Humphreys BD, Duffield JS (2012) Pericyte TIMP3 and ADAMTS1 modulate vascular stability after kidney injury. J Am Soc Nephrol 23:868–883
Sehnert AJ, Huq A, Weinstein BM, Walker C, Fishman M, Stainier DY (2002) Cardiac troponin T is essential in sarcomere assembly and cardiac contractility. Nat Genet 31:106–110
Serluca FC, Fishman MC (2001) Pre-pattern in the pronephric kidney field of zebrafish. Development 128:2233–2241
Serluca FC, Drummond IA, Fishman MC (2002) Endothelial signaling in kidney morphogenesis: a role for hemodynamic forces. Curr Biol 12:492–497
Seufert DW, Brennan HC, DeGuire J, Jones EA, Vize PD (1999) Developmental basis of pronephric defects in Xenopus body plan phenotypes. Dev Biol 215:233–242
Shimozono S, Iimura T, Kitaguchi T, Higashijima S, Miyawaki A (2013) Visualization of an endogenous retinoic acid gradient across embryonic development. Nature 496:363–366
Singhal PC, Sagar S, Garg P (1996) Simulated glomerular pressure modulates mesangial cell 72 kDa metalloproteinase activity. Connect Tissue Res 33:257–263
Skouloudaki K, Puetz M, Simons M, Courbard JR, Boehlke C, Hartleben B, Engel C, Moeller MJ, Englert C, Bollig F, Schafer T, Ramachandran H, Mlodzik M, Huber TB, Kuehn EW, Kim E, Kramer-Zucker A, Walz G (2009) Scribble participates in Hippo signaling and is required for normal zebrafish pronephros development. Proc Natl Acad Sci USA 106:8579–8584
Slanchev K, Putz M, Schmitt A, Kramer-Zucker A, Walz G (2011) Nephrocystin-4 is required for pronephric duct-dependent cloaca formation in zebrafish. Hum Mol Genet 20:3119–3128
Spemann H, Mangold H (1923) Induction of embryonic primordia by implantation of organizers from a different species. Int J Dev Biol 45:13–38
Sullivan-Brown J, Schottenfeld J, Okabe N, Hostetter CL, Serluca FC, Thiberge SY, Burdine RD (2008) Zebrafish mutations affecting cilia motility share similar cystic phenotypes and suggest a mechanism of cyst formation that differs from pkd2 morphants. Dev Biol 314:261–275
Sun Z, Hopkins N (2001) vhnf1, the MODY5 and familial GCKD-associated gene, regulates regional specification of the zebrafish gut, pronephros, and hindbrain. Genes Dev 15:3217–3229
Sun Z, Amsterdam A, Pazour GJ, Cole DG, Miller MS, Hopkins N (2004) A genetic screen in zebrafish identifies cilia genes as a principal cause of cystic kidney. Development 131:4085–4093
Takeda T, Go WY, Orlando RA, Farquhar MG (2000) Expression of podocalyxin inhibits cell-cell adhesion and modifies junctional properties in Madin-Darby canine kidney cells. Mol Biol Cell 11:3219–3232
Tena JJ, Neto A, de la Calle-Mustienes E, Bras-Pereira C, Casares F, Gomez-Skarmeta JL (2007) Odd-skipped genes encode repressors that control kidney development. Dev Biol 301:518–531
Thisse B, Thisse C (2004) Fast release clones: a high throughput expression analysis. ZFIN Direct Data Submission. http://zfin.org
Thisse C, Thisse B (2005) High throughput expression analysis of ZF-models consortium clones. ZFIN Direct Data Submission. http://zfin.org
Thisse B, Pflumio S, Fürthauer M, Loppin B, Heyer V, Degrave A, Woehl R, Lux A, Steffan T, Charbonnier XQ, Thisse C (2001) Expression of the zebrafish genome during embryogenesis. ZFIN Direct Data Submission. http://zfin.org
Tian Y, Kolb R, Hong JH, Carroll J, Li D, You J, Bronson R, Yaffe MB, Zhou J, Benjamin T (2007) TAZ promotes PC2 degradation through a SCFbeta-Trcp E3 ligase complex. Mol Cell Biol 27:6383–6395

Tomar R, Mudumana SP, Pathak N, Hukreide NA, Drummond IA (2014) osr1 is required for podocyte development downstream of wt1a. J Am Soc Nephrol ASN. 2013121327

Torres M, Gomez-Pardo E, Dressler GR, Gruss P (1995) Pax-2 controls multiple steps of urogenital development. Development 121:4057–4065

Tsujikawa M, Malicki J (2004) Intraflagellar transport genes are essential for differentiation and survival of vertebrate sensory neurons. Neuron 42:703–716

Tucker JA, Mintzer KA, Mullins MC (2008) The BMP signaling gradient patterns dorsoventral tissues in a temporally progressive manner along the anteroposterior axis. Dev Cell 14:108–119

Vasilyev A, Liu Y, Mudumana S, Mangos S, Lam PY, Majumdar A, Zhao J, Poon KL, Kondrychyn I, Korzh V, Drummond IA (2009) Collective cell migration drives morphogenesis of the kidney nephron. PLoS Biol 7:e9

Vasilyev A, Liu Y, Hellman N, Pathak N, Drummond IA (2012) Mechanical stretch and PI3K signaling link cell migration and proliferation to coordinate epithelial tubule morphogenesis in the zebrafish pronephros. PLoS One 7:e39992

Wang Q, Lan Y, Cho ES, Maltby KM, Jiang R (2005) Odd-skipped related 1 (Odd 1) is an essential regulator of heart and urogenital development. Dev Biol 288:582–594

Wanner A, Salathe M, O'Riordan TG (1996) Mucociliary clearance in the airways. Am J Respir Crit Care Med 154:1868–1902

Warga RM, Nusslein-Volhard C (1999) Origin and development of the zebrafish endoderm. Development 126:827–838

Warga RM, Mueller RL, Ho RK, Kane DA (2013) Zebrafish Tbx16 regulates intermediate mesoderm cell fate by attenuating Fgf activity. Dev Biol 383:75–89

Whitesell TR, Kennedy RM, Carter AD, Rollins EL, Georgijevic S, Santoro MM, Childs SJ (2014) An alpha-smooth muscle actin (acta2/alphasma) zebrafish transgenic line marking vascular mural cells and visceral smooth muscle cells. PLoS One 9:e90590

Wingert R, Davidson A (2008) The zebrafish pronephros: a model to study nephron segmentation. Kidney Int 73:1120–1127

Wingert RA, Davidson AJ (2011) Zebrafish nephrogenesis involves dynamic spatiotemporal expression changes in renal progenitors and essential signals from retinoic acid and irx3b. Dev Dyn 240:2011–2027

Wingert RA, Selleck R, Yu J, Song H-D, Chen Z, Song A, Zhou Y, Thisse B, Thisse C, McMahon AP (2007) The cdx genes and retinoic acid control the positioning and segmentation of the zebrafish pronephros. PLoS Genet 3:1922–1938

Woo K, Fraser SE (1995) Order and coherence in the fate map of the zebrafish nervous system. Development 121:2595–2609

Xu PF, Houssin N, Ferri-Lagneau KF, Thisse B, Thisse C (2014) Construction of a vertebrate embryo from two opposing morphogen gradients. Science (New York, NY) 344:87–89

Yasuda T, Kondo S, Homma T, Harris RC (1996) Regulation of extracellular matrix by mechanical stress in rat glomerular mesangial cells. J Clin Invest 98:1991–2000

Zhang Q, Liu Q, Austin C, Drummond I, Pierce EA (2012) Knockdown of ttc26 disrupts ciliogenesis of the photoreceptor cells and the pronephros in zebrafish. Mol Biol Cell 23:3069–3078

Zhang J, Yuan S, Vasilyev A, Amin Arnaout M (2015) The transcriptional coactivator Taz regulates proximodistal patterning of the pronephric tubule in zebrafish. Mech Dev 138:328–335

Zust B, Dixon KE (1975) The effect of u.v. irradiation of the vegetal pole of Xenopus laevis eggs on the presumptive primordial germ cells. J Embryol Exp Morpholog 34:209–220

Chapter 3
Zebrafish as a Model of Kidney Disease

Elvin E. Morales and Rebecca A. Wingert

Abstract Animal models have been an invaluable means to advance biomedical research as they provide experimental avenues for cellular and molecular investigations of disease pathology. The zebrafish (*Danio rerio*) is a good alternative to mammalian models that can be used to apply powerful genetic experimental methods normally used in invertebrates to answer questions about vertebrate development and disease. In the case of the kidney, the zebrafish has proven itself to be an applicable and versatile experimental system, mainly due to the simplicity of its pronephros, which contains two nephrons that possess conserved structural and physiological aspects with mammalian nephrons. Numerous genes that were not previously related to kidney conditions have now been linked to renal diseases by applying genetic screening with the zebrafish. In fact, a large collection of mutations that affect nephron formation and function were generated through phenotype-based forward screens. Complementary reverse genetic approaches have also been insightful, with methods spanning the use of antisense morpholino oligonucleotides to genome editing approaches such as the CRISPR/Cas9 system, to selectively knock down or knock out genes of interest to see if they produce kidney phenotypes. Acute kidney injury (AKI) has also been easily modeled in the zebrafish by injecting nephrotoxins, directly inducing damage through surgical intervention, or by generating transgenic lines that express compounds in a tissue-specific manner that when exposed to certain drugs promote an apoptotic response within cells. In this chapter, we provide an overview of these various approaches as well as discuss many of the contributions that have been achieved through the use of zebrafish to model kidney disease.

E.E. Morales • R.A. Wingert (✉)
Department of Biological Sciences, Center for Stem Cells and Regenerative Medicine, Center for Zebrafish Research, University of Notre Dame, 100 Galvin Life Sciences, Notre Dame, IN 46556, USA
e-mail: rwingert@nd.edu

R.K. Miller (ed.), *Kidney Development and Disease*, Results and Problems in Cell Differentiation 60, DOI 10.1007/978-3-319-51436-9_3

3.1 Introduction

Animal models are the cornerstones of biomedical research. Over the last two decades, the zebrafish (*Danio rerio*) has received an unprecedented amount of attention within the scientific community (Pickart and Klee 2014). Although mammalian models have been a central means of studying human afflictions due to anatomical and evolutional similarities, there are also limiting factors with their applications. For example, rodent models are not always the optimal organisms to model certain conditions, due to factors such as inherent biological differences, the economical and space costs associated with the care and husbandry of the animals, or the molecular biology tools available for researching a particular question. A large degree of functional and genetic conservation between humans and nonmammalian organisms means that many diseases can be accurately modeled at the molecular level in a more cost-effective manner by using less complex organisms, such as flies or worms. With regard to this chapter, the genetic conservation between zebrafish and humans (Howe et al. 2013) has made them a premier model for efficacious basic and translational research and provided many prospects for novel ongoing studies (Lieschke and Currie 2007; Santoriello and Zon 2012).

For the purposes of kidney disease modeling, the zebrafish provides a particularly useful system to apply genetic approaches that have been traditionally used in the aforementioned invertebrate models to answer questions about cellular and developmental processes within vertebrates (Drummond 2005; Ebarasi et al. 2011). A suite of traits makes zebrafish amenable for organ development and disease studies in the embryo. For example, zebrafish development occurs ex utero, and the embryos are optically clear, enabling researchers to readily observe processes in real time within the context of the whole animal (Kimmel et al. 1995; Laale 1997). Most importantly, all common vertebrate organs are visible by 120–144 hours post-fertilization (hpf), with many apparent even earlier, at 24 hpf, such as the heart, eyes, and kidney (Rubenstein 2003). Thus, the development as well as the onset of metabolic functions for these common organs is not only fast but can be observed through the transparent embryos.

The zebrafish embryonic kidney, or pronephros, is composed of two blood-filtering glomeruli fused at the midline that are connected to two epithelial tubules (each divided into proximal and distal domains) and followed by paired bilateral pronephric ducts that join at the cloaca, which is where wastes exit the body (Drummond et al. 1998; Gerlach and Wingert 2013). The pronephros becomes functional beginning around 48 hpf (Drummond et al. 1998), and the structural composition of the zebrafish pronephros, as well as the segmentation pattern of the tubules, is highly conserved to that of the mammalian kidney (Wingert et al. 2007; Wingert and Davidson 2008, 2011). This high degree of structural conservation makes the zebrafish a useful counterpart to studies of mammalian kidney development (Fig. 3.1) (also discussed in Chap. 2).

The adult kidney is also useful to model advanced developmental stages as well as renal afflictions that affect mature tissues. At around 12–14 days post-

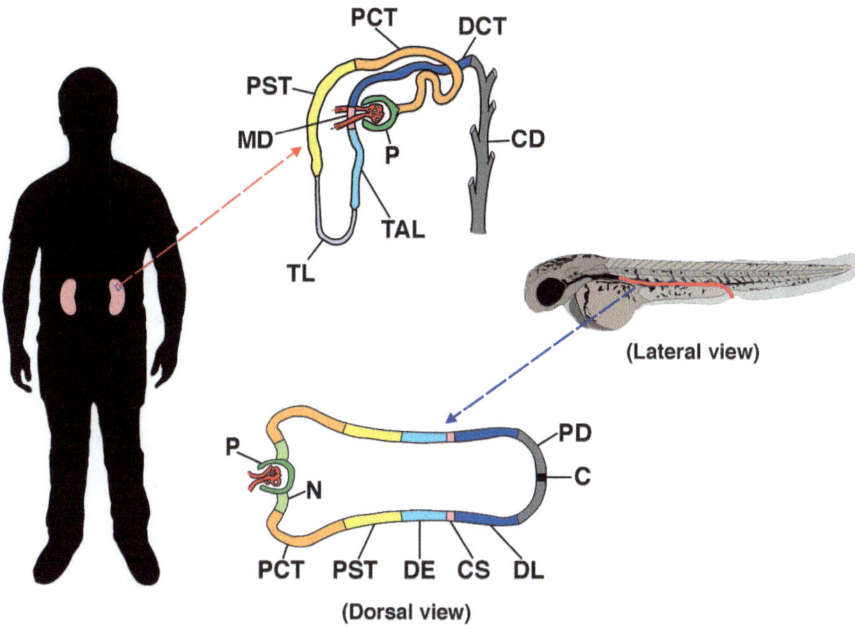

Fig. 3.1 Comparison between human nephron and zebrafish pronephros. There is a high degree of similarity between the nephron segmentation patterns of both organisms. Abbreviations in the human nephron: *P* podocytes, *PCT* proximal convoluted tubule, *PST* proximal straight tubule, *TL* thin limb, *TAL* thick ascending limb, *MD* macula densa, *DCT* distal convoluted tubule, *CD* collecting duct. Abbreviations in the zebrafish pronephros: *P* podocytes, *N* neck, *PCT* proximal convoluted tubule, *PST* proximal straight tubule, *DE* distal early, *CS* corpuscle of Stannius, *DL* distal late, *PD* pronephric duct, *C* cloaca

fertilization (dpf), the adult form, known as the mesonephros, begins to develop, with the progressive addition of new nephrons to the existing bilateral pair (Diep et al. 2015). When mesonephros generation is complete, the zebrafish possesses a more complex branched arrangement of nephron units within the kidney, but the nephrons show the same segmental composition (McCampbell et al. 2014, 2015). Mammals contain a fixed number of nephrons after birth and cannot regenerate or produce more after they are destroyed (Stocum 2012). Interestingly enough, in organisms such as the zebrafish or the goldfish, new nephrons keep being added to the existing blood-filtering array throughout the lifetime of the animal in response to a naturally increasing biomass (Reimschuessel et al. 1990; Reimschuessel and Williams 1995; Zhou et al. 2010; Diep et al. 2011). This phenomenon known as neonephrogenesis also occurs after kidney injury (Reimschuessel 2001). Because of the fundamental similarities between the mammalian and zebrafish kidney structures, the mesonephros can also be used to study the onset, progression, and recovery from kidney injury. Although complex in its anatomy, the mesonephros is relatively conducive to analysis when compared to arrangements of several thousand nephrons present in the rodent adult kidney (Kroeger and Wingert 2014).

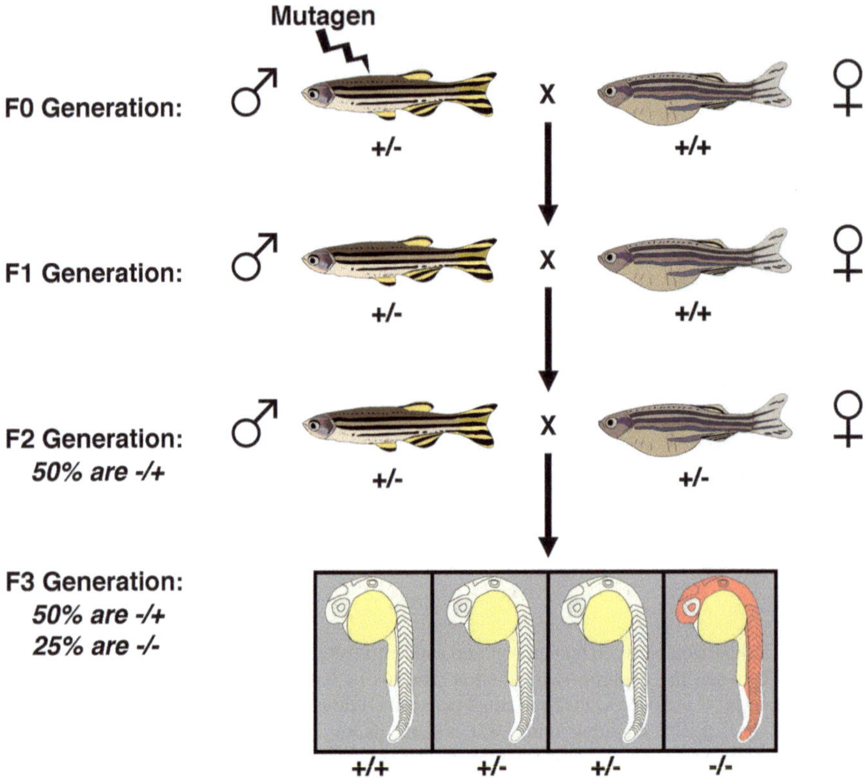

Fig. 3.2 Diagram of forward genetic screen approaches in the zebrafish model. F0 Generation: Mutagenized males are crossed with WT females to generate F1 families. F1 Generation: Heterozygotes are selected based on their phenotype and then outcrossed to expand the heterozygote population and generate F2 families. F2 Generation: Heterozygotes of the F2 generation are incrossed to bring the desired mutant phenotype to homozygosity. F3 Generation: Homozygous mutant is analyzed to identify the genetic cause of the observed mutant phenotype

In this chapter, we will discuss the versatility and advantages of the zebrafish pronephros and mesonephros kidney as models for renal disease, along with the various tools that have been implemented to study these systems in recent years.

3.2 Forward Genetic Approaches

One of the ways that zebrafish researchers can assess developmental dynamics is through the use of forward genetics, which involves identifying genes responsible for a specific phenotype. Because of the recently discussed advantages of zebrafish development, this organism can be subjected to large-scale forward genetic screens that produce a large number of mutants for analysis (Fig. 3.2).

3.2.1 Mutagenesis Screens

In a mutagenesis screen, adult males are exposed to an agent such as the chemical mutagen ethylnitrosourea (ENU), which induces random point mutations in their spermatogonial stem cells (Solnica-Krezel et al. 1994; Driever et al. 1996). The mutagenized males are then crossed with wild-type (WT) females to transmit the mutations onto their progeny. The F_1 generation is then outcrossed with WT females in order to expand the heterozygote population, after which the resulting heterozygotes of the F_2 generation are incrossed in order to produce homozygous mutants for analysis. Random mutagenesis screens can also be performed through exogenous DNA insertion. Although less efficient than chemical mutagenesis, the insertional mutagenesis method can greatly accelerate the identification of mutant genes (Amsterdam et al. 1999). This method involves inserting an exogenous DNA sequence, such as a retroviral vector, into the zebrafish genome to produce random mutations (Amsterdam and Hopkins 2006). The insertions can then serve as tags that can be used to identify the mutated gene in question. Both chemical and insertional mutagenesis screens can produce a multitude of mutants, where pertinent ones are then selected based on a phenotype of interest. Within the nephrology field, zebrafish mutants have been particularly useful for the study of cystic diseases.

Polycystic kidney disease (PKD) is a genetic condition characterized by the formation of multiple liquid-filled cysts within the kidney (Halvorson et al. 2010). This condition has been connected to mutations in the genes *POLYCYSTIN-1* (*PKD1*) and *POLYCYSTIN-2* (*PKD2*), as well as to overproliferation of the renal epithelium (Nadasy et al. 1995). PKD can be subdivided into two types: autosomal dominant (ADPKD) and autosomal recessive (ARPKD). ADPKD is one of the most common monogenic human disorders, affecting anywhere from 1/400 to 1/1000 live births, so the importance of understanding how this condition arises cannot be overstated. Using zebrafish mutants obtained through forward genetic mutagenesis screens, researchers have been able to develop various models of these conditions that have contributed to the understanding of disease onset and enabled further cell biological studies.

In a study by Drummond et al., researchers analyzed data obtained from a previously performed ENU mutagenesis screen for the purposes of identifying mutations affecting pronephric development (Drummond et al. 1998). Selection of mutants was based upon a common phenotype of fluid-filled cysts in place of the normal pronephric tubule. The group was able to identify 18 independent recessive mutations that affected all of the identifiable parts of the pronephros: the glomerulus, the tubules, and the duct. One of these identified mutants named *double bubble* showed glomerular expansion due to fluid buildup at around 40 hpf. By 56 hpf, the glomerulus was severely distorted and loose. Researchers stated that the aberrant glomerular morphology suggests that cyst formation was most likely due to early defects in glomerular structure, specifically with defects that affect filtration pressure. This study was able to link cyst formation with defective glomerulogenesis, providing some insight into the primary defects that can be responsible for cyst formation, which could lead to further insights about PKD pathogenesis.

A subsequent study used insertional mutagenesis to produce zebrafish embryos containing a similar polycystic renal phenotype (Sun et al. 2004). From this insertional mutagenesis screen, researchers were able to identify 12 genes whose disrupted expression causes cysts within the developing zebrafish kidney, one of which was *pkd2*, the zebrafish orthologue of human *PKD2*. Researchers found that although the *pkd2* mutant produced by this screen did not exhibit kidney cysts before death, this was most likely due to maternal contribution of *pkd2* transcript that carries the embryo through early development. Subsequent morpholino-mediated knockdown of *pkd2* (method discussed in the next section), a process that targets both endogenous and maternally contributed transcripts, resulted in glomerular cyst formation, which speaks to the relevance these mutants have to human PKD.

Interestingly, three of the identified genes are homologues of genes encoding components of intraflagellar transport (IFT) particles, which are important in cilia formation. Cilia are microtubule-based organelles present on the cells in most tissues that serve a variety of metabolic and signaling functions. They have been shown to function as environmental sensors of mechanical stimuli (Vincensini et al. 2011) as well as signaling centers for various developmental pathways such as Hedgehog and Wnt (Lancaster and Gleeson 2009). Ciliopathies are conditions stemming from mutations in genes encoding and/or that process ciliary proteins, which result in ciliary dysfunction in the form of altered cilia structure or movement (Tobin and Beales 2008). Renal disease is a common aspect of ciliopathies, including, but not limited to, cyst formation, fusion of the lobules, and clubbing. For example, previous studies demonstrate that mutations in a human core intraflagellar transport gene, *IFT80*, can cause Jeune syndrome, a condition that among its characteristics features renal cysts (Beales et al. 2007). This work by Sun et al. (2004) similarly exemplifies how a forward genetic screen was useful to create zebrafish PKD models, and how mutants obtained through these screens can lead researchers to find previously unknown links between kidney cystogenesis and other cellular processes, providing a clearer picture of the genetic cause of PKD.

3.2.2 Future Directions

Various zebrafish mutants that resemble a number of human kidney-related conditions, as well as developmental defects, are still being discovered from the data produced by large-scale mutagenesis screens, and used to further elucidate the mechanisms of disease progression. Some such examples, the *locke*, *switch hitter* (*swt*), and *kurly* mutants (Brand et al. 1996; Haffter et al. 1996), which are characterized by a distinct "curly tail down" phenotype as well as cystic dilations of the pronephric tubules, were used to study the earlier cellular defects of cyst formation in the kidney. Sullivan-Brown et al. (2008) observed that *locke*, *swt*, *kurly* mutants develop tubular kidney cysts due to ciliary defects ranging from immotile cilia to cilia that beat irregularly. Analyzing the cilia of *pkd2* morphants revealed that they beat in a coordinated fashion similar to that of WT embryos, which meant that the glomerular cysts resulting from *pkd2* knockdown did not arise due to defects in cilia motility, but rather another, as of yet unknown mechanism. In doing this, researchers were able to observe a new mechanism for cyst formation that differed from what was seen in *pkd2* morphants.

Combining the before-mentioned methods with the organism's ex utero development, high fecundity, and the large number of embryos produced after mating, the zebrafish can be a powerful tool for in vivo drug discovery, looking for potential therapeutic agents that can ameliorate kidney conditions like PKD. Using a technique called "chemical genetics," researchers expose embryos to a small molecule and then analyze the resulting phenotype within the context of a genetic pathway or disease of interest to assess the effects on gene expression or the disease phenotype.

For example, Cao et al. (2009) performed a chemical screen for compounds that ameliorated body curvature and laterality defects, aberrant phenotypes that are closely associated with kidney cyst formation, in previously identified zebrafish *pkd2* and *ift172* mutants. After applying the compounds to the mutant embryos, researchers were able to identify trichostatin A (TSA), a pan HDAC (histone deacetylase) inhibitor, as a compound that could modulate both body curvature and laterality defects. Additionally, they were able to demonstrate how this compound, as well as valproic acid (VPA), a Class I-specific HDAC inhibitor that is structurally unrelated to TSA that can also modulate body curvature and laterality defects, was able to suppress kidney cyst formation in *pkd2* morphants. Through the use of a chemical genetic screen, Cao et al. (2009) were able to identify HDAC inhibitors as drug candidates for potential PKD treatments. Because of this and many other examples, forward genetic screens are a very promising approach that takes full advantage of the experimental versatility of this animal model.

3.3 Reverse Genetic Approaches

Although forward genetics is great at identifying multiple loci responsible for disease onset, random mutagenesis makes it hard to produce disease models for conditions that result from abnormal expression of specific genes because identification of mutants is based purely on the aberrant phenotype. Additionally, trying to identify the specific gene responsible for the observed mutant phenotype through methods like positional cloning is very work intensive and requires a large amount of effort. In cases like this, researchers turn to the use of reverse genetics, defined as the analysis of phenotypical consequences resulting from disruptions in genes of interest. Simply disrupt or mutate the gene, and then observe how the organism is affected.

3.3.1 Morpholino oligos

A standard reverse genetic technique performed on zebrafish is morpholino oligo-mediated gene knockdown. Morpholino oligos (MOs) are modified, antisense oligomer molecules that bind to a desired RNA transcript to prevent access of cell components to a target site on said transcript (Summerton 1999). MOs possess a nonionic backbone at physiological pH composed of 6-membered morpholine rings connected through phosphorodiamidate linkages. Because of the inherent nature of this chemical structure, MOs are highly soluble in aqueous solutions, highly stable in vivo, and resistant to a variety of enzymes and biological fluids (Janson and During 2006; Amantana and Iversen 2005; Hudziak et al. 2009). MOs work transiently in the embryo, where they act to block the ATG start site preventing translation, or they can block splice sites, which prevents processing of pre-mRNA into mature mRNA. Using this type of compound does not cause degradation of the RNA but prevents the biological activity of the RNA until it is degraded naturally. The process of using MOs involves microinjecting one-cell stage zebrafish embryos to deliver the oligonucleotide ultimately to all of the cells of the organism (Fig. 3.3a). This is possible because of the intrinsic high solubility of MOs, which allows them to diffuse readily throughout the one-cell stage embryo such that the resulting cells of the organism can subsequently inherit MO molecules. The embryos are then allowed to develop until a desired growth stage, at which point the specimen can be studied through any number of protocols to interrogate the phenotype.

The binding affinity of the MO to a desired target sequence regulates its efficacy; however, MO concentration can be a limiting factor as it could become too diluted when development has progressed to a sufficient degree, making the MO-mediated knockdown a transient gene knockdown. Nevertheless, MO effects can be typically seen within the first three days of development but have also been reported after 5 dpf (Bill et al. 2009). To achieve genetic knockdowns over longer intervals, some researchers have supplemented the MO levels with more MO into individual or a small group of cells through techniques like electroporation which involves applying an electric field to cells in order to increase their permeabilization and introduce drugs or other molecules into the cell (Eisen and Smith 2008). Although not a complete gene knockout, this tool can still serve to substantially alter the normal expression of specific genes and can therefore provide a valuable approach to generate specific zebrafish models of disease.

One category of renal affliction that has benefited from the use of MOs has been ciliopathies. Researchers used MOs in zebrafish to generate models for this and other ciliopathies such as Jeune syndrome, Bardet–Biedl syndrome, Meckel syndrome, oro-facial-digital syndrome, and nephronophthisis, screen them for kidney phenotypes, and then identify compounds that might ameliorate said phenotypes. Representative gene candidates for all of these conditions were knocked down and each subsequent morphant developed kidney cysts, random body situs, and a downturned body axis, all features characteristic of ciliary mutants. These

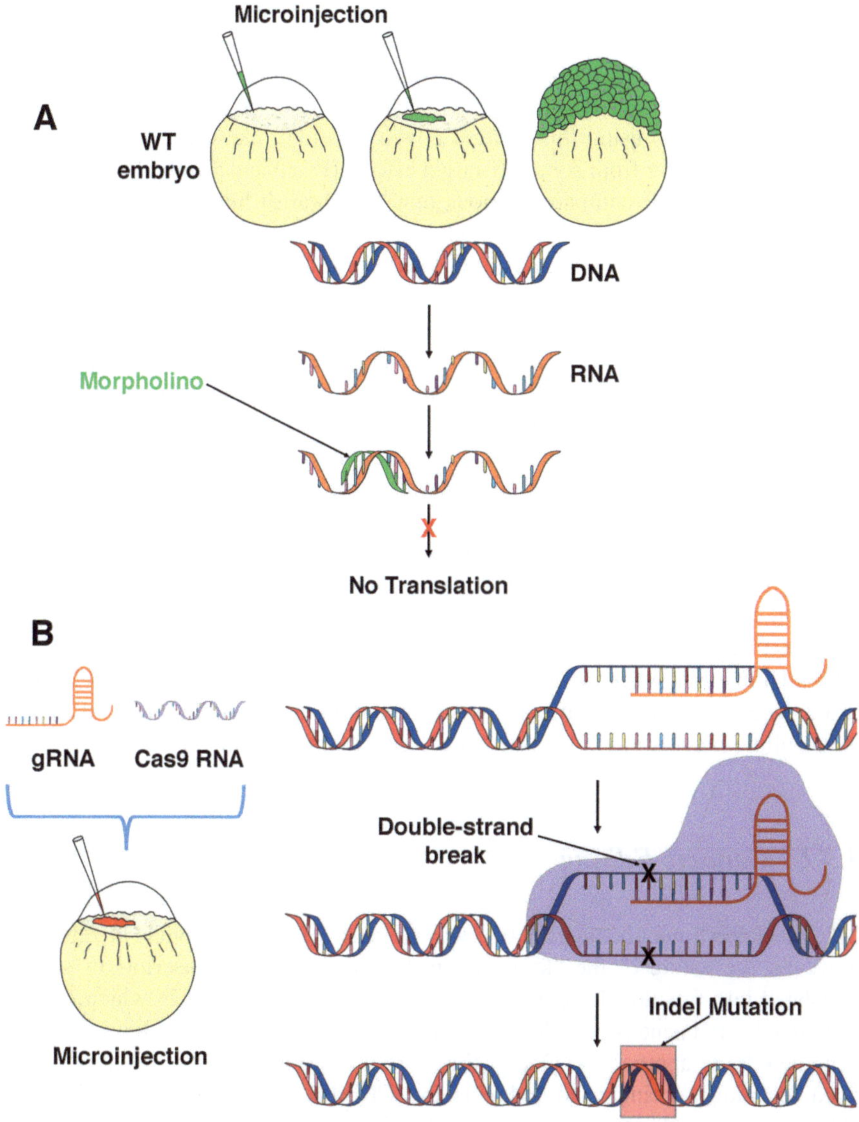

Fig. 3.3 Diagrams of reverse genetic techniques used in the zebrafish model. (**a**) Morpholino oligonucleotide is microinjected into the one-cell stage zebrafish embryo where it targets and binds to a specific mRNA strand. This prevents translation of the mRNA strand, drastically decreasing the amount of functional protein available. (**b**) CRISPR-Cas9 system. A guide RNA (gRNA) that specifically targets a gene of interest, as well as Cas9 endonuclease mRNA, is microinjected into the one-cell stage embryo. The gRNA binds to its target site, where it "guides" the binding of Cas9 which makes a double-strand cut in the DNA sequence. An insertion/deletion (indel) mutation is then produced through nonhomologous end joining

ciliopathy models were then used to assay the effects of rapamycin and roscovitine, two compounds that have been shown to retard or reverse the formation of kidney cysts (Tobin and Beales 2008). The individual compounds were able to rescue the renal edema caused by the cysts.

One drawback to using this method however is that, like most antisense technologies, MOs suffer from some off-target effects (Eisen and Smith 2008). A study by Kok et al. (2015) compares various mutant zebrafish lines generated through site-specific genome editing (discussed in next section) to published morpholino phenotypes. Researchers saw that out of the 20 genes they created mutant lines for, mutants for 10 of them did not recapitulate the morpholino phenotype, suggesting that these previously published morphant phenotypes were the result of morpholino off-target effects. In order to assess the off-target effects of an MO, appropriate controls are needed, and one of the best ways to do this is to compare the morphant phenotype to a mutant phenotype, should a mutation in the gene of interest be available. Validating the effects of one morpholino by analyzing the effects of another independent morpholino targeting the same gene to see if they are the same can also serve to assess off-target effects. Additionally, rescue studies can be performed to assess whether the morphant phenotype can be ameliorated through co-injection of the morpholino and a form of the targeted RNA (Eisen and Smith 2008). If the RNA for the affected gene is unable to rescue the phenotype, this suggests that the phenotype is most likely due to off-target effects (Eisen and Smith 2008). Ultimately, when the appropriate controls are applied, these works demonstrate the effectiveness of MOs as tools for renal disease modeling, as well as how the resulting models might be used for the development of therapeutics.

3.3.2 Genome Editing

MO knockdown may be an effective way to generate disease models in zebrafish; however, the nature of the "knockdown" allows for some transcripts to still be translated into functional protein. To create "knockout" models, in which expression of a desired gene is completely abrogated, zebrafish researchers have turned to genome editing approaches. One of the more promising and more recently developed techniques being used today is the CRISPR/Cas9 system. CRISPRs, an abbreviation for the term "Clustered Regularly Interspaced Short Palindromic Repeats," are short segments of prokaryotic DNA that serve as part of the bacterial immune system to protect it from foreign DNA elements (Barrangou et al. 2007). By recognizing and binding to the foreign DNA sequences, they recruit CRISPR-associated protein 9 (Cas9), an RNA-guided endonuclease enzyme, to degrade DNA sequence the CRISPR is bound to. Scientists took advantage of this naturally occurring bacterial immune system to edit the genome of other animal models, such as the zebrafish (Cong et al. 2013). This type of genome editing begins by designing a guide RNA (gRNA) that will target a gene of interest and then "guide" the Cas9 protein to that location. Next, a cocktail of gRNA and Cas9 mRNA are microinjected into the one-cell stage zebrafish embryo, where the gRNA will bind to the desired target sequence and the Cas9 mRNA will be translated to produce the endonuclease enzyme. The Cas9 will then recognize the gRNA, bind to it, and produce a double-stranded break in the DNA. This break will then be repaired through nonhomologous end joining, which will result in random indel (insertion/deletion) mutations that can abrogate gene expression (Fig. 3.3b).

Using this system, researchers can successfully knock out genes that have been linked to various renal conditions (Anderson et al. 2015). Coding variants of human aplipoprotein L1 (*APOL1*) have been attributed to an increased risk of focal glomerular segmental sclerosis (FSGS) in African Americans; however, very little evidence exists as to the role of *APOL1* in the kidney. Anderson et al. (2015) used the CRISPR/Cas9 system to create an *apoll* knockout zebrafish mutant as a means of supplementing MO-mediated knockdown studies that try to analyze the role of this gene in renal function. Analysis of the mutant and morphant phenotypes revealed the presence of a pericardial edema at 3 dpf, indicative of glomerular filtration defects. Additionally, analysis of the organization and patterning of mutant and morphant glomeruli revealed an altered glomerular ultrastructure in both morphant and mutant embryos that was conducive to a poor glomerular filtration barrier. Using the CRISPR-Cas9 system, in conjunction with MO-mediated knockdown researchers were able to establish a solid, well-evidenced relationship between disease onset and variants of *APOL1*.

Like most molecular biology techniques, genome editing systems also have their drawbacks. A study by Rossi et al. (2015) revealed that mutants generated by TAL effector nucleases (TALENs), an alternative genome editing technique to the CRISPR-Cas9 system, could potentially be activating compensatory networks to shield against harmful mutations. Researchers generated mutants for *egfl7*, a gene that lacks obvious phenotypes when knocked out in mice and demonstrates severe vascular tube formation defects when knocked down in zebrafish, frog, and human cells, and then compared those mutants to *egfl7* morphants. They observed that the severe vascular tube formation defects present in the morphants were either very mild or not present at all in the mutants. After confirming that the severe vascular phenotype in the morphants was not due to off-target effects, they proceeded to analyze both the transcriptome and proteome of morphant and mutant embryos and saw that *emilin3a*, *emilin3b*, and *emilin2a* transcripts and Emilin3a protein were upregulated in *egfl7* mutants but not in morphant embryos. The *emilin* genes have an EMI domain, a key unit of Egfl7 function, and co-injection of *emilin* gene mRNA with *egfl7* MO revealed that these transcripts could actually rescue *egfl7* morphants. These results suggest that the upregulation of these genes within the *egfl7* mutants could serve to compensate for the lack of *egfl7*. The possibility of these compensatory networks makes interpreting mutants generated via genome editing somewhat difficult which is why future studies should utilize both MO-mediated knockdown and mutants generated through genome editing techniques so as to compensate the drawbacks of each.

3.3.3 Future Directions

In addition to the MO and CRISPR/Cas9 systems, there are a number of other methods that are being used in zebrafish that have yet to be applied to the zebrafish renal field. One such example would be the Cre/loxP system, which has been widely used in mice to model diseases. Wang et al. (2008) demonstrated how efficient this system is to investigate genetic pathways in a tissue-specific manner by driving Cre recombinase under the control of the zebrafish *lmo2* promoter, a tissue-specific promoter expressed in primitive hematopoietic stem cells as well as vascular endothelial cells. Not only were they able to show that Cre could effectively be expressed only in specific tissues, but they were also able to show that Cre expression faithfully recapitulated the endogenous *lmo2* expression pattern. Additionally this system can be adapted to allow for temporal control of a transgene by expressing a modified version of Cre fused to a mutated hormone-binding domain of the estrogen receptor (CreER), which can only be activated in the presence of tamoxifen, a synthetic estrogen receptor ligand (Hans et al. 2009). This study serves to emphasize the usefulness of this method and denotes a possible way to study other, more late-onset renal diseases such as adult-onset PKD and renal cancer in the much simpler zebrafish renal system. A modified version of the CRISPR system can also be used for this purpose (inducible CRISPR or iCRISPR), by taking advantage of the Tetracycline-On system. This system uses a mutated version of the tetracycline repressor fused with virion protein 16 (VP16) called reverse tetracycline-controlled transactivator (rtTA) to activate expression of a transgene cassette when exposed to doxycycline, a member of the tetracycline class of antibiotics. Three expression cassettes are developed for this system: rtTA under the control of a constitutive promoter, the gRNA under the control of a ubiquitous promoter, and the *Cas9* gene under the control of a tetracycline-dependent promoter. These tetracycline-dependent promoters consist of a tetracycline response element (TRE) upstream of a minimal promoter. When added, doxycycline binds with rtTA and changes the rtTA protein conformation, which now means that rtTA can bind to the TRE, effectively promoting expression of Cas9. The Cas9 then interacts with the gRNA molecules to produce targeted indel mutation (González et al. 2014; Dow et al. 2015). This system can also be modified for tissue-specific knockout by putting expression of rtTA under the control of a constitutively expressed, tissue-specific promoter like that of *cadherin17*, which is expressed exclusively in the zebrafish kidney from 24 hpf to adulthood.

An example of an alternate knockdown system would be the introduction of synthetic microRNA (miRNA) cassettes into the zebrafish genome to knockdown a desired gene (Dong et al. 2009; Giacomotto et al. 2015). RNA interference (RNAi) using synthetic small hairpin RNAs (shRNAs) has been shown to be an efficient method of posttranscriptional gene silencing in many model organisms (Dong et al. 2009). In combination with the miR-30 miRNA backbone, higher amounts of shRNAs can be produced. Dong et al. (2009) took the zebrafish homologues of miR-30 and miR-155 miRNA precursors (pri-miR-30e and pri-miR-155, respectively), cloned them into a pCS2 vector, and then replaced the endogenous miRNA stem-loop sequence with a 24 nucleotide long linker sequence containing two Bbs I restriction sites. This would allow insertion of a synthetic shRNA stem loop specific for a desired gene, effectively making a miRNA/shRNA hybrid construct (mir-shRNA). After determining that targeting the 3′ untranslated region (UTR) resulted in a more potent knockdown of reporter genes, researchers created miR-shRNA constructs for the genes *chordin* and *α-catenin* that, when injected into embryos, induced efficient knockdown of the two developmentally important genes. Transgenic zebrafish lines were also successfully created containing these miR-shRNA constructs under the control of the lineage-specific *gata1* promoter, meaning that not only is this miRNA-mediated knockdown effect heritable but tissue specific as well. Giacomotto et al. (2015) also induced effective heritable gene knockdown of *smn1* using this approach, effectively reproducing spinal muscular atrophy (SMA) in zebrafish, an autosomal recessive human disease that is characterized by motor neuron loss, progressive muscle weakness, and in severe cases death.

Notably, there has been some controversy regarding the use of RNAi-mediated knockdown. Zhao et al. (2008) demonstrated how injecting zebrafish embryos with small interrupting RNAs (siRNAs) specific for *six3a* and *eri1* inhibited endogenous miR-430 processing, which resulted in many off-target effects such as tail truncations and distinct morphological features at the mesencephalic–mentencephalic boundary due to saturation of the miRNA pathway. However, recent studies have shown that generating transgenic lines that stably express integrated miRNA constructs for a desired gene of interest can circumvent these off-target effects, which speaks to the effectiveness of this knockdown method (Dong et al. 2009; Giacomotto et al. 2015).

Targeted Induced Local Lesions In Genomes, or TILLING, is probably one of the most recognized reverse genetic techniques currently in use across multiple fields. It involves screening DNA of randomly mutagenized fish for point mutations in a gene of interest. Once a mutation has been discovered, either the corresponding male is bred or its cryopreserved sperm are implanted in a female through in vitro fertilization until the mutation is brought to homozygosity (Draper et al. 2004). To date, there has been little to no data obtained using this technique pertaining to the study of renal disease; however, TILLING is still a good alternative to more common techniques, such as MOs or CRISPR, as a means to identify kidney mutants. These zebrafish research technologies are being developed and improved with each passing day, making said organism an increasingly attractive animal model for kidney disease.

3.4 Inducible Damage Models

Acute kidney injury (AKI) is a type of disorder that encompasses a wide range of factors, with a 50–70% mortality rate within intensive care patients, a rate that has not changed over the last few decades (Thadhani et al. 1996; Chertow et al. 2005). Treatments for AKI are extremely limited for patients with more severe cases, where renal replacement and dialysis therapies are the only treatments shown to be effective in humans (Thadhani et al. 1996). Mammalian models, such as rats or mice, of AKI are often difficult to interpret because of a plethora of challenges that include inaccessibility of the kidney and poor visualization of the renal tubules. Some alternative models like cell culture have come close but still cannot faithfully recapitulate the in vivo environment or produce adequate three-dimensional cultures in order to model injury.

Thus, researchers have turned to the zebrafish as a model system to study AKI. As previously stated, not only does the zebrafish kidney have conserved nephron structure and segmentation to that of mammalian nephrons, but their embryos and larvae are translucent, which makes observing changes along the entire length of the kidney under a microscope quite simple. Additionally, zebrafish have a relatively complex mesonephros that provides a convenient and useful adult setting for renal studies. Combined with the fact that they produce a large number of offspring when they breed, and that they have an incredible regenerative capacity after injury (McCampbell et al. 2015), zebrafish are uniquely qualified to help better elucidate the mechanisms of AKI progression and recovery.

3.4.1 Nephrotoxins

One of the ways that researchers model AKI in zebrafish is by injecting them with gentamicin and/or cisplatin, antibiotics that are also known nephotoxins (Fig. 3.4a). In a study led by Hentschel et al. (2004), the researchers characterized and quantified the effects of gentamicin injection in zebrafish larvae, as evidenced by histological and functional changes similar to aminoglycoside toxicity in mammals. Gentamicin injury was characterized by pronephric tubule flattening of the brush border, distension of the tubules and glomeruli, lysosomal phospholipidosis, accumulation of leukocytes, as well as debris in the tubular lumen. Similar effects were observed in cisplatin-injected zebrafish kidneys. These changes resulted in decreased renal function and an inability to maintain fluid homeostasis, resulting in the development of a pericardial edema.

Because of this ability to recapitulate AKI, nephrotoxins have also been used to characterize regeneration within the adult zebrafish kidney (Diep et al. 2011; McCampbell et al. 2015). In particular, McCampbell et al. (2015) analyzed and mapped the temporal progression of the cellular and molecular dynamics of

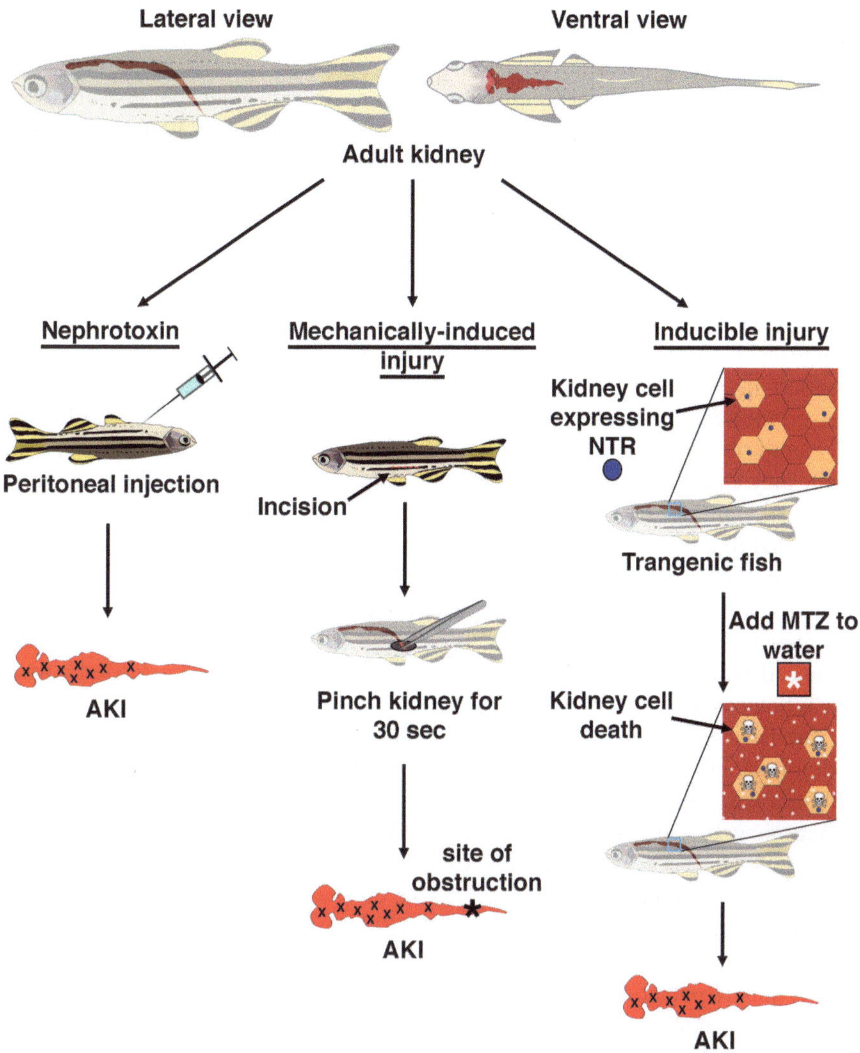

Fig. 3.4 Assortment of techniques used to generate acute kidney injury (AKI) in the adult zebrafish kidney. Diagram of peritoneal nephrotoxin injection, mechanically induced ischemic injury, and the MTZ-NTR-inducible transgenic model. "X" on kidney diagrams represents AKI

regeneration after gentamicin injury, establishing a regeneration time course damage induction to complete regeneration of proximal epithelial nephron tubules.

3.4.2 Mechanically Induced Kidney Insults

Some drawbacks to using nephrotoxins on both larval and adult zebrafish are likely to include leakage of toxins out of the animal because solubilized liquid compounds are injected into an organism that lives in an aquatic environment. This may partially account for variability in the degree of AKI in fish exposed to the same dosage of antibiotic (Kamei et al. 2015). Consequently, there is no true way to quantify how much of the drug is actually affecting the fish kidney. As an alternative, inducing AKI through mechanical means can circumvent this difficulty. In a study done by Hellman et al. (2010), researchers mechanically obstructed the zebrafish kidney fluid flow for the purpose of understanding regulation of cilia function in response to injury by looking at *foxj1a* expression, an important transcription factor required for motile ciliogenesis. Researchers took anesthetized 12-month-old fish and made an incision at the level of the distal collecting system. Next, they used tweezers to pinch off the distal collecting system for 30 s which produced damage in the form of dilated nephron tubules. After which they proceeded to allow the fish to recover, along with providing antibiotics to prevent infection, for a period of 12 h overnight (Fig. 3.4b). Using this mechanical means of inducing kidney injury, they were able to determine that inducible *foxj1a* is essential for cilia beat rate maintenance over a 24-h time period, suggesting that enhanced cilia function may be an important, previously uncharacterized part of organ homeostasis.

Another type of mechanical insult used in the field today is laser-mediated cell ablation (Johnson et al. 2011). Palmyre et al. (2014) utilized focused violet (405 nm) laser photoablation on a 20–100 μm stretch of pronephric epithelium in transgenic zebrafish embryos expressing GFP under the control of kidney-specific promoters (Fig. 3.4c). The rationale behind the use of the 405 nm laser on GFP-expressing zebrafish is twofold: (1) the GFP would allow for a more focused and intense beam on a specific area, in this case the fluorescent tissue, and (2) GFP absorbs light around 405 nm, meaning that the fluorescent protein would serve as an energy sink to potentiate cell injury. Using this targeted and highly specific method for inducing injury in the zebrafish embryo, researchers were able to analyze the mechanisms of renal repair, noting how collective kidney epithelial cell migration is an early response to injury and how cell proliferation occurs after migration.

3.4.3 Spatial/Temporally Controlled Methods of Kidney Injury

One of the most significant advances in the last couple of years has been the development of a noninvasive, tissue-specific method of inducing kidney injury that is both quantifiable and able to recapitulate more specific and subtle forms of kidney injury, such as podocyte ablation, which leads to glomerulosclerosis. The injury models in question take advantage of bacterial nitroreductase (NTR) and its ability to convert the prodrug metronidazole (MTZ) into a cytotoxic metabolite (Zhou and Hildebrandt 2012; Huang et al. 2013). Researchers generated transgenic zebrafish expressing NTR fused to a fluorescent reporter under the control of the *podocin* promoter, so as to only induce injury in the podocytes. After which they added a specific amount of MTZ to the fish tanks, which was taken in by the fish through their gills (Fig. 3.4d). Using this system, the researchers were able to create animal models of glomerulosclerosis to characterize podocyte injury progression, as well as the mechanisms of recovery (Zhou and Hildebrandt 2012; Huang et al. 2013).

3.4.4 Future Directions

Zebrafish carry with them the immense possibility for answering many questions pertaining to AKI pathophysiology, as well as injury recovery and regeneration. Understanding both global and specific kidney damage using the methods described previously can help physicians provide better care for patients with AKI. Additionally, zebrafish have incredible regenerative capacity after AKI, completely recovering from damage 14–21 days after injury (McCampbell et al. 2015). This is of significant importance because even though mammals can regenerate their kidney tubule epithelium, they cannot regenerate nephrons after they are destroyed (Stocum 2012), by either disease or AKI. By understanding the cellular mechanisms of zebrafish kidney regeneration potential, cell therapies can be developed that can ameliorate kidney damage.

Another way zebrafish kidney damage models can be used to develop kidney therapies is when they are used in tandem with chemical genetic screens to find compounds that can ameliorate kidney damage, enhance kidney injury recovery, or post-injury fibrosis after AKI. de Groh et al. (2010) performed a chemical screen to identify compounds that would expand the renal progenitor cell field, explaining that compounds that expand renal progenitor cell number might also enhance recovery after injury. In this screen, they identified the HDAC inhibitor methyl-4-(phenylthio) butanoate (m4PTB) as a compound that expanded the renal progenitor cell field in a proliferation-dependent manner. Based on the results of this screen, Cianciolo et al. (2013) induced AKI in both zebrafish larvae and adult mice through gentamicin injection and ischemic reperfusion (IR) AKI, respectively, and then treated the organisms with m4PTB to see if this would affect their recovery. m4PTB treatment after AKI proved to accelerate recovery and survivorship in both zebrafish and mice, making this compound a potentially viable approach to treating AKI. These methods of inducible injury, combined with an incredible regenerative potential and a large molecular biology tool box, have made the zebrafish an incredible tool for disease study and therapy development.

3.5 Conclusion

From forward genetic screens to genome editing and inducible injuries approaches, this chapter has discussed various ways in which the zebrafish can be used to recapitulate kidney disease and has discussed how these can be used to further understand disease onset to provide better treatment, and hopefully create new therapies for patients with renal disease. The question of "how" zebrafish models of kidney disease and injury can further development of therapeutics, which is the ultimate goal of understanding any human disease, is one that is constantly being discussed however.

That being said, no disease model is perfect. There will be instances in which gene function will diverge between zebrafish and humans and others in which kidney physiology is too dissimilar in order to create an adequate disease phenotype, an example of which would be the lack of intermediate tubule segments in the zebrafish nephron. Nevertheless, the organism's experimental versatility, limited in many ways only by a researcher's imagination, is vast. The best way to take advantage of this animal model is to use it in conjunction with mammalian ones in order to get a clearer picture of the cellular processes taking place during kidney injury and disease. From a pet store novelty to an important model organism, the zebrafish has become an invaluable tool for scientific discovery.

References

Amantana A, Iversen PL (2005) Pharmacokinetics and biodistribution of phosphorodiamidate morpholino antisense oligomers. Curr Opin Pharmacol 5:550–555

Amsterdam A, Hopkins N (2006) Mutagenesis strategies in zebrafish for identifying genes involved in development and disease. Trends Genet 22:473–478

Amsterdam A, Burgess S, Golling G et al (1999) A large-scale insertional mutagenesis screen in zebrafish. Genes Dev 13:2713–2724

Anderson BR, Howell DN, Soldano K et al (2015) In vivo modeling implicates APOL1 in nephropathy: evidence for dominant negative effects and epistasis under anemic stress. PLoS Genet 11:e1005349

Barrangou R, Fremaux C, Deveau H et al (2007) CRISPR provides acquired resistance against viruses in prokaryotes. Science 315:1709–1712

Beales PL, Balnd E, Tobin JL et al (2007) IFT80, which encodes a conserved intraflagellar transport protein, is mutated in Jeune asphyxiating thoracic dystrophy. Nat Genet 39:727–729

Bill BR, Petzold AM, Clark KJ et al (2009) A primer for morpholino use in zebrafish. Zebrafish 6:69–77

Brand M, Heisenberg CP, Warga RM et al (1996) Mutations affecting development of the midline and general body shape during zebrafish embryogenesis. Development 123:129–142

Cao Y, Semanchik N, Lee SH et al (2009) Chemical modifier screen identifies HDAC inhibitors as suppressors of PKD models. Proc Natl Acad Sci 106:21819–21824

Chertow GM, Burdick E, Honour M et al (2005) Acute kidney injury, mortality, length of stay, and costs in hospitalized patients. J Am Soc Nephrol 16:3365–3370

Cianciolo CC, Skrypnyk NI, Brilli LL et al (2013) Histone deacetylase inhibitor enhances recovery after AKI. J Am Soc Nephrol 24:943–953

Cong L, Ran FA, Cox D et al (2013) Multiplex genome engineering using CRISPR/Cas systems. Science 339:819–823
de Groh ED, Swanhart LM, Cosentino CC et al (2010) Inhibition of histone deacetylase expands the renal progenitor cell population. J Am Soc Nephrol 21:794–802
Diep CQ, Ma D, Deo RC et al (2011) Identification of adult nephron progenitors capable of kidney regeneration in zebrafish. Nature 470:95–100
Diep CQ, Peng Z, Ukah TK et al (2015) Development of the zebrafish mesonephros. Genesis 53:257–269
Dong M, Fu YF, Du TT et al (2009) Heritable and lineage-specific gene knockdown in zebrafish embryo. PLoS One 4:e6125
Dow LE, Fisher J, O'Rourke KP et al (2015) Inducible *in vivo* genome editing with CRISPR-Cas9. Nat Biotechnol 33:390–394
Draper BW, McCallum CM, Stout JL et al (2004) A high-throughput method for identifying N-ethyl-N-nitrosourea (ENU)-induced point mutations in zebrafish. Methods Cell Biol 77:91–112
Driever W, Solnica-Krezel L, Schier AF et al (1996) A genetic screen for mutations affecting embryogenesis in zebrafish. Development 123:37–46
Drummond IA (2005) Kidney development and disease in the zebrafish. J Am Soc Nephrol 16:299–304
Drummond IA, Majumdar A, Hentschel H et al (1998) Early development of the zebrafish pronephros and analysis of mutations affecting pronephric function. Development 125:4655–4667
Ebarasi L, Oddsson A, Hultenby K et al (2011) Zebrafish: a model system for the study of vertebrate renal development, function, and pathophysiology. Curr Opin Nephrol Hypertens 20:416–424
Eisen JS, Smith JC (2008) Controlling morpholino experiments: don't stop making antisense. Development 135:1735–1743
Gerlach GF, Wingert RA (2013) Kidney organogenesis in the zebrafish: insights into vertebrate morphogenesis and regeneration. Wiley Interdiscip Rev Dev Biol 2:559–585
Giacomotto J, Rinkwitz S, Becker TS (2015) Effective heritable gene knockdown in zebrafish using synthetic microRNAs. Nat Commun 6:7378
González F, Zhu Z, Shi Z et al (2014) An iCRISPR platform for rapid, multiplexable, and inducible genome editing in human pluripotent stem cells. Cell Stem Cell 15:215–226
Haffter P, Granato M, Brand M et al (1996) The identification of genes with unique and essential functions in the development of the zebrafish, *Danio rerio*. Development 123:1–36
Halvorson CR, Bremmer MS, Jacobs SC (2010) Polycystic kidney disease: inheritance, pathophysiology, prognosis, and treatment. Int J Nephrol Renovasc Dis 3:69–83
Hans S, Jaslin J, Freudenreich D et al (2009) Temporally–controlled site-specific recombination in zebrafish. PLoS One 4(2):e4640
Hellman NE, Liu Y, Merkel E et al (2010) The zebrafish foxj1a transcription factor regulates cilia function in response to injury and epithelial stretch. Proc Natl Acad Sci 107:18499–18504
Hentschel DM, Park KM, Cilenti L et al (2004) Acute renal failure in zebrafish-a novel system to study a complex disease. Am J Physiol Renal Physiol 288:F923–F929
Howe K, Clark MD, Torroja CF et al (2013) The zebrafish reference genome sequence and its relationship to the human genome. Nature 496:498–503
Huang J, McKee M, Huang HD et al (2013) A zebrafish model of conditional targeted podocyte ablation and regeneration. Kidney Int 83:1193–1200
Hudziak RM, Barofsky E, Barofsky DF et al (2009) Resistance of morpholino phosphorodiamidate oligomers to enzymatic degradation. Antisense Nucleic Acid Drug Dev 6:267–272
Janson CG, During MJ (2006) Peptide nucleic acids, morpholinos and related antisense biomolecules. Morpholinos and PNAs compared. Landes Bioscience and Kluwer Academic, New York, pp. 89–113

Johnson CS, Holzemer NF, Wingert RA (2011) Laser ablation of the zebrafish pronephros to study renal epithelial regeneration. J Vis Exp 54:e2839
Kamei CN, Liu Y, Drummond IA (2015) Kidney regeneration in adult zebrafish by gentamicin induced injury. J Vis Exp 102:e51912
Kimmel CB, Ballard WW, Kimmel SR et al (1995) Stages of embryonic development of the zebrafish. Dev Dyn 203:253–310
Kok FO, Shin M, Ni CW, Gupta A, Grosse AS, van Impel A, Kirchmaier BC, Peterson-Maduro J, Kourkoulis G, Male I, DeSantis DF, Sheppard-Tindell S, Ebarasi L, Betsholtz C, Schulte-Merker S, Wolfe SA, Lawson ND (2015) Reverse genetic screening reveals poor correlation between morpholinoinduced and mutant phenotypes in zebrafish. Dev Cell 32:97–108
Kroeger PT Jr, Wingert RA (2014) Using zebrafish to study podocyte genesis during kidney development and regeneration. Genesis 52:771–792
Laale HW (1997) The biology and use of zebrafish. Brachydanio rerio in fisheries research. A literature review. J Fish Biol 10:121–173
Lancaster MA, Gleeson JG (2009) The primary cilium as a cellular signaling center: lessons from disease. Curr Opin Genet Dev 19:220–229
Lieschke GJ, Currie PD (2007) Animal models of human disease: zebrafish swim into view. Nat Rev Genet 8:353–367
McCampbell KK, Springer KN, Wingert RA (2014) Analysis of nephron composition and function in the adult zebrafish kidney. J Vis Exp 90:e51644
McCampbell KK, Springer K, Wingert RA (2015) Atlas of cellular dynamics during zebrafish adult kidney regeneration. Stem Cells Int 2015:547636
Nadasdy T, Laszik Z, Lajoie G et al (1995) Proliferative activity of cyst epithelium in human renal cystic diseases. J Am Soc Nephrol 5:1462–1468
Palmyre A, Lee J, Ryklin G et al (2014) Collective epithelial migration drives kidney repair after acute injury. PLoS One 9:e101304
Pickart MA, Klee EW (2014) Zebrafish approaches enhance the translational research tackle box. Transl Res 163:65–78
Reimschuessel R (2001) A fish model of renal regeneration and development. ILAR J 42:285–291
Reimschuessel R, Williams D (1995) Development of new nephrons in adult kidneys following gentamicin-induced nephrotoxicity. Ren Fail 17:101–106
Reimschuessel R, Bennett RO, May EB et al (1990) Renal tubular cell regeneration, cell proliferation and chronic nephrotoxicity in the goldfish *Carassius auratus* following exposure to a single sublethal dose of hexachlorobutadiene. Dis Aquat Org 8:211–224
Rossi A, Kontarakis Z, Gerri C et al (2015) Genetic compensation induced by deleterious mutations but not gene knockdowns. Nature 524:230–233
Rubinstein AL (2003) Zebrafish: from disease modeling to drug discovery. Curr Opin Drug Discov Devel 6:218–223
Santoriello C, Zon LI (2012) Hooked! modeling human disease in zebrafish. J Clin Invest 122:2337–2343
Solnica-Krezel L, Schier AF, Driever W (1994) Efficient recovery of ENU-induced mutations from the zebrafish germline. Genetics 136:1401–1420
Stocum, DL (2012) Regeneration of digestive respiratory and urinary tissues. Regenerative biology and medicine. 2nd edn. Burlington: Elsevier Academic. Pg. 118
Sullivan-Brown J, Schottenfeld J, Okabe N et al (2008) Zebrafish mutations affecting cilia motility share similar cystic phenotypes and suggest a mechanism of cyst formation that differs from pkd2 morphants. Dev Biol 314:261–275
Summerton J (1999) Morpholino antisense oligomers: the case for an RNase H-independent structural type. Biochim Biophys Acta 1489:141–158
Sun Z, Amsterdam A, Pazour GJ et al (2004) A genetic screen in zebrafish identifies cilia genes as a principal cause of cystic kidney. Development 131:4085–4093
Thadhani R, Pascual M, Bonventure JV (1996) Acute renal failure. N Engl J Med 334:1448–1460
Tobin JL, Beales PL (2008) Restoration of renal function in zebrafish models of ciliopathies. Pediatr Nephrol 23:2095–2099

Vincensini L, Blisnick T, Bastin P (2011) 1001 model organisms to study cilia and flagella. Biol Cell 103:109–130
Wang L, Zhang Y, Zhou T et al (2008) Functional characterization of Lmo2-Cre transgenic zebrafish. Dev Dyn 237:2139–2146
Wingert RA, Davidson AJ (2008) The zebrafish pronephros: a model to study nephron segmentation. Kidney Int 73:1120–1127
Wingert RA, Davidson AJ (2011) Zebrafish nephrogenesis involves dynamic spatiotemporal expression changes in renal progenitors and essential signals from retinoic acid and irx3b. Dev Dyn 240:2011–2027
Wingert RA, Selleck R, Yu J et al (2007) The cdx genes and retinoic acid control the positioning and segmentation of the zebrafish pronephros. PLoS Genet 3:1922–1938
Zhou W, Hildebrandt F (2012) Inducible podocyte injury and proteinuria in transgenic zebrafish. J Am Soc Nephrol 23:1039–1047
Zhou X, Fjose A, Larsen N et al (2008) Treatment with small interfering RNA affects the microRNA pathway and causes unspecific defects in zebrafish embryos. FEBS J 275:2177–2184
Zhou W, Boucher RC, Bollig F et al (2010) Characterization of mesonephric development and regeneration using transgenic zebrafish. Am J Physiol Renal Physiol 299:F1040–F1047

Chapter 4
Use of *Xenopus* Frogs to Study Renal Development/Repair

Shoshoni T. Droz and Kelly A. McLaughlin

Abstract The *Xenopus* genus includes several members of aquatic frogs native to Africa but is perhaps best known for the species *Xenopus laevis* and *Xenopus tropicalis*. These species were popularized as model organisms from as early as the 1800s and have been instrumental in expanding several biological fields including cell biology, environmental toxicology, regenerative biology, and developmental biology. In fact, much of what we know about the formation and maturation of the vertebrate renal system has been acquired by examining the intricate genetic and morphological patterns that epitomize nephrogenesis in *Xenopus*. From these numerous reports, we have learned that the process of kidney development is as unique among organs as it is conserved among vertebrates. While development of most organs involves increases in size at a single location, development of the kidney occurs through a series of three increasingly complex nephric structures that are temporally distinct from one another and which occupy discrete spatial locales within the body. These three renal systems all serve to provide homeostatic, osmoregulatory, and excretory functions in animals. Importantly, the kidneys in amphibians, such as *Xenopus,* are less complex and more easily accessed than those in mammals, and thus tadpoles and frogs provide useful models for understanding our own kidney development. Several descriptive and mechanistic studies conducted with the *Xenopus* model system have allowed us to elucidate the cellular and molecular mediators of renal patterning and have also laid the foundation for our current understanding of kidney repair mechanisms in vertebrates. While some species-specific responses to renal injury have been observed, we still recognize the advantage of the *Xenopus* system due to its distinctive similarity to mammalian wound healing, reparative, and regenerative responses. In addition, the first evidence of renal regeneration in an amphibian system was recently demonstrated in *Xenopus laevis*. As genetic and molecular tools continue to advance, our appreciation for and utilization of this amphibian model organism can only intensify and

S.T. Droz
Northeastern University, Boston, MA, USA

K.A. McLaughlin (✉)
Tufts University, Medford, MA, USA
e-mail: kelly.mclaughlin@tufts.edu

R.K. Miller (ed.), *Kidney Development and Disease*, Results and Problems in Cell Differentiation 60, DOI 10.1007/978-3-319-51436-9_4

will certainly provide ample opportunities to further our understanding of renal development and repair.

4.1 Introduction

The kidney is a vital organ that performs essential homeostatic functions in animals, thus allowing for survival of a wide variety of species in diverse aquatic and terrestrial environments. This concept of homeostasis was introduced by the nineteenth-century French physiologist Claude Bernard, who first coined the term "*milieu intérieur.*" Bernard's commentary on the internal environment had a huge impact on the American physiologist Homer W. Smith, who extensively studied the kidney in the 1930s. Smith is known for his eloquence with regard to the importance of the renal system, and is famously quoted as saying, "*Perhaps more than any other organ, the kidney has allowed the survival and evolution of organisms in new environments ranging from seawater to dry land. The universal demand for tissue fluid homeostasis and osmoregulation in diverse habitats has led to the development of a variety of kidney forms.*" Smith summarizes his sentiments by writing, "*It was the view of this physiologist that we achieve a free and independent life, mentally and physically, because of the constancy of the composition of our blood. Recognizing that we have the kind of blood we have because we have the kind of kidneys that we have, we must acknowledge that our kidneys constitute the major foundation of our physiological freedom*" (Smith 1953).

Indeed, the kidney is unique among other organs in the body in its sequential development through three distinct phases: the pronephros, the mesonephros, and the metanephros. All vertebrate kidneys progress through two or more of these stages, which differ temporally and spatially during organogenesis. This successive development of the renal system allows an organism to survive and function in vastly different environs during its maturation. For example, the mammalian fetus develops in utero, but, after parturition, it must survive external to the mother. In other amniotes (birds and reptiles), embryonic kidneys function differently in the egg as compared to the mature kidneys found in hatchlings and adults. Similarly, dramatic changes also occur in kidney development during maturation of both fully aquatic anamniotes (fish and some amphibians) and semiaquatic amphibians. These changes are especially essential for amphibians that must transition from aquatic to terrestrial habitats after metamorphosis as the issue of water conservation becomes even more critical. This is achieved by the successive nature of renal development, which replaces each kidney with a more elaborate and multiunit organ. In general, a more convoluted kidney is capable of accommodating a larger volume of blood (due to the increased size of a growing organism) and retaining a greater amount of water during the excretory process (Kardong 2014).

Due to its physiological importance, several organisms, including both vertebrates and invertebrates, have developed reparative processes to counteract the detrimental effects of injuring or losing renal tissue. Regrettably, while these repair phenomena have been observed and studied for several decades, the mechanisms regulating them are still poorly understood. In general, every incidence of tissue injury must be addressed by some degree of wound healing, with the initial inflammatory response largely determining whether an event will proceed toward regeneration or result in repair dominated by cicatrization (Harty et al. 2003). Both regenerative and non-regenerative wounds are closed by re-epithelialization; however, non-regenerative tissues tend to experience excessive matrix deposition (Raghow 1994; Wynn 2008; Liu 2011). The resulting scar, primarily comprised of collagen (Stichel 1999; King et al. 2003; Huang et al. 2011), serves to reestablish the anatomical continuity of the damaged tissue but is unable to restore lost function. In contrast, wound healing during regenerative events involves re-epithelialization in the absence of fibroproliferation (Tanaka and Galliot 2009). Consequently, successful regeneration is characterized by the scar-free restoration of functional tissues or organs and, in some circumstances, of entire body plans (Tsonis 2000; Sánchez Alvarado and Tsonis 2006), after damage, disease, casting, autotomy, or fission. It should be noted that the term "regeneration" is also used to describe the homeostatic turnover of short-lived cell types such as those found in blood (Zon 2008), skin (Martin 1997), bone (Huang and Ogawa 2010; Panetta et al. 2010), and the intestinal epithelium (Ishizuya-Oka 2007; Barker et al. 2008). However, cell replacement during these turnover events is a separate context from repair and thus is not discussed within this chapter.

The mammalian kidney has a limited capacity for restoring structural components after severe damage (Boti et al. 1982; Nonclercq et al. 1992; Imgrund et al. 1999), and, as seen in other non-regenerative organs, its repair is epitomized by fibrosis caused by chronic inflammation (Hewitt et al. 2008; Wynn 2008; Liu 2011). This process occurs via the dysregulation of innate and adaptive immune responses (Wynn and Ramalingam 2012). In contrast to the mammalian system, fibroproliferation has not been described in regeneration of piscine kidneys (Elger et al. 2003; Watanabe et al. 2009; Zhou et al. 2010; Diep et al. 2011) or *Xenopus laevis* pronephroi (Caine and McLaughlin 2013), and was only briefly observed after newt mesonephrectomy (Scadding and Liversage 1974). These phylogenetically basal members of the infraphylum Gnathostomata (jawed vertebrates; Phylum: Chordata, Subphylum: Vertebrata) have retained a more ancestral immunity (Mescher et al. 2007) and thus are more likely to have an attenuated inflammatory response to renal damage as compared to their mammalian counterparts. Rather than become fibrotic, damaged renal structures in tadpoles (Caine and McLaughlin 2013) and fish (Reimschuessel and Williams 1995; Augusto et al. 1996; Salice et al. 2001; Liu et al. 2002; Elger et al. 2003) tend to display a greater propensity for regeneration.

Notably, all vertebrate kidneys share important characteristics including highly conserved developmental programs, anatomical similarities (throughout the many stages of organogenesis as well as during maturity), and physiological function. These shared characteristics are likely to extend, at least in part, to cellular and molecular responses initiated after tissue damage. In fact, much of our current understanding of kidney repair and regeneration has been gleaned from investigating nonmammalian renal systems, and some of the most seminal research in this field has been conducted on juvenile and adult amphibians (Howland 1916; Swingle 1919; Babaeva 1964; Chopra and Simnett 1969, 1970; Scadding and Liversage 1974).

In general, there is considerable interspecies variation among reparative and regenerative capacities, with invertebrates typically exhibiting superior proficiency to vertebrates (Monks 1903, 1904; Sánchez Alvarado 2000). For this reason, the regenerative biology community has devoted considerable effort to elucidating mechanisms of repair with the use of several model organisms (Carlson 1978; Tsonis 2000; Brockes and Kumar 2002; Tsonis 2002; Slack 2003; Sánchez Alvarado and Tsonis 2006; Birnbaum and Sánchez Alvarado 2008). While model animals have historically been selected based on local availability and personal preference (Nieuwkoop 1996), the recent development of innovative molecular tools and advances in genetic approaches allow researchers new opportunities to address complex questions in a wide number of model systems, including amphibians (Stoick-Cooper et al. 2007; Birnbaum and Sánchez Alvarado 2008; Jewhurst et al. 2014; Bonasio 2015). Among vertebrates, members of the Amphibia class are outstanding in their ability to regenerate several larval and adult tissues. For this reason, both orders of amphibians—Urodela (amphibians that retain their tails after metamorphosis such as newts and salamanders) and Anura (amphibians that resorb their tails during metamorphosis such as frogs and toads)—have produced a number of popular model organisms used in regenerative studies (Tsonis 2000).

Xenopus laevis, commonly known as the African clawed/claw-toed frog (Fig. 4.1), has long been a favorite among laboratory research animals due to its experimental tractability and rapid speed of embryonic and larval development, as well as the relative ease with which researchers can culture embryos, raise juveniles, and care for adults (Nieuwkoop 1996). Additionally, unlike in mammalian system, *Xenopus* embryos and larvae develop independently of the mother and as such are available for manipulations even at the earliest stages of development. These characteristics, which have been especially appealing to developmental biologists, have also allowed *Xenopus* to emerge as a leading model in regenerative biology (Beck et al. 2009).

To date, *Xenopus* has been instrumental in elucidating reparative mechanisms of several structures including fore- and hindlimb buds (Singer 1951; Dent 1962; Muneoka et al. 1986; Endo et al. 1997, 2000; Yokoyama et al. 2000; Hayashi et al. 2015), tail (Beck et al. 2003; Gargioli and Slack 2004; Sugiura et al. 2004; Tazaki et al. 2005; Mochii et al. 2007; Mondia et al. 2011; Tseng et al. 2011; Love et al. 2013), intestines (reviewed in Ishizuya-Oka 2007), brain (Endo et al. 2007; Bernardini et al. 2010), olfactory epithelium (Frontera et al. 2015), spinal cord (Muñoz et al. 2015), lens (Yoshii et al. 2007; Malloch et al. 2009; Fukui and Henry 2011; Hamilton et al. 2016), and retina (Sakaguchi et al. 1997; Moshiri et al. 2004; Yoshii et al. 2007; Lee et al. 2013). Notably, *Xenopus* has also been a valuable study system for characterizing the conserved events in kidney development (Vize et al. 1997; Brennan et al. 1998; Brändli 1999; Carroll et al. 1999a, b; Hensey et al. 2002; Jones 2005; Wessely and Tran 2011) and for investigating the mechanisms of kidney repair (Chopra and Simnett 1969, 1970; Caine and McLaughlin 2013).

Fig. 4.1 Adult *Xenopus laevis* frog. Photo credit: Hallie Gluk and Elizabeth Card

4.2 Development of the *Xenopus* Renal System

Organogenesis in the frog *Xenopus laevis* involves a dramatic ballet of morphological changes that occur shortly after fertilization and proceed through metamorphosis to maturation. During early embryogenesis, organogenesis follows the establishment of the three primary germ layers (gastrulation) to confer shape and function to developing organs. For most tissues, this process involves cellular differentiation and morphological rearrangements at the terminal anatomical site. Kidney development in *X. laevis* is unique in that it occurs through a series of two distinct forms, the pronephros and the mesonephros, which vary in location along the anterior–posterior axis at different stages of embryogenesis (Saxén 1987). Both kidneys are derived from the intermediate mesoderm (Vize et al. 1995; Dressler 2006; Attia et al. 2012) and develop in response to similar inductive signals between mesenchymal and epithelial tissues (Vize et al. 1997; Carroll et al. 1999a; Kuure et al. 2000). They develop as bilaterally symmetrical organ pairs and share the basic functional unit of all vertebrate kidneys, the nephron. To accommodate the increasing size and requirements of the developing organism, each successive kidney form increases in nephron number and morphological complexity (Fig. 4.2).

4.2.1 Overview of Sequential Development During Vertebrate Nephrogenesis

The pronephros develops first and is the most anterior of all vertebrate kidneys (Vize et al. 1995). It is the functional kidney found in embryos and larvae of fish (Armstrong 1932; Drummond and Davidson 2010), urodelean amphibians (Howland 1916; O'Connor 1940), and anuran amphibians (Jaffee 1954; Møbjerg et al. 2000) including *Xenopus* (Vize et al. 1995). In juvenile animals, the pronephroi exist as a pair of single nephrons located on either side of the body (Drummond et al. 1998; Carroll et al. 1999b). The pronephros also functions as part

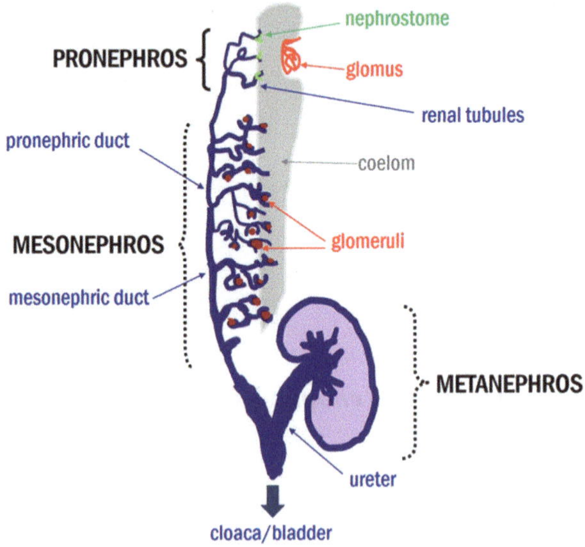

Fig. 4.2 Sequential kidney development in vertebrates. Development of the pronephros, mesonephros, and metanephros occurs at distinct spatiotemporal intervals, with the pronephros being established first at the most anterior location within the body (as depicted). As development progresses, each kidney degenerates and is replaced by a larger, multi-nephron organ. The mesonephros functions as the adult kidney in fish and amphibians, while the metanephros serves this function in reptiles, birds, and mammals. All three kidney forms contain several renal tubules that connect either directly to a glomerulus or indirectly to a glomus, via the nephrostomes and coelom. Filtrate flows through the tubules to the ducts and is finally excreted by the cloaca (in fish, amphibians, birds, and some reptiles) or the bladder (in mammals and some reptiles)

of the adult excretory system in extant species of the primitive Agnathan (jawless) class of fish such as lampreys and hagfish (Ellis and Youson 1989). In more recently evolved taxa (reptiles, birds, mammals), the pronephros is functionally vestigial and exists for a very transient period during embryogenesis (Bremer 1916; Saxén 1987). Despite its impermanence, development of the pronephros is critical for the formation of the mesonephros and, ultimately, the metanephros (O'Connor 1940; Vize et al. 1995, 1997; Jones 2005).

As fish and amphibian juveniles develop, eventually the pronephros degenerates and all urinary and homeostatic functions are taken over by the mesonephros (Bremer 1916; Carroll et al. 1999a). Mesonephric kidneys are typically comprised of multiple nephrons (Kingsley 1917; Vize et al. 1997) and develop caudally to the pronephros (Lipschutz 1998). In adult anamniotes, the mesonephros functions as the primary excretory organ and is then referred to as the opisthonephros (Barch et al. 1966; Fedorova et al. 2008). In contrast, the mesonephros in amniotes functions briefly (a few days to many weeks depending on the organism) during intrauterine development (Tiedemann and Wettstein 1980; Moritz and Wintour 1999). The mesonephros then regresses (Pole et al. 2002) as the terminal metanephric kidney develops. The metanephros forms posteriorly to the mesonephros

(Fig. 4.2), is comprised of up to a million nephrons per kidney (Saxén 1987), and is retained throughout adulthood (Lipschutz 1998).

4.2.2 Gene Patterning During Pronephric Kidney Development in Xenopus

The amphibian pronephros was first identified in the mid-1800s and has previously been described in great detail (Fox 1963). The anatomical and morphological characteristics of the *Xenopus* pronephros, in particular, have been meticulously outlined by Nieuwkoop and Faber (1994) and by Vize et al. (1997). Compared to the multi-nephron networks of the mesonephric and metanephric kidneys, the architecture of the pronephros is relatively simple and development is completed by approximately 2 days after fertilization, which corresponds to NF stage 37/38 in *Xenopus laevis* (Nieuwkoop and Faber 1994). The single nephron of the pronephros is comprised of three main components: the pronephric corpuscle (consisting of the glomus and coelom), the pronephric tubules, and the pronephric duct (Fox 1963; Vize et al. 1997; Brändli 1999) (Fig. 4.2). All three components are also present in the pronephros that forms during early mammalian renal development and are histologically, morphologically, and functionally analogous to the nephrons that comprise mature mammalian metanephric kidneys. As such, the simple *Xenopus* pronephros provides a useful model for understanding development, tissue maintenance, and reparative mechanisms in more complex kidneys such as those found in humans.

In fact, several signaling cascades that contribute to pronephric kidney development have previously been identified in *Xenopus laevis* and appear to be well conserved during mammalian metanephric development as well (reviewed in Lienkamp 2016). One of the earliest known genetic markers of kidney development is *lhx1* (also known as lim1), a member of the LIM family of homeobox genes (Taira et al. 1994). In *Xenopus laevis*, expression of this transcription factor is critical for specification of cells to a renal fate (Carroll and Vize 1999; Cirio et al. 2011) and for the promotion of tubule growth and elongation as development progresses (Chan et al. 2000). This phenomenon is also observed in the murine system in which *lim1* is essential for multiple steps in morphogenesis of epithelial tubules in metanephric kidney development (Kobayashi et al. 2005). Similarly, the *paired homeobox 2* (*pax2*) gene also encodes a transcription factor that is essential for development of both pronephric tubules (Drummond et al. 1998) and metanephric tubules (Dressler 1996; Davies and Fisher 2002). *Pax2* is detectable in the pronephric region in 1-day-old *Xenopus laevis* embryos (NF stage 23; Nieuwkoop and Faber 1994) after the completion of specification/determination (Heller and Brändli 1997; Carroll and Vize 1999). This finding suggests a role for *pax2* in pronephric differentiation and/or morphogenesis.

Additionally, conserved genes and signaling pathways have also been found to coordinate the process of tubulogenesis during kidney formation in both amphibians and mammals. Among these is the *Wnt4* gene, which encodes a secreted glycoprotein (Stark et al. 1994). *Wnt4* is transiently expressed throughout the pronephric tubule and duct anlagen and then becomes restricted to the dorsolateral region that will ultimately form the functional pronephric tubules (Vize et al. 2003). Notably, morpholino-mediated knockdown of *wnt4* resulted in the complete loss of pronephric tubules (Saulnier et al. 2002). Similar outcomes were seen in knockout studies in murine metanephroi (Stark et al. 1994; Kispert et al. 1998). These studies indicate that Wnt4 is not required for early inductive activities but does play a critical role in inducing the mesenchymal-to-epithelial transitions that epitomize tubulogenesis.

As renal development continues to progress, markers of differentiated cells are expressed in distinct regions of the pronephros. Most prominent among these are a variety of ion channels that coordinate specific homeostatic and osmoregulatory functions including sodium/glucose cotransporters (Zhou and Vize 2004), sodium/potassium ATPases (Eid and Brändli 2001; Rahman et al. 2015), calcium-binding senescence marker protein 30 (Sato et al. 2000), sodium bicarbonate transporters (Zhou and Vize 2004), chloride channel proteins (Vize et al. 2003), sodium/potassium/chloride cotransporters (Zhou and Vize 2004), and several members of the solute carrier gene family (Raciti et al. 2008).

4.2.3 *Structure and Function of the* Xenopus *Pronephric Kidney*

Circulating blood is supplied to the kidney by branches of the dorsal aorta, which connect to each pronephros via a glomus, the vascularized portion of the pronephric corpuscle. Pronephric gloma are composed of fenestrated endothelial cells and podocytes (Drummond and Majumdar 2003), which filter nitrogenous waste products from the blood. In fish, a large, single (fused) glomus, located at the embryo midline, is closely integrated with the nephric tubules and so is referred to as a glomerulus (Drummond et al. 1998). In *Xenopus* and other amphibians, the glomus on either side of the animal connects to a coelomic cavity called the nephrocoele via the pronephric capsule (the nonvascular component of the pronephric corpuscle; Brändli 1999). This coelom, which is lined with squamous mesothelial cells (Wrobel and Süß 2000), is also associated with the mesonephric glomeruli in fish but is absent in mammalian metanephric kidneys (as depicted in Fig. 4.2).

The glomal filtrate is swept forward by cilia lining the cuboidal epithelial cells of the nephrostomes (Jaffee 1963). Each thin nephrostome funnels filtrate into the wider lumen of a branch of the proximal tubules. These tubules are comprised of columnar epithelial cells that are extensively convoluted at their basolateral membranes to facilitate transepithelial transport, which is also seen in the proximal tubules in mammals (Aronson 1989). The apical surfaces of these tubule cells end in dense microvilli (Møbjerg et al. 2000), and this brush border is responsible for the majority of osmoregulatory functions performed by the pronephric kidney. These function include solute reabsorption and pH regulation (Zhou and Vize 2004; Christensen et al. 2008). Resorbed nutrients, salts, and water are returned to the blood via the venous system surrounding the tubules (Vize et al. 1997). The actual number of nephrostomes, and corresponding number of proximal tubule branches, varies among organisms (Vize et al. 2003). In *Xenopus laevis* tadpoles, there are three nephrostomes connected to three proximal tubule branches (Brennan et al. 1998). These dorso-anterior branches connect directly to a common branch of the proximal tubules called the broad tubule.

Once water and nutrient reabsorption has occurred in the proximal tubules, a more concentrated urine flows through the proximal broad tubules into the adjoining distal tubules, which are located more posteriorly. The cuboidal epithelial cells of the distal tubules lack an apical brush border but have extensive basal membrane folding (Møbjerg et al. 2000). Nitrogenous waste flows through the narrow lumen of the distal tubules and into the pronephric duct. In *Xenopus laevis*, this waste predominantly contains ammonia with smaller amounts of urea (Balinsky and Baldwin 1961). The duct, otherwise known as the connecting tubule, is comprised of a least two distinct types of low cylindrical cells that lack both apical microvilli and basal membrane folding (Møbjerg et al. 2000). This tube is responsible for carrying urine to the rectal diverticulum and ultimately to the cloaca (Drawbridge et al. 2003), from where it is excreted.

While only two cell types are known to comprise the pronephric duct (Raciti et al. 2008), the pronephric tubules are more extensively segmented. A decade ago, the proximal and distal tubules in *Xenopus* were divided into so-called "early" regions (connected to the nephrostomes) and "late" regions (connected to the pronephric duct) as described by Zhou and Vize (2004; Fig. 4.3a). However, through large-scale gene expression mapping, Raciti et al. (2008) have provided the *Xenopus* community with a new model for nephron organization that closely resembles the segmentation seen in mammalian kidneys (Maunsbach and Christensen 1992). This model introduces the "intermediate tubule" segment (Fig. 4.3b), which expresses several kidney genes not found in the proximal and/or distal tubules that adjoin it on either side (Raciti et al. 2008).

Differential gene expression patterns in the nephric regions correspond to a division of labor among the functional domains of the nephron. Specifically, expression of several members of the solute carrier gene family indicates that specialized ion transport channels localize to one or more of the four pronephric regions (proximal tubule, intermediate tubule, distal tubule, and pronephric duct/collecting tubule; Raciti et al. 2008). While this hugely impressive undertaking has provided us with an updated nephron model and many new renal gene expression profiles, it has also expanded the way we must think about kidney repair. Since damaged renal tubules in several injury models are often replaced by surviving tubule cells (Cuppage and Tate 1967; Cuppage et al. 1972; Boti et al. 1982; Nonclercq et al. 1992; Gobé et al. 1995; Gobé and Buttyan 2002; Bonventre

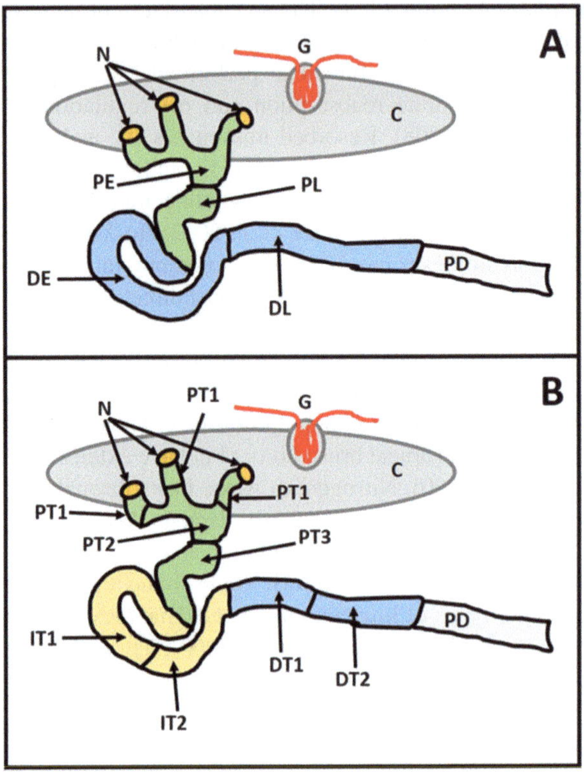

Fig. 4.3 Pronephric tubule segmentation models. The Zhou and Vize (2004) model of tubule segmentation **(a)** designates early (E) and late (L) segments of the proximal (P, *green*) and distal (D, *blue*) tubules in the pronephric kidney. These segments indicate functional domains of the pronephric tubules. A newer model introduced by Raciti et al. (2008) **(b)** demonstrates that pronephric tubules are more highly segmented than previously thought, with differential functional domains established along the proximal tubule segments (PT1, PT2, PT3) and distal tubule segments (DT1, DT2). This model also introduces a third tubule component, the intermediate tubule, which is also further segmented (IT1, IT2) by differences in gene expression patterns. This newer model is similar to the segmentation observed in mammalian kidneys (see Maunsbach and Christensen 1992). The coelom (C), glomus (G), nephrostomes (N), and pronephric duct (PD) are also present in the pronephros

2003; Nony and Schnellmann 2003), identifying the source of these cells in any renal repair or regenerative pathway is paramount to determining the mechanisms involved.

4.3 *Xenopus laevis* as a Model for Kidney Repair

The position of the pronephros, located dorso-laterally just beneath the transparent *Xenopus* tadpole skin, makes this system particularly appealing as a model for kidney repair research. Its visibility and accessibility have previously provided the unique advantages of observing patterns of gene expression in situ (Brändli 1999), examining morphological changes in embryos (Nieuwkoop and Faber 1994; Vize et al. 1997), and directly observing kidney development in vivo at subcellular resolution (Lienkamp et al. 2012). Recently, these advantages have also proven to be useful during mechanical injury studies involving partial pronephrectomies (Caine and McLaughlin 2013). Moreover, since the genetic programs and cellular mechanisms that coordinate tissue regeneration are frequently found to be similar to those that direct early organogenesis (Goldin and Fabian 1978; Sánchez Alvarado 2004; Birnbaum and Sánchez Alvarado 2008), the pattern of renal development established in the *Xenopus* model system may be especially beneficial for elucidating the mechanisms of renal repair in this system (and, by extension, perhaps all vertebrate kidneys). Although some other visceral organs do appear to have very different programs for development and regeneration (Dor and Stanger 2007), it is still useful to infer from mechanisms of organ ontogeny to guide new inquiries into tissue restoration. This type of comparative investigation is highly productive when conducted with a model organism in which the development of the organ of interest has been as well characterized as it has with the *Xenopus* renal system.

Unfortunately, the pseudo-allotetraploid genome of *Xenopus laevis* has previously caused a considerable delay in the development of genetic tools, unlike those currently available for studying another well-established, nonmammalian model system, the zebrafish (*Danio rerio*) (Poss et al. 2003). However, recent advances in the creation of chemical genetic screening systems (Wheeler and Brändli 2009; Tomlinson et al. 2012; Wheeler and Liu 2012), as well as targeted gene disruption via editing technologies including clustered regularly interspaced short palindromic repeats (CRISPR)/CRISPR-associated protein 9 (CRISPR/Cas9; Wang et al. 2015), and transcription activator-like effector nucleases/zinc-finger nucleases (TALENs/ZFNs; Lei et al. 2012; Suzuki et al. 2013; Nadake et al. 2015; Liu et al. 2016), provide new tools for research with *Xenopus*. Additionally, the optimization of optogenetic tools to induce organ damage in *Xenopus laevis* (Jewhurst et al. 2014) has further expanded the suitability of this system in repair studies. Importantly, the *Xenopus* and mammalian genomes are quite similar (Hellsten et al. 2010), which may account for the comparable regenerative capacities observed in these two systems (Gargioli and Slack 2004; Slack et al. 2004). Taken together, these advantages make the *Xenopus* system a uniquely suitable model organism with which to study renal repair.

4.3.1 *Renal Injury Models:* Xenopus, *Other Amphibians, Fish, Mammals*

Several vertebrate models of kidney disease have been used to investigate the reparability of nephric structures, most often after damage to the tubule or glomerular basement membrane. Mammalian metanephric kidneys undergo epithelial replacement, with adjacent surviving cells migrating and proliferating to repopulate the denuded basement membrane of injured nephrons (Cuppage and Tate 1967; Cuppage et al. 1972; Boti et al. 1982; Nonclercq et al. 1992; Gobé et al. 1995; Gobé and Buttyan 2002; Bonventre 2003; Nony and Schnellmann 2003). Compensatory hypertrophism also occurs in both the injured kidney (Wesson 1989) and the contralateral kidney (Sheridan and Bonventre 2000; Menè et al. 2003) after renal damage.

In contrast, neonephrogenesis after renal injury has only been observed in species of fish (reviewed in Davidson 2011). While acute injury to fish mesonephric tubules sometimes results in reestablishment of tubule integrity via local cell proliferation (Reimschuessel and Williams 1995; Salice et al. 2001), both teleost fish (bony skeletons) and elasmobranch fish (cartilaginous skeletons, e.g., sharks and rays) also possess the ability to initiate de novo nephrogenesis after chemotoxic mesonephric injury. This phenomenon has been observed in the zebrafish *Danio rerio* (Salice et al. 2001; Liu et al. 2002; Diep et al. 2011; McCampbell et al. 2015), goldfish *Carassius auratus* (Reimschuessel and Williams 1995), Japanese rice fish (also known as the medaka; Watanabe et al. 2009), catfish, rainbow trout, tilapia, toadfish, and tomcod (Reimschuessel 2001). These fish studies have contributed immensely to our understanding of the repair response of mesonephroi to chemically induced nephrotoxicity. However, the fish system is less conducive to mechanical injury and as such is less well represented in the current literature (Elger et al. 2003). Partial nephrectomy of the fish pronephros is complicated by the presence of the centrally fused glomerulus, which lies ventral to the dorsal aorta, and which is shared by both kidneys (Drummond 2000; Fig. 4.4). Conversely, amphibian gloma are completely independent of each other and tubules are easily accessible via the skin (Fig. 4.4).

Although the pronephros is a transient embryonic structure, it is essential for the survival of aquatic juvenile organisms, who must maintain osmotic balance despite their surrounding environment. This was established a century ago when Howland demonstrated that bilateral nephrectomy in the spotted salamander, *Amblystoma punctatum*, results in rapid death due to edema (1916). Unilateral nephrectomy, however, does not result in certain death and thus has been performed repeatedly in amphibian model systems in an effort to understand the repair response after severe renal damage in vertebrates (Swingle 1919; Babaeva 1964; Chopra and Simnett 1969, 1970; Scadding and Liversage 1974; Caine and McLaughlin 2013). Importantly, partial nephrectomy of pronephric tubules is a useful mechanical injury model for understanding kidney damage caused by obstructive injury or severe infection (Becker et al. 1999; Mantur et al. 2000). This model is also useful for

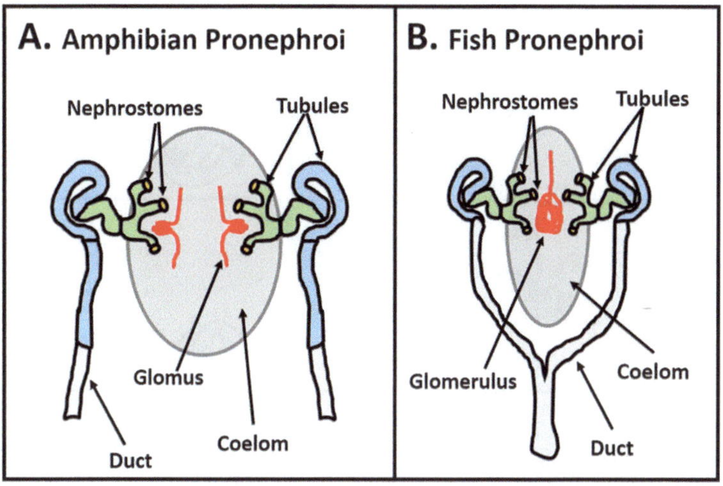

Fig. 4.4 The comparative anatomies of pronephric kidneys found in amphibians such as *Xenopus laevis* (**a**) and fish such as *Danio rerio* (**b**). Each glomus in the amphibian pronephric kidney system exists independently of the other, and neither glomus is directly integrated with the pronephric tubules. In contrast, each pair of fish pronephroi associates with a singular glomerular unit

investigating the destruction of nephric tissue experienced by patients with acute kidney injury or chronic kidney disease. Thus, the repair responses observed after nephrectomy may provide essential insight into the response of human kidneys to common causes of renal dysfunction.

In a series of articles published over 40 years ago, Chopra and Simnett described their studies of proliferation in the *Xenopus laevis* kidney and revealed the ability of both the juvenile pronephros and the adult mesonephros to undergo contralateral hypertrophy in response to unilateral partial nephrectomy (Chopra and Simnett 1969, 1970, 1971; Simnett and Chopra 1969). However, these studies were limited in their ability to establish the molecular mechanisms mediating this contralateral phenomenon. In addition, the restorative capacity of the ipsilateral (injured) kidney was never examined.

Four decades after this exploration of the compensatory hypertrophic response, Droz (previously published as Caine) and McLaughlin examined the cellular and molecular events that coordinate repair of the ipsilateral kidney after unilateral nephrectomy in *Xenopus laevis* tadpoles (Caine and McLaughlin 2013). To avoid the inaccessibility and complexity issues inherent with the frog mesonephric kidney, partial nephrectomy was performed on one pronephros in each pre-metamorphic, free-swimming, NF stage 37/38 (Nieuwkoop and Faber 1994) *Xenopus laevis* tadpole after the completion of pronephric development (Fig. 4.5; Caine and McLaughlin 2013). The nephrectomy undoubtedly disrupts the pronephric sinus (capillary system) surrounding the tubules, but is performed with negligible damage to other nephric components (pronephric duct and glomus) and to

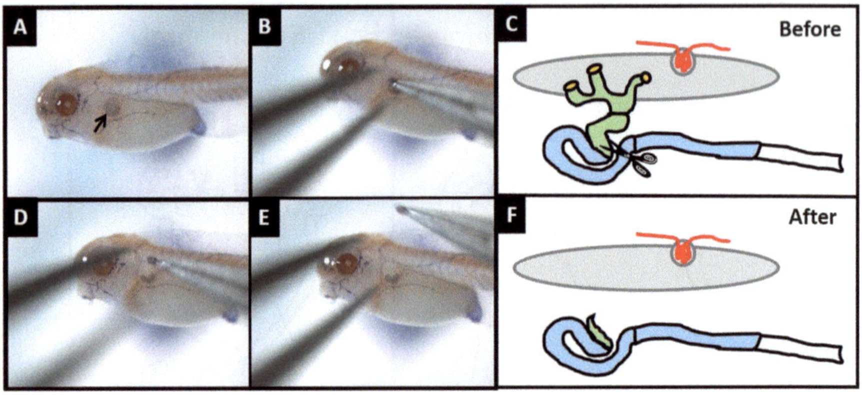

Fig. 4.5 Partial unilateral nephrectomy is performed on pronephric proximal tubules in free-swimming, premetamorphic *Xenopus laevis* tadpoles. Animals were briefly anesthetized with 0.04% tricaine and nephrectomized as depicted in panels (**a**–**f**). However, to better visualize the semitransparent pronephric tubules typically removed during this technique, the tadpole photographed in this figure was euthanized, fixed in formaldehyde, and assayed for Na,K-ATPase α5 protein expression via immunohistochemistry (**a**, *black arrow*). A small incision is made adjacent to the gill slit, in the skin covering the proximal tubules (**b**), and a large portion of the proximal tubules is excised (**c**–**f**). For more details, see Caine and McLaughlin (2013)

adjacent structures that flank the pronephros dorsally (somites) and ventrally (hypaxial muscles). Combined these results demonstrate that unilateral nephrectomy can be used to create mechanical damage to a specific component of the pronephric kidney in *Xenopus laevis* tadpoles, while maintaining the integrity of neighboring tissues (Caine and McLaughlin 2013).

4.3.2 Renal Repair in Xenopus *Requires a Permissive Microenvironment*

Droz (previously published as Caine) and McLaughlin (2013) postulate that a permissive microenvironment must be established in the renal milieu after injury and must be sustained throughout the repair process in order to achieve successful restoration of lost structures in *Xenopus laevis*. Recently, the authors described the apoptotic (Fig. 4.6) and extracellular matrix (ECM) remodeling patterns (Fig. 4.7) that follow renal damage and provided the first evidence of tissue regeneration in an amphibian kidney (Fig. 4.8; Caine and McLaughlin 2013).

The discovery that regenerated *Xenopus laevis* pronephric proximal tubules were restored with structural and functional fidelity after massive tissue damage (Fig. 4.8) provides renewed hope that all vertebrate kidneys may possess a similar potential for restoration under the correct balance of permissive conditions in the renal microenvironment. This is supported by the propensity of mammalian proximal tubule cells to proliferate and re-epithelialize denuded tubular basement

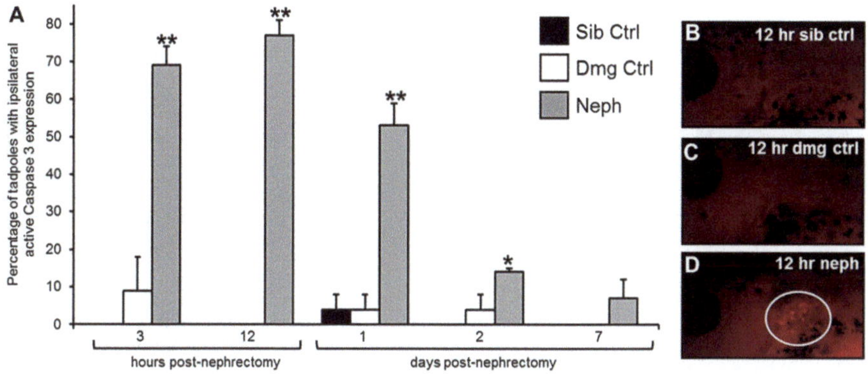

Fig. 4.6 Nephrectomized tadpoles have increased levels of apoptosis during the first 24 h post-surgery. A significant increase in the number of nephrectomized tadpoles expressing active Caspase 3 protein (as compared to both damage controls and unoperated sibling controls) was observed throughout the first day post-surgery (**a**). This number decreased significantly by the second day and remained at low levels throughout the remainder of the first week post-surgery. 25–70 animals from each treatment group were assayed at each time-point examined. Error bars indicate standard error among 2–3 replicates/group. *Asterisks* denote single-factor ANOVA tests among different groups at the same time-point: $^{*}p < 0.05$; $^{**}p < 0.001$. Photographs are lateral views (anterior to the *left*) of an unoperated sibling control (**b**), damage control (**c**), and nephrectomized tadpole (**d**), at 12 h post-surgery. *White circle* indicates renal area with active Caspase 3-expressing cells. Figure reproduced with permission from Caine and McLaughlin (2013)

membranes under permissive conditions (Cuppage and Tate 1967; Cuppage et al. 1972; Boti et al. 1982; Nonclercq et al. 1992; Gobé et al. 1995; Gobé and Buttyan 2002; Bonventre 2003; Nony and Schnellmann 2003). However, since it is still unknown which molecular mediators are governing these repair events, it is critical that further analysis of the extracellular matrix components and remodeling events be performed in future kidney repair studies.

At this time, it appears that early ECM remodeling, as indicated by a significant upregulation in *mmp9* (previously *Xmmp9*) expression, promotes tubule regeneration (Caine and McLaughlin 2013). Since the expression of *mmp9* also increases at later stages in the renal repair process, it is likely that ECM remodelers may play dualistic roles in the vertebrate renal repair response (Fig. 4.7). Interestingly, the ECM has also been shown to be critical during renal development, where it provides both a scaffold for spatial organization and a source of growth factors necessary for inductive signaling (Lelongt and Ronco 2003). Taken together, these data further emphasize the delicate balance that must be achieved within a permissive tissue microenvironment in order to successfully accomplish both renal development and repair/regeneration of pronephric kidney structures.

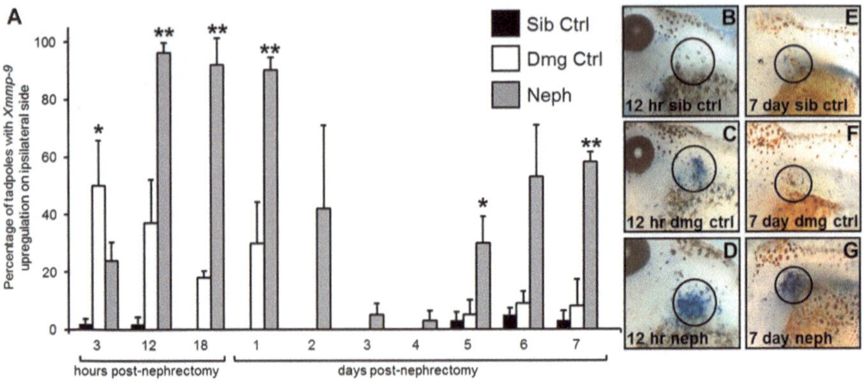

Fig. 4.7 *Xenopus laevis* tadpoles express *mmp9* in a biphasic pattern after unilateral nephrectomy. In situ hybridization analysis demonstrated that *mmp9* (*Xmmp-9*) expression was dramatically upregulated in both nephrectomized and damage control groups within 3 h of injury. Significantly more nephrectomized tadpoles displayed this phenotype between 12 and 24 h post-surgery. This number decreased by 2 days post-nephrectomy (dpn) and remained low until 5 dpn, at which time a second phase of *mmp9* upregulation began on the ipsilateral side of only nephrectomized tadpoles. This phase persisted through 7 dpn (**a**). 25–70 animals from each treatment group were assayed at time-points examined. Error bars indicate standard error among 3–5 replicates/group. *Asterisks* denote single-factor ANOVA tests among different groups at the same time-point: $^{*}p = 0.05$; $^{**}p < 0.005$. Photographs are lateral views (anterior to the *left*) of unoperated sibling controls (**b** and **e**), damage controls (**c** and **f**), and nephrectomized tadpoles (**d** and **g**), at 12 h (**b**–**d**) or 7 days (**e**–**g**) post-surgery. *Black circles* indicate renal area in each tadpole. Figure reproduced with permission from Caine and McLaughlin (2013)

4.3.3 Renal Repair in Xenopus *and the Inflammatory Response to Damage*

Importantly, a remarkable regenerative capacity has been described in several organs in amphibians (Brockes and Kumar 2002; Nye et al. 2003; Sánchez Alvarado and Tsonis 2006). These findings support a prevailing hypothesis in regenerative biology, which posits that the evolution of adaptive immunity correlates, and perhaps contributes, to the restriction of regenerative ability (Mescher and Neff 2004). Further evidence of this is seen in invertebrates, which completely lack adaptive immunity and have outstanding regenerative capacities (Monks 1904; Bryant 1971; French and Domican 1982; Carnevali et al. 1995; Kobayashi et al. 1999; Mito et al. 2002; Bode 2003; Scimone et al. 2011; Wenemoser et al. 2012).

Adaptive immunity originated with jawed vertebrates (Flajnik and Du Pasquier 2004) but is less evolved in anamniotes as compared to mammals (Robert and Cohen 1998; Menger et al. 2010). Additionally, adaptive immunocompetence varies among amphibians in many ways: between anurans and urodeles (Mescher and Neff 2004), between larval and adult stages of anurans (Robert and Cohen 1998), and between different stages during anuran larvae development (Fukazawa et al. 2009). For these reasons, amphibian model systems provide a unique

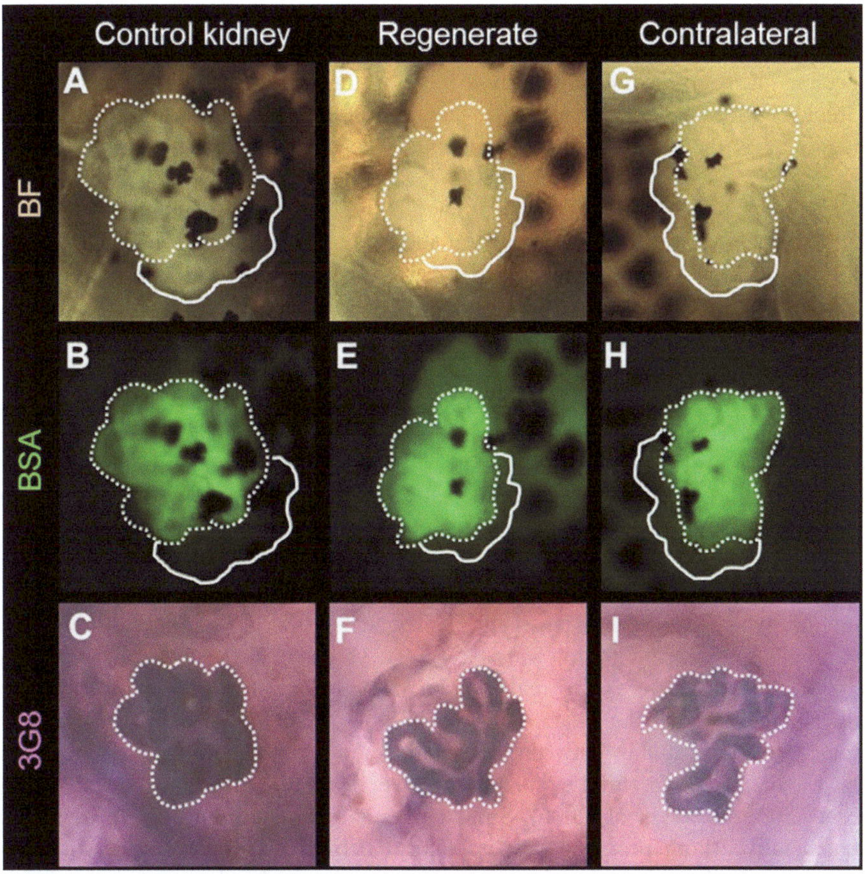

Fig. 4.8 Regenerated proximal tubules are functional 3 weeks after pronephrectomy. All unoperated control kidneys reabsorbed fluorescently tagged bovine serum albumin (BSA) into proximal tubules (**a–b**), as confirmed by the expression of the proximal tubule-specific protein, 3G8 (**c**). Some nephrectomized tadpoles regenerated coiled tubules, which reabsorbed fluorescently tagged BSA (**d–e**) and expressed 3G8 (**f**) as well. Comparable coiled morphology, BSA reuptake capability, and 3G8 expression were also seen in all unoperated contralateral proximal tubules (**g–i**). Photographs are lateral views of proximal tubules in representative tadpoles, anterior to the *left* in (**a–f**), and to the *right* in (**g–i**). *BF* bright field. *Broken outlines*: proximal tubules. *Solid outlines*: distal tubules. Figure reproduced with permission from Caine and McLaughlin (2013)

opportunity to demonstrate the relationship between adaptive immunity and reparative capacity.

Urodeles, which are immunodeficient compared to anurans (Mescher and Neff 2004), retain the ability to regenerate from juvenile stages through to adulthood (Graver 1978; Ferretti et al. 1991; Thouveny et al. 1991; Simon et al. 1995; Mescher 1996; Ghosh et al. 1996; Torok et al. 1998; Gardiner et al. 1999; Flink 2002; Imokawa and Brockes 2003; Bettencourt-Dias et al. 2003; Lévesque et al.

2005). Anurans like *Xenopus*, however, simultaneously develop adaptive immunity and lose regenerative capacity as they approach metamorphosis. This regenerative decline has been most clearly demonstrated in the froglet/frog limb (Thorton and Shields 1945; Muneoka et al. 1986; Suzuki et al. 2006). Premetamorphic anuran tadpoles, on the other hand, have remarkable regenerative abilities (Singer 1951; Dent 1962; Endo et al. 1997, 2000, 2007; Sakaguchi et al. 1997; Yokoyama et al. 2000; Beck et al. 2003; Gargioli and Slack 2004; Tazaki et al. 2005; Ishizuya-Oka 2007; Mochii et al. 2007) and concomitantly rely primarily on their innate immunity throughout these larval stages (Robert and Cohen 1998; Mescher and Neff 2004).

Among anurans, *Xenopus laevis* (larvae and adults) remain the preferred nonmammalian model for studying the ontogeny and phylogeny of the immune system (Du Pasquier et al. 1989; Goyos and Robert 2009; Robert and Ohta 2009; Robert and Cohen 2011; Nedelkovska et al. 2013). Fortunately, *Xenopus* also provides a useful system in which to investigate the relationship between immunity and regeneration. Adult frogs are typically limited in their regenerative capacity (Bertolotti et al. 2013), with some evidence suggesting that ontogenetic loss of regeneration correlates to development of the adaptive immune system (reviewed in Godwin and Rosenthal 2014). *Xenopus* tadpoles have a greater capacity for regeneration, which has been extensively investigated in the tail and larval limb buds (Christen et al. 2003; Gargioli and Slack 2004; Mochii et al. 2007; Tseng and Levin 2008). These structures are regeneration incompetent during specific stages of development (Muneoka et al. 1986; Slack et al. 2004; Mescher et al. 2013), and high-throughput studies between regenerative and non-regenerative hindlimb buds have indicated differential expression of several immunomodulatory genes (King et al. 2003; Grow et al. 2006). Similarly, tails amputated during refractory and regenerative (both pre- and post-refractory) periods displayed distinct immune responses, including a loss of regenerative capacity as development of the T-cell population progressed (Fukazawa et al. 2009).

These studies clearly demonstrate an intimate relationship between adaptive immunity and regeneration in tadpole limbs, but it is not yet known if a similar pattern is found in visceral organs such as the kidney. However, we can use this knowledge of the role of immunity in tadpole repair to inform future investigation on pronephric regeneration. Specifically, since the adaptive immunity is not active in *Xenopus* tadpoles until 12 days post-fertilization (Smith et al. 2002; Tomlinson et al. 2008), which corresponds to NF stage 49 (Nieuwkoop and Faber 1994), renal damage can be induced earlier as was done by Droz (previously published as Caine) and McLaughlin (2013) with tadpoles at NF stage 37/38 (Nieuwkoop and Faber 1994). Presumably, the microenvironment that is created after this type of damage will involve an inflammatory response dominated by innate immune cells. These cells are likely members of the phagocytic myeloid lineage, which include neutrophils and macrophages (Smith et al. 2002; Robert and Ohta 2009). Macrophage recruitment, in particular, may be responsible for mediating early cellular events that are required for successful pronephric proximal tubule regeneration. Preliminary data from the McLaughlin laboratory at Tufts University suggest that these cells may be expressing the critical matrix metalloproteinase *mmp9* previously observed during a very early window of pronephric repair (Fig. 4.7) and that the proteolytic activity of this ECM remodeler is crucial for tubule regeneration in *Xenopus laevis* (personal communication).

While chemical signaling and structural scaffolds established during inflammation and wound healing may strongly influence the repatterning of tissues during renal regeneration, successful restoration is also dependent on a source of cells to replace lost structures. In fact, with a few exceptions (such as morphallaxis and metaplasia), regeneration typically requires a source of undifferentiated cells in order to rebuild damaged tissues. This source may be attained by one (or a combination) of three mechanisms: local de- or transdifferentiation of cells that survive the injury, dedifferentiation of undamaged neighboring cells that will contribute to the regenerate, and/or proliferation of a reserve progenitor cell population, which often retains some levels of oligo- or multipotency (Tanaka and Reddien 2011).

In mammalian kidneys, surviving renal cells are the primary cellular source for repopulating damaged tubular basement membranes (Cuppage and Tate 1967; Cuppage et al. 1972; Nonclercq et al. 1992; Gobé et al. 1995; Gobé and Buttyan 2002; Bonventre 2003; Nony and Schnellmann 2003). In several different injury models, the absence of surviving tubule cells has been shown to severely attenuate renal repair and diminish renal function (Cuppage et al. 1972; Venkatachalam et al. 1978; Kovacs et al. 1982; Wallin et al. 1992; Witzgall et al. 1994; Kays and Schnellmann 1995; Nadasdy et al. 1995; Humphreys et al. 2008). Recently, it has also been shown that a complete loss of proximal tubules after nephrectomy is correlated with ablation of the regenerative capacity in the *Xenopus* pronephric kidney (Caine and McLaughlin 2013). At this time, it is unknown whether surviving pronephric tubule cells dedifferentiate and revert to a more mesenchymal state prior to reconstruction of the tubule, as is seen in mice (Humphreys et al. 2006).

4.4 Conclusion

Elucidating the mechanisms of kidney development and repair certainly requires an appreciation for the formation, growth, maintenance, and function of the renal system. However, to discover these mechanisms, researchers must also methodically explore both the nephric components themselves as well as the surrounding milieu that supports this system. As described herein, *Xenopus* as a model organism is particularly suitable for this type of investigation due to its accessibility, relatively simple renal system, and unique position within the evolutionary history of the vertebrate lineage (i.e., comparable to regenerative urodele amphibians and non-regenerative mammals).

The anatomical and morphological changes that epitomize renal development have fascinated scientists for decades and have been eloquently described in *Xenopus* (Nieuwkoop and Faber 1994; Vize et al. 1997). These reports were later expounded by advances in molecular and genetic tools that allowed researchers to investigate the inducers and effectors, which coordinate the distinctive developmental patterns involving the three different organ forms: the pronephros, mesonephros, and metanephros. Knowledge gained through these studies has contributed greatly to the fields of renal physiology and nephrology and has laid the foundation for the study of renal repair and regeneration.

Older repair research done in amphibians involved unilateral induction of mechanical damage via pronephrectomy. These studies mainly focused on the morphological response of the unmanipulated kidney and demonstrated that this contralateral kidney becomes swollen and hypertrophic to compensate for the loss of its counterpart (Howland 1916; Swingle 1919; Fox 1963; Babaeva 1964; Chopra and Simnett 1969, 1970; Scadding and Liversage 1974). More recently, the reparability of the injured pronephros was determined, both morphologically/histologically and with the use of molecular tools to characterize the gene expression profiles of cells comprising both the wound site and the regenerate (Caine and McLaughlin 2013). Moreover, Droz (previously published as Caine) and McLaughlin demonstrate for the first time that *Xenopus laevis* tadpoles are capable of regenerating functional pronephric proximal tubules and that this process may be completed within three weeks of partial unilateral nephrectomy in a subpopulation of tadpoles.

Matrix metalloproteinase 9, a key component of the amphibian wound healing and regenerative response (Yang et al. 1999; Carinato et al. 2000), appears to play a critical role during the early stages of renal repair observed in *Xenopus laevis* (Caine and McLaughlin 2013). Interestingly, this tightly regulated secreted member of the MMP family of proteases has also been implicated in several kidney disease models (McMillan et al. 1996; González-Avila et al. 1998; Lelongt et al. 2001a; Yang et al. 2002; Chromek et al. 2003; Bengatta et al. 2009; Liu et al. 2009; Yao et al. 2009; Wang et al. 2010; Kunugi et al. 2011). The difficulty with studying MMP9 is that this enzyme has several substrates, including both ECM and non-ECM components, and thus the active protein has several potential (and actual) roles in the wound environment (Legallicier et al. 2001; Lelongt et al. 2001b; Bengatta et al. 2009). MMP9 has been shown to be involved with both the synthesis and the degradation of the ECM and can function in both anti-fibrotic and pro-inflammatory contexts. Most of these roles have been described in mammalian metanephric kidneys, which can repopulate denuded basement membranes during acute renal injury (Boti et al. 1982) but experience fibrosis (renal scarring) under chronic conditions (McMillan et al. 1996; Liu et al. 2009; Wang et al. 2010).

However, the multifaceted (and often unpredictable) nature of MMP9 in renal repair is likely not limited to re-epithelialization or fibrotic events in mammalian kidneys, but rather may also play a critical role during *Xenopus laevis* pronephric proximal tubule regeneration (Caine and McLaughlin 2013). Although the function of MMP9 during *Xenopus laevis* renal repair is still unclear, this study indicates that early protease activity after injury promotes regeneration in this system. It is possible that MMP9 may be mitigating the cellular processes that cause fibroproliferation, thereby creating a microenvironment at the wound site that is more permissive for subsequent regeneration, as has been described in both regenerative pathways (Ingber and Levin 2007) and cancer progression (Noël et al. 2008). The characteristics of such a microenvironment remain unknown, and thus downstream effects of this *mmp9* expression and MMP9 activity need to be further investigated.

Elucidating the roles of MMP9 and other mediators of renal repair/regeneration in *Xenopus* may reveal potential therapeutic targets for enhancing the regenerative capacity of mammalian metanephric kidneys. Humans, like most mammals, have a very limited regenerative capacity, relying instead on localized wound healing with minimal restoration of lost structures. Tissue regeneration would be an incredibly beneficial therapeutic tool to address the pathology associated with several human diseases, specifically those with current treatments limited to invasive organ/tissue transplant. As such, determining the role for this protease in the *Xenopus laevis* pronephric model is paramount to understanding how MMP9 may mediate vertebrate renal repair in general. Just as the *Xenopus* system has afforded researchers insight into the developmental programs that coordinate renal patterning, it is also likely that this model organism will be valuable for revealing important events in renal regeneration that may be induced in the human kidney. Determining the cellular mechanisms that regulate these reparative and regenerative events would be beneficial for the development of renal disease therapies and would also contribute to distinguishing between species-specific adaptations in kidney repair mechanisms.

Moving forward in the genomic era, researchers are taking advantage of the ability to examine gene function using new and innovative approaches. For example, customizable genome editing technologies such as TALENs and ZFNs (Lei et al. 2012; Suzuki et al. 2013; Nadake et al. 2015; Liu et al. 2016) and, more recently, CRISPR/Cas9 (Wang et al. 2015) have been successfully used in *Xenopus* to create precise, targeted changes to the genome of living cells (reviewed in Tandon et al. 2016). Future studies that take advantage of this versatile, externally developing vertebrate using genome-edited *Xenopus* in combination with established injury models such as the partial unilateral pronephrectomy will greatly enhance our ability to understand gene function during development, repair, and regeneration of the kidney.

References

Armstrong PB (1932) The embryonic origin of function in the pronephros through differentiation and parenchyma-vascular association. Am J Anat 51:157–188

Aronson PS (1989) The renal proximal tubule: a model for diversity of anion exchangers and stilbene-sensitive anion transporters. Annu Rev Physiol 51:419–441

Attia L, Yelin R, Schultheiss TM (2012) Analysis of nephric duct specification in the avian embryo. Development 139:4143–4151

Augusto J, Smith B, Smith S, Robertson J, Reimschuessel R (1996) Gentamicin-induced nephrotoxicity and nephroneogenesis in *Oreochromis nilotica*, a tilapian fish. Dis Aquat Org 26:49–58

Babaeva AG (1964) Regeneration of the kidney in the red-bellied toad (*Bombina bombina*). Bull Exp Biol Med 57:99–103

Balinsky JB, Baldwin E (1961) The mode of excretion of ammonia and urea in *Xenopus laevis*. J Exp Biol 38:695–705

Barch SH, Shaver JR, Wilson GB (1966) An electron microscopic study of the nephric unit in the frog. Trans Am Microsc Soc 85:350–359

Barker N, De Wetering MV, Clevers H (2008) The intestinal stem cell. Genes Dev 22:1856–1864

Beck CW, Christen B, Slack JMW (2003) Molecular pathways needed for regeneration of spinal cord and muscle in a vertebrate. Dev Cell 5:429–439

Beck CW, Izpisúa Belmont JC, Christen B (2009) Beyond early development: *Xenopus* as an emerging model for the study of regenerative mechanisms. Dev Dyn 238:1226–1248

Becker JL, Miller F, Nuovo GJ, Josepovitz C, Schubach WH, Nord EP (1999) Epstein-Barr virus infection of renal proximal tubule cells: possible role in chronic interstitial nephritis. J Clin Invest 104:1673–1681

Bengatta S, Arnould C, Letavernier E, Monge M, De Préneuf HM, Werb Z, Ronco P, Lelongt B (2009) MMP9 and SCF protect from apoptosis in acute kidney injury. J Am Soc Nephrol 20:787–797

Bernardini S, Gargioli C, Cannata SM, Filoni S (2010) Neurogenesis during optic tectum regeneration in *Xenopus laevis*. Develop Growth Differ 52:365–376

Bertolotti E, Malagoli D, Franchini A (2013) Skin wound healing in different aged *Xenopus laevis*. J Morphol 274:956–964

Bettencourt-Dias M, Mittnacht S, Brockes JP (2003) Heterogeneous proliferative potential in regenerative adult newt cardiomyocytes. J Cell Sci 116:4001–4009

Birnbaum KD, Sánchez Alvarado A (2008) Slicing across kingdoms: regeneration in plants and animals. Cell 132:697–710

Bode HR (2003) Head regeneration in hydra. Dev Dyn 226:225–236

Bonasio R (2015) The expanding epigenetic landscape of non-model organisms. J Exp Biol 218:114–122

Bonventre JV (2003) Dedifferentiation and proliferation of surviving epithelial cells in acute renal failure. J Am Soc Nephrol 14:S55–S61

Boti Z, Kobor J, Ormos J (1982) Activity of glucose-6-phosphatase in regenerating tubular epithelium in rat kidney after necrosis induced with mercuric chloride: A light and electronmicroscopical study. Br J Exp Pathol 63:615–624

Brändli AW (1999) Towards a molecular anatomy of the *Xenopus* pronephric kidney. Int J Dev Biol 43:381–395

Bremer JL (1916) The interrelations of the mesonephros, kidney and placenta in different classes of animals. Am J Anat 19:179–209

Brennan HC, Nijjar S, Jones EA (1998) The specification of the pronephric tubules and duct in *Xenopus laevis*. Mech Dev 75:127–137

Brockes JP, Kumar A (2002) Plasticity and reprogramming of differentiated cells in amphibian regeneration. Nat Rev Mol Cell Biol 3:566–574

Bryant PJ (1971) Regeneration and duplication following operations *in situ* on the imaginal discs of *Drosophila melanogaster*. Dev Biol 26:637–651

Caine ST, McLaughlin KA (2013) Regeneration of functional pronephric proximal tubules after partial nephrectomy in *Xenopus laevis*. Dev Dyn 242:219–229

Carinato ME, Walter BE, Henry JJ (2000) *Xenopus laevis* gelatinase B (Xmmp-9): development, regeneration, and wound healing. Dev Dyn 217:377–387

Carlson BM (1978) Types of morphogenetic phenomena in vertebrate regenerating systems. Integr Comp Biol 18:869–882

Carnevali MD, Bonasoro F, Lucca E, Thorndyke MC (1995) Pattern of cell proliferation in the early stages of arm regeneration in the feather star *Antedon mediterranea*. J Exp Zool 272:464–474

Carroll TJ, Vize PD (1999) Synergism between Pax-8 and lim-1 in embryonic kidney development. Dev Biol 214:46–59

Carroll T, Wallingford J, Seufert D, Vize PD (1999a) Molecular regulation of pronephric development. Curr Top Dev Biol 44:67–100

Carroll TJ, Wallingford JB, Vize PD (1999b) Dynamic patterns of gene expression in the developing pronephros of *Xenopus laevis*. Dev Genet 24:199–207

Chan TC, Takahashi S, Asashima M (2000) A role for Xlim-1 in pronephros development in *Xenopus laevis*. Dev Biol 228:256–269

Chopra DP, Simnett JD (1969) Changes in mitotic rate during compensatory renal growth in *Xenopus laevis* tadpoles after unilateral pronephrectomy. J Embryol Exp Morphol 21:539–548

Chopra DP, Simnett JD (1970) Stimulation of cell division in pronephros of embryonic grafts following partial nephrectomy in the host (*Xenopus laevis*). J Embryol Exp Morphol 24:525–533

Chopra DP, Simnett JD (1971) Stimulation of cell division in larval kidney (*Xenopus laevis*) by rat kidney antiserum. Exp Cell Res 64:396–402

Christen B, Beck CW, Lombardo A, Slack JMW (2003) Regeneration-specific expression pattern of three posterior hox genes. Dev Dyn 226:349–355

Christensen EI, Raciti D, Reggiani L, Verroust PJ, Brändli AW (2008) Gene expression analysis defines the proximal tubule as the compartment for endocytic receptor-mediated uptake in the *Xenopus* pronephric kidney. Pflugers Arch - Eur J Physiol 456:1163–1176

Chromek M, Tullus K, Hertting O, Jaremko G, Khalil A, Li Y, Brauner A (2003) Matrix metalloproteinase-9 and tissue inhibitor of metalloproteinases-1 in acute pyelonephritis and renal scarring. Pediatr Res 53:698–705

Cirio MC, Hui Z, Haldin CE, Cosentino CC, Stuckenholz C, Chen X, Hong S, Dawid IB, Hukriede NA (2011) Lhx1 is required for specification of the renal progenitor cell field. PLoS One 6: e18858

Cuppage FE, Tate A (1967) Repair of the nephron following injury with mercuric chloride. Am J Pathol 51:405–429

Cuppage FE, Chiga M, Tate A (1972) Cell cycle studies in the regenerating rat nephron following injury with mercuric chloride. Lab Investig 26:122–126

Davidson AJ (2011) Uncharted waters: nephrogenesis and renal regeneration in fish and mammals. Pediatr Nephrol 26:1435–1443

Davies JA, Fisher CE (2002) Genes and proteins in renal development. Exp Nephrol 10:102–113

Dent JN (1962) Limb regeneration in larvae and metamorphosing individuals of the South African clawed toad. J Morphol 110:61–77

Diep CQ, Ma D, Deo RC, Holm TM, Naylor RW, Arora N, Wingert RA, Bollig F, Djordjevic G, Lichman B, Zhu H, Ikenaga T, Ono F, Englert C, Cowan CA, Hukriede NA, Handin RI, Davidson AJ (2011) Identification of adult nephron progenitors capable of kidney regeneration in zebrafish. Nature 470:95–101

Dor Y, Stanger BZ (2007) Regeneration in liver and pancreas: time to cut the umbilical cord? Sci STKE 414:pe66

Drawbridge J, Meighan CM, Lumpkins R, Kite ME (2003) Pronephric duct extension in amphibian embryos: migration and other mechanisms. Dev Dyn 226:1–11

Dressler GR (1996) Pax-2, kidney development, and oncogenesis. Med Pediatr Oncol 27:440–444

Dressler GR (2006) The cellular basis of kidney development. Annu Rev Cell Dev Biol 22:509–529

Drummond IA (2000) The zebrafish pronephros: a genetic system for studies of kidney development. Pediatr Nephrol 14:428–435

Drummond IA, Davidson AJ (2010) Zebrafish kidney development. Methods Cell Biol 100:233–260

Drummond IA, Majumdar A (2003) The pronephric glomus and vasculature. In: Vize PD, Woolf AS, Bard JBL (eds) The kidney: from normal development to congenital disease. Elsevier Science, San Diego, p 61

Drummond IA, Majumdar A, Hentschel H, Elger M, Solnica-Krezel L, Schier AF, Neuhauss SCF, Stemple DL, Zwartkruis F, Rangini Z, Driever W, Fishman MC (1998) Early development of the zebrafish pronephros and analysis of mutations affecting pronephric function. Development 125:4655–4667

Du Pasquier L, Schwager J, Flajnik MF (1989) The immune system of *Xenopus*. Annu Rev Immunol 7:251–275

Eid SR, Brändli AW (2001) *Xenopus* Na, K-ATPase: primary sequence of the β2 subunit and *in situ* localization of α1, β1, and γ expression during pronephric kidney development. Differentiation 68:115–125

Elger M, Hentschel H, Litteral J, Wellner M, Kirsch T, Luft FC, Haller H (2003) Nephrogenesis is induced by partial nephrectomy in the elasmobranch *Leucoraja erinacea*. J Am Soc Nephrol 14:1506–1518

Ellis LC, Youson JH (1989) Ultrastructure of the pronephric kidney in upstream migrant sea lamprey, *Petromyzon marinus* L. Am J Anat 185:429–443

Endo T, Yokoyama H, Tamura K, Ide H (1997) Shh expression in developing and regenerating limb buds of *Xenopus laevis*. Dev Dyn 209:227–232

Endo T, Tamura K, Ide H (2000) Analysis of gene expressions during *Xenopus* forelimb regeneration. Dev Biol 220:296–306

Endo T, Yoshino J, Kado K, Tochinai S (2007) Brain regeneration in anuran amphibians. Develop Growth Differ 49:121–129

Fedorova S, Miyamoto R, Harada T, Isogai S, Hashimoto H, Ozato K, Wakamatsu Y (2008) Renal glomerulogenesis in medaka fish, *Oryzias latipes*. Dev Dyn 237:2342–2352

Ferretti P, Brockes JP, Brown R (1991) A newt type II keratin restricted to normal and regenerating limbs and tails is responsive to retinoic acid. Development 111:497–507

Flajnik MF, Du Pasquier L (2004) Evolution of innate and adaptive immunity: can we draw a line? Trends Immunol 25:640–644

Flink IL (2002) Cell cycle reentry of ventricular and atrial cardiomyocytes and cells within the epicardium following amputation of the ventricular apex in the axolotl, *Amblystoma mexicanum*: confocal microscopic immunofluorescent image analysis of bromodeoxyuridine-labeled nuclei. Anat Embryol 205:235–244

Fox H (1963) The amphibian pronephros. Q Rev Biol 38:1–25

French V, Domican J (1982) The regeneration of supernumerary cockroach antennae. J Embryol Exp Morphol 67:153–165

Frontera JL, Cervino AS, Jungblut LD, Paz DA (2015) Brain-derived neurotrophic factor (BDNF) expression in normal and regenerating olfactory epithelium of *Xenopus laevis*. Ann Anat 198:41–48

Fukazawa T, Naora Y, Kunieda T, Kubo T (2009) Suppression of the immune response potentiates tadpole tail regeneration during the refractory period. Development 136:2323–2327

Fukui L, Henry JJ (2011) FGF signaling is required for lens regeneration in *Xenopus laevis*. Biol Bull 221:137–145

Gardiner DM, Carlson MRJ, Roy S (1999) Towards a functional analysis of limb regeneration. Semin Cell Dev Biol 10:385–393

Gargioli C, Slack JMW (2004) Cell lineage tracing during *Xenopus* tail regeneration. Development 131:2669–2679

Ghosh S, Thorogood P, Ferretti P (1996) Regeneration of lower and upper jaws in urodeles is differentially affected by retinoic acid. Int J Dev Biol 40:1161–1170

Gobé GC, Buttyan R (2002) Apoptosis in the pathogenesis of renal disease with a focus on tubulointerstitial injury. Nephrology 7:287–293

Gobé GC, Buttyan R, Wyburn KRL, Etheridge MR, Smith PJ (1995) Clusterin expression and apoptosis in tissue remodeling associated with renal regeneration. Kidney Int 47:411–420

Godwin JW, Rosenthal N (2014) Scar-free wound healing and regeneration in amphibians: Immunological influences on regenerative success. Differentiation 87:66–75

Goldin G, Fabian B (1978) The regulation of growth in the mesonephric kidney of adult *Xenopus laevis* by an endogenous inhibitor of proliferation. Dev Biol 66:529–538

González-Avila G, Iturria C, Vadillo-Ortega F, Ovalle C, Montaño M (1998) Changes in matrix metalloproteinases during the evolution of interstitial renal fibrosis in a rat experimental model. Pathobiology 66:196–204

Goyos A, Robert J (2009) Tumorigenesis and anti-tumor immune responses in *Xenopus*. Front Biosci 14:167–176

Graver HT (1978) Re-regeneration of lower jaws and the dental lamina in adult urodeles. J Morphol 157:269–276

Grow M, Neff AW, Mescher AL, King MW (2006) Global analysis of gene expression in *Xenopus* hindlimbs during stage-dependent complete and incomplete regeneration. Dev Dyn 235:2667–2685

Hamilton PW, Sun Y, Henry JJ (2016) Lens regeneration from the cornea requires suppression of Wnt/β-catenin signaling. Exp Eye Res 145:206–215

Harty M, Neff AW, King MW, Mescher AL (2003) Regeneration or scarring: an immunologic perspective. Dev Dyn 226:268–279

Hayashi S, Kawaguchi A, Uchiyama I, Kawasumi-Kita A, Kobayashi T, Nishide H, Tsutsumi R, Tsuru K, Inoue T, Ogino H, Agata K, Tamura K, Yokoyama H (2015) Epigenetic modification maintains intrinsic limb-cell identity in *Xenopus* limb bud regeneration. Dev Biol 406:271–282

Heller N, Brändli AW (1997) *Xenopus* Pax-2 displays multiple splice forms during embryogenesis and pronephric kidney development. Mech Dev 69:83–104

Hellsten U, Harland RM, Gilchrist MJ, Hendrix D, Jurka J, Kapitonov V, Ovcharenko I, Putnam NH, Shu S, Taher L, Blitz IL, Blumberg B, Dichmann DS, Dubchak L, Amaya E, Detter JC, Fletcher R, Gerhard DS, Goodstein D, Graves T, Grigoriev IV, Grimwood J, Kawashima T, Lindquist E, Lucas SM, Mead PE, Mitros T, Ogino H, Ohta Y, Poliakov AV, Pollet N, Robert J, Salamov A, Sater AK, Schmutz J, Terry A, Vize PD, Warren WC, Wells D, Wills A, Wilson RK, Zimmerman LB, Zorn AM, Grainger R, Grammer T, Khokha MK, Richardson PM, Rokhsar DS (2010) The genome of the western clawed frog *Xenopus tropicalis*. Science 328:633–636

Hensey C, Dolan V, Brady HR (2002) The *Xenopus* pronephros as a model system for the study of kidney development and pathophysiology. Nephrol Dial Transplant 17:73–74

Hewitt IK, Zucchetta P, Rigon L, Maschio F, Molinari PP, Tomasi L, Toffolo A, Pavanello L, Crivellaro C, Bellato S, Montini G (2008) Early treatment of acute pyelonephritis in children fails to reduce renal scarring: data from the Italian renal infection study trials. Pediatrics 122:486–490

Howland RB (1916) On the effect of removal of the pronephros of the amphibian embryo. Proc Natl Acad Sci USA 2:231–234

Huang C, Ogawa R (2010) Mechanotransduction in bone repair and regeneration. FASEB J 24:3625–3632

Huang Y, Chen M, Chiu N, Chou H, Lin K, Chiou Y (2011) Adjunctive oral methylprednisolone in pediatric acute pyelonephritis alleviates renal scarring. Pediatrics 128:e496–e504

Humphreys BD, Duffield JS, Bonventre JV (2006) Renal stem cells in recovery from acute kidney injury. Minerva Urol Nefrol 58:329–337

Humphreys BD, Valerius MT, Kobayashi A, Mugford JW, Soeung S, Duffield JS, McMahon AP, Bonventre JV (2008) Intrinsic epithelial cells repair the kidney after injury. Cell Stem Cell 2:284–291

Imgrund M, Gröne E, Gröne H, Kretzler M, Holzman L, Schlöndorff D, Rothenpieler UW (1999) Re-expression of the developmental gene pax-2 during experimental acute tubular necrosis in mice. Kidney Int 56:1423–1431

Imokawa Y, Brockes JP (2003) Selective activation of thrombin is a critical determinant for vertebrate lens regeneration. Curr Biol 13:877–881

Ingber DE, Levin M (2007) What lies at the interface of regenerative medicine and development? Development 134:2541–2547

Ishizuya-Oka A (2007) Regeneration of the amphibian intestinal epithelium under the control of stem cell niche. Develop Growth Differ 49:99–107

Jaffee OC (1954) Morphogenesis of the pronephros of the leopard frog (*Rana pipiens*). J Morphol 95:109–123

Jaffee OC (1963) Cellular differentiation in the anuran pronephros. Anat Rec 145:179–182

Jewhurst K, Levin M, McLaughlin KA (2014) Optogenetic control of apoptosis in targeted tissues of *Xenopus laevis* embryos. J Cell Death 13:25–31

Jones EA (2005) *Xenopus*: A prince among models for pronephric kidney development. J Am Soc Nephrol 16:313–321

Kardong KV (ed) (2014) Vertebrates: comparative anatomy, function, evolution, 7th edn. McGraw Hill Higher Education, Boston, 816 p

Kays SE, Schnellmann RG (1995) Regeneration of renal proximal tubule cells in primary culture following toxicant injury: response to growth factors. Toxicol Appl Pharmacol 132:273–280

King MW, Nguyen T, Calley J, Harty MW, Muzinich MC, Mescher AL, Chalfant C, N'Cho M, McLeaster K, McEntire J, Stocum D, Smith RC, Neff AW (2003) Identification of genes expressed during *Xenopus laevis* limb regeneration by using subtractive hybridization. Dev Dyn 226:398–409

Kingsley JS (ed) (1917) Outlines of comparative anatomy of vertebrates, 2nd edn. Revised. P. Blakiston Son's, Philadelphia, 449 p

Kispert A, Vainio S, McMahon AP (1998) Wnt-4 is a mesenchymal signal for epithelial transformation of metanephric mesenchyme in the developing kidney. Development 125:4225–4234

Kobayashi C, Watanabe K, Agata K (1999) The process of pharynx regeneration in planarians. Dev Biol 211:27–38

Kobayashi A, Kwan KM, Carroll TJ, McMahon AP, Mendelsohn CL, Behringer RR (2005) Distinct and sequential tissue-specific activities of the LIM-class homeobox gene Lim1 for tubular morphogenesis during kidney development. Development 132:2809–2823

Kovacs CJ, Braunschweiger PG, Schenken LL, Burholt DR (1982) Proliferative defects in renal and intestinal epithelium after cis-dichlorodiammine platinum (II). Br J Cancer 45:286–294

Kunugi S, Shimizu A, Kuwahara N, Du X, Takahashi M, Terasaki Y, Fujita E, Mii A, Nagasaka S, Akimoto T, Masuda Y, Fukuda Y (2011) Inhibition of matrix metalloproteinases reduces ischemia-reperfusion acute kidney injury. Lab Investig 91:170–180

Kuure S, Vuolteenaho R, Vainio S (2000) Kidney morphogenesis: cellular and molecular regulation. Mech Dev 92:31–45

Lee DC, Hamm LM, Moritz OL (2013) *Xenopus laevis* tadpoles can regenerate neural retina lost after physical excision but cannot regenerate photoreceptors lost through targeted ablation. Invest Ophthalmol Vis Sci 54:1859–1867

Legallicier B, Trugnan G, Murphy G, Lelongt B, Ronco P, Delauche M, Fontanges P (2001) Expression of the type IV collagenase system during mouse kidney development and tubule segmentation. J Am Soc Nephrol 12:2358–2369

Lei Y, Guo X, Liu Y, Cao Y, Deng Y, Chen X, Cheng CH, Dawid IB, Chen Y, Zhao H (2012) Efficient targeted gene disruption in *Xenopus* embryos using engineered transcription activator-like effector nucleases (TALENs). Proc Natl Acad Sci USA 109:17484–17489

Lelongt B, Ronco P (2003) Role of extracellular matrix in kidney development and repair. Pediatr Nephrol 18:731–742

Lelongt B, Bengatta S, Delauche M, Lund LR, Werb Z, Ronco PM (2001a) Matrix metalloproteinase 9 protects mice from anti-glomerular basement membrane nephritis through its fibrinolytic activity. J Exp Med 193:793–802

Lelongt B, Legallicier B, Piedagnel R, Ronco PM (2001b) Do matrix metalloproteinases MMP-2 and MMP-9 (gelatinases) play a role in renal development, physiology and glomerular diseases? Curr Opin Nephrol Hypertens 10:7–12

Lévesque M, Guimond J, Pilote M, Leclerc S, Moldovan F, Roy S (2005) Expression of heat-shock protein 70 during limb development and regeneration in the axolotl. Dev Dyn 233:1525–1534

Lienkamp SS (2016) Using *Xenopus* to study genetic kidney diseases. Semin Cell Dev Biol S1084-9521:30045–30043

Lienkamp SS, Liu K, Karner CM, Carroll TJ, Ronneberger O, Wallingford JB, Walz G (2012) Vertebrate kidney tubules elongate using a planar cell polarity-dependent, rosette-based mechanism of convergent extension. Nat Genet 44:1382–1387

Lipschutz JH (1998) Molecular development of the kidney: a review of the results of gene disruption studies. Am J Kidney Dis 31:383–397

Liu Y (2011) Cellular and molecular mechanisms of renal fibrosis. Nat Rev Nephrol 7:684–696

Liu M, Reimschuessel R, Hassel BA (2002) Molecular cloning of the fish interferon stimulated gene, 15 kDa (ISG15) orthologue: a ubiquitin-like gene induced by nephrotoxic damage. Gene 298:129–139

Liu W, Tang N, Zhang Q (2009) Could mycophenolate mofetil combined with benazapril delay tubulointerstitial fibrosis in 5/6 nephrectomized rats? Chin Med J 122:199–204

Liu Y, Zhao H, Cheng CH (2016) Mutagenesis in *Xenopus* and Zebrafish using TALENs. Methods Mol Biol 1338:207–227

Love NR, Chen Y, Ishibashi S, Kritsiligkou P, Lea R, Koh Y, Gallop JL, Dorey K, Amaya E (2013) Amputation-induced reactive oxygen species are required for successful *Xenopus* tadpole tail regeneration. Nat Cell Biol 15:222–228

Malloch EL, Perry KJ, Fukui L, Johnson VR, Wever J, Beck CW, King MW, Henry JJ (2009) Gene expression profiles of lens regeneration and development in *Xenopus laevis*. Dev Dyn 238:2340–2356

Mantur M, Kemona H, Dabrowska M, Dabrowska J, Sobolewski S, Prokopowicz J (2000) α1-microglobulin as a marker of proximal tubular damage in urinary tract infection in children. Clin Nephrol 53:283–287

Martin P (1997) Wound healing – aiming for perfect skin regeneration. Science 276:75–81

Maunsbach AB, Christensen EI (1992) Functional ultrastructure of the proximal tubule. Compr Physiol (Online, 2011). Supplement 25: Handbook of Physiology, Renal Physiol 41–107

McCampbell KK, Springer KN, Wingert RA (2015) Atlas of cellular dynamics during zebrafish adult kidney regeneration. Stem Cells Int 547636

McMillan JI, Riordan JW, Couser WG, Pollock AS, Lovett DH (1996) Characterization of a glomerular epithelial cell metalloproteinase as matrix metalloproteinase-9 with enhanced expression in a model of membranous nephropathy. J Clin Invest 97:1094–1101

Menè P, Polci R, Festuccia F (2003) Mechanisms of repair after kidney injury. J Nephrol 16:186–195

Menger B, Vogt PM, Kuhbier JW, Reimers K (2010) Applying amphibian limb regeneration to human wound healing: a review. Ann Plast Surg 65:504–510

Mescher AL (1996) The cellular basis of limb regeneration in urodeles. Int J Dev Biol 40:785–795

Mescher AL, Neff AW (2004) Loss of regenerative capacity: a trade-off for immune specificity? Cell Sci Rev 1:1–10

Mescher AL, Wolf WL, Moseman EA, Hartman B, Harrison C, Nguyen E, Neff AW (2007) Cells of cutaneous immunity in *Xenopus*: studies during larval development and limb regeneration. Dev Comp Immunol 31:383–393

Mescher AL, Neff AW, King MW (2013) Changes in the inflammatory response to injury and its resolution during the loss of regenerative capacity in developing *Xenopus* limbs. PLoS One 8: e80477

Mito T, Inoue Y, Kimura S, Miyawaki K, Niwa N, Shinmyo Y, Ohuchi H, Noji S (2002) Involvement of hedgehog, wingless, and dpp in the initiation of proximodistal axis formation during the regeneration of insect legs, a verification of the modified boundary model. Mech Dev 114:27–35

Møbjerg N, Larsen EH, Jespersen Å (2000) Morphology of the kidney in larvae of *Bufo viridis* (Amphibia, Anura, Bufonidae). J Morphol 245:177–195

Mochii M, Taniguchi Y, Shikata I (2007) Tail regeneration in the *Xenopus* tadpole. Dev Growth Differ 49:155–161

Mondia JP, Levin M, Omenetto FG, Orendorff RD, Branch MR, Adams DS (2011) Long-distance signals are required for morphogenesis of the regenerating *Xenopus* tadpole tail. PLoS One 6: e24953

Monks SP (1903) Regeneration of the body of a starfish. Proc Acad Natl Sci Phila 55:351

Monks SP (1904) Variability and autotomy of *Phataria*. Proc Acad Natl Sci Phila 56:596–600

Moritz KM, Wintour EM (1999) Functional development of the meso- and metanephros. Pediatr Nephrol 13:171–178

Moshiri A, Close J, Reh TA (2004) Retinal stem cells and regeneration. Int J Dev Biol 48:1003–1014

Muneoka K, Holler-Dinsmore G, Bryant SV (1986) Intrinsic control of regenerative loss in *Xenopus laevis* limbs. J Exp Zool 240:47–54

Muñoz R, Edwards-Faret G, Moreno M, Zuñiga N, Cline H, Larraín J (2015) Regeneration of *Xenopus laevis* spinal cord requires Sox2/3 expressing cells. Dev Biol 408:229–243

Nadake S, Sakuma T, Sakane Y, Hara Y, Kurabayashi A, Kashiwago K, Yamamoto T, Obara M (2015) Homeolog-specific targeted mutagenesis in *Xenopus* using TALENs. In Vitro Cell Dev Biol Anim 51:879–884

Nadasdy T, Laszik Z, Blick KE, Johnson DL, Burst-Singer K, Nast C, Cohen AH, Ormos J, Silva FG (1995) Human acute tubular necrosis: a lectin and immunohistochemical study. Hum Pathol 26:230–239

Nedelkovska H, Edholm ES, Haynes N, Robert J (2013) Effective RNAi-mediated beta2-microglobulin loss of function by transgenesis in *Xenopus laevis*. Biol Open 2:335–342

Nieuwkoop PD (1996) What are the key advantages and disadvantages of urodele species compared to anurans as a model system for experimental analysis of early development? Int J Biol 40:617–619

Nieuwkoop PD, Faber J (eds) (1994) Normal table of *Xenopus laevis* (Daudin). Garland Publishing, New York, 252 p

Noël A, Jost M, Maquoi E (2008) Matrix metalloproteinases at cancer tumor-host interface. Semin Cell Dev Biol 19:52–60

Nonclercq D, Wrona S, Toubeau G, Zanen J, Heuson-Stiennon J, Schaudies RP, Laurent G (1992) Tubular injury and regeneration in the rat kidney following acute exposure to gentamicin: a time-course study. Renal Fail 14:507–521

Nony PA, Schnellmann RG (2003) Mechanisms of renal cell repair and regeneration after acute renal failure. J Pharmacol Exp Ther 304:905–912

Nye HLD, Cameron JA, Chernoff EAG, Stocum DL (2003) Regeneration of the urodele limb: a review. Dev Dyn 226:280–294

O'Connor RJ (1940) The evolutionary significance of the embryology of the amphibian nephric system. J Anat 75:95–101

Panetta NJ, Gupta DM, Longaker MT (2010) Bone regeneration and repair. Curr Stem Cell Res Ther 5:122–128

Pole RJ, Qi BQ, Beasley SW (2002) Patterns of apoptosis during degeneration of the pronephros and mesonephros. J Urol 167:269–271

Poss KD, Keating MT, Nechiporuk A (2003) Tales of regeneration in zebrafish. Dev Dyn 226:202–210

Raciti D, Reggiani L, Geffers L, Jiang Q, Bacchion F, Subrizi AE, Clements D, Tindal C, Davidson DR, Kaissling B, Brändli AW (2008) Organization of the pronephric kidney revealed by large-scale gene expression mapping. Genome Biol 9:R84

Raghow R (1994) The role of extracellular matrix in postinflammatory wound healing and fibrosis. FASEB J 8:823–831

Rahman MM, Tae HJ, Cho HS, Shin GW, Park BY (2015) Developmental expression analysis of Na, K-ATPase α subunits in *Xenopus*. Dev Genes Evol 225:105–111

Reimschuessel R (2001) A fish model of renal regeneration and development. ILAR J 42:285–291

Reimschuessel R, Williams D (1995) Development of new nephrons in adult kidneys following gentamicin-induced nephrotoxicity. Ren Fail 17:101–106

Robert J, Cohen N (1998) Evolution of immune surveillance and tumor immunity: studies in *Xenopus*. Immunol Rev 166:231–243

Robert J, Cohen N (2011) The genus *Xenopus* as a multispecies model for evolutionary and comparative immunobiology of the 21st century. Dev Comp Immunol 35:916–923

Robert J, Ohta Y (2009) Comparative and developmental study of the immune system in *Xenopus*. Dev Dyn 238:1249–1270

Sakaguchi DS, Janick LM, Reh TA (1997) Basic fibroblast growth factor (FGF-2) induced transdifferentiation of retinal pigment epithelium: generation of retinal neurons and glia. Dev Dyn 209:387–398

Salice CJ, Rokous JS, Kane AS, Reimschuessel R (2001) New nephron development in goldfish (*Carassius auratus*) kidneys following repeated gentamicin-induced nephrotoxicosis. Comp Med 51:56–59

Sánchez Alvarado A (2000) Regeneration in the metazoans: why does it happen? BioEssays 22:578–590

Sánchez Alvarado A (2004) Regeneration and the needs for simpler model organisms. Philos Trans R Soc Lond Ser B Biol Sci 359:759–763

Sánchez Alvarado A, Tsonis PA (2006) Bridging the regeneration gap: genetic insights from diverse animal models. Nat Rev Genet 7:873–884

Sato A, Asashima M, Yokota T, Nishinakamura R (2000) Cloning and expression pattern of a *Xenopus* pronephros-specific gene, XSMP-30. Mech Dev 92:273–275

Saulnier DME, Ghanbari H, Brändli AW (2002) Essential function of Wnt-4 for tubulogenesis in the *Xenopus* pronephric kidney. Dev Biol 248:13–28

Saxén L (1987) Organogenesis of the kidney. Cambridge University Press, Cambridge, 173 p

Scadding SR, Liversage RA (1974) Studies on the response of the adult newt kidney to partial nephrectomy. Am J Anat 140:349–368

Scimone ML, Srivastava M, Bell GW, Reddien PW (2011) A regulatory program for excretory system regeneration in planarians. Development 138:4387–4398

Sheridan AM, Bonventre JV (2000) Cell biology and molecular mechanisms of injury in ischemic acute renal failure. Curr Opin Nephrol Hypertens 9:427–434

Simnett JD, Chopra DP (1969) Organ specific inhibitor of mitosis the amphibian kidney. Nature 222:1189–1190

Simon H, Nelson C, Goff D, Laufer E, Morgan BA, Tabin C (1995) Differential expression of myogenic regulatory genes and msx-1 during dedifferentiation and redifferentiation of regenerating amphibian limbs. Dev Dyn 202:1–12

Singer M (1951) Induction of regeneration of forelimb of the frog by augmentation of the nerve supply. Proc Soc Exp Biol Med 76:413–416

Slack JMW (2003) Regeneration research today. Dev Dyn 226:162–166

Slack JMW, Beck CW, Gargioli C, Christen B (2004) Cellular and molecular mechanisms of regeneration in *Xenopus*. Philos Trans R Soc Lond Ser B Biol Sci 359:745–751

Smith HW (1953) From fish to philosopher. Little, Brown, Boston, 304 p

Smith SJ, Kotecha S, Towers N, Latinkic BV, Mohun TJ (2002) XPOX2-peroxidase expression and the XLURP-1 promoter reveal the site of embryonic myeloid cell development in *Xenopus*. Mech Dev 117:173–186

Stark K, Vainio S, Vassileva G, McMahon AP (1994) Epithelial transformation metanephric mesenchyme in the developing kidney regulated by Wnt-4. Nature 372:679–683

Stichel CC (1999) Inhibition of collagen IV deposition promotes regeneration of injured CMS axons. Eur J Neurosci 11:632–646

Stoick-Cooper CL, Moon RT, Weidinger G (2007) Advances in signaling in vertebrate regeneration as a prelude to regenerative medicine. Genes Dev 21:1292–1315

Sugiura T, Taniguchi Y, Tazaki A, Ueno N, Watanabe K, Mochii M (2004) Differential gene expression between the embryonic tail bud and regenerating larval tail in *Xenopus laevis*. Develop Growth Differ 46:97–105

Suzuki M, Yakushiji N, Nakada Y, Satoh A, Ide H, Tamura K (2006) Limb regeneration in *Xenopus laevis* froglet. Sci World J 6:26–37

Suzuki KT, Isoyama Y, Kashiwagi K, Sakuma T, Ochiai H, Sakamoto N, Furuno N, Kashiwagi A, Yamamoto T (2013) High efficiency TALENs enable F0 functional analysis by targeted gene disruption in *Xenopus laevis* embryos. Biol Open 2:448–452

Swingle WW (1919) On the experimental production of edema by nephrectomy. J Genet Physiol 1:509–514

Taira M, Otani H, Jamrich M, Dawid IB (1994) Expression of the LIM class homeobox gene Xlim-1 in pronephros and CNS cell lineages of *Xenopus* embryos is affected by retinoic acid and exogastrulation. Development 120:1525–1536

Tanaka E, Galliot B (2009) Triggering the regeneration and tissue repair programs. Development 136:349–353

Tanaka E, Reddien P (2011) The cellular basis for animal regeneration. Dev Cell 21:172–185

Tandon P, Conlon F, Furlow JD, Horb ME (2016) Expanding the genetic toolkit in *Xenopus*: approaches and opportunities for human disease modeling. Dev Biol (in press). doi:10.1016/j.ydbio.2016.04.009

Tazaki A, Kitayama A, Terasaka C, Watanabe K, Ueno N, Mochii M (2005) Macroarray-based analysis of tail regeneration in *Xenopus laevis* larvae. Dev Dyn 233:1394–1404

Thorton CS, Shields TW (1945) Five cases of atypical regeneration in the adult frog. Am Soc Ichthyol Herpetol 1945:40–42

Thouveny YR, Komorowski TE, Arsanto JP, Carlson BM (1991) Early innervation of skeletal muscle during tail regeneration in urodele amphibians. J Exp Zool 260:354–370

Tiedemann K, Wettstein R (1980) The mature mesonephric nephron of the rabbit embryo I. SEM studies. Cell Tissue Res 209:95–109

Tomlinson ML, Garcia-Morales C, Abu-Elmagd M, Wheeler GN (2008) Three matrix metalloproteinases are required in vivo for macrophage migration during embryonic development. Mech Dev 125:1059–1070

Tomlinson ML, Hendry AE, Wheeler GN (2012) Chemical genetics and drug discovery in *Xenopus*. Methods Mol Biol 917:155–166

Torok MA, Gardiner DM, Shubin NH, Bryant SV (1998) Expression of HoxD genes in developing and regenerating axolotl limbs. Dev Biol 200:225–233

Tseng AS, Levin M (2008) Tail regeneration in *Xenopus laevis* as a model for understanding tissue repair. J Dent Res 87:806–816

Tseng AS, Carneiro K, Lemire JM, Levin M (2011) HDAC activity is required during *Xenopus* tail regeneration. PLoS One 6:e26382

Tsonis PA (2000) Regeneration in vertebrates. Dev Biol 221:273–284

Tsonis PA (2002) Regenerative biology: the emerging field of tissue repair and restoration. Differentiation 70:397–409

Venkatachalam MA, Bernard DB, Donohoe JF, Levinsky NG (1978) Ischemic damage and repair in the rat proximal tubule: differences among the S1, S2, and S3 segments. Kidney Int 14:31–49

Vize PD, Jones EA, Pfister R (1995) Development of the *Xenopus* pronephric system. Dev Biol 171:531–540

Vize PD, Seufert DW, Carroll TJ, Wallingford JB (1997) Model systems for the study of kidney development: use of the pronephros in the analysis of organ induction and patterning. Dev Biol 188:189–204

Vize PD, Carroll TJ, Wallingford JB (2003) Induction, development, and physiology of the pronephric tubules, pp 19–50. In: Vize PD, Woolf AS, JBL B (eds) The kidney: from normal development to congenital disease. Elsevier Science, California, p 519

Wallin A, Zhang G, Jones TW, Jaken S, Stevens JL (1992) Mechanism of the nephrogenic repair response: Studies on proliferation and vimentin expression after 35S-1,2-dichlorovinyl-L-cysteine nephrotoxicity in vivo and in cultured proximal tubule epithelial cells. Lab Investig 66:474–484

Wang X, Zhou Y, Tan R, Xiong M, He W, Fang L, Wen P, Jiang L, Yang J (2010) Mice lacking the matrix metalloproteinase-9 gene reduce renal interstitial fibrosis in obstructive nephropathy. Am J Physiol Renal Physiol 299:F973–F982

Wang F, Shi Z, Cui Y, Guo X, Shi YB, Chen Y (2015) Targeted gene disruption in *Xenopus laevis* using CRISPR/Cas 9. Cell Biosci 5:15

Watanabe N, Kato M, Suzuki N, Inoue C, Fedorova S, Hashimoto H, Maruyama S, Matsuo S, Wakamatsu Y (2009) Kidney regeneration through nephron neogenesis in medaka. Develop Growth Differ 51:135–143

Wenemoser D, Lapan SW, Wilkinson AW, Bell GW, Reddien PW (2012) A molecular wound response program associated with regeneration initiation in planarians. Genes Dev 26:988–1002

Wessely O, Tran U (2011) *Xenopus* pronephros development – past, present, and future. Pediatr Nephrol 26:1545–1551

Wesson LG (1989) Compensatory growth and other growth responses of the kidney. Nephron 51:149–184

Wheeler GN, Brändli AW (2009) Simple vertebrate models for chemical genetics and drug discovery screens: lessons from zebrafish and *Xenopus*. Dev Dyn 238:1287–1308

Wheeler GN, Liu KJ (2012) *Xenopus*: An ideal system for chemical genetics. Genesis 50:207–218

Witzgall R, Brown D, Schwarz C, Bonventre JV (1994) Localization of proliferating cell nuclear antigen, vimentin, c-fos, and clusterin in the postischemic kidney. Evidence for a heterogenous genetic response among nephron segments, and a large pool of mitotically active and dedifferentiated cells. J Clin Invest 93:2175–2188

Wrobel K, Süß F (2000) The significance of rudimentary nephrostomial tubules for the origin of the vertebrate gonad. Anat Embryol 201:273–290

Wynn TA (2008) Cellular and molecular mechanisms of fibrosis. J Pathol 214:199–210

Wynn TA, Ramalingam TR (2012) Mechanisms of fibrosis: therapeutic translation for fibrotic disease. Nat Med 18:1028–1040

Yang EV, Gardiner DM, Carlson MRJ, Nugas CA, Bryant SV (1999) Expression of Mmp-9 and related matrix metalloproteinase genes during axolotl limb regeneration. Dev Dyn 216:2–9

Yang J, Shultz RW, Mars WM, Wegner RE, Li Y, Dai C, Nejak K, Liu Y (2002) Disruption of tissue-type plasminogen activator gene in mice reduces renal interstitial fibrosis in obstructive nephropathy. J Clin Invest 110:1525–1538

Yao X, Ye S, Chen Y, Zai Z, Li X, Wang Y, Chen K (2009) Rosiglitazone protects diabetic rats against kidney injury through the suppression of renal matrix metalloproteinase-9 expression. Diabetes Obes Metab 11:519–522

Yokoyama H, Yonei-Tamura S, Endo T, Izpisúa Belmonte JC, Tamura K, Ide H (2000) Mesenchyme with fgf-10 expression is responsible for regenerative capacity in *Xenopus* limb buds. Dev Biol 219:18–29

Yoshii C, Ueda Y, Okamoto M, Araki M (2007) Neural retinal regeneration in the anuran amphibian *Xenopus laevis* post-metamorphosis: transdifferentiation of retinal pigmented epithelium regenerates the neural retina. Dev Biol 303:45–56

Zhou X, Vize PD (2004) Proximo-distal specialization of epithelial transport processes within the *Xenopus* pronephric kidney tubules. Dev Biol 271:322–338

Zhou W, Boucher RC, Bollig F, Englert C, Hildebrandt F (2010) Characterization of mesonephric development and regeneration using transgenic zebrafish. Am J Physiol Renal Physiol 299: F1040–F1047

Zon LI (2008) Intrinsic and extrinsic control of haematopoietic stem-cell self-renewal. Nature 453:306–313

Part II
Cell Renewal, Differentiation and Fate in Nephrogenesis

Chapter 5
Tissue-Specific Functions of p53 During Kidney Development

Zubaida Saifudeen

Abstract p53 is best identified as a tumor suppressor for its transcriptional control of genes involved in cell cycle progression and apoptosis. Beyond its irrefutable involvement in restraining unchecked cell proliferation, research over the past several years has indicated a requirement for p53 function in sustaining normal development. Here I summarize the role of p53 in embryonic development, with a focus on knowledge gained from p53 loss and overexpression during kidney development. In contrast to its classical role in suppressing proliferative pathways, p53 positively regulates nephron progenitor cell (NPC) renewal. Emerging evidence suggests p53 may control cell fate decisions by preserving energy metabolism homeostasis of progenitors in the nephrogenic niche. Maintaining a critical level of p53 function appears to be a prerequisite for optimal nephron endowment. Defining the molecular networks targeted by p53 in the NPC may well provide new targets not only for regenerative medicine but also for cancer treatment.

5.1 Introduction

TP53 is the most commonly mutated gene in cancer with nearly half of all human cancers showing protein loss or mutation (Olivier et al. 2010). Somatic frameshift and nonsense mutations account for loss of wild-type protein. More prevalent, however, are the somatic missense point mutations resulting in a single amino acid substitution that encode a stable mutant protein with a gain-of-function phenotype (Olivier et al. 2010; Muller Patricia and Vousden 2014). Of the cancers that do not have mutations in the p53 gene locus, a vast majority exhibit mutations or altered levels of Mdm2, the negative regulator of p53 (Onel and Cordon-Cardo 2004; Zhang and Lu 2009). p53 is best known for maintaining genome stability and as an inhibitor of cell proliferation pathways; however, research in recent years has expounded the role of p53 as much

Z. Saifudeen, Ph.D. (✉)
Department of Pediatrics, Section of Pediatric Nephrology, Tulane University School of Medicine, 1430 Tulane Avenue, SL37, New Orleans, LA 70112, USA
e-mail: zubisaif@tulane.edu

R.K. Miller (ed.), *Kidney Development and Disease*, Results and Problems in Cell Differentiation 60, DOI 10.1007/978-3-319-51436-9_5

more than a tumor suppressor (Molchadsky et al. 2010; Bondar and Medzhitov 2010; Cicalese et al. 2009; Balaburski et al. 2010; Bensaad et al. 2006; Donehower 1996; Park et al. 2009). Here, I discuss the roles of p53 and Mdm2 in kidney development and their impact on cell fate in nephrogenesis.

5.2 p53 Gene Expression and Protein Stabilization

p53 expression is ubiquitous early in mouse embryogenesis until nearly mid-gestation (Molchadsky et al. 2010; Schmid et al. 1991; Komarova et al. 1997), after which time expression is restricted to specific tissues during organogenesis. Full-length human p53 is a 393 aa (390 aa mice) protein, expressed from a gene with 11 exons (Brady and Attardi 2010). Twelve isoforms have been identified that occur from alternative splicing, alternate promoter usage, and alternate translation start sites (Khoury and Bourdon 2011).

p53 is a short-lived protein with a half-life of 5–20 min in most cell types (Giaccia and Kastan 1998; Liu et al. 1994). p53 levels as well as its activity are primarily regulated by stimuli-induced post-translational modifications (PTMs) (Giaccia and Kastan 1998; Loughery 2013; Dai and Gu 2010; Kruse and Gu 2008, 2009; Vousden and Prives 2009; Meek and Anderson 2009). The number and combinations of the PTMs—the PTM signature—drive context-specific pathway activation. Protein stability and function are controlled by phosphorylation (ATM, CHK1/2, CK, AMPK, p38MAPK), acetylation (histone acetyltransferases or HATs—CBP, PCAF, etc.), poly-ubiquitination (MDM2/4, PIRH2, etc.), sumoylation (PIAS), neddylation (MDM2, FBXO11), and methylation (PRMT5, G9a, etc.) (Dai and Gu 2010; Meek and Anderson 2009; Knights et al. 2006; Ivanov et al. 2007; An et al. 2004; Jansson et al. 2008; Schmidt and Müller 2002) [Fig. 5.1,

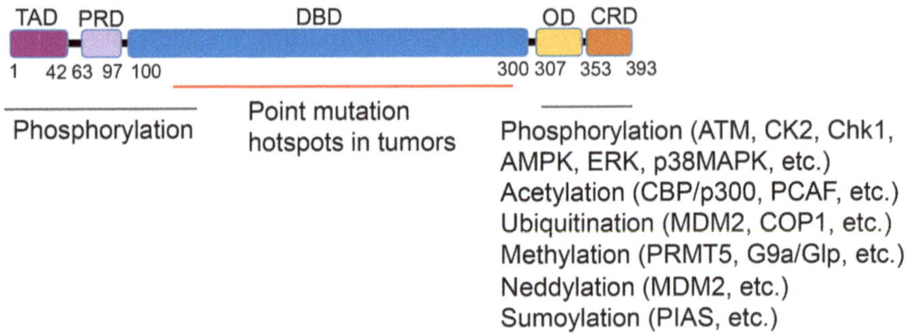

Fig. 5.1 p53 protein functional domains—*TAD* Transcription Activation Domain, *PRD* Proline-Rich Domain, *DBD* DNA Binding Domain, *OD* Oligomerization Domain, *CRD* C-terminal regulatory domain. Regions of the protein that undergo various PTMs are shown along with examples of the modifying enzymes in parentheses. The hotspots for point mutations found in cancers mostly occur in the DBD [From (Dai and Gu 2010; Meek and Anderson 2009; Vousden and Lane 2007)]

from (Dai and Gu 2010; Meek and Anderson 2009; Vousden and Lane 2007)]. Given the critical role of the PTMs in controlling stability and function, phosphatases (PP1, PP2A, etc.), deacetylases (HDAC1, KMD1, SIRT1), and deubiquitinases (HAUSP, PDCD5, etc.) are essential factors in determining duration of activity (Meek and Anderson 2009; Jenkins et al. 2012; Brooks and Gu 2006; Park et al. 2015). Tight control of p53 activity is essential for maintaining cellular and tissue homeostasis. Unmodified p53 is a target for MDM2 which binds to p53, blocks its transcriptional activity, and promotes p53 degradation through the ubiquitin proteasome system. Phosphorylation sites overlap Mdm2 binding site in the N-terminus of p53; thus, modified p53 is stable because it is refractory to binding and ubiquitination by Mdm2 (Meek and Anderson 2009).

Modified p53 binds to DNA and transcriptionally regulates expression of genes in diverse pathways that impact cell fate—cell cycle arrest, apoptosis, senescence, autophagy, cell migration, metabolism, differentiation, and self-renewal (Bensaad et al. 2006; Brady and Attardi 2010; Gadea et al. 2007; Gottlieb and Vousden 2010; Hwang et al. 2011; Schoppy et al. 2010; Tasdemir et al. 2008; Xu 2005; Tang et al. 2006; Milyavsky et al. 2010; Menendez et al. 2009; El-Deiry et al. 1993; Kastan et al. 1992). The inherent structural plasticity of the DNA-binding domain (DBD) enables binding to a consensus binding site that is highly degenerate and often gene context specific (Brady and Attardi 2010; Menendez et al. 2009; Joerger and Fersht 2008). The 20 bp binding site consists of two head-to-head or head-to-tail decameric palindromes separated by 0–13 bp (Kruse and Gu 2009; Ho et al. 2006; el-Deiry et al. 1992). Each decamer is an inverted repeat of the pentamer 5′-RRRCW/WGYYY-3′, each of which binds a p53 monomer; thus, p53 is bound as a tetramer, multimerizing via the oligomerization domain (Fig. 5.1). The PTMs contribute to site selectivity based on the DNA sequence and also possibly by facilitating interactions with auxiliary proteins activated in response to specific stimuli (Kruse and Gu 2009). The N-terminus contains the transcription activation domain (TAD) (Fig. 5.1); by undergoing disordered-to-ordered transitions, the TAD enables interaction of p53 with numerous target proteins with high specificity to assist transcriptional regulation (Kruse and Gu 2009; Joerger and Fersht 2008). The PTMs dictate interactions of proteins with p53, enabling stimuli-specific cellular output (Fig. 5.2).

Mdm2 tightly regulates p53 levels such that baseline protein levels are low with a short half-life unless stabilized upon stress-induced PTMs as discussed above (Moll and Petrenko 2003; Iwakuma and Lozano 2003). Loss of Mdm2 is embryonic lethal due to unchecked transcriptional and apoptotic activities of p53 (Montes de Oca Luna et al. 1995). Although p53-independent functions of Mdm2 are starting to emerge (Bouska and Eischen 2009; Wienken et al. 2016), the rescue of embryonic lethality of Mdm2-null mice by p53 deletion indicates that a major function for Mdm2 is keeping p53 activity in check.

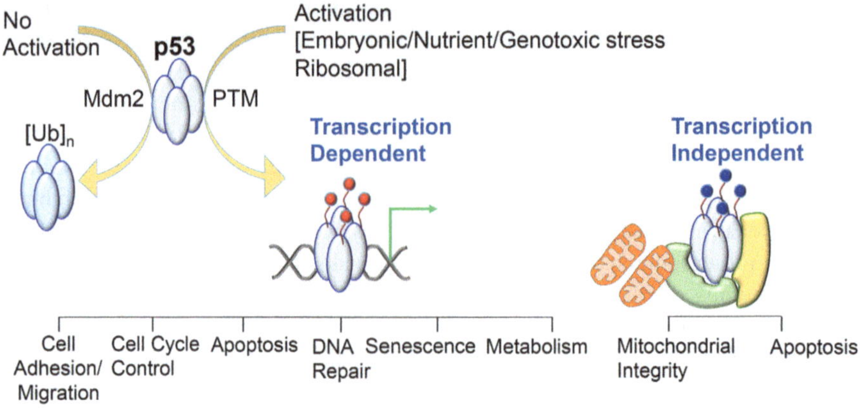

Fig. 5.2 Regulation of p53 stability and function by post-translational modifications in response to various stimuli. Combinations of modifications (*red circles*)—the PTM signature—will dictate the appropriate transcriptional response that will translate to a phenotypic outcome. Transcription-independent response requires translocation of mono-ubiquitinated (*blue circles*) p53 to the mitochondria. Localization at the mitochondrial membrane and interaction with anti-apoptotic Bcl proteins stimulate apoptosis. Translocation into the mitochondrial matrix requires interactions with proteins (denoted in *green* and *yellow*) where it interacts with mitochondrial proteins to preserve mitochondrial integrity. See text for details

5.3 p53 Function

5.3.1 Transcription-Dependent Effects

In its canonical role as a tumor suppressor, p53 plays a key role in cell-fate decisions by transcriptional regulation of genes that control cell cycle arrest, DNA repair, apoptosis, senescence, and autophagy, working to limit the propagation of cells with damaged genomes (Gottlieb and Vousden 2010; Hwang et al. 2011; Schoppy et al. 2010; Tasdemir et al. 2008). Besides restricting proliferation arising from genotoxic stress, p53 also regulates genes involved in cellular homeostasis. p53 is a crucial regulator of metabolic homeostasis assisting cells in adapting to adverse conditions by promoting catabolic pathways such as fatty acid oxidation (FAO) in times of nutrient stress by activating expression of multiple genes in the FAO pathway (Zhang et al. 2010; Puzio-Kuter 2011). In line with its role as a tumor-suppressor p53 represses glycolysis, the favored pathway of glucose metabolism by proliferating cells, while promoting oxidative phosphorylation (Bensaad et al. 2006; Matoba et al. 2006). However, p53 may also enhance some steps in glycolysis via transcriptional activation of glycolysis pathway genes in a cell context-dependent manner (Maddocks and Vousden 2011).

One key aspect of p53-mediated transcription involves recruitment of chromatin-modifying enzymes to either stimulate or repress transcription. Besides recruiting chromatin modifiers to regulate individual genes, p53 can regulate global gene expression by regulating expression of chromatin modifiers, such as Ezh2

(Tang et al. 2004). In addition to protein-coding genes, p53 also regulates transcription of microRNAs (miRNAs) and long noncoding RNAs (lncRNAs), both important regulators of gene expression (Vousden and Prives 2009; Grossi et al. 2016).

5.3.2 *Transcription-Independent Effects of p53 Function*

Apoptosis induction by p53 is a well-established (patho)physiological pathway in response to extreme genotoxic and oxidative stress as a means of tumor suppression and for tissue homeostasis during embryonic development (Erster and Moll 2005; Mihara et al. 2003). In addition to transcriptionally activating pro-apoptotic genes in response to extreme stress, non-nuclear mono-ubiquitinated p53 translocates to the mitochondrial surface, where it binds to the anti-apoptotic BclX and Bcl2 proteins and initiates mitochondrial membrane permeabilization, cytochrome c release, and cell death (Moll et al. 2005; Vaseva and Moll 2009).

A non-apoptotic role for p53 is also indicated inside the mitochondria in the matrix, the site of mitochondrial DNA localization (Park et al. 2009; Gupta et al. 2013; Xavier et al. 2014). Translocation to the mitochondria requires interactions with proteins such as RECQL4 (Gupta et al. 2013). Mitochondrial p53 physically interacts with the mitochondrial transcription factor A (TFAM), an essential protein for mitochondrial DNA transcription, replication, and repair (Park et al. 2009). Interaction of mitochondrial p53 with mitochondria-specific DNA polymerase gamma (Polγ) increases the enzyme's processivity, signifying a role for p53 in maintaining mitochondrial DNA stability (Fig. 5.2); accordingly, decreased mitochondrial DNA content and/or mitochondrial DNA mutations are detected in fibroblasts from Li-Fraumeni patients harboring p53 gene mutations (Gupta et al. 2013). These observations of p53 as a critical regulator of mitochondrial size and shape, mitochondrial genomic integrity, mitophagy, aerobic metabolism, and cellular redox state vastly widen the reach of p53 in regulating cellular homeostasis (Lebedeva et al. 2009; Park et al. 2016; Wang et al. 2014).

5.3.3 *p53 in Embryonic Development*

Given the pleiotropy of p53 functions, it was surprising that p53 loss in embryonic life did not result in a more severe outcome. This led to the erroneous conclusion that p53 is not required for development and is only required at times of genotoxic or other stress (Donehower et al. 1992; Hickman and Helin 2000). p53 levels are elevated ubiquitously during early embryonic life and then in an organ-specific manner and decrease postnatally to follow its more recognized expression pattern of stabilization under cellular stress (Molchadsky et al. 2010; Schmid et al. 1991; Gottlieb et al. 1997). Subsequent studies informed that genetic background differences are a dominant effector of phenotype penetrance. Thus, germline deletion on a 129S background produces an apparently normal offspring, while deletion on a CBA background results in embryonic lethality (Dey et al. 2000). Likewise, p53 deletion on a 129/Sv background produced exencephaly at a higher rate than on a 129/Sv-C57BL6 mixed background (Sah et al. 1995). Absence of wild-type p53 activity in Xenopus is embryonic lethal due to inhibition of mesoderm differentiation (Wallingford et al. 1997; Cordenonsi et al. 2003). Interestingly, injection of mutant p53 mRNA into Xenopus blastomeres destined to give rise to the kidney resulted in inhibition of pronephric tubule differentiation (Wallingford et al. 1997). The presence of p53 family members p63/p73 during mammalian embryogenesis but not in Xenopus and possibly other genetic redundancies may protect murine development from p53 loss (Cordenonsi et al. 2003; Danilova et al. 2008). Presence of mutant p53 in humans with Li-Fraumeni syndrome predisposes individuals to cancer (Wang et al. 2013; Li et al. 1988); however, complete absence of p53 function has not been reported in humans perhaps due to embryonic nonviability.

p53 is required for mesoderm induction in Xenopus (Cordenonsi et al. 2003). FGF-RTK-Ras-MAPK-CK1ε/δ-induced p53 phosphorylation and subsequent interaction with TGF-β-activated Smads was demonstrated to drive mesoderm determination in Xenopus (Cordenonsi et al. 2007). P-p53/P-Smads cooperatively induced expression of mesodermal genes. These studies demonstrate the role of p53 as an integrator of growth factor signaling in development, independent of its functions in genotoxic stress response.

In mouse embryonic stem cells (mESC), p53 represses expression of pluripotency factors and activates expression of genes linked to differentiation (Li et al. 2012). In contrast, another study in mESC showed that p53 induces expression of Wnt ligands and receptors and Lef1 expression and promotes proliferation of self-renewing progenitor populations (Schoppy et al. 2010; Xu 2005; Lee et al. 2010). p53 also regulates the self-renewal potential of tissue-specific stem cells including those from mesenchymal, mammary, hematopoietic, and neuronal lineages (Cicalese et al. 2009; Liu et al. 2009; Meletis et al. 2006). p53 loss in these cells enhances their self-renewal and consequently the stem cell pool. In the hematopoietic system, p53 mediates cell competition in the stem and progenitor compartments, a process by which fittest cells are selected for renewal and the “weakest” cells are removed from the renewing pool (Bondar and Medzhitov 2010).

5.4 p53 in Kidney Development

The mammalian kidney develops via inductive signaling between two intermediate mesoderm derivatives, the ureteric bud (UB) and the metanephric mesenchyme (MM) (Little and McMahon 2012; Costantini and Kopan 2010). The MM condenses around the UB tip forming the cap mesenchyme (CM) which houses the

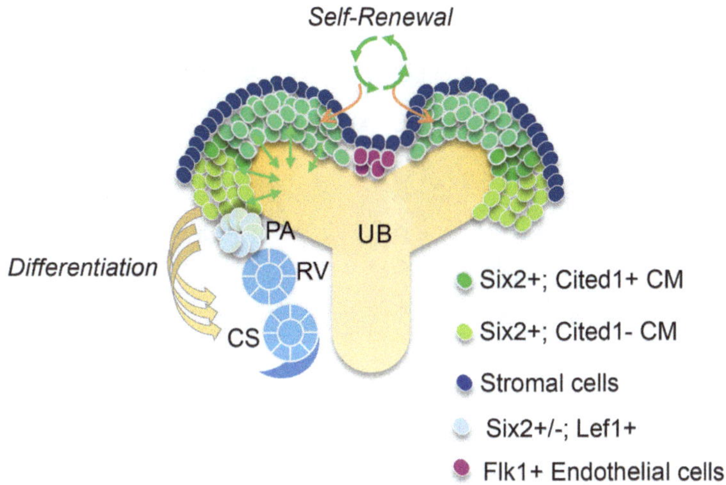

Fig. 5.3 Organization of the nephron progenitor cells (NPCs) in the cap mesenchyme (CM) in the embryonic kidney. The Cited1+/Six2+ NPC (dark green) are the self-renewing NPCs, resistant to Wnt/β-catenin-induced differentiation. Loss of Cited1 expression (*light green*) renders the Six2+/Cited1-NPC inducible to differentiation by signaling from the UB (*green arrows*); the induced cells form the pretubular aggregate (PA) which expresses Wnt4, Fgf8, Pax8, and decreasing levels of Six2. The PA epithelializes to form the renal vesicle (RV) which expresses epithelial markers ZO1 and E-Cadherin. The RV will progress through nascent nephron structures (*CS* comma-shaped body) which gradually mature to functional nephrons

nephron progenitor cells (NPCs) (Fig. 5.3). The NPCs located distal to the UB tip express Cited1 and Six2—these are the self-renewing multipotent progenitor cells that will undergo nephrogenesis (Boyle et al. 2008; Kobayashi et al. 2008). Six2 expression is essential for maintaining NPC in stem-like state, and deletion of Six2 results in premature differentiation of the NPC pool (Self et al. 2006). Wnt9b from the UB maintains the Cited1+/Six2+ NPC in progenitor state, while inducing the Cited1−/Six2+ NPC to undergo differentiation (Carroll et al. 2005; Karner et al. 2011). The differentiating NPCs are lateral to the UB tip and will form the pretubular aggregate (PA, Fig. 5.3) before undergoing mesenchyme-to-epithelial transition that results in the formation of the renal vesicle (RV). The PA expresses Wnt4, Fgf8, Pax8, and decreasing levels of Six2; these cells have exited the progenitor pool (El-Dahr et al. 2008; Georgas et al. 2009). The RV expresses epithelial markers ZO1 and E-Cadherin. The RV will progress through nascent nephron structures (CS, comma-shaped body) which gradually mature to functional nephrons (Georgas et al. 2009).

p53 expression is developmentally regulated in the kidney, showing approximately a fourfold decline in p53 mRNA expression from embryonic age (E)15.5 to adulthood (Saifudeen et al. 2009); protein expression shows a similar profile with a more pronounced decrease in protein levels in adult kidneys (Aboudehen et al. 2012). Notably, RNA-Seq only identified the full-length p53 transcript expression in the Six2+ nephron progenitor cells (Li et al. 2015); however, the prevalence of

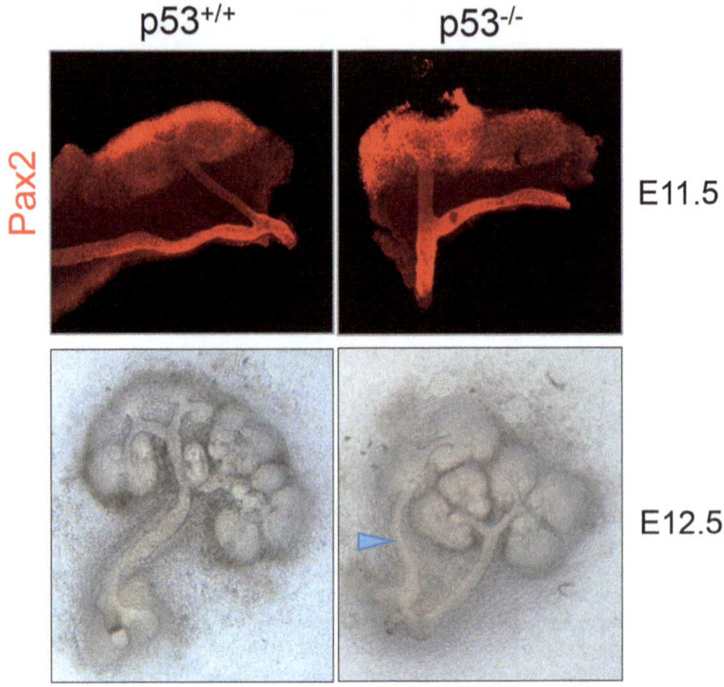

Fig. 5.4 Renal phenotype after germline p53 deletion. Upper Panels - Whole-mount Pax2 immunostain (*red*) shows hypoplastic metanephric mesenchyme in mutant kidney at E11.5 (unpublished data, Z. Saifudeen and S. El-Dahr). Lower Panels - Duplex ureters (*blue arrowhead*) are observed in 33% p53-null embryos (Saifudeen et al. 2009)

alternate isoforms from alternate translation initiation site usage remains a possibility.

Modified p53 phosphorylated at S6, S9, S15, and S392 showed a significant decline from embryo to adult, as did acetylated p53 at K373 and K382 (Aboudehen et al. 2012). DNA-binding activity also decreased in E13.5 to adult kidney nuclear extracts. p53 transcript is detected in all cell types of the embryonic kidney (Saifudeen et al. 2009). Significantly, differentially modified p53 showed specific spatial distribution, with p53-P-S392 localized in differentiated proximal tubule cells but not in proliferating cells and acetylated p53–K373/K382/K386 showing predominant localization in the nephrogenic zone that houses the ureteric tips, the nephron progenitor cells, and cortical stroma (Aboudehen et al. 2012; Saifudeen et al. 2012).

A spectrum of developmental kidney and urinary tract defects are observed with germline p53 deletion in mice on a C57BL6 background, most frequently double ureters and renal hypoplasia (Fig. 5.4 and Table 5.1) (Saifudeen et al. 2009; Saifudeen et al. 2002). Whole-mount immunofluorescence with Pax2 antibody shows a smaller metanephric mesenchyme from a $p53^{-/-}$ embryo at E11.5 (Fig. 5.4; Z. Saifudeen and S. El-Dahr, unpublished data). Further characterization

Table 5.1 Renal abnormalities in germline and conditional p53-null mice

	Germline p53 deletion		Conditional p53 deletion	
	−/−	+/+	$UB^{p53-/-}$	$MM^{p53-/-}$
Hypoplasia	7/7 (100%)	0/10	8/11 (73%)	17/19 (90%)
Duplex	3/9 (33%)	0/21	4/86 (5%)	0/19

From Saifudeen et al. (2009, 2012), Li et al. (2015), and unpublished data

of the renal defect in this and additional conditional gene deletion models on B6SJL and FVB mixed backgrounds showed hypoplasia in all deletion models, whereas duplex occurrence is only observed when p53 function is deleted or impaired in the nephric duct/UB lineage (Table 5.1) (Saifudeen et al. 2009; Li et al. 2015). Since nephrogenesis is dependent on signaling from the ureteric tips, the impact of conditional p53 deletion or overexpression from both mesenchymal and ureteric lineages will be discussed. The contribution of p53/Mdm2 function in the stroma is not yet characterized.

5.4.1 p53 Loss in the Nephric Duct Lineage

Abolishing p53 function either by epithelial-specific p53 gene deletion (Ksp- or Hoxb7-Cre) or by expression of dominant-negative p53 in the UB lineage (Hoxb7-DNp53) recapitulated the duplex ureter phenotype, albeit with lower penetrance than observed after germline p53 deletion (Saifudeen et al. 2009). The duplex ureter phenotype often demonstrates incomplete penetrance. Decreased penetrance may be from inefficient Cre-mediated recombination resulting in incomplete p53 deletion. Alternatively, the duplex phenotype might require p53 deletion from both the UB and the mesenchyme surrounding the UB as is the case with germline deletion. However, since Hoxb7-DNp53 recapitulates the duplex phenotype at the penetrance observed after germline deletion, the former possibility is more likely. Furthermore, modulation of nephric duct sensitivity to GDNF-induced budding by pharmacological p53 inhibition or activation supports a role for a nephric duct cell-autonomous role for p53 function. Hypersensitivity to low-dose GDNF treatment was observed in isolated nephric duct cultures, provided the ducts had either genetic or pharmacological loss of p53 function. Conversely, p53 stabilization by treating ND with Nutlin, a small molecule inhibitor of Mdm2–p53 interaction and thereby p53 degradation, prevents budding even at high GDNF concentrations (Saifudeen et al. 2009).

Mechanistically, p53 inactivation by siRNA in UB cells potentiated phosphatidylinositol-3 kinase activation by GDNF, measured as increase in phospho-Akt, providing biochemical evidence that p53 antagonizes the GDNF → c-Ret → PI3K pathway (Saifudeen et al. 2009). However, since increased proliferation of cells was not observed in p53-null nephric ducts, other

Akt-mediated pathways such as cell migration may likely be altered. The underlying mechanisms for $ND^{p53-/-}$ hypersensitivity to GDNF remain to be elucidated. Whether the increased sensitivity of $ND^{p53-/-}$ also extends to other growth factors (FGFs) is not known. Unlike other models of UB ectopia that exhibit increased receptor tyrosine kinase (RTK) signaling from decreased levels of the RTK signaling antagonist Sprouty-1 (Spry1) or anterior expansion of the GDNF domain, neither showed a change in the p53-null models (Saifudeen et al. 2009). Chimera experiments with GFP-labeled p53-null cells in the UB lineage revealed neither a preponderance nor exclusion of mutant cells at the leading edge of the bud or UB tips (Z. Saifudeen and F. Costantini, unpublished data). An attractive possibility that remains to be tested is that p53-P-Smad cooperate to inhibit ectopic budding in the nephric duct by activating a cytostatic pathway; interestingly, heterozygous deletion of Bmp4 results in CAKUT, including duplex ureters (Miyazaki et al. 2000). Alternatively, modulation of regulators of heparin sulfate such as heparanase and extracellular heparin sulfate 6-*O*-endosulfatases, which are known p53 targets, may deregulate heparin sulfate proteoglycans (HSPG) that control effective growth factor (GDNF, FGFs, BMPs) concentrations in a microenvironment, thereby contributing to ectopic signaling and bud formation (Chau et al. 2009; Baraz et al. 2006; Shah et al. 2011). Also, the involvement of other factors such as modifier genes cannot be discounted in regulating UB induction.

5.4.2 p53 Loss in the Metanephric Mesenchyme

Conditional deletion of p53 from the metanephric mesenchyme (MM) using a Six2Cre driver results in kidney hypoplasia with a sparse, disorganized cap mesenchyme (CM) [Fig. 5.5, A–B′ (Li et al. 2015)]. A marked paucity of differentiating nephrons is observed in P0 kidney sections. Glomerular dysfunction is indicated in PAS-stained kidneys displaying proteinaceous or hyaline material in Bowman's space. Immunofluorescent staining with CM markers Amphiphysin, Six2, and Cited1 demonstrates a diminished and less compact CM at E14.5–15.5 (Fig. 5.5, C–E′). The reduced Cited1 domain is lost by P0; however, Six2+ cells persist in a visibly smaller CM (Li et al. 2015). Although a simple decrease in proliferation might explain the fewer Six2+ cells, this would not explain the progressive loss and premature absence of Cited1+ cells observed in this p53-deficiency model. The progressive loss of Cited1+ cells suggests a decreased capacity for self-renewal because, as per our current knowledge, Cited1 marks the self-renewing population. Quantitatively, $Six2^{p53-/-}$ kidneys have 30% less Six2+ cells. Neural cell adhesion molecule (NCAM1), present at intercellular domains of CM and nascent nephrons, is greatly decreased in the mutant CM (Li et al. 2015). NCAM1 is required for cell–cell and cell–matrix interactions during development and differentiation. Reduced Pax2 staining was also observed in both germline and conditional $Six2^{p53-/-}$ kidneys. In vivo and in vitro reporter analysis demonstrated Pax2 is a p53 target

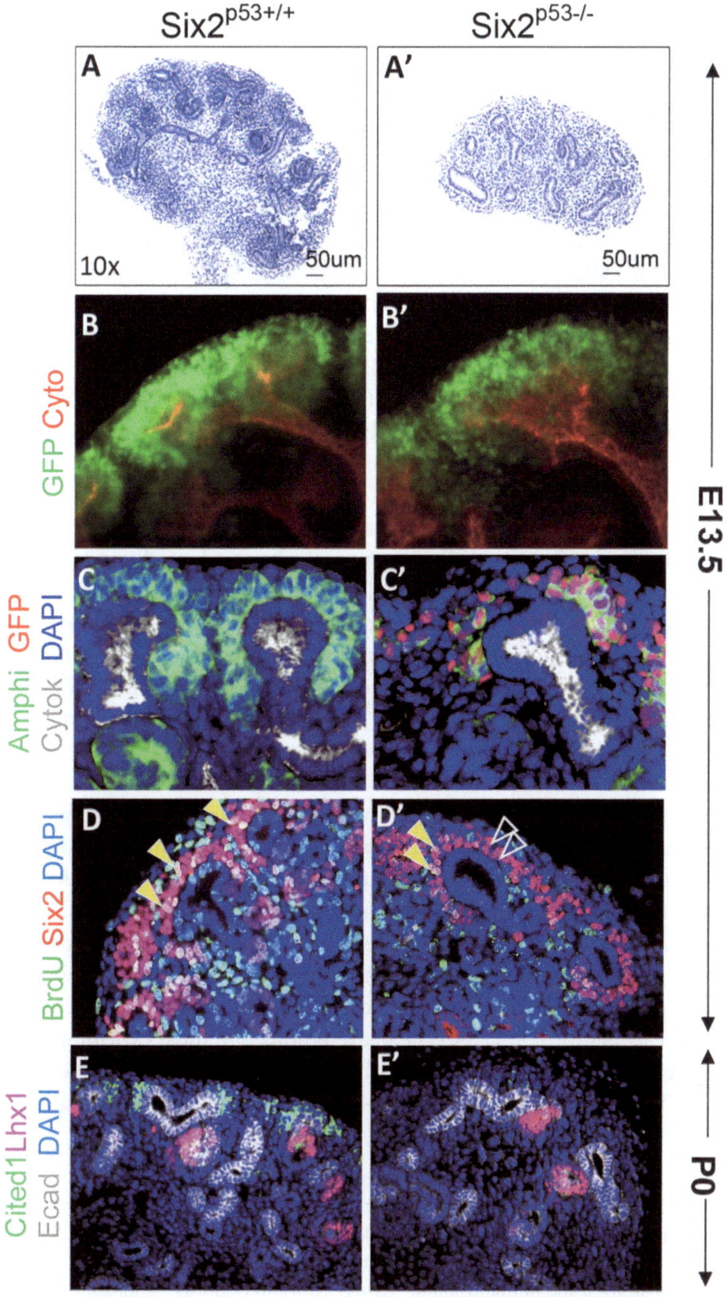

Fig. 5.5 Phenotype after conditional deletion of p53 from the Six2+ mesenchyme (Li et al. 2015). (**A**, **A′**)—H&E staining shows hypoplasia, branching defects, dysmorphic CM, and paucity of nascent nephrons in E13.5 mutant kidneys. (**B**, **B′**)—*Six2*$^{p53-/-}$ cells are dispersed and loosely organized around the ureteric tip, in contrast to *Six2*$^{p53+/+}$ cells arranged compactly around the ureteric tip. Six2+ cells are GFP+ (*green*); cytokeratin for UB (*red*). (**C**, **C′**)—Decreased expression of cap marker amphiphysin in mutant kidneys. Note disorganization of the mutant

gene (Saifudeen et al. 2012), implicating p53 in regulating the expression of a key kidney development protein.

No increase in either apoptosis or senescence was observed in Six2$^{p53-/-}$ cells, monitored by PARP or active caspase-3 and senescence marker SA-βgal, respectively (Li et al. 2015). Moreover, γH2Ax staining—a gauge of DNA damage—was also not increased, indicating that p53 is not exclusively involved in the maintenance of genomic stability in the developing kidney. Unexpectedly, after p53 loss the proliferation index of the CM is significantly lower compared to wild-type CM at E15.5. More Six2$^{p53-/-}$ cells are in G0/G1 phases of the cell cycle, and a significantly smaller fraction are in S and G2/M phases (Li et al. 2015). Furthermore, in contrast to the wild-type CM, BrdU incorporation was detected predominantly in the laterally located Six2+ cells that are the Wnt/β-catenin-inducible CM rather than in the self-renewing cells ventral to the ureteric tip (Fig. 5.5, D–D′). This explains why Six2+ cells are not entirely lost by P0. Gene expression analysis of Six2$^{p53-/-}$ CM surprisingly showed decreased, rather than increased, levels of transcripts of cell cycle inhibitors p21 (*Cdkn1a*) and p57 (*Cdkn1c*), indicative of an alternate explanation for decreased proliferation.

Similar to the p53 chromatin occupancy observed at diverse genes in the developing kidney beyond the canonical p53 targets (Li et al. 2013), transcription analysis also showed dysregulation of genes beyond classical pathways associated with p53 regulation, including development and morphology, cell adhesion and migration, cell survival, metabolism, and ion transport (Li et al. 2015). In accord with the loosely organized CM in Six2$^{p53-/-}$ kidneys, expression of multiple genes in cell adhesion and migration pathways is decreased. These include genes encoding collagens, which contribute to the extracellular matrix, as well as genes involved in cytoskeletal remodeling (Table 5.2). Although the functions of many of the products of these genes in the NPC are unknown, disruption of biophysical and biomechanical cues from the extracellular matrix to cells within the niche can influence renewal and differentiation programs (Cox and Erler 2011).

Genes involved in glucose metabolism, the pentose phosphate pathway, and components of the electron transport chain (ETC) (Table 5.3) or genes required to maintain a proton gradient for ATP production during oxidative respiration are dysregulated in Six2$^{p53-/-}$ cells. These are among the most downregulated genes and include phosphoenol pyruvate carboxykinase 1 (Pck1, ~ −14-fold), cytochrome P450 2D26 (Cyp2d26, −13.0-fold), fructose bisphosphatase 1 (Fbp1,

---

Fig. 5.5 (continued) GFP-stained CM. Markers—UB (cytokeratin—cyto, *white*) and CM (amphiphysin—amphi, *green*); nuclear stain—DAPI, *blue*. (**D, D′**)—Proliferative index measured by BrdU administration and immunostaining. BrdU incorporation was significantly lower in Six2+ cells in E13.5 Six2$^{p53-/-}$ kidneys (*p* < 0.001). BrdU was detected mostly in lateral section of the Six2+ cap of the mutant kidney (*solid yellow arrowheads*), whereas the medial Six2+ cells dorsal to the ureteric tip had few BrdU+ cells (*empty arrowheads*). In contrast, wild-type kidney showed BrdU incorporation in all areas of the Six2+ cap (*yellow arrowheads*). (**E, E′**)—Cited1 domain is completely lost by P0. Images from (Li et al. 2015), with permission

Table 5.2 Adhesion and migration genes' expression in Six2$^{p53-/-}$ CM [TPM, transcripts per million; from (Li et al. 2015)]

Gene symbol	Fold change	WT TPM	KO TPM	Function
Ntn	−4.5	19.0	4.2	Netrin is thought to be involved in axon guidance, cell adhesion, and cell migration during development (Sun et al. 2011)
Cdc42ep1	−2.5	5.5	2.2	Cdc42 effector proteins (Cdc42ep or CEPs) bind Cdc42 to induce cytoskeletal remodeling and cell shape changes (Hirsch et al. 2001)
Cdc42ep2	−2.0	2.2	1.1	
Cdc42ep5	−2.1	8.6	4.1	
Rock1	1.3	12.1	16.1	Protein serine/threonine kinases. Key regulators of actin remodeling, cell polarity, focal adhesion formation, cell adhesion, and motility (Shimizu et al. 2005)
Rock2	1.3	13.6	17.1	
Twist1	−2.6	3.6	1.4	Transcription factors that regulate genes involved in EMT, mesenchymal cell lineage determination, and differentiation (Ansieau et al. 2008)
Twist2	−2.7	3.5	1.3	
Dpp4	−4.1	6.6	1.6	Ubiquitous, membrane-bound enzyme that can modify growth regulators such as chemokines, cytokines, and colony-stimulating factors in hematopoietic stem/progenitor cells (Ou et al. 2013)
Arhgap24	−1.9	10.5	5.6	A Rho GTPase-activating protein involved in actin remodeling, cell migration and cell polarity (Lavelin and Geiger 2005)
Gng11	−1.9	9.8	5.1	Is a G-protein, gamma subunit; participates in signal transduction and regulation of enzymes and ion channels (Ray et al. 1995)
Col1a1	−4.5	96.2	21.4	Component of the extracellular matrix (ECM) (Cox and Erler 2011)
Col1a2	−3.5	104.5	29.7	
Col6a3	−4.3	10.4	2.4	
Col5a2	−2.1	13.3	6.3	
Col3a1	−3.9	169.0	43.4	

~12.0-fold), and aldolase b (Aldob, −12.0-fold) (Li et al. 2015). Moreover, functional analysis showed decreased ROS and ATP levels in Six2$^{p53-/-}$ cells, indicative of mitochondrial dysfunction. Thus, it appears that the p53-deficient NPCs undergo deregulation of metabolic homeostasis that compromises the metabolic fitness of the Cited1+/Six2+ NPC, thereby impeding their self-renewal. It needs to be emphasized, however, that decreased ATP (and ROS) are not the only products of deregulated energy metabolism that impact cell fate. Alteration in energy pathways results in changes in metabolites as well, which are key regulators of epigenetic modifications that control gene expression.

The Six2+(p53-null) cells are capable of differentiating into nascent nephrons, although a nephron deficit is observed in mutant kidneys in embryonic and P0 kidney sections. Induction of the Wnt/β-catenin pathway to induce mesenchyme–epithelial transition (MET) in an in vitro assay on isolated Six2+ CM cells revealed a differentiation defect in p53-null cells (Li et al. 2015). The mutant cells were unable to form complex structures observed in wild-type cells and show decreased conversion to epithelia as determined by E-cadherin staining. Fgf8 expression

Table 5.3 Energy metabolism genes' expression in Six2$^{p53-/-}$ CM (from (Li et al. 2015))

Gene symbol	Fold change	Process	Function
Pck1	−14.3	Gluconeogenesis	Catalyzes the conversion of oxaloacetate (OAA) to phosphoenolpyruvate (PEP), the rate-limiting step in the metabolic pathway that produces glucose from lactate (Yu et al. 1993)
H6pd	−1.4	Pentose phosphate pathway	Catalyzes the first two reactions of the PPP, thus generating NADPH (Hewitt et al. 2005). H6PD KO mice have anomalies in glucose homeostasis and ER stress (Zielinska et al. 2011)
Gpd1	−2.1	Glycolysis, glycerol phosphate shunt, transfer of H+ to mitochondria	Forms the glycerol phosphate shuttle with mitochondrial GPD2 which facilitates the transfer of reducing equivalents (H+) from the cytosol to mitochondria to maintain the proton gradient across the mitochondrial membrane (Ansell et al. 1997)
Pfkl	−1.2	Glycolysis	Key regulatory enzyme in glycolysis—catalyzes the irreversible conversion of fructose-6-phosphate to fructose-1,6-bisphosphate (Sola-Penna et al. 2010). Encodes liver isoform
Aldob	−12.1		Catalyzes the conversion of fructose-1,6-bisphosphate to glyceraldehyde 3-phosphate and dihydroxyacetone phosphate and participates in gluconeogenesis (Yañez et al. 2005)
Idh1	−1.4	Cytoplasmic NADPH production	Catalyzes the oxidative decarboxylation of isocitrate into α-ketoglutarate used in the TCA cycle, with NADP+ as the electron acceptor (Kloosterhof et al. 2011)
Pdk1	−1.3	Pyruvate activation and consumption	Reversibly inactivates the mitochondrial PDH complex by phosphorylation; may thereby alter the metabolite flux by altering the activity of pyruvate dehydrogenase (PDH), which converts pyruvate to acetyl-CoA (Kaplon et al. 2013)
Dld	1.3		Component of the pyruvate dehydrogenase complex and the alpha-ketoglutarate dehydrogenase complex (Brown et al. 1994)
Ndufs5	−1.5	Complex I (ETC)	Belong to first enzyme complex of the ETC
Ndufs6	−1.3		
Uqcr11	−1.2	Complex III (ETC)	Ubiquinol-cytochrome c reductase (Islam et al. 1997)

(continued)

Table 5.3 (continued)

Gene symbol	Fold change	Process	Function
Cox5b	−1.5	Complex IV (ETC)	Belong to the terminal enzyme complex IV of the ETC that transfers electrons from reduced cytochrome c to oxygen, thus generating a proton electrochemical gradient across the inner mitochondrial membrane that is critical for ATP generation (Li et al. 2006)
Cox6b2	−1.9		
Cox7c	−1.2		
Atp5e	−1.3	Complex V (ETC, ATP synthase)	Essential subunit in the biosynthesis and assembly of the F1 part of the mitochondrial ATP synthase Complex V that catalyzes ATP synthesis from ADP, utilizing an electrochemical gradient of protons across the inner membrane during oxidative phosphorylation (Mayr et al. 2010)

(typically observed in the PA/RV; see Fig. 5.3) is greatly diminished in the Six2$^{p53-/-}$ kidneys. However, Pax8 and Wnt4 expression did not show a marked change. Treatment of Six2$^{p53-/-}$ kidneys ex vivo with Fgf8 beads did not demonstrate any localized rescue of MET in comparison to the control BSA beads, indicative of additional epithelialization defects in the p53-null CM. Thus, it appears that the nephron deficit in Six2$^{p53-/-}$ kidneys is a result of at least two developmental defects: (1) self-renewal defect of Cited1+/Six2+ NPC resulting in reduced availability of progenitors for nephrogenesis and (2) improper differentiation, either from an inefficient response to Wnt/β-catenin signaling or defective MET or both in Six2+(p53-null) cells (Fig. 5.6) (Li et al. 2015).

Expression of NPC maintenance and survival regulators Six2, Fgf20, Bmp7, Sall1, WT1, and Mdm2 was unchanged in Six2$^{p53-/-}$ CM (Li et al. 2015). Also, Cited1+/Six2+ progenitor loss upon p53 deletion is not a result of precocious differentiation as was shown with the loss of Six2, Fgf9/20, or Sall1 (Self et al. 2006; Barak et al. 2012; Basta et al. 2014). The nephron deficit results from impaired progenitor cell renewal, secondary UB branching defects, and loss of organized niche architecture, which disrupts the CM microenvironment and the signaling pathways between the CM and UB lineages and between the CM and stroma lineages that are crucial for NPC renewal and differentiation (Das et al. 2013; Fetting et al. 2014). Six2$^{p53-/-}$ mice have increased interstitial stroma, visualized by immunostaining with antibody against the stromal cell marker Meis1/2 (Saifudeen et al. 2012). Stromal cells are interspersed among the loosely dispersed CM cells and often in close apposition to the UB, disrupting the clear CM-stroma demarcation that is present in wild-type kidneys (Li et al. 2015).

Surprisingly, a consistent observation in all the models studied has been the decrease in cell proliferation with p53 loss (Saifudeen et al. 2009, 2012; Li et al. 2015), suggesting that p53 mediates divergent pathways in kidney development than in cancer. This idea is further ratified upon comparison of p53 occupancy on chromatin from embryonic kidneys or cancer-derived cell lines (Li et al. 2013). Not

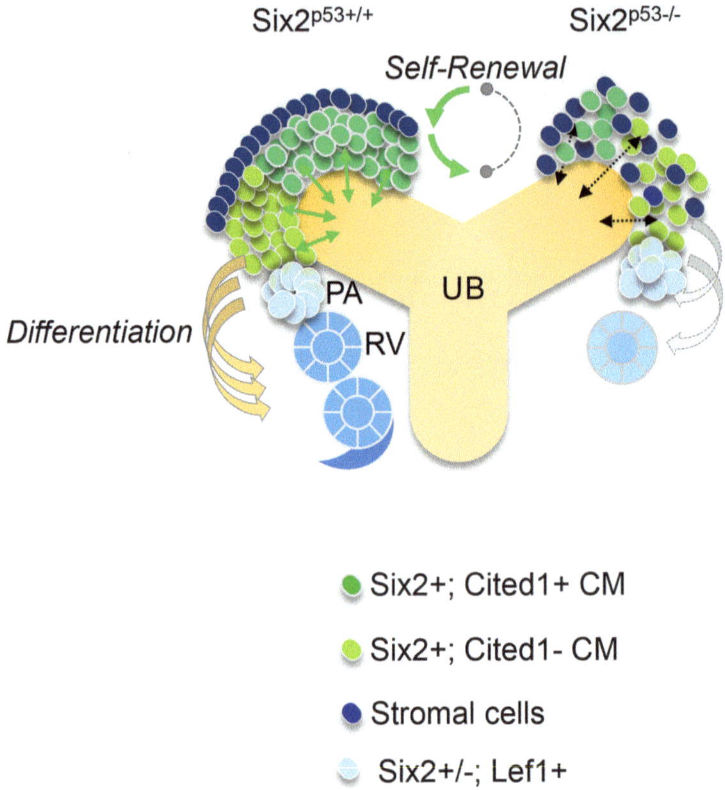

Fig. 5.6 Model for p53 function in NPC self-renewal and nephrogenesis. Loss of self-renewing cells in Six2$^{p53-/-}$ CM. Cited1+ cells are lost (*dark green circles*) in mutant CM. A Cited1-/Six2+ domain (*light green circles*) is present, but with defects in organization and integrity loss. Stromal cells (*navy blue circles*) are interspersed in between the Six2+ cap cells. Thus, a distinct CM–stroma and CM–UB interface that is present in wild-type kidneys is lost in the Six2-p53-null kidneys. Disorganization of the niche would result in impaired signaling and cross talk between the different lineages (denoted by *dashed arrows* in the Six2$^{p53-/-}$ CM). Improper differentiation (*dotted arrow*) results in decreased nascent nephrons (*PA* pretubular aggregate, *RV* renal vesicle) either from an inefficient response to Wnt/β-catenin signaling, or defective MET or both in Six2+(p53-null) cells. Endothelial cells are not shown [adapted from (Li et al. 2015)]

only is the pattern of chromatin occupancy different between embryonic kidney and cancer cells (promoter versus non-promoter/intergenic regions, respectively) but also the number and variety of genes that are enriched (Li et al. 2013). p53 gene occupancy in embryonic kidneys was noted beyond cell cycle and apoptosis genes—expected in cells with genotoxic stress-induced p53—in development and morphology, chromatin-modifying enzymes and genes in metabolic pathways (Li et al. 2013).

5.4.3 p53 Overexpression by Mdm2 Deletion

As stated above, Murine Double Minute-2 (Mdm2) controls cellular p53 protein levels and thereby its activity by ubiquitinating the protein that marks it for proteasomal degradation (Brooks and Gu 2006; Montes de Oca Luna et al. 1995). Germline Mdm2 deletion results in early embryonic death from unabated p53 levels and activity that can be rescued by p53 deletion (Montes de Oca Luna et al. 1995). Inordinate levels of p53 protein activate growth arrest and apoptosis pathways at the peri-implantation stage (Montes de Oca Luna et al. 1995). In the embryonic kidney, Mdm2 mRNA and protein expectedly follow the temporal and spatial expression pattern of p53 (Hilliard et al. 2011). Mice with conditional deletion of Mdm2 from the UB epithelium ($UB^{Mdm2-/-}$) show increased p53 protein levels by immunostaining, display severe renal hypodysplasia, and die soon after birth (Hilliard et al. 2011). The kidneys show defective UB branching and a marked decrease in UB tips. However, surprisingly, c-Ret and Wnt11 continue to be expressed. $UB^{Mdm2-/-}$ cells showed reduced proliferation rate and enhanced apoptosis. The nephrogenic zone is underdeveloped with decreased Wnt9b, Lhx-1, and Pax-2 expression levels. Rescue of the mutant phenotype by UB-specific p53 deletion indicates that the renal phenotype is largely p53 dependent (Hilliard et al. 2011).

Mdm2 deletion in the Six2+ nephron progenitor cells ($NPC^{Mdm2-/-}$) also results in perinatal lethality (Hilliard et al. 2014). The mutant kidneys are hypodysplastic. The CM is markedly thinner with fewer Six2+ cells and reduced or lost progenitor markers Amphiphysin, Cited1, Sall1, and Pax2. Increased p53 levels drive increased apoptosis and reduced proliferation observed in $NPC^{Mdm2-/-}$, along with reduced renal parenchyma and increased stroma. Expression of Eya1, Pax2, and Bmp7 is reduced, whereas Wnt4, Lhx1, and Pax8 expression is unchanged. As with the germline and $UB^{Mdm2-/-}$ mutants, p53 deletion rescued the progenitor depletion and hypodysplasia and restored normal renal development and postnatal survival of mice (Hilliard et al. 2014). These results demonstrate a critical and cell-autonomous role for Mdm2 in the UB and MM lineages. Mdm2-mediated inhibition of p53 activity is a prerequisite for renal organogenesis and plays an essential role in the renewal and survival of NPCs. Loss of functional Mdm2 results in stable, constitutively active p53 that is responsible for elevated apoptosis and premature depletion of the Six2 cap cells.

5.5 Conclusions and Future Prospects

Contrary to its classical role in suppressing proliferative pathways, data from $Six2^{p53-/-}$ mice indicate that p53 positively regulates NPC renewal. Both loss and gain of p53 protein levels result in NPC depletion. However, the mechanisms leading to this phenotype are quite different in both cases (Fig. 5.7). Whereas p53 loss

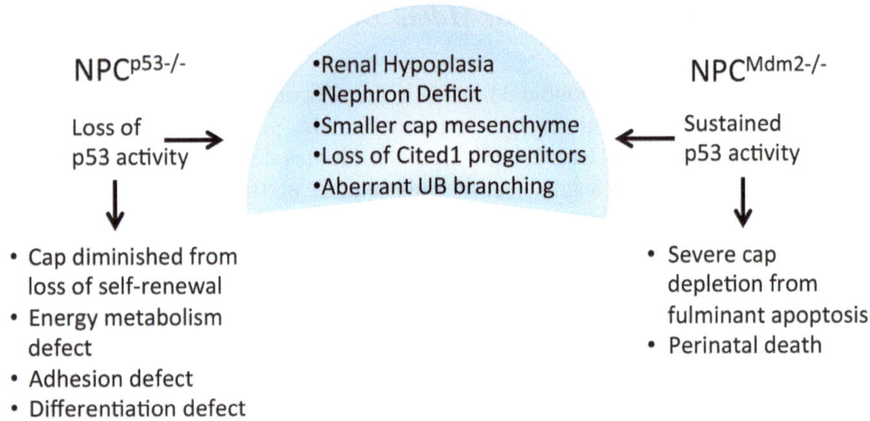

Fig. 5.7 Distinct mechanisms of nephron deficit in kidneys with p53 loss or excess. Both loss and gain of p53 protein levels result in NPC depletion. However, the mechanisms leading to this phenotype are quite different in both cases. Whereas p53 loss decreases NPC self-renewal possibly from disrupted cellular metabolism and adhesion, p53 overexpression causes NPC depletion by decreasing proliferation and increasing apoptosis

decreases NPC self-renewal possibly from disrupted cellular metabolism and adhesion, p53 overexpression causes NPC depletion by decreasing proliferation and increasing apoptosis. Maintaining a critical level of p53 function appears to be a prerequisite for optimal nephron endowment. p53-controlled energy metabolism homeostasis in the NPC is emerging as a crucial regulator of NPC fate, not only from the point of energy generation, but more importantly as a regulator of metabolite availability for macromolecule biosynthesis and epigenetic modifications which drive gene expression that controls cell fate. p53 regulates multiple metabolic pathways including glycolysis, oxidative phosphorylation, the pentose phosphate pathway, and fatty acid metabolism. Further mechanistic studies will shed light on which of these pathways are key regulatory points in p53-mediated NPC self-renewal and differentiation. Given the robust impact of p53 activity modulation on NPC fate, how p53 interacts with other important regulators of NPC fate such as miRNAs, chromatin modifiers (Ho et al. 2011; Chen and El-Dahr 2013; McLaughlin et al. 2013), and critical growth factor signaling pathways (Lindstrom et al. 2014; Muthukrishnan et al. 2015) remains to be explored. Knowledge gained by defining the molecular networks targeted by p53 in the NPC may well provide new targets not only for regenerative medicine but also for cancer treatment.

Acknowledgements The author thanks Dr. Samir El-Dahr for helpful discussions and suggestions and Dr. Yuwen Li for help with the figures and acknowledges funding support from the NIDDK (R56DK104779) and the Tulane Hypertension and Renal Center of Excellence.

References

Aboudehen K, Hilliard S, Saifudeen Z, El-Dahr SS (2012) Mechanisms of p53 activation and physiological relevance in the developing kidney. Am J Physiol Renal Physiol 302(8):F928–F940. doi:10.1152/ajprenal.00642.2011 (PubMed PMID: 22237799; PMCID: PMC3330719)

An W, Kim J, Roeder RG (2004) Ordered cooperative functions of PRMT1, p300, and CARM1 in transcriptional activation by p53. Cell 117(6):735–748. doi:10.1016/j.cell.2004.05.009

Ansell R, Granath K, Hohmann S, Thevelein JM, Adler L (1997) The two isoenzymes for yeast NAD+-dependent glycerol 3-phosphate dehydrogenase encoded by GPD1 and GPD2 have distinct roles in osmoadaptation and redox regulation. EMBO J 16(9):2179–2187. doi:10.1093/emboj/16.9.2179 (PubMed PMID: 9171333; PMCID: 1169820)

Ansieau S, Bastid J, Doreau A, Morel A-P, Bouchet BP, Thomas C, Fauvet F, Puisieux I, Doglioni C, Piccinin S, Maestro R, Voeltzel T, Selmi A, Valsesia-Wittmann S, Caron de Fromentel C, Puisieux A (2008) Induction of EMT by twist proteins as a collateral effect of tumor-promoting inactivation of premature senescence. Cancer Cell 14(1):79–89. doi:10.1016/j.ccr.2008.06.005

Balaburski GM, Hontz RD, Murphy ME (2010) p53 and ARF: unexpected players in autophagy. Trends Cell Biol 20(6):363–369. doi:10.1016/j.tcb.2010.02.007

Barak H, Huh SH, Chen S, Jeanpierre C, Martinovic J, Parisot M, Bole-Feysot C, Nitschke P, Salomon R, Antignac C, Ornitz DM, Kopan R (2012) FGF9 and FGF20 maintain the stemness of nephron progenitors in mice and man. Dev Cell 22(6):1191–1207. doi:10.1016/j.devcel.2012.04.018 (PubMed PMID: 22698282; PMCID: 3376351)

Baraz L, Haupt Y, Elkin M, Peretz T, Vlodavsky I (2006) Tumor suppressor p53 regulates heparanase gene expression. Oncogene 25(28):3939–3947

Basta JM, Robbins L, Kiefer SM, Dorsett D, Rauchman M (2014) Sall1 balances self-renewal and differentiation of renal progenitor cells. Development 141(5):1047–1058. doi:10.1242/dev.095851 (PubMed PMID: 24550112; PMCID: 3929412)

Bensaad K, Tsuruta A, Selak MA, Vidal MNC, Nakano K, Bartrons R, Gottlieb E, Vousden KH (2006) TIGAR, a p53-inducible regulator of glycolysis and apoptosis. Cell 126(1):107–120. doi:10.1016/j.cell.2006.05.036

Bondar T, Medzhitov R (2010) p53-mediated hematopoietic stem and progenitor cell competition. Cell Stem Cell 6(4):309–322. doi:10.1016/j.stem.2010.03.002

Bouska A, Eischen CM (2009) Mdm2 affects genome stability independent of p53. Cancer Res 69(5):1697–1701. doi:10.1158/0008-5472.can-08-3732

Boyle S, Misfeldt A, Chandler KJ, Deal KK, Southard-Smith EM, Mortlock DP, Baldwin HS, de Caestecker M (2008) Fate mapping using Cited1-CreERT2 mice demonstrates that the cap mesenchyme contains self-renewing progenitor cells and gives rise exclusively to nephronic epithelia. Dev Biol 313(1):234–245. doi:10.1016/j.ydbio.2007.10.014 (PubMed PMID: 18061157; PMCID: PMC2699557)

Brady CA, Attardi LD (2010) p53 at a glance. J Cell Sci 123(Pt 15):2527–2532. doi:10.1242/jcs.064501 (PubMed PMID: 20940128; PMCID: PMC2912460)

Brooks CL, Gu W (2006) p53 ubiquitination: Mdm2 and beyond. Mol Cell 21(3):307–315. doi:10.1016/j.molcel.2006.01.020

Brown GK, Otero LJ, LeGris M, Brown RM (1994) Pyruvate dehydrogenase deficiency. J Med Genet 31(11):875–879 (PubMed PMID: 7853374; PMCID: 1016663)

Carroll TJ, Park JS, Hayashi S, Majumdar A, McMahon AP (2005) Wnt9b plays a central role in the regulation of mesenchymal to epithelial transitions underlying organogenesis of the mammalian urogenital system. Dev Cell 9(2):283–292. doi:10.1016/j.devcel.2005.05.016 (PubMed PMID: 16054034)

Chau BN, Diaz RL, Saunders MA, Cheng C, Chang AN, Warrener P, Bradshaw J, Linsley PS, Cleary MA (2009) Identification of SULF2 as a novel transcriptional target of p53 by use of integrated genomic analyses. Cancer Res 69(4):1368–1374. doi:10.1158/0008-5472.CAN-08-2742 (PubMed PMID: 19190338)

Chen S, El-Dahr SS (2013) Histone deacetylases in kidney development: implications for disease and therapy. Pediatr Nephrol 28(5):689–698. doi:10.1007/s00467-012-2223-8 (PubMed PMID: 22722820)

Cicalese A, Bonizzi G, Pasi CE, Faretta M, Ronzoni S, Giulini B, Brisken C, Minucci S, Di Fiore PP, Pelicci PG (2009) The tumor suppressor p53 regulates polarity of self-renewing divisions in mammary stem cells. Cell 138(6):1083–1095. doi:10.1016/j.cell.2009.06.048

Cordenonsi M, Dupont S, Maretto S, Insinga A, Imbriano C, Piccolo S (2003) Links between tumor suppressors: p53 is required for TGF-β gene responses by cooperating with smads. Cell 113(3):301–314. doi:10.1016/S0092-8674(03)00308-8

Cordenonsi M, Montagner M, Adorno M, Zacchigna L, Martello G, Mamidi A, Soligo S, Dupont S, Piccolo S (2007) Integration of TGF-ß and Ras/MAPK signaling through p53 phosphorylation. Science 315(5813):840–843. doi:10.1126/science.1135961

Costantini F, Kopan R (2010) Patterning a complex organ: branching morphogenesis and nephron segmentation in kidney development. Dev Cell 18(5):698–712. doi:10.1016/j.devcel.2010.04.008 (PubMed PMID: 20493806; PMCID: PMC2883254)

Cox TR, Erler JT (2011) Remodeling and homeostasis of the extracellular matrix: implications for fibrotic diseases and cancer. Dis Model Mech 4(2):165–178. doi:10.1242/dmm.004077 (PubMed PMID: PMC3046088)

Dai C, Gu W (2010) p53 post-translational modification: deregulated in tumorigenesis. Trends Mol Med 16(11):528–536. doi:10.1016/j.molmed.2010.09.002 (PubMed PMID: 20932800; PMCID: PMC2978905)

Danilova N, Sakamoto KM, Lin S (2008) p53 family in development. Mech Dev 125 (11–12):919–931. doi:10.1016/j.mod.2008.09.003

Das A, Tanigawa S, Karner CM, Xin M, Lum L, Chen C, Olson EN, Perantoni AO, Carroll TJ (2013) Stromal-epithelial crosstalk regulates kidney progenitor cell differentiation. Nat Cell Biol 15(9):1035–1044. doi:10.1038/ncb2828 (PubMed PMID: PMC3891676)

Dey DC, Bronson RP, Dahl J, Carroll JP, Benjamin TL (2000) Accelerated development of polyoma tumors and embryonic lethality: different effects of p53 loss on related mouse backgrounds. Cell Growth Differ 11(5):231–237

Donehower LA (1996) The p53-deficient mouse: a model for basic and applied cancer studies. Semin Cancer Biol 7(5):269–278. doi:10.1006/scbi.1996.0035 (PubMed PMID: 9110404)

Donehower LA, Harvey M, Slagle BL, McArthur MJ, Montgomery CA Jr, Butel JS, Bradley A (1992) Mice deficient for p53 are developmentally normal but susceptible to spontaneous tumours. Nature 356(6366):215–221. doi:10.1038/356215a0 (PubMed PMID: 1552940)

El-Dahr SS, Aboudehen K, Saifudeen Z (2008) Transcriptional control of terminal nephron differentiation. Am J Physiol Renal Physiol 294(6):F1273–F1278. doi:10.1152/ajprenal.00562.2007 (PubMed PMID: 18287399; PMCID: PMC2606293)

el-Deiry WS, Kern SE, Pietenpol JA, Kinzler KW, Vogelstein B (1992) Definition of a consensus binding site for p53. Nat Genet 1(1):45–49. doi:10.1038/ng0492-45 (PubMed PMID: 1301998)

El-Deiry WS, Tokino T, Velculescu VE, Levy DB, Parsons R, Trent JM, Lin D, Mercer WE, Kinzler KW, Vogelstein B (1993) WAF1, a potential mediator of p53 tumor suppression. Cell 75(4):817–825. doi:10.1016/0092-8674(93)90500-P

Erster S, Moll UM (2005) Stress-induced p53 runs a transcription-independent death program. Biochem Biophys Res Commun 331(3):843–850. doi:10.1016/j.bbrc.2005.03.187 (PubMed PMID: 15865940)

Fetting JL, Guay JA, Karolak MJ, Iozzo RV, Adams DC, Maridas DE, Brown AC, Oxburgh L (2014) FOXD1 promotes nephron progenitor differentiation by repressing decorin in the embryonic kidney. Development 141(1):17–27. doi:10.1242/dev.089078 (PubMed PMID: 24284212; PMCID: 3865747)

Gadea G, de Toledo M, Anguille C, Roux P. Loss of p53 promotes RhoA-ROCK-dependent cell migration and invasion in 3D matrices. J Cell Biol. 2007;178(1):23-30. PubMed PMID: 17606864.

Georgas K, Rumballe B, Valerius MT, Chiu HS, Thiagarajan RD, Lesieur E, Aronow BJ, Brunskill EW, Combes AN, Tang D, Taylor D, Grimmond SM, Potter SS, McMahon AP, Little MH (2009) Analysis of early nephron patterning reveals a role for distal RV proliferation in fusion to the ureteric tip via a cap mesenchyme-derived connecting segment. Dev Biol 332 (2):273–286. doi:10.1016/j.ydbio.2009.05.578 (PubMed PMID: 19501082)

Giaccia AJ, Kastan MB (1998) The complexity of p53 modulation: emerging patterns from divergent signals. Genes Dev 12(19):2973–2983. doi:10.1101/gad.12.19.2973

Gottlieb E, Vousden KH (2010) p53 regulation of metabolic pathways. Cold Spring Harb Perspect Biol 2(4):a001040. doi:10.1101/cshperspect.a001040

Gottlieb E, Haffner R, King A, Asher G, Gruss P, Lonai P, Oren M (1997) Transgenic mouse model for studying the transcriptional activity of the p53 protein: age- and tissue-dependent changes in radiation-induced activation during embryogenesis. EMBO J 16(6):1381–1390. doi:10.1093/emboj/16.6.1381 (PubMed PMID: 9135153; PMCID: PMC1169735)

Grossi E, Sánchez Y, Huarte M (2016) Expanding the p53 regulatory network: LncRNAs take up the challenge. Biochim Biophys Acta 1859(1):200–208. doi:10.1016/j.bbagrm.2015.07.011

Gupta S, De S, Srivastava V, Hussain M, Kumari J, Muniyappa K, Sengupta S (2013) RECQL4 and p53 potentiate the activity of polymerase γ and maintain the integrity of the human mitochondrial genome. Carcinogenesis. doi:10.1093/carcin/bgt315

Hewitt KN, Walker EA, Stewart PM (2005) Minireview: hexose-6-phosphate dehydrogenase and redox control of 11β-hydroxysteroid dehydrogenase type 1 activity. Endocrinology 146 (6):2539–2543. doi:10.1210/en.2005-0117 (PubMed PMID: 15774558)

Hickman ES, Helin K (2000) The p53 tumour suppressor protein. Biotechnol Genet Eng Rev 17:179–211 (PubMed PMID: 11255666)

Hilliard S, Aboudehen K, Yao X, El-Dahr SS (2011) Tight regulation of p53 activity by Mdm2 is required for ureteric bud growth and branching. Dev Biol 353(2):354–366. doi:10.1016/j.ydbio.2011.03.017 (PubMed PMID: 21420949; PMCID: 3086838)

Hilliard SA, Yao X, El-Dahr SS (2014) Mdm2 is required for maintenance of the nephrogenic niche. Dev Biol 387(1):1–14. doi:10.1016/j.ydbio.2014.01.009 (PubMed PMID: 24440154; PMCID: 3951515)

Hirsch DS, Pirone DM, Burbelo PD (2001) A new family of Cdc42 effector proteins, CEPs, function in fibroblast and epithelial cell shape changes. J Biol Chem 276(2):875–883. doi:10.1074/jbc.M007039200

Ho WC, Fitzgerald MX, Marmorstein R (2006) Structure of the p53 core domain dimer bound to DNA. J Biol Chem 281(29):20494–20502. doi:10.1074/jbc.M603634200

Ho J, Pandey P, Schatton T, Sims-Lucas S, Khalid M, Frank MH, Hartwig S, Kreidberg JA (2011) The pro-apoptotic protein Bim is a MicroRNA target in kidney progenitors. J Am Soc Nephrol 22(6):1053–1063. doi:10.1681/ASN.2010080841 (PubMed PMID: PMC3103725)

Hwang C-I, Matoso A, Corney DC, Flesken-Nikitin A, Körner S, Wang W, Boccaccio C, Thorgeirsson SS, Comoglio PM, Hermeking H, Nikitin AY (2011) Wild-type p53 controls cell motility and invasion by dual regulation of MET expression. Proc Natl Acad Sci 108 (34):14240–14245. doi:10.1073/pnas.1017536108

Islam MM, Suzuki H, Yoneda M, Tanaka M (1997) Primary structure of the smallest (6.4-kDa) subunit of human and bovine ubiquinol-cytochrome c reductase deduced from cDNA sequences. Biochem Mol Biol Int 41(6):1109–1116 (PubMed PMID: 9161705)

Ivanov GS, Ivanova T, Kurash J, Ivanov A, Chuikov S, Gizatullin F, Herrera-Medina EM, Rauscher F, Reinberg D, Barlev NA (2007) Methylation-acetylation interplay activates p53 in response to DNA damage. Mol Cell Biol 27(19):6756–6769. doi:10.1128/mcb.00460-07

Iwakuma T, Lozano G (2003) MDM2, an introduction. Mol Cancer Res 1(14):993–1000

Jansson M, Durant ST, Cho E-C, Sheahan S, Edelmann M, Kessler B, La Thangue NB (2008) Arginine methylation regulates the p53 response. Nat Cell Biol 10(12):1431–1439 (http://www.nature.com/ncb/journal/v10/n12/suppinfo/ncb1802_S1.html)

Jenkins LMM, Durell SR, Mazur SJ, Appella E (2012) p53 N-terminal phosphorylation: a defining layer of complex regulation. Carcinogenesis 33(8):1441–1449. doi:10.1093/carcin/bgs145

Joerger AC, Fersht AR (2008) Structural biology of the tumor suppressor p53. Annu Rev Biochem 77:557–582. doi:10.1146/annurev.biochem.77.060806.091238 (PubMed PMID: 18410249)

Kaplon J, Zheng L, Meissl K, Chaneton B, Selivanov VA, Mackay G, van der Burg SH, Verdegaal EM, Cascante M, Shlomi T, Gottlieb E, Peeper DS (2013) A key role for mitochondrial gatekeeper pyruvate dehydrogenase in oncogene-induced senescence. Nature 498 (7452):109–112. doi:10.1038/nature12154 (PubMed PMID: 23685455)

Karner CM, Das A, Ma Z, Self M, Chen C, Lum L, Oliver G, Carroll TJ (2011) Canonical Wnt9b signaling balances progenitor cell expansion and differentiation during kidney development. Development 138(7):1247–1257. doi:10.1242/dev.057646 (PubMed PMID: 21350016; PMCID: PMC3050658)

Kastan MB, Zhan Q, El-Deiry WS, Carrier F, Jacks T, Walsh WV, Plunkett BS, Vogelstein B, Fornace AJ Jr (1992) A mammalian cell cycle checkpoint pathway utilizing p53 and GADD45 is defective in ataxia-telangiectasia. Cell 71(4):587–597. doi:10.1016/0092-8674(92)90593-2

Khoury MP, Bourdon JC (2011) p53 isoforms: an intracellular microprocessor? Genes Cancer 2 (4):453–465. doi:10.1177/1947601911408893 (PubMed PMID: 21779513; PMCID: PMC3135639)

Kloosterhof NK, Bralten LBC, Dubbink HJ, French PJ, van den Bent MJ (2011) Isocitrate dehydrogenase-1 mutations: a fundamentally new understanding of diffuse glioma? Lancet Oncol 12(1):83–91. doi:10.1016/S1470-2045(10)70053-X

Knights CD, Catania J, Giovanni SD, Muratoglu S, Perez R, Swartzbeck A, Quong AA, Zhang X, Beerman T, Pestell RG, Avantaggiati ML (2006) Distinct p53 acetylation cassettes differentially influence gene-expression patterns and cell fate. J Cell Biol 173(4):533–544. doi:10.1083/jcb.200512059

Kobayashi A, Valerius MT, Mugford JW, Carroll TJ, Self M, Oliver G, McMahon AP (2008) Six2 defines and regulates a multipotent self-renewing nephron progenitor population throughout mammalian kidney development. Cell Stem Cell 3(2):169–181. doi:10.1016/j.stem.2008.05.020 (PubMed PMID: 18682239; PMCID: PMC2561900)

Komarova EA, Chernov MV, Franks R, Wang K, Armin G, Zelnick CR, Chin DM, Bacus SS, Stark GR, Gudkov AV (1997) Transgenic mice with p53-responsive lacZ: p53 activity varies dramatically during normal development and determines radiation and drug sensitivity in vivo. EMBO J 16(6):1391–1400. doi:10.1093/emboj/16.6.1391 (PubMed PMID: 9135154; PMCID: PMC1169736)

Kruse J-P, Gu W (2008) SnapShot: p53 posttranslational modifications. Cell 133(5):930–9e1. doi:10.1016/j.cell.2008.05.020

Kruse J-P, Gu W (2009) Modes of p53 regulation. Cell 137(4):609–622. doi:10.1016/j.cell.2009.04.050

Lavelin I, Geiger B (2005) Characterization of a novel GTPase-activating protein associated with focal adhesions and the actin cytoskeleton. J Biol Chem 280(8):7178–7185. doi:10.1074/jbc.M411990200

Lebedeva MA, Eaton JS, Shadel GS (2009) Loss of p53 causes mitochondrial DNA depletion and altered mitochondrial reactive oxygen species homeostasis. Biochim Biophys Acta 1787 (5):328–334. doi:10.1016/j.bbabio.2009.01.004

Lee KH, Li M, Michalowski AM, Zhang X, Liao H, Chen L, Xu Y, Wu X, Huang J (2010) A genomewide study identifies the Wnt signaling pathway as a major target of p53 in murine embryonic stem cells. Proc Natl Acad Sci USA 107(1):69–74. doi:10.1073/pnas.0909734107 (PubMed PMID: 20018659; PMCID: PMC2806696)

Li FP, Fraumeni JF, Mulvihill JJ, Blattner WA, Dreyfus MG, Tucker MA, Miller RW (1988) A cancer family syndrome in twenty-four kindreds. Cancer Res 48(18):5358–5362

Li Y, Park JS, Deng JH, Bai Y (2006) Cytochrome c oxidase subunit IV is essential for assembly and respiratory function of the enzyme complex. J Bioenerg Biomembr 38(5–6):283–291. doi:10.1007/s10863-006-9052-z (PubMed PMID: 17091399; PMCID: 1885940)

Li M, He Y, Dubois W, Wu X, Shi J, Huang J (2012) Distinct regulatory mechanisms and functions for p53-activated and p53-repressed DNA damage response genes in embryonic stem cells. Mol Cell 46(1):30–42. doi:10.1016/j.molcel.2012.01.020

Li Y, Liu J, McLaughlin N, Bachvarov D, Saifudeen Z, El-Dahr SS (2013) Genome-wide analysis of the p53 gene regulatory network in the developing mouse kidney. Physiol Genomics 45 (20):948–964. doi:10.1152/physiolgenomics.00113.2013 (PubMed PMID: 24003036; PMCID: 3798767)

Li Y, Liu J, Li W, Brown A, Baddoo M, Li M, Carroll T, Oxburgh L, Feng Y, Saifudeen Z (2015) p53 Enables metabolic fitness and self-renewal of nephron progenitor cells. Development 142 (7):1228–1241. doi:10.1242/dev.111617 (PubMed PMID: 25804735; PMCID: PMC4378244)

Lindstrom NO, Lawrence ML, Burn SF, Johansson JA, Bakker ER, Ridgway RA, Chang CH, Karolak MJ, Oxburgh L, Headon DJ, Sansom OJ, Smits R, Davies JA, Hohenstein P (2014) Integrated beta-catenin, BMP, PTEN, and Notch signalling patterns the nephron. Elife 3: e04000. doi:10.7554/eLife.04000 (PubMed PMID: 25647637; PMCID: PMC4337611)

Little MH, McMahon AP (2012) Mammalian kidney development: principles, progress, and projections. Cold Spring Harb Perspect Biol 4(5):a008300. doi:10.1101/cshperspect.a008300

Liu M, Dhanwada KR, Birt DF, Hecht S, Pelling JC (1994) Increase in p53 protein half-life in mouse keratinocytes following UV-B irradiation. Carcinogenesis 15(6):1089–1092 (PubMed PMID: 8020138)

Liu Y, Elf SE, Miyata Y, Sashida G, Liu Y, Huang G, Di Giandomenico S, Lee JM, Deblasio A, Menendez S, Antipin J, Reva B, Koff A, Nimer SD (2009) p53 regulates hematopoietic stem cell quiescence. Cell Stem Cell 4(1):37–48. doi:10.1016/j.stem.2008.11.006

Loughery JDM (2013) Switching on p53: an essential role for protein phosphorylation? BioDiscovery 8:1–20. doi:10.7750/BioDiscovery.2013.8.1

Maddocks ODK, Vousden KH (2011) Metabolic regulation by p53. J Mol Med 89(3):237–245. doi:10.1007/s00109-011-0735-5 (PubMed PMID: PMC3043245)

Matoba S, Kang J-G, Patino WD, Wragg A, Boehm M, Gavrilova O, Hurley PJ, Bunz F, Hwang PM (2006) p53 regulates mitochondrial respiration. Science 312(5780):1650–1653. doi:10.1126/science.1126863

Mayr JA, Havlíčková V, Zimmermann F, Magler I, Kaplanová V, Ješina P, Pecinová A, Nůsková H, Koch J, Sperl W, Houštěk J (2010) Mitochondrial ATP synthase deficiency due to a mutation in the ATP5E gene for the F1 ε subunit. Hum Mol Genet 19(17):3430–3439. doi:10.1093/hmg/ddq254

McLaughlin N, Yao X, Li Y, Saifudeen Z, El-Dahr SS (2013) Histone signature of metanephric mesenchyme cell lines. Epigenetics 8(9):970–978. doi:10.4161/epi.25753 (PubMed PMID: 23867747; PMCID: PMC3883774)

Meek DW, Anderson CW (2009) Posttranslational modification of p53: cooperative integrators of function. Cold Spring Harb Perspect Biol 1(6):a000950. doi:10.1101/cshperspect.a000950

Meletis K, Wirta V, Hede S-M, Nistér M, Lundeberg J, Frisén J (2006) p53 suppresses the self-renewal of adult neural stem cells. Development 133(2):363–369. doi:10.1242/dev.02208

Menendez D, Inga A, Resnick MA (2009) The expanding universe of p53 targets. Nat Rev Cancer 9(10):724–737 (http://www.nature.com/nrc/journal/v9/n10/suppinfo/nrc2730_S1.html)

Mihara M, Erster S, Zaika A, Petrenko O, Chittenden T, Pancoska P, Moll UM (2003) p53 has a direct apoptogenic role at the mitochondria. Mol Cell 11(3):577–590. doi:10.1016/S1097-2765(03)00050-9

Milyavsky M, Gan OI, Trottier M, Komosa M, Tabach O, Notta F, Lechman E, Hermans KG, Eppert K, Konovalova Z, Ornatsky O, Domany E, Meyn MS, Dick JE (2010) A distinctive DNA damage response in human hematopoietic stem cells reveals an apoptosis-independent role for p53 in self-renewal. Cell Stem Cell 7(2):186–197. doi:10.1016/j.stem.2010.05.016 (PubMed PMID: 20619763)

Miyazaki Y, Oshima K, Fogo A, Hogan BLM, Ichikawa I (2000) Bone morphogenetic protein 4 regulates the budding site and elongation of the mouse ureter. J Clin Invest 105(7):863–873. doi:10.1172/JCI8256

Molchadsky A, Rivlin N, Brosh R, Rotter V, Sarig R (2010) p53 is balancing development, differentiation and de-differentiation to assure cancer prevention. Carcinogenesis 31 (9):1501–1508. doi:10.1093/carcin/bgq101

Moll UM, Petrenko O (2003) The MDM2-p53 interaction. Mol Cancer Res 1(14):1001–1008

Moll UM, Wolff S, Speidel D, Deppert W (2005) Transcription-independent pro-apoptotic functions of p53. Curr Opin Cell Biol 17(6):631–636. doi:10.1016/j.ceb.2005.09.007

Montes de Oca Luna R, Wagner DS, Lozano G (1995) Rescue of early embryonic lethality in mdm2-deficient mice by deletion of p53. Nature 378(6553):203–206

Muller Patricia AJ, Vousden KH (2014) Mutant p53 in cancer: new functions and therapeutic opportunities. Cancer Cell 25(3):304–317. doi:10.1016/j.ccr.2014.01.021

Muthukrishnan SD, Yang X, Friesel R, Oxburgh L (2015) Concurrent BMP7 and FGF9 signalling governs AP-1 function to promote self-renewal of nephron progenitor cells. Nat Commun 6:10027. doi:10.1038/ncomms10027 (PubMed PMID: 26634297; PMCID: PMC4686668)

Olivier M, Hollstein M, Hainaut P (2010) TP53 mutations in human cancers: origins, consequences, and clinical use. Cold Spring Harb Perspect Biol 2(1):a001008. doi:10.1101/cshperspect.a001008 (PubMed PMID: PMC2827900)

Onel K, Cordon-Cardo C (2004) MDM2 and prognosis. Mol Cancer Res 2(1):1–8 (PubMed PMID: 14757840)

Ou X, O'Leary HA, Broxmeyer HE (2013) Implications of DPP4 modification of proteins that regulate stem/progenitor and more mature cell types. Blood 122(2):161–169. doi:10.1182/blood-2013-02-487470

Park JY, Wang PY, Matsumoto T, Sung HJ, Ma W, Choi JW, Anderson SA, Leary SC, Balaban RS, Kang JG, Hwang PM (2009) p53 improves aerobic exercise capacity and augments skeletal muscle mitochondrial DNA content. Circ Res 105(7):705–712. doi:10.1161/CIRCRESAHA.109.205310 (11 p following 12, PubMed PMID: 19696408; PMCID: PMC2761626)

Park J-H, Zhuang J, Li J, Hwang PM (2016) p53 as guardian of the mitochondrial genome. FEBS Lett. doi:10.1002/1873-3468.12061.

Park S-Y, Choi H-K, Choi Y, Kwak S, Choi K-C, Yoon H-G (2015) Deubiquitinase OTUD5 mediates the sequential activation of PDCD5 and p53 in response to genotoxic stress. Cancer Lett 357:419–427.

Puzio-Kuter AM (2011) The role of p53 in metabolic regulation. Genes Cancer 2(4):385–391. doi:10.1177/1947601911409738 (PubMed PMID: PMC3135642)

Ray K, Kunsch C, Bonner LM, Robishaw JD (1995) Isolation of cDNA clones encoding eight different human G protein γ subunits, including three novel forms designated the γ4, γ10, and γ11 subunits. J Biol Chem 270(37):21765–21771

Sah VP, Attardi LD, Mulligan GJ, Williams BO, Bronson RT, Jacks T (1995) A subset of p53-deficient embryos exhibit exencephaly. Nat Genet 10(2):175–180. doi:10.1038/ng0695-175 (PubMed PMID: 7663512)

Saifudeen Z, Dipp S, El-Dahr SS (2002) A role for p53 in terminal epithelial cell differentiation. J Clin Invest 109(8):1021–1030. doi:10.1172/JCI13972 (PubMed PMID: 11956239; PMCID: PMC150944)

Saifudeen Z, Dipp S, Stefkova J, Yao X, Lookabaugh S, El-Dahr SS (2009) p53 regulates metanephric development. J Am Soc Nephrol 20(11):2328–2337. doi:10.1681/ASN.2008121224 (PubMed PMID: 19729440; PMCID: 2799183)

Saifudeen Z, Liu J, Dipp S, Yao X, Li Y, McLaughlin N, Aboudehen K, El-Dahr SS (2012) A p53-Pax2 pathway in kidney development: implications for nephrogenesis. PLoS One 7(9): e44869. doi:10.1371/journal.pone.0044869 (PubMed PMID: 22984579; PMCID: 3440354)

Schmid P, Lorenz A, Hameister H, Montenarh M (1991) Expression of p53 during mouse embryogenesis. Development 113(3):857–865 (PubMed PMID: 1821855)

Schmidt D, Müller S (2002) Members of the PIAS family act as SUMO ligases for c-Jun and p53 and repress p53 activity. Proc Natl Acad Sci 99(5):2872–2877. doi:10.1073/pnas.052559499

Schoppy DW, Ruzankina Y, Brown EJ (2010) Removing all obstacles: a critical role for p53 in promoting tissue renewal. Cell Cycle 9(7):1313–1319 (PMCID: PMC2980577)

Self M, Lagutin OV, Bowling B, Hendrix J, Cai Y, Dressler GR, Oliver G (2006) Six2 is required for suppression of nephrogenesis and progenitor renewal in the developing kidney. EMBO J 25 (21):5214–5228. doi:10.1038/sj.emboj.7601381

Shah MM, Sakurai H, Gallegos TF, Sweeney DE, Bush KT, Esko JD, Nigam SK (2011) Growth factor-dependent branching of the ureteric bud is modulated by selective 6-O sulfation of heparan sulfate. Dev Biol 356(1):19–27. doi:10.1016/j.ydbio.2011.05.004 (PubMed PMID: 21600196; PMCID: PMC3130836)

Shimizu Y, Thumkeo D, Keel J, Ishizaki T, Oshima H, Oshima M, Noda Y, Matsumura F, Taketo MM, Narumiya S (2005) ROCK-I regulates closure of the eyelids and ventral body wall by inducing assembly of actomyosin bundles. J Cell Biol 168(6):941–953. doi:10.1083/jcb.200411179

Sola-Penna M, Da Silva D, Coelho WS, Marinho-Carvalho MM, Zancan P (2010) Regulation of mammalian muscle type 6-phosphofructo-1-kinase and its implication for the control of the metabolism. IUBMB Life 62(11):791–796. doi:10.1002/iub.393

Sun KLW, Correia JP, Kennedy TE (2011) Netrins: versatile extracellular cues with diverse functions. Development 138(11):2153–2169. doi:10.1242/dev.044529

Tang X, Milyavsky M, Shats I, Erez N, Goldfinger N, Rotter V (2004) Activated p53 suppresses the histone methyltransferase EZH2 gene. Oncogene 23(34):5759–5769

Tang Y, Luo J, Zhang W, Gu W (2006) Tip60-dependent acetylation of p53 modulates the decision between cell-cycle arrest and apoptosis. Mol Cell 24(6):827–839. doi:10.1016/j.molcel.2006.11.021

Tasdemir E, Maiuri MC, Galluzzi L, Vitale I, Djavaheri-Mergny M, D'Amelio M, Criollo A, Morselli E, Zhu C, Harper F, Nannmark U, Samara C, Pinton P, Vicencio JM, Carnuccio R, Moll UM, Madeo F, Paterlini-Brechot P, Rizzuto R, Szabadkai G, Pierron G, Blomgren K, Tavernarakis N, Codogno P, Cecconi F, Kroemer G (2008) Regulation of autophagy by cytoplasmic p53. Nat Cell Biol 10(6):676–687 (http://www.nature.com/ncb/journal/v10/n6/suppinfo/ncb1730_S1.html)

Vaseva AV, Moll UM (2009) The mitochondrial p53 pathway. Biochim Biophys Acta 1787 (5):414–420. doi:10.1016/j.bbabio.2008.10.005

Vousden KH, Lane DP (2007) p53 in health and disease. Nat Rev Mol Cell Biol 8(4):275–283. doi:10.1038/nrm2147 (PubMed PMID: 17380161)

Vousden KH, Prives C (2009) Blinded by the light: the growing complexity of p53. Cell 137 (3):413–431. doi:10.1016/j.cell.2009.04.037

Wallingford JB, Seufert DW, Virta VC, Vize PD (1997) P53 activity is essential for normal development in Xenopus. Curr Biol 7(10):747–757. doi:10.1016/S0960-9822(06)00333-2

Wang P-Y, Ma W, Park J-Y, Celi FS, Arena R, Choi JW, Ali QA, Tripodi DJ, Zhuang J, Lago CU, Strong LC, Talagala SL, Balaban RS, Kang J-G, Hwang PM (2013) Increased oxidative metabolism in the Li–Fraumeni syndrome. N Engl J Med 368(11):1027–1032. doi:10.1056/NEJMoa1214091 (PubMed PMID: 23484829)

Wang DB, Kinoshita C, Kinoshita Y, Morrison RS (2014) p53 and mitochondrial function in neurons. Biochim Biophys 1842(8):1186–1197. doi:10.1016/j.bbadis.2013.12.015

Wienken M, Dickmanns A, Nemajerova A, Kramer D, Najafova Z, Weiss M, Karpiuk O, Kassem M, Zhang Y, Lozano G, Johnsen Steven A, Moll Ute M, Zhang X, Dobbelstein M (2016) MDM2 associates with polycomb repressor complex 2 and enhances stemness-promoting chromatin modifications independent of p53. Mol Cell 61(1):68–83. doi:10.1016/j.molcel.2015.12.008

Xavier JM, Morgado AL, Sola S, Rodrigues CM (2014) Mitochondrial translocation of p53 modulates neuronal fate by preventing differentiation-induced mitochondrial stress. Antioxid Redox Signal 21(7):1009–1024. doi:10.1089/ars.2013.5417 (PubMed PMID: 24329038; PMCID: PMC4123470)

Xu Y (2005) A new role of p53 in maintaining genetic stability in embryonic stem cells. Cell Cycle 4(3):363–364

Yañez AJ, Ludwig HC, Bertinat R, Spichiger C, Gatica R, Berlien G, Leon O, Brito M, Concha II, Slebe JC (2005) Different involvement for aldolase isoenzymes in kidney glucose metabolism:

aldolase B but not aldolase A colocalizes and forms a complex with FBPase. J Cell Physiol 202 (3):743–753. doi:10.1002/jcp.20183

Yu H, Thun R, Chandrasekharappa S, Trent JM, Zhang J, Meisler MH (1993) Human PCK1 encoding phosphoenolpyruvate carboxykinase is located on chromosome 20q13.2. Genomics 15(1):219–221. doi:10.1006/geno.1993.1040

Zhang Y, Lu H (2009) Signaling to p53: ribosomal proteins find their way. Cancer Cell 16 (5):369–377. doi:10.1016/j.ccr.2009.09.024 (PubMed PMID: 19878869; PMCID: PMC4369769)

Zhang X-D, Qin Z-H, Wang J (2010) The role of p53 in cell metabolism. Acta Pharmacol Sin 31 (9):1208–1212

Zielinska AE, Walker EA, Stewart PM, Lavery GG (2011) Biochemistry and physiology of hexose-6-phosphate knockout mice. Mol Cell Endocrinol 336(1-2):213–218. doi:10.1016/j.mce.2010.12.004 (PubMed PMID: 21146583)

Chapter 6
Growth Factor Regulation in the Nephrogenic Zone of the Developing Kidney

Leif Oxburgh, Sree Deepthi Muthukrishnan, and Aaron Brown

Abstract New nephrons are induced by the interaction between mesenchymal progenitor cells and collecting duct tips, both of which are located at the outer edge of the kidney. This leading edge of active nephron induction is known as the nephrogenic zone. Cell populations found within this zone include collecting duct tips, cap mesenchyme cells, pretubular aggregates, nephrogenic zone interstitium, hemoendothelial progenitor cells, and macrophages. The close association of these dynamic progenitor cell compartments enables the intricate and synchronized patterning of the epithelial and the vascular components of the nephron. Understanding signaling interactions between the distinct progenitor cells of the nephrogenic zone are essential to determining the basis for new nephron formation, an important goal in regenerative medicine. A variety of technologies have been applied to define essential signaling pathways, including organ culture, mouse genetics, and primary cell culture. This chapter provides an overview of essential signaling pathways and discusses how these may be integrated.

6.1 Introduction

The kidney fills several essential physiological functions, and it has been heavily modified during the course of evolution to adapt to a variety of demands (Vize and Smith 2004). One of the simpler examples is the Malpighian tubule of the fruitfly, an epithelial tube connected to the gut that filters hemolymph (Sözen et al. 1997). In the zebrafish, kidney structures are more comparable to our own, with the filtering tubule capped by a tuft of glomerular capillaries through which blood is filtered. These nephrons are contained within the mesonephros, which continues growing throughout the life of the fish and has the capacity to regenerate in response to injury (Diep et al. 2011; Diep et al. 2015). While humans form a mesonephros during the course of embryonic development, this structure is transient. Instead, the adult human kidney is the metanephros, and its organogenesis obeys distinct rules. The central principle of

L. Oxburgh (✉) • S.D. Muthukrishnan • A. Brown
Center for Molecular Medicine, Maine Medical Center Research Institute, 81 Research Drive, Scarborough, ME 04074, USA
e-mail: Oxburl@mmc.org

R.K. Miller (ed.), *Kidney Development and Disease*, Results and Problems in Cell Differentiation 60, DOI 10.1007/978-3-319-51436-9_6

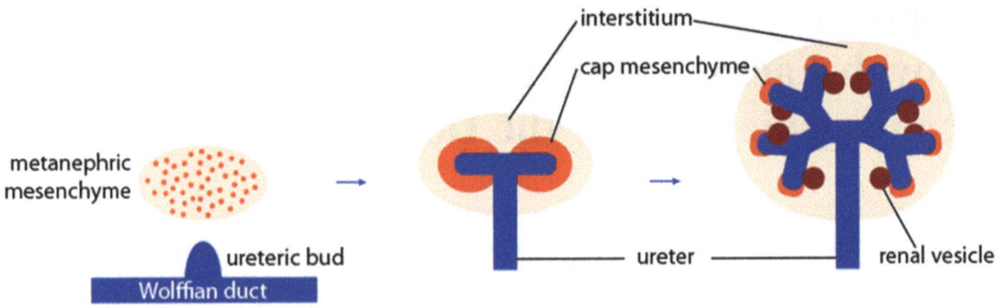

Fig. 6.1 Ureteric bud outgrowth and branching. Schematic showing outgrowth of the ureteric bud from the Wolffian duct and subsequent branching of the collecting duct system. In this largely dichotomous branching process, cap mesenchyme is displaced radially at collecting duct tips, while a subset of cap mesenchyme cells that have differentiated are deposited as renal vesicles which will form the nephrons of the kidney

metanephric development is reciprocal epithelium–mesenchyme induction leading to radial branching of a central collecting duct tree and induction of new nephrons by collecting duct tips at the outer edge of the developing organ (Fig. 6.1). Iteration of this basic program throughout the period of active nephrogenesis determines nephron endowment of the adult human kidney, which is between 250,000 and 2.5 million (Bertram et al. 2011). There is no capacity for generation of new nephrons in the adult, and the number of nephrons in the human is decided during fetal life, with nephrogenesis ending at approximately 36 weeks of gestation. A large body of data shows that the fundamental organogenetic principles of human development are similar in the mouse. Notable exceptions are that the final nephron number is lower, in the range of 9,000–13,000 per kidney, and that the period of developmental nephrogenesis extends to the second day after birth (Cullen-McEwen et al. 2003; Baldelomar et al. 2016; Hartman et al. 2007). Based on analyses of morphology and gene expression of human embryonic kidneys, it is assumed that fundamental cellular interactions and signaling events underlying nephrogenesis are conserved between human and mouse, although detailed studies have shown some subtle and intriguing differences that may explain morphological diversity between the two species (O'Brien et al. 2016). An excellent description of the morphology of the embryonic mouse kidney is available at the genito-urinary development microanatomy project site (GUDMAP).

6.2 Anatomy and Function of the Nephrogenic Zone

New nephrons are induced by the interaction between mesenchymal progenitor cells and collecting duct tips, both of which are located at the outer edge of the kidney. This leading edge of active nephron induction is known as the nephrogenic

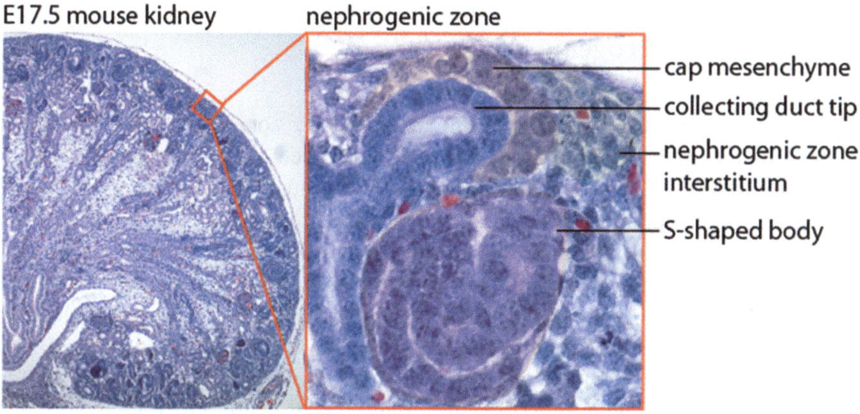

Fig. 6.2 The nephrogenic zone of an E17.5 mouse kidney

zone. According to current anatomical ontology, this region lies between the kidney capsule and the first layer of epithelialized nascent nephrons known as renal vesicles (Fig. 6.2) (Little et al. 2007). The metanephros appears at approximately embryonic day (E) 10.5 in the mouse, the first wave of renal vesicles becomes evident at approximately E12, and the nephrogenic zone is represented in the embryonic kidney from this point until cessation of nephrogenesis. The major cell populations within this zone are: collecting duct tips, cap mesenchyme cells [nephron progenitor cells (NPCs)], pretubular aggregates (aggregates of induced NPCs), and nephrogenic zone interstitium. Less conspicuous cell populations include hemoendothelial progenitor cells and macrophages (Schmidt-Ott et al. 2006; Rae et al. 2007; Hu et al. 2015). On the one hand, the nephrogenic zone can be viewed as a collection of discrete microanatomical structures, but on the other hand it can be viewed as a dynamic collection of progenitor cells that are closely associated to enable formation of the nephron, which depends on intricate and synchronized patterning of structures derived from these distinct progenitors. The tubular epithelium, podocytes, and cells of the Bowman's capsule derive from the cap mesenchyme, whereas the glomerular endothelium and the capillary bed surrounding the nephron derive from endothelial progenitor cells (Kobayashi et al. 2008; Hu et al. 2015). The glomerular mesangium and pericytes associated with the capillary surrounding the nephron originate from the nephrogenic zone interstitium (Kobayashi et al. 2014). The collecting duct tip is itself a highly dynamic progenitor cell compartment that grows radially while contributing cells to the trunk of the collecting duct tree (Riccio et al. 2016). These progenitor cell compartments are closely integrated, and one particularly intriguing question is to what degree they control each other's renewal and differentiation.

6.3 Experimental Systems for the Study of the Nephrogenic Zone

The composition and dynamics of the nephrogenic zone have primarily been studied in the mouse, although some work has been done on human kidneys (Metsuyanim et al. 2009; O'Brien et al. 2016). Because of the scarcity of tissue, few mechanistic studies of the human nephrogenic zone have been reported. The few reports available are, however, generally supportive of similarities in cell behavior between mouse and human nephrogenic zones, for example showing progenitor properties of nephron progenitor cells expressing the same molecular markers in both species (Harari-Steinberg et al. 2013; Pode-Shakked et al. 2016). The mouse has been a highly versatile model for human kidney development for two major reasons. The first is that the organ can be cultured ex vivo; groundbreaking studies in organogenesis reported in the 1950s demonstrated not only that kidney organogenesis could be maintained for several days in culture, but also that the mesenchymal component of the kidney could be cultured in isolation from the epithelial component (Grobstein 1953a, b, 1956). These important studies made it possible to study effects of proteins, neutralizing antibodies and small molecules specifically on the mesenchyme from which the nephrogenic zone arises. Lauri Saxén, who played a major role in exploiting this system to understand the cellular and molecular bases of nephron induction and epithelial–mesenchymal interactions in general, summarized many of these early studies in his influential monograph "Organogenesis of the kidney," which remains essential reading (Saxén 1987). The second reason for the dominance of the mouse as a model system for human kidney development is the potential for genetic modification; the discoveries of transgenesis and gene targeting in the 1980s ushered in a new era in which the basic organogenetic rules established through culture-based experiments could be critically tested in vivo. It is important to point out that the genetic era did not displace organ culture-based inquiry and that these complementary technologies have led to some of the most important advances in our mechanistic understanding of kidney organogenesis when used together.

The strength of the in vitro organogenesis system is of course that it accurately mimics the complexity of the organ in vivo. However, this has been a limitation for studies aiming to understand the importance of signal transduction pathways for specific cell types. Signaling interactions between cells of the nephrogenic zone have of course been the focus of a great deal of interest because it is the region in which the earliest decisions are made with regard to formation of the nephron. For this reason, there has been substantial interest in development of cell culture systems. Although immortalized cells have been versatile tools for studies of signaling biochemistry and basic cellular functions, their aberrant behavior in comparison to the cells from which they were generated have generally precluded their use in modeling signaling and cellular behaviors that determine function in vivo. Primary cell systems that are cultured in environments similar to the developing organ enable accurate modeling of key events in nephrogenesis such as turnover of progenitor cells in the undifferentiated state and epithelial differentiation. Purification and culture systems are now available for mixed cells of the nephrogenic zone as well as nephron progenitor cells (Blank et al. 2009; Brown et al. 2011a, b, 2013, 2015). Work is in progress in several research laboratories to define conditions for the growth of primary nephrogenic zone interstitial cells, collecting duct tip cells, and primary endothelial progenitors derived from the nephrogenic zone. When available, this complete toolkit will enable sophisticated investigations of the signaling relationships between the distinct cell types contained within the nephrogenic zone.

Finally, an influential experimental system that has been developed in only the past few years is the directed differentiation of human pluripotent stem cells to kidney tissue. Progress in this area has seen remarkably rapid development, a testament to its potential utility in regenerative medicine (Mae et al. 2013; Lam et al. 2014; Taguchi et al. 2014; Takasato et al. 2014, 2015; Morizane et al. 2015; Sharmin et al. 2015). For the first time, we now have a tractable experimental system with which we can observe and manipulate human nephrogenesis. This technology is still in its infancy and many aspects need to be refined, but studies are already being reported that highlight its enormous potential to understand molecular pathways underlying human nephrogenesis and congenital kidney disease (Freedman et al. 2015). It is now clear that all of the different cell types of the nephrogenic zone can be differentiated from pluripotent stem cells, and with further refinement it appears feasible to generate an experimental system specifically for the study of this niche (Takasato et al. 2015).

6.4 Nephron Progenitor Cells

6.4.1 Cellular Compartments and Signaling Within the Cap Mesenchyme

NPCs are arranged into a series of compartments within the cap mesenchyme, and cells within each of these compartments express a particular combination of transcriptional activators and co-activators (Mugford et al. 2009). While the biological meaning of this compartmentalization is still being unraveled, one hypothesis with significant experimental support is that it reflects a segregation of NPCs based on their ability to respond to differentiation signals (Fig. 6.3). Dividing the NPC population into cells that are receptive to differentiation signals and cells that are refractory is an important feature of a progenitor population such as the cap mesenchyme that differentiates to form new nephrons but must also be carried forward to give rise to the next round of nephrons. Progressive differentiation through nephron progenitor compartments provides a mechanism that balances renewal and differentiation, preventing NPCs from premature depletion and protecting the organism from low nephron endowment. This essential balance

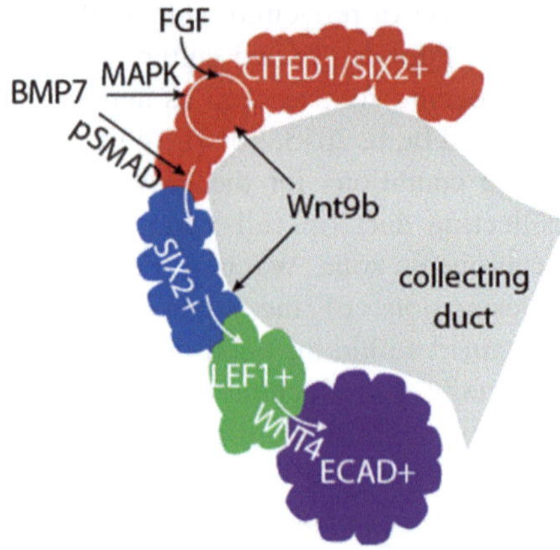

Fig. 6.3 Schematic of cap mesenchyme and collecting duct tip. Compartments of the cap mesenchyme and the predominant signaling pathways that regulate them are shown. Adapted from Brown et al. (2013)

between self-renewal and differentiation appears largely to be governed by the interplay between BMP, FGF, and WNT signaling pathways.

The least differentiated compartment of NPCs is marked by the transcriptional co-activator CITED1 and the transcription factor SIX2 (Boyle et al. 2008). CITED1 +/SIX2+ progenitors require FGF9/20 signaling for proliferation and maintenance, and loss of FGF9/20 leads to their premature differentiation and apoptosis (Brown et al. 2011a, b; Barak et al. 2012). Chemical inhibitor studies using cultured nephrogenic zone cells have shown that FGF-promoted renewal is most likely mediated through the RAS/PI3K signaling pathway (Brown et al. 2011a, b). Inhibitor studies on organ explants support the central role of PI3K signaling in NPC self-renewal and indicate that it acts as a convergence point for multiple signaling pathways including BMP and β-catenin (Lindström et al. 2015). While CITED1 is dispensable for function of the cap mesenchyme (Boyle et al. 2007), SIX2 cooperates with collecting duct derived WNT9b to promote proliferation of undifferentiated NPCs and also prevents their differentiation by limiting inductive WNT/β-catenin activity (Karner et al. 2011; Park et al. 2012; Self et al. 2006). The CITED1 compartment can be further subdivided based upon the expression of the transcription factors *Meox1* and *Dpf3* (Mugford et al. 2009), which are coexpressed in a domain directly adjacent to the CITED1-/SIX2+ compartment and thus appear primed for loss of CITED1 expression. While the function of *Meox1* and *Dpf3* is unknown, their expression appears to be co-regulated with *Cited1* through FGF signaling (Brown et al. 2011a, b).

BMP signaling has a bimodal function in the cap mesenchyme, stimulating proliferation of CITED1+/SIX2+ cells through activation of the TAK1/JNK map kinase pathway and promoting transition to the more differentiated CITED1-/SIX2 + compartment through SMAD activation (Blank et al. 2009; Brown et al. 2013; Muthukrishnan et al. 2015). NPCs that transition to the CITED1-/SIX2+ compartment display reduced SIX2 expression and become susceptible to ureteric bud derived WNT9b, which activates β-catenin signaling, causing transition into a loosely associated structure called the pretubular aggregate (PTA) in which SIX2 expression is strongly downregulated and the transcription factor LEF1 is expressed (Mugford et al. 2009; Park et al. 2012; Carroll et al. 2005). Primary cell studies suggest that the switch from JNK to SMAD signaling in response to BMP7 is required for NPCs to become susceptible to the differentiating effect of WNT/β-catenin (Brown et al. 2013). Molecular mechanisms that determine the balance between BMP/JNK and BMP/SMAD activation by BMP7 in NPCs remain undefined, but one interesting possibility is that the balance may be controlled by other cells in the nephrogenic zone that interact with the CITED1+/SIX2+/Meox1+ compartment. For example, nephrogenic zone interstitial cell expression of the transcription factor *Foxd1* and the protocadherin Fat4 regulate the capacity of NPCs to differentiate to the pretubular aggregate stage (Hatini et al. 1996; Das et al. 2013; Bagherie-Lachidan et al. 2015; Mao et al. 2015). Finally, WNT9b/β-catenin-induced expression of WNT4 in the LEF1-expressing compartment promotes epithelialization of the PTA to form a renal vesicle through noncanonical JNK and calcium-dependent WNT pathways (Kispert et al. 1998, Osafune et al. 2006, Tanigawa et al. 2011; Burn et al. 2011).

6.5 Recapitulation of the Signaling Niche for NPC Propagation in Culture

Our sophisticated knowledge of the signaling events that regulate self-renewal versus differentiation of NPCs has facilitated development of an in vitro niche for their undifferentiated propagation (Brown et al. 2015). Simultaneous modulation of BMP, FGF, WNT, and Rho kinase pathways with recombinant protein ligands and small molecules allows CITED1+/SIX2+ NPCs to be passaged and expanded over one hundred thousand fold in serum free monolayer culture, while still retaining the ability to differentiate into tubular segments of the nephron. As predicted, modulation of the BMP signaling pathway is essential for propagation of NPCs; in the absence of the BMP/SMAD inhibitor LDN-193189, NPCs signal through phospho-SMAD1/5 and lose expression of undifferentiated NPC markers and competency to undergo differentiation. In the absence of BMP ligand, NPCs failed to expand, indicating that the TAK1 branch of the BMP signaling pathway is necessary for NPC self-renewal and expansion in culture. While FGF9 appears to be the optimal FGF ligand for NPC maintenance of the undifferentiated state, FGF1 and 2 can also be used as substitutes in progenitor expansion medium, in agreement with previous studies that show these two ligands promote CITED1 and SIX2 expression to similar degree (Brown et al. 2011a, b). Interestingly, FGF20, a known natural ligand for NPC maintenance, does not cause a strong proliferative response in cultured NPCs alone (unpublished observation), which may be related to the quality of the recombinant protein itself, or the fact that FGF20 works in conjunction with FGF9 in vivo. Because of difficulties in obtaining recombinant Wnt protein, the WNT/β-catenin pathway is activated using the GSK3β inhibitor CHIR99021.

Titration of this inhibitor down to concentrations at which it mimics the low level of WNT/β-catenin seen in renewing cells of the cap mesenchyme in vivo is essential for in vitro propagation of NPCs (Karner et al. 2011). Interestingly, Rho kinase inhibition is necessary to maintain NPC identity as well as for cell survival during passage. The basis for this effect may be that Rho signaling is required for the nuclear activities of the Hippo transcription factors Yap and Taz that are known to promote stem cell renewal (Ohgushi et al. 2015). Alternately, the requirement for Rho kinase inhibition could simply stem from the cytoskeletal adaptations that NPCs must undergo in order to renew on a tissue culture surface (Watanabe et al. 2007).

NPCs from embryonic day 13.5 (E13.5) to postnatal day 1 (P1) can be propagated using the in vitro niche described by Brown et al (Brown et al. 2015). However, NPCs derived from embryonic stages prior to establishment of the nephrogenic zone (E11.5) display poor capacity for expansion in these conditions. The signaling requirements for these early NPCs appears to differ significantly, and several groups have taken advantage of aggregate culture of whole metanephric mesenchymes to expand progenitors from E11.5 mouse and E13.5 rat kidneys (Tanigawa et al. 2015; Yuri et al. 2015). While modulation of the Notch signaling pathway in cultured NPCs isolated from E13.5-P1 showed negligible effects on propagation, consistent with loss of function phenotypes in which effects are limited to NPC differentiation and determination of proximal versus distal tubule fate, inhibiting Notch appears to aid in the expansion of E11.5 NPCs (Bonegio et al. 2011; Cheng et al. 2007; Brown et al. 2015). Differentiated cells within metanephric mesenchyme cultures interfere with NPC renewal, an effect that can be abrogated by addition of the Notch inhibitor DAPT (Yuri et al. 2015). Notch inhibition also promotes renewal of purified E11.5 NPCs, suggesting that there may also be cell autonomous Notch signaling in NPCs at this early developmental stage (Brown et al. 2015). Studies using a serum free medium containing FGF2, TGFα, LIF, and the ROCK inhibitor Y-27632 demonstrate the preferential expansion of SIX2+ NPCs isolated from E11.5 mouse and E13.5 rat metanephric mesenchymes grown in aggregate culture (Tanigawa et al. 2015). In this study, Y-27632 was shown to attenuate LIF-induced MET and maintain the undifferentiated state of NPCs through activation of YAP.

6.5.1 NPCs Derived from Pluripotent Stem Cells

A number of groups have succeeded in differentiating human pluripotent stem cells (PSCs) through the intermediate mesoderm lineage to NPCs. Interestingly, in these differentiation schemes, many of the same signaling pathways that control NPC proliferation and renewal during embryonic development in mice are manipulated to promote differentiation of human NPCs from pluripotent stem cells, including FGF, BMP and WNT (Mae et al. 2013; Lam et al. 2014; Taguchi et al. 2014; Takasato et al. 2014, 2015; Morizane et al. 2015). One experimental hurdle that remains is to obtain human pluripotent cell-derived NPCs in large quantities for both biochemical and tissue engineering studies. Data on the signaling pathways necessary for expansion of these human NPCs in their undifferentiated state is limited, and progress toward human NPC expansion has been hampered by the lack of a well-defined surface marker for their isolation and enrichment. Using the in vitro niche for NPC propagation described above, it is possible to culture PSC-derived NPCs, but contaminating non-NPCs that arise during the differentiation process rapidly take over these cultures (Brown et al. 2015). A recent report identifies integrin α8 as a surface marker using which NPCs can be isolated from human fetal kidneys, opening up the possibility of purifying PSC-derived NPCs by fluorescent sorting in order to formulate conditions for their propagation (O'Brien et al. 2016).

6.6 Collecting Duct Cells

Formation of the collecting duct starts with the outgrowth of the ureteric bud from the nephric (Wolffian) duct within the intermediate mesoderm. Ingrowth to the metanephric mesenchyme and the establishment of the reciprocal signaling interchange between collecting ducts and cap mesenchyme ensure that a program of branching is maintained throughout the course of metanephric development. The tubular epithelium of the collecting duct can be divided into trunk and tips, and it is the tips that are located within the nephrogenic zone. In the basic signaling axis that drives collecting duct growth and branching, GDNF expressed by the cap mesenchyme activates the Ret receptor complex, which is expressed at collecting duct tips (Moore et al. 1996; Pichel et al. 1996; Sánchez et al. 1996; Pachnis et al. 1993; Sainio et al. 1997; Schuchardt et al. 1994; Vega et al. 1996). The expression of Ret is limited to tips, and the process of collecting duct growth and branching is determined in great part by signaling interactions within the nephrogenic zone. In fact, the tip consists of a self-renewing progenitor population, the daughter cells of which are either retained in the tip for ongoing outward growth or deposited in the trunk (Riccio et al. 2016).

Far more is known about signaling that controls ureteric bud outgrowth than the signaling within the nephrogenic zone that determines ongoing branching and growth throughout development. However, animals with reduced rather than abrogated *Gdnf* and *Ret* expression form hypomorphic kidneys, indicating reduced branching following ureteric bud formation and ingrowth into the metanephric mesenchyme (Pichel et al. 1996; de Graaff et al. 2001). Also, organ culture studies of collecting duct branching in isolation from cap mesenchyme reveal a dependence on GDNF addition to the culture medium for iterative branching (Qiao et al. 1999). Based on these findings, it seems clear that the GDNF-Ret axis functions within the nephrogenic zone throughout development (Fig. 6.4). It also seems likely that other signaling pathways known to control ureteric bud formation also regulate growth and branching in the nephrogenic zone throughout kidney development. Studies of

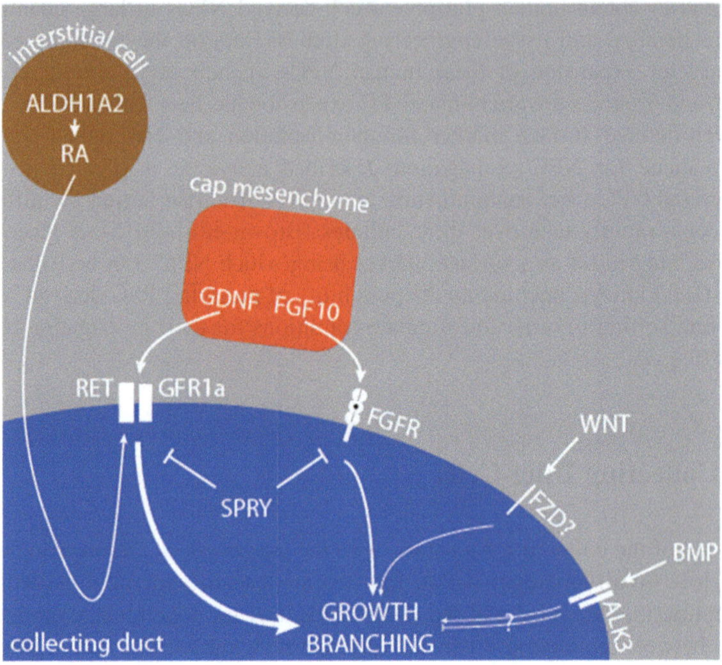

Fig. 6.4 Signals from cap mesenchyme and interstitial cells that regulate collecting duct growth and branching

the receptor tyrosine kinase inhibitor Sprouty 1 have led to intriguing findings regarding signaling redundancy in the GDNF-Ret pathway; compound genetic inactivation experiments reveal that *Gdnf* and *Ret* are dispensable for ureteric bud outgrowth in the *Spry1* null background (Basson et al. 2005; Michos et al. 2010). The redundant signal that activates receptor tyrosine kinase signaling in the collecting duct tip of the *Gdnf/Spry1* double mutant is FGF10, which is produced by the mesenchyme. *Fgf10* remains expressed in the cap mesenchyme throughout nephrogenesis, and one can extrapolate that signaling through the FGF10:FGFR pathway acts redundantly with GDNF:Ret through the course of collecting duct morphogenesis. The retinoic acid signaling pathway is also essential for ureteric bud outgrowth and likely operates later in collecting duct development. Reduction of retinoic acid signaling results in reduced Ret expression and associated impairment of ureteric bud outgrowth, which can be overcome by forced expression of Ret in the collecting duct (Mendelsohn et al. 1999; Batourina et al. 2001). This effect can be mimicked by expression of a dominant negative retinoic acid receptor in the collecting duct, indicating that retinoic acid signaling acts directly on collecting duct tips to maintain Ret expression (Rosselot et al. 2010). The source of the retinoic acid is predicted to be the nephrogenic zone interstitial cells, which express the retinoic acid processing enzyme retinaldehyde dehydrogenase 2 (*Aldh1A2*).

This highlights the integration and interdependence of the three major cell types within the nephrogenic zone, the collecting duct, nephrogenic zone interstitium, and cap mesenchyme.

WNT and BMP signaling are also thought to be involved in collecting duct growth and branching, although in both cases it has been difficult to assign growth factor-receptor relationships because of extensive redundancy of expression. Conditional inactivation of β-catenin specifically in the ureteric bud/collecting duct lineage interrupts growth and branching of the collecting duct already at E11.5 in the mouse, leading to aplasia or severe hypoplasia (Bridgewater et al. 2008; Marose et al. 2008). Although it has not been confirmed that this requirement for β-catenin in cells of the collecting duct is maintained throughout nephrogenesis, in vivo reporters for the β-catenin-associated Tcf/Lef transcription factors show that signaling is strong from the beginning of metanephric development through to the postnatal period (Iglesias et al. 2007; Bridgewater et al. 2008). Several ligands that are known to activate the β-catenin signaling pathway are expressed in or immediately adjacent to the nephrogenic zone, including Wnt7b and Wnt9b which are expressed in the collecting duct itself (Carroll et al. 2005; Yu et al. 2009). Conditional inactivation of the BMP receptor Alk3 in the ureteric bud/collecting duct results in a reduction of collecting duct growth and branching (Hartwig et al. 2008). Interestingly, analysis of these kidneys in culture reveals that the lack of BMP receptor causes an increase in branching early in kidney development, but a reduction later in development. The molecular basis for this biphasic requirement for BMP signaling in collecting duct morphogenesis is not known. The identification of ligands responsible for these effects is complicated by redundant expression; *Bmp7* is expressed in both the collecting duct and cap mesenchyme, whereas *Bmp2* is expressed in the pretubular aggregate, and *Bmp3* is expressed in the kidney capsule (Dudley and Robertson 1997).

6.7 Vascular Endothelial Progenitor Cells

The kidney is highly vascularized and receives approximately 20% of the cardiac output in the adult. Renal function relies on intimate integration of blood vessels and nephrons, which is established during the embryonic period. Both the vasculature of the glomerulus and the capillary network that feeds the tubular nephron are morphogenetically highly complex. Kidneys transferred from embryos at a developmental stage preceding the first overt signs of vascularization to an adult host are vascularized without ingrowth of host-derived endothelial cells, indicating that renal vasculature arises largely from intrinsic endothelial progenitors (Robert et al. 1996; Sequeira Lopez et al. 2001). These C-kit+/SCL+ endothelial progenitors are interspersed with *Foxd1*+ interstitial progenitors throughout the nephrogenic zone (Dekel et al. 2004; Schmidt-Ott et al. 2006; Hu et al. 2015). The angiogenic factor VegfA acts chemotactically on endothelial cells of the kidney, ensuring their directional migration (Tufró 2000). Vegf is expressed in cap mesenchyme under the control of the transcription factor WT1, which provides an explanation for the placement of endothelial progenitor cells around the cap mesenchyme (Gao et al. 2005). This position is important because endothelial cell invasion of the nascent nephron that develops from aggregated cap mesenchyme cells is essential for formation of the glomerular capillary. Strictly speaking, this invasion process occurs immediately outside the nephrogenic zone, but the range of the chemotactic signal emanating from the newly formed nephron may extend into the nephrogenic zone. Once the cap mesenchyme has aggregated and undertaken its mesenchyme-to-epithelium transition, it forms a polarized S-shaped body with a vascular cleft at the proximal pole. This is the site of podocyte differentiation, and these early podocytes express *Vegf* very strongly (Robert et al. 2000). Conditional inactivation of Vegf in early podocytes leads to a marked decrease in endothelial cell recruitment and loss of formation of glomerular vasculature (Eremina et al. 2003). Thus, VEGF secretion by podocytes is responsible for recruitment of endothelial cells into the forming glomerulus. Endothelial cells in their turn recruit mesangial cells at this very early stage in glomerulus formation through expression of PDGFB, a process that is described in the section on *Foxd1*+ interstitial cells.

6.8 Nephrogenic Zone Interstitial Cells

The interstitial cell compartment, which is otherwise referred to as the stroma, consists of all cell types except cap mesenchyme, collecting duct epithelium, and endothelial cells. In the nephrogenic zone, this includes the *Foxd1*-expressing progenitor for vascular mural cells and fibroblasts as well as macrophages.

6.8.1 The Progenitor for Vascular Mural Cells and Fibroblasts

The majority of the cells between the cap mesenchyme and the capsule and between cap mesenchymes of distinct tips are multipotential progenitors expressing the transcription factor FOXD1. These cells can differentiate to a number of different cell types including the pericytes that line capillaries, vascular smooth muscle cells that line larger vessels, renin-producing cells, mesangial cells of the glomerulus, and interstitial fibroblasts of the kidney, which encompass the erythropoietin-producing cells (Kobayashi et al. 2014, 2016; Sequeira-Lopez et al. 2015). Interestingly, there appear to be distinct origins of this multipotent progenitor population in different regions of the kidney. The majority of these cells arise from an *Osr1*-expressing intermediate mesoderm progenitor which is bipotential for *Six2*+ NPCs and *Foxd1*+ interstitial cells (Mugford et al. 2008). However, on the medial aspect of the organ which surrounds the entry point of the ureter, these cells are derived from a distinct lineage of *Tbx18*+ intermediate mesoderm cells whose main function is to give rise to the mesenchyme of the ureter (Bohnenpoll et al. 2013). Although they differ in their ontogenies, nephrogenic zone interstitial cells of these distinct origins appear to follow the same differentiation pathways, and it seems most likely that cell signaling processes driving these fates are conserved between the two. We know little about which signaling pathways are responsible for maintenance/proliferation of nephrogenic zone interstitial cells. In contrast to the cap mesenchyme, which we know depends on multiple growth factor pathways for its maintenance, studies in interstitial progenitor cells have not yielded any clear results. Perhaps the best candidates pathways for growth control of this population are FGF, WNT/β-catenin, Notch, and BMP; because they are essential for maintenance of the neighboring cap mesenchyme, we assume that pathway ligands are available in the niche. Surprisingly, neither inactivation of β-catenin, compound inactivation of FGF receptors, or inactivation of numerous components of the Notch signaling pathway using the *Foxd1cre* deleter strain results in a noticeable reduction in abundance of these cells (Yu et al. 2009; Boivin et al. 2015; Boyle et al. 2014; Walker et al. 2016). Investigations into the role of BMP and other signaling pathways aimed at understanding if growth factor signaling indeed is essential for maintenance of this cell population are ongoing.

It has been clear for some time that the interstitial cell population of the nephrogenic zone provides important signals for neighboring cells; inactivation of the gene encoding the FOXD1 transcription factor resulted in accumulation of cap mesenchyme cells with a lack of nephron differentiation (Hatini et al. 1996). This suggested that there may be signaling directly from interstitial cells to cap mesenchyme cells. In one informative approach to try to resolve this question, investigators have investigated the consequences of removing these cells by diphtheria toxin-mediated ablation (Das et al. 2013; Hum et al. 2014). Cap mesenchymes are enlarged, and there is a paucity of nephron differentiation in kidneys lacking the *Foxd1*+ interstitial cell population, similar to the *Foxd1* null. NPCs are blocked in their differentiation, resulting in accumulation of cells in the cap mesenchyme compartment. A complex mechanism involving components of the Hippo pathway has been proposed to regulate cross-talk between *Foxd1*+ interstitial cells and NPCs; Fat4 signaling from interstitial cells causes exclusion of the Yap transcription factor from the nucleus. Yap modulates the transcriptional response to Wnt9b from the collecting duct and when Yap is nuclear, the Wnt9b/β-catenin response activates targets that are required to maintain the cell in its progenitor state. However, when Yap is excluded from the nucleus by Fat4 signaling, Wnt9b/β-catenin activates targets that promote NPC differentiation (Das et al. 2013). Thus, in mice lacking *Foxd1*+ interstitial cells, Yap is constitutively nuclear in the NPC, preventing it from differentiating. This mechanism for interstitial control of NPC differentiation has been challenged by a series of compound genetic studies (Bagherie-Lachidan et al. 2015; Mao et al. 2015). These studies confirm the essential role that Fat4 expression in *Foxd1*+ interstitial cells plays in controlling NPC differentiation, and reveals Dachsous 1 (Dchs1) as the cognate partner for Fat4 in cells of the cap mesenchyme, but show that compound inactivation of Fat4 and

Yap in cap mesenchyme does not affect NPC differentiation. Collectively, these studies reveal an essential role for the interstitial cell population of the nephrogenic zone in controlling the balance between renewal and differentiation of the cap mesenchyme through Fat4–Dchs1 interaction, but the molecular mechanism underlying this remains a work in progress.

In addition to the Fat4–Dchs1 axis, *Foxd1+* interstitial cells also modulate the extracellular signaling environment for NPCs. A screen for genes controlled by the FOXD1 transcription factor uncovered the small leucine rich protein decorin (*Dcn*) as a target (Fetting et al. 2014). FOXD1 acts as a repressor, and nephrogenic zones of kidneys from *Foxd1* null mice display abnormal accumulation of DCN protein. DCN inhibits BMP/Smad signaling, thus reducing a pro-differentiation signal to NPCs, providing an explanation for the accumulation of NPCs seen in the *Foxd1* null. Compound inactivation of *Foxd1* and *Dcn* results in partial rescue of the *Foxd1* null phenotype, showing that *Dcn* plays an essential role but also indicating that additional factors contribute to the *Foxd1* null kidney phenotype. Interestingly, *Dcn* is also under the control of the transcription factor SALL1 which is expressed in interstitial cells of the nephrogenic zone. Inactivation of *Sall1* in *Foxd1+* interstitial cells results in a similar phenotype to the *Foxd1* null with cap mesenchyme expansion and elevated *Dcn* expression in the nephrogenic zone (Ohmori et al. 2015). The *Foxd1+* interstitium thus plays an essential role in regulating the extracellular signaling environment for the cap mesenchyme (Fig. 6.5).

Pathways that control differentiation of *Foxd1+* interstitial cells themselves are partially understood, although enough knowledge is not yet available to comprehensively understand how cells choose to commit to one of the many differentiation pathways available to them. Signaling events controlling mesangial differentiation

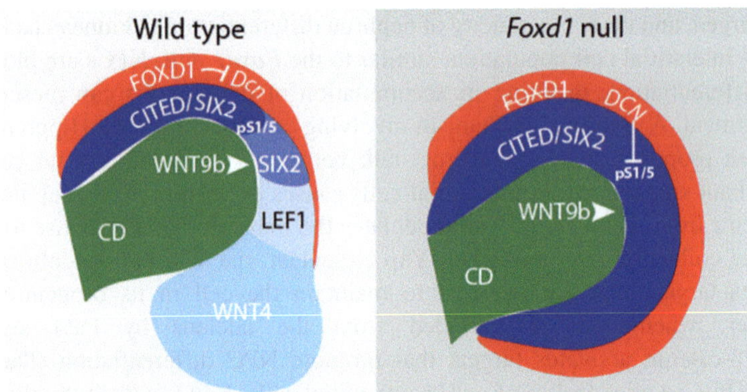

Fig. 6.5 Schematic showing signaling interactions in the cap mesenchyme that are controlled by Foxd1 in nephrogenic zone interstitial cells. Decorin (DCN) acts as an antagonist of BMP/SMAD signaling and is normally repressed by the FOXD1 transcription factor in the nephrogenic zone interstitium. Following inactivation of *Foxd1*, DCN antagonism of SMAD activation by BMP7 (pS1/5) prevents progression of cap mesenchyme cells to a state in which they are susceptible for epithelial induction by Wnt9b leading to accumulation of undifferentiated cap mesenchyme. Adapted from Fetting et al. (2014)

have been the most actively studied, although progress has been hampered by the lack of mesangial-specific markers and mesangial-specific cre deleter strains. Development of the *Foxd1cre* strain facilitated systematic study of pathways that control these cells, but because it is relatively new a limited number of studies have so far been conducted (Kobayashi et al. 2014). Inactivation of the genes encoding PDGFB or the PDGF receptor β results in failure of mesangial investment in newly formed glomeruli (Levéen et al. 1994; Soriano 1994). PDGFB expressed by differentiating endothelial cells signals through PDGFRβ expressed on cells of the *Foxd1*+ interstitial cell lineage, providing a recruitment signal (Lindahl et al. 1998). Notch signaling is also involved in the decision of *Foxd1*+ cells to take on the mesangial fate; inactivation of the gene encoding the RBPjk transcription factor that mediates Notch signaling with *Foxd1cre* specifically prevents entry into the mesangial lineage but does not affect the capacity of cells to contribute to pericyte or vascular smooth muscle fates (Boyle et al. 2014; Lin et al. 2014). The Notch and PDGFB pathways appear to be interconnected in specification of mesangial cells; PDGFRβ is expressed at a low level in *Foxd1*+ cells within the nephrogenic zone, but PDGFRβ expression increases in those cells neighboring the vascular clefts into which they are to be recruited as mesangial cells. The switch from low- to high-level PDGFRβ expression is regulated by the transcription factor PBX1 which represses *Pdgfrβ* in *Foxd1*+ cells and is downregulated as they differentiate (Hurtado et al. 2015). The elevation of PDGFRβ expression that preempts recruitment of mesangial cells does not occur in the *Foxd1cre*; *RBPjk* conditional mutant. A pathway for the differentiation of mesangial cells from their *Foxd1*+ progenitors appears to be emerging, although many questions remain to be answered. For example, is *Pbx1* a target of Notch signaling, and what exactly is the function of PDGFB signaling in mesangial recruitment?

6.8.2 Macrophages

In the developing embryo, yolk sac-derived macrophages/phagocytic cells infiltrate all solid organs and become tissue resident macrophages. They play vital roles during organogenesis including phagocytic clearance of dead cells associated with morphogenesis and tissue remodeling, production of trophic factors that stimulate growth and differentiation, and promotion of angiogenesis. Macrophages are also known to promote development and branching morphogenesis of mammary gland, lung, pancreas, and kidney (Pollard and Hennighausen 1994; Jones et al. 2013; Banaei-Bouchareb et al. 2004; Rae et al. 2007). In addition to resident macrophages, monocytes derived from the circulation traffic to developing organs. As development proceeds, hematopoiesis and production of macrophages shift from the yolk sac to the liver, and subsequently to the bone marrow. Bone marrow-derived monocytes infiltrate tissues and differentiate into resident macrophages. Thus, solid organs contain both yolk sac- and bone marrow-derived resident macrophages that persist throughout life. Mice lacking macrophage colony stimulating factor 1 (*Csf1*) exhibit numerous developmental defects in the skeleton, brain, and lung and show impaired growth and fertility (Marks and Lane 1976; Michaelson et al. 1996; Cohen et al. 1996; Pollard et al. 1991; Harris et al. 2012). CSF-1 signals through its high affinity tyrosine-kinase receptor CSF1R to control survival, proliferation, and differentiation of macrophages, and targeted inactivation of *Csf1r* recapitulates the *Csf1* loss of function phenotype in mice (Dai et al. 2002). Transgenic mice expressing EGFP driven by the *Csf1r* promoter revealed the presence of macrophages around E11.5-E12.5 in the embryonic kidney (Rae et al. 2007). At E11.5, macrophages are localized near the vascular bed between the metanephric mesenchyme and metanephric duct. During later stages of kidney development, they reside between the interstitial cells surrounding the UB and CM in the nephrogenic zone and in close apposition to the developing nephrons. Their trophic role and functional significance to embryonic kidney development is not well understood, although reports suggest that they may provide important developmental cues. Organ culture experiments revealed that addition of CSF-1 to E11.5 kidney explants promotes ureteric bud growth and branching and nephron formation (Rae et al. 2007), and administration of CSF-1 to neonatal mice promotes kidney growth (Alikhan et al. 2011). The finding that *Csf1* null mice do not have kidney defects argues against a role for CSF-1 signaling in normal kidney development (Stanley et al. 1994; Wiktor-Jedrzejczak et al. 1994). However, the possibility that macrophage secretion of other trophic factors such as PDGFs, TGFβs, and WNTs may be involved in regulation of kidney development remains to be explored.

6.8.3 Growth Factor Signaling Integration

Reciprocal signaling interaction between the distinct cellular compartments of the nephrogenic zone is essential for normal nephron endowment of the adult kidney. As summarized above, we know the identities of many of the growth factors and cytokines essential in this process. However, we are only really beginning to understand how these signaling pathways integrate at the molecular level. BMP7 and FGF9 signaling pathways interact to promote self-renewal of NPCs; while both *Bmp7* and *Fgf9* are expressed in cap mesenchyme and collecting ducts, conditional inactivation of *Fgf9* in the collecting duct reveals that it is essential for cap mesenchyme maintenance (Barak et al. 2012). Cooperative regulation of AP-1 transcriptional activity has emerged as a key intersection between BMP7 and FGF9 pathways, partially explaining their synergistic regulation of NPC maintenance and renewal. The mechanism involves phosphorylation and activation of the AP-1 transcription factors JUN and FOS distinctly by BMP7 and FGF9 respectively, resulting in collaborative regulation of AP-1 transcriptional activity and activation of downstream targets involved in cell cycle and proliferation (Fig. 6.6) (Muthukrishnan et al. 2015). A number of well-characterized AP-1 transcriptional targets are key regulators of the cell cycle, apoptosis, and proliferation including

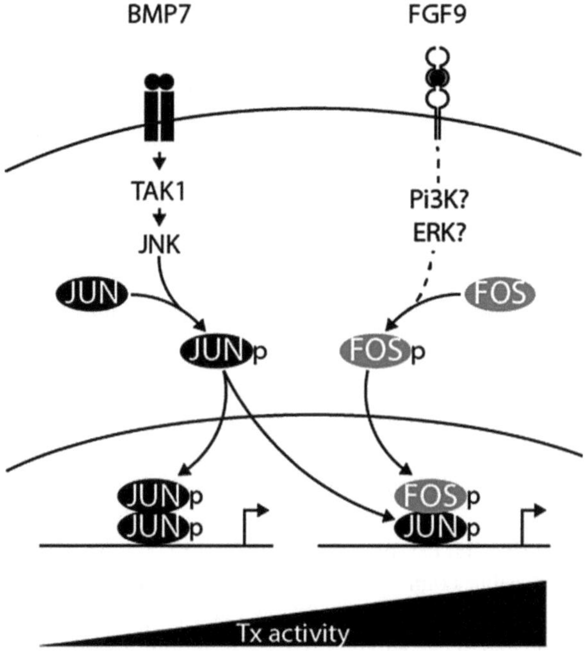

Fig. 6.6 Model for interaction between BMP and FGF signaling in maintenance of the nephron progenitor cell. BMP signaling activates the JUN transcription factor while FGF9 activates FOS. Although the AP1 transcription factor that promotes proliferation of these cells can be assembled from JUN:JUN homodimers, the transcriptional activity of the JUN:FOS heterodimer is stronger. AP1 activates a number of cell cycle regulators

Ccnd1, *p53*, *Myc*, *Bim*, *Bcl-2*, *p21* (Shaulian and Karin 2001). Of these targets, *Bcl-2*, *Myc*, and *p53* have already been shown to be essential for NPC survival, renewal, and differentiation and normal kidney development (Sorenson et al. 1995; Couillard and Trudel 2009; Saifudeen et al. 2012).

In other cellular contexts, WNT/β-catenin signaling has been shown to regulate *Jun* and AP-1 target genes such as *Myc, Ccnd1*, and *Ccnd2* (Nateri et al. 2005; Yochum et al. 2008; Toualbi et al. 2007). WNT9b/β-catenin signaling could also converge with BMP7 and FGF9 to govern AP-1 transcription in NPCs. Understanding how AP-1 function is cooperatively controlled by these distinct NPC renewal pathways will be important for integration of the growth factor signaling inputs governing nephron formation. Although the importance of BMP7-MAPK and WNT9b/β-catenin in NPCs has been established in vivo, intracellular signal transduction mechanisms utilized by FGFs 9/20 remains to be elucidated. In vitro findings suggest that RAS and PI3K pathways are essential for maintenance of NPCs (Brown et al. 2011a, b). A recent study reported that PI3K has dose-dependent effects on NPC renewal and differentiation through regulation of β-catenin signaling (Lindström et al. 2015). However, the function of the ERK/MAPK pathway is unknown. Future studies should aim to generate conditional mutants of PI3K and ERK pathway components to elucidate their role in regulation of AP-1 transcription in NPCs. Interestingly, YAP expression is regulated by BMP7-SMAD signaling in embryonic neural stem cells, and YAP phosphorylation is regulated by JNK-JUN in MEFs (Danovi et al. 2008; Tomlinson et al. 2010; Yao et al. 2014). YAP interacts with both SMAD and AP-1 transcription factors, and the intersection of FAT4-DCHS1/2 signaling with both BMP7-MAPK and BMP7-SMAD is an interesting possibility (Estaras et al. 2015; Zanconato et al. 2015). Mapping the intracellular signal transduction mechanisms utilized by these distinct pathways will be required to build a unified model of NPC renewal and differentiation.

6.9 Changes in the Nephrogenic Zone Associated with Cessation of Nephrogenesis

The formation of new nephrons ceases on approximately the second day after birth in the mouse. Morphologically, this arrest of nephrogenesis is associated with a loss of *Cited1*+ cells in a final wave of differentiation (Hartman et al. 2007). Although nephrogenesis ends abruptly, there is a gradual change in the morphology of the nephrogenic zone preceding cessation, with a reduction in size of both collecting duct tips and cap mesenchymes over the course of several days (Short et al. 2014). A possible connection between the gradual thinning of the cap mesenchyme and cessation of nephrogenesis has been proposed; theoretical modeling predicts that once cap mesenchyme size falls below a certain threshold, the final wave of differentiation and cessation will ensue (Zubkov et al. 2015). Alterations in key signaling pathways responsible for cap mesenchyme maintenance have been associated with these terminal alterations in the nephrogenic zone. BMP/SMAD signaling increases within the cap mesenchyme over the last three days of kidney development (Brown et al. 2015). The balance between MAPK and SMAD signaling downstream of BMP7 plays an important role in balancing renewal and differentiation of NPCs; increased SMAD signaling is predicted to increase the sensitivity of these cells to WNT/β-catenin mediated differentiation, thus depleting the cap mesenchyme more rapidly. Heterochronic engraftment studies in which postnatal cap mesenchyme cells were engrafted into cap mesenchymes of E12.5 kidneys revealed that the older cells are more prone to exit the cap mesenchyme and differentiate than the E12.5 cells (Chen et al. 2015). The aging cap mesenchyme thus appears fundamentally altered in its capacity for renewal, and engraftment studies with mutant kidneys indicates that reduced FGF expression may be associated with this age-related change. FGF9 and 20 are essential for the maintenance of the cap mesenchyme in its undifferentiated state, and a reduction in FGF20 is predicted to tip the balance in favor of differentiation and more rapid exit of cells from the cap mesenchyme. Whether there is a connection between elevated BMP/SMAD signaling and reduced FGF20 remains to be determined.

6.10 Conclusion

With this overview, we hope to have given an introduction to the vast number of signaling interactions within the nephrogenic zone. Much of this signaling is going on simultaneously in a relatively small space and one key feature that has not really been explored is how the different cells within this niche insulate themselves from unwanted signaling inputs. We are now starting to acquire the tools to study the different cell types within the nephrogenic zone in isolation, and we expect that many intriguing questions such as this will be answered within the coming years.

Acknowledgements This work was supported by National Institutes of Diabetes and Digestive and Kidney Disease.

Grant R01DK078161 (L.O.). Literature included in this chapter represents a selection of reports from this vigorous research field, and we apologize to the many colleagues whose work we could not cite due to space constraints.

References

Alikhan MA, Jones CV, Williams TM, Beckhouse AG, Fletcher AL, Kett MM, Sakkal S, Samuel CS, Ramsay RG, Deane JA, Wells CA, Little MH, Hume DA, Ricardo SD (2011) Colony-stimulating factor-1 promotes kidney growth and repair via alteration of macrophage responses. Am J Pathol 179(3):1243–1256. doi:10.1016/j.ajpath.2011.05.037 (Epub 2011 Jul 14. PubMed PMID: 21762674; PubMed Central PMCID: PMC3157188)

Bagherie-Lachidan M, Reginensi A, Pan Q, Zaveri HP, Scott DA, Blencowe BJ, Helmbacher F, McNeill H (2015) Stromal Fat4 acts non-autonomously with Dchs1/2 to restrict the nephron progenitor pool. Development 142(15):2564–2573. doi:10.1242/dev.122648 (Epub 2015 Jun 26. PubMed PMID: 26116661)

Baldelomar EJ, Charlton JR, Beeman SC, Hann BD, Cullen-McEwen L, Pearl VM, Bertram JF, Wu T, Zhang M, Bennett KM (2016) Phenotyping by magnetic resonance imaging nondestructively measures glomerular number and volume distribution in mice with and without nephron reduction. Kidney Int 89(2):498–505 (PubMed PMID: 26535998; PubMed Central PMCID: PMC4854807)

Banaei-Bouchareb L, Gouon-Evans V, Samara-Boustani D, Castellotti MC, Czernichow P, Pollard JW, Polak M (2004) Insulin cell mass is altered in Csf1op/Csf1op macrophage-deficient mice. J Leukoc Biol 76(2):359–367. doi:10.1189/jlb.1103591

Barak H, Huh SH, Chen S, Jeanpierre C, Martinovic J, Parisot M, Bole-Feysot C, Nitschké P, Salomon R, Antignac C, Ornitz DM, Kopan R (2012) FGF9 and FGF20 maintain the stemness of nephron progenitors in mice and man. Dev Cell 22(6):1191–1207. doi:10.1016/j.devcel.2012.04.018 (PubMed PMID: 22698282; PubMed Central PMCID: PMC3376351)

Basson MA, Akbulut S, Watson-Johnson J, Simon R, Carroll TJ, Shakya R, Gross I, Martin GR, Lufkin T, McMahon AP, Wilson PD, Costantini FD, Mason IJ, Licht JD (2005) Sprouty1 is a critical regulator of GDNF/RET-mediated kidney induction. Dev Cell 8(2):229–239 (PubMed PMID: 15691764)

Batourina E, Gim S, Bello N, Shy M, Clagett-Dame M, Srinivas S, Costantini F, Mendelsohn C (2001) Vitamin A controls epithelial/mesenchymal interactions through Ret expression. Nat Genet 27(1):74–78 (PubMed PMID: 11138002)

Bertram JF, Douglas-Denton RN, Diouf B, Hughson MD, Hoy WE (2011) Human nephron number: implications for health and disease. Pediatr Nephrol 26(9):1529–1533. doi:10.1007/s00467-011-1843-8 (Epub 2011 May 22. Review. PubMed PMID: 21604189)

Blank U, Brown A, Adams DC, Karolak MJ, Oxburgh L (2009) BMP7 promotes proliferation of nephron progenitor cells via a JNK-dependent mechanism. Development 136(21):3557–3566. doi:10.1242/dev.036335 (Epub 2009 Sep 30. PubMed PMID: 19793891; PubMed Central PMCID: PMC2761106)

Bohnenpoll T, Bettenhausen E, Weiss AC, Foik AB, Trowe MO, Blank P, Airik R, Kispert A (2013) Tbx18 expression demarcates multipotent precursor populations in the developing urogenital system but is exclusively required within the ureteric mesenchymal lineage to suppress a renal stromal fate. Dev Biol 380(1):25–36. doi:10.1016/j.ydbio.2013.04.036 (Epub 2013 May 15. PubMed PMID: 23685333)

Boivin FJ, Sarin S, Lim J, Javidan A, Svajger B, Khalili H, Bridgewater D (2015) Stromally expressed β-catenin modulates Wnt9b signaling in the ureteric epithelium. PLoS One 10(3): e0120347. doi:10.1371/journal.pone.0120347 (eCollection 2015. PubMed PMID: 25803581; PubMed Central PMCID: PMC4372213)

Bonegio RG, Beck LH, Kahlon RK, Lu W, Salant DJ (2011) The fate of Notch-deficient nephrogenic progenitor cells during metanephric kidney development. Kidney Int 79 (10):1099–1112. doi:10.1038/ki.2010.553 (Epub 2011 Jan 26. PubMed PMID: 21270765)

Boyle S, Shioda T, Perantoni AO, de Caestecker M (2007) Cited1 and Cited2 are differentially expressed in the developing kidney but are not required for nephrogenesis. Dev Dyn 236 (8):2321–2330 (PubMed PMID: 17615577)

Boyle S, Misfeldt A, Chandler KJ, Deal KK, Southard-Smith EM, Mortlock DP, Baldwin HS, de Caestecker M (2008) Fate mapping using Cited1-CreERT2 mice demonstrates that the cap mesenchyme contains self-renewing progenitor cells and gives rise exclusively to nephronic epithelia. Dev Biol 313(1):234–245 (Epub 2007 Oct 24. PubMed PMID: 18061157; PubMed Central PMCID: PMC2699557)

Boyle SC, Liu Z, Kopan R (2014) Notch signaling is required for the formation of mesangial cells from a stromal mesenchyme precursor during kidney development. Development 141 (2):346–354. doi:10.1242/dev.100271 (Epub 2013 Dec 18. PubMed PMID: 24353058; PubMed Central PMCID: PMC4074211)

Bridgewater D, Cox B, Cain J, Lau A, Athaide V, Gill PS, Kuure S, Sainio K, Rosenblum ND (2008) Canonical WNT/beta-catenin signaling is required for ureteric branching. Dev Biol 317 (1):83–94. doi:10.1016/j.ydbio.2008.02.010 (Epub 2008 Feb 21. PubMed PMID: 18358465)

Brown AC, Adams D, de Caestecker M, Yang X, Friesel R, Oxburgh L (2011a) FGF/EGF signaling regulates the renewal of early nephron progenitors during embryonic development. Development 138(23):5099–5112. doi:10.1242/dev.065995 (Epub 2011 Oct 26. PubMed PMID: 22031548; PubMed Central PMCID: PMC3210493)

Brown AC, Blank U, Adams DC, Karolak MJ, Fetting JL, Hill BL, Oxburgh L (2011b) Isolation and culture of cells from the nephrogenic zone of the embryonic mouse kidney. J Vis Exp (50). pii: 2555. doi:10.3791/2555 (PubMed PMID: 21540822; PubMed Central PMCID: PMC3169285)

Brown AC, Muthukrishnan SD, Guay JA, Adams DC, Schafer DA, Fetting JL, Oxburgh L (2013) Role for compartmentalization in nephron progenitor differentiation. Proc Natl Acad Sci USA 110(12):4640–4645. doi:10.1073/pnas.1213971110 (Epub 2013 Mar 4. PubMed PMID: 23487745; PubMed Central PMCID: PMC3607044)

Brown AC, Muthukrishnan SD, Oxburgh L (2015) A synthetic niche for nephron progenitor cells. Dev Cell 34(2):229–241. doi:10.1016/j.devcel.2015.06.021 (Epub 2015 Jul 16. PubMed PMID: 26190145; PubMed Central PMCID: PMC4519427)

Burn SF, Webb A, Berry RL, Davies JA, Ferrer-Vaquer A, Hadjantonakis AK, Hastie ND, Hohenstein P (2011) Calcium/NFAT signalling promotes early nephrogenesis. Dev Biol 352 (2):288–298. doi:10.1016/j.ydbio.2011.01.033

Carroll TJ, Park JS, Hayashi S, Majumdar A, McMahon AP (2005) Wnt9b plays a central role in the regulation of mesenchymal to epithelial transitions underlying organogenesis of the mammalian urogenital system. Dev Cell 9(2):283–292 (PubMed PMID: 16054034)
Chen S, Brunskill EW, Potter SS, Dexheimer PJ, Salomonis N, Aronow BJ, Hong CI, Zhang T, Kopan R (2015) Intrinsic age-dependent changes and cell-cell contacts regulate nephron progenitor lifespan. Dev Cell 35(1):49–62. doi:10.1016/j.devcel.2015.09.009 (PubMed PMID: 26460946; PubMed Central PMCID: PMC4615609)
Cheng HT, Kim M, Valerius MT, Surendran K, Schuster-Gossler K, Gossler A, McMahon AP, Kopan R (2007) Notch2, but not Notch1, is required for proximal fate acquisition in the mammalian nephron. Development 134(4):801–811 (Epub 2007 Jan 17. PubMed PMID: 17229764; PubMed Central PMCID: PMC2613851)
Cohen PE, Chisholm O, Arceci RJ, Stanley ER, Pollard JW (1996) Absence of colony-stimulating factor-1 in osteopetrotic (csfmop/csfmop) mice results in male fertility defects. Biol Reprod 55 (2):310–317
Couillard M, Trudel M (2009) C-myc as a modulator of renal stem/progenitor cell population. Dev Dyn 238(2):405–414. doi:10.1002/dvdy.21841
Dai XM, Ryan GR, Hapel AJ, Dominguez MG, Russell RG, Kapp S, Sylvestre V, Stanley ER (2002) Targeted disruption of the mouse colonystimulating factor 1 receptor gene results in osteopetrosis, mononuclear phagocyte deficiency, increased primitive progenitor cell frequencies, and reproductive defects. Blood 99(1):111–120
Danovi SA, Rossi M, Gudmundsdottir K, Yuan M, Melino G, Basu S (2008) Yes-associated protein (YAP) is a critical mediator of c-Jun-dependent apoptosis. Cell Death Differ 15 (1):217–219. doi:10.1038/sj.cdd.4402226
Cullen-McEwen LA, Kett MM, Dowling J, Anderson WP, Bertram JF (2003) Nephron number, renal function, and arterial pressure in aged GDNF heterozygous mice. Hypertension 41 (2):335–340 (PubMed PMID: 12574104)
Das A, Tanigawa S, Karner CM, Xin M, Lum L, Chen C, Olson EN, Perantoni AO, Carroll TJ (2013) Stromal-epithelial crosstalk regulates kidney progenitor cell differentiation. Nat Cell Biol 15(9):1035–1044. doi:10.1038/ncb2828 (Epub 2013 Aug 25. Erratum in: Nat Cell Biol. 2013 Oct;15(10):1260. PubMed PMID: 23974041; PubMed Central PMCID: PMC3891676)
de Graaff E, Srinivas S, Kilkenny C, D'Agati V, Mankoo BS, Costantini F, Pachnis V (2001) Differential activities of the RET tyrosine kinase receptor isoforms during mammalian embryogenesis. Genes Dev 15(18):2433–2444 (PubMed PMID: 11562352; PubMed Central PMCID: PMC312785)
Dekel B, Hochman E, Sanchez MJ, Maharshak N, Amariglio N, Green AR, Izraeli S (2004) Kidney, blood, and endothelium: developmental expression of stem cell leukemia during nephrogenesis. Kidney Int 65(4):1162–1169 (Erratum in: Kidney Int 2004 Jun;65(6):2465-6. PubMed PMID: 15086455)
Diep CQ, Ma D, Deo RC, Holm TM, Naylor RW, Arora N, Wingert RA, Bollig F, Djordjevic G, Lichman B, Zhu H, Ikenaga T, Ono F, Englert C, Cowan CA, Hukriede NA, Handin RI, Davidson AJ (2011) Identification of adult nephron progenitors capable of kidney regeneration in zebrafish. Nature 470(7332):95–100. doi:10.1038/nature09669 (Epub 2011 Jan 26. PubMed PMID: 21270795; PubMed Central PMCID: PMC3170921)
Diep CQ, Peng Z, Ukah TK, Kelly PM, Daigle RV, Davidson AJ (2015) Development of the zebrafish mesonephros. Genesis 53(3-4):257–269. doi:10.1002/dvg.22846 (Epub 2015 Mar 14. PubMed PMID: 25677367; PubMed Central PMCID: PMC4412820)
Dudley AT, Robertson EJ (1997) Overlapping expression domains of bone morphogenetic protein family members potentially account for limited tissue defects in BMP7 deficient embryos. Dev Dyn 208(3):349–362 (PubMed PMID: 9056639)
Eremina V, Sood M, Haigh J, Nagy A, Lajoie G, Ferrara N, Gerber HP, Kikkawa Y, Miner JH, Quaggin SE (2003) Glomerular-specific alterations of VEGF-A expression lead to distinct congenital and acquired renal diseases. J Clin Invest 111(5):707–716 (PubMed PMID: 12618525; PubMed Central PMCID: PMC151905)

Estaras C, Benner C, Jones KA (2015) SMADs and YAP compete to control elongation of beta-catenin:LEF-1-recruited RNAPII during hESC differentiation. Mol Cell 58(5):780–793. doi:10.1016/j.molcel.2015.04.001

Fetting JL, Guay JA, Karolak MJ, Iozzo RV, Adams DC, Maridas DE, Brown AC, Oxburgh L (2014) FOXD1 promotes nephron progenitor differentiation by repressing decorin in the embryonic kidney. Development 141(1):17–27. doi:10.1242/dev.089078 (Epub 2013 Nov 27. PubMed PMID: 24284212; PubMed Central PMCID: PMC3865747)

Freedman BS, Brooks CR, Lam AQ, Fu H, Morizane R, Agrawal V, Saad AF, Li MK, Hughes MR, Werff RV, Peters DT, Lu J, Baccei A, Siedlecki AM, Valerius MT, Musunuru K, McNagny KM, Steinman TI, Zhou J, Lerou PH, Bonventre JV (2015) Modelling kidney disease with CRISPR-mutant kidney organoids derived from human pluripotent epiblast spheroids. Nat Commun 6:8715. doi:10.1038/ncomms9715 (PubMed PMID: 26493500; PubMed Central PMCID: PMC4620584)

Gao X, Chen X, Taglienti M, Rumballe B, Little MH, Kreidberg JA (2005) Angioblast-mesenchyme induction of early kidney development is mediated by Wt1 and Vegfa. Development 132(24):5437–5449 (Epub 2005 Nov 16. PubMed PMID: 16291795)

Grobstein C (1953a) Morphogenetic interaction between embryonic mouse tissues separated by a membrane filter. Nature 172(4384):869–870 (PubMed PMID: 13111219)

Grobstein C (1953b) Inductive epitheliomesenchymal interaction in cultured organ rudiments of the mouse. Science 118(3054):52–55 (PubMed PMID: 13076182)

Grobstein C (1956) Trans-filter induction of tubules in mouse metanephrogenic mesenchyme. Exp Cell Res 10(2):424–440 (PubMed PMID: 13317909)

Harari-Steinberg O, Metsuyanim S, Omer D, Gnatek Y, Gershon R, Pri-Chen S, Ozdemir DD, Lerenthal Y, Noiman T, Ben-Hur H, Vaknin Z, Schneider DF, Aronow BJ, Goldstein RS, Hohenstein P, Dekel B (2013) Identification of human nephron progenitors capable of generation of kidney structures and functional repair of chronic renal disease. EMBO Mol Med 5 (10):1556–1568. doi:10.1002/emmm.201201584 (Epub 2013 Sep 2. PubMed PMID: 23996934; PubMed Central PMCID: PMC3799579)

Harris SE, MacDougall M, Horn D, Woodruff K, Zimmer SN, Rebel VI, Fajardo R, Feng JQ, Gluhak-Heinrich J, Harris MA, Abboud Werner S (2012) Meox2Cre-mediated disruption of CSF-1 leads to osteopetrosis and osteocyte defects. Bone 50(1):42–53. doi:10.1016/j.bone.2011.09.038

Hartman HA, Lai HL, Patterson LT (2007) Cessation of renal morphogenesis in mice. Dev Biol 310(2):379–387 (Epub 2007 Aug 16. PubMed PMID: 17826763; PubMed Central PMCID: PMC2075093)

Hartwig S, Bridgewater D, Di Giovanni V, Cain J, Mishina Y, Rosenblum ND (2008) BMP receptor ALK3 controls collecting system development. J Am Soc Nephrol 19(1):117–124. doi:10.1681/ASN.2007010080 (PubMed PMID: 18178801; PubMed Central PMCID: PMC2391036)

Hatini V, Huh SO, Herzlinger D, Soares VC, Lai E (1996) Essential role of stromal mesenchyme in kidney morphogenesis revealed by targeted disruption of Winged Helix transcription factor BF-2. Genes Dev 10(12):1467–1478 (PubMed PMID: 8666231)

Hu Y, Li M, Göthert JR, Gomez RA, Sequeira-Lopez ML (2015) Hemovascular progenitors in the kidney require sphingosine-1-phosphate receptor 1 for vascular development. J Am Soc Nephrol (pii: ASN.2015060610. [Epub ahead of print] PubMed PMID: 26534925)

Hum S, Rymer C, Schaefer C, Bushnell D, Sims-Lucas S (2014) Ablation of the renal stroma defines its critical role in nephron progenitor and vasculature patterning. PLoS One 9(2): e88400. doi:10.1371/journal.pone.0088400 (eCollection 2014. PubMed PMID: 24505489; PubMed Central PMCID: PMC3914987)

Hurtado R, Zewdu R, Mtui J, Liang C, Aho R, Kurylo C, Selleri L, Herzlinger D (2015) Pbx1-dependent control of VMC differentiation kinetics underlies gross renal vascular patterning. Development 142(15):2653–2664. doi:10.1242/dev.124776 (Epub 2015 Jul 2. PubMed PMID: 26138478; PubMed Central PMCID: PMC4529034)

Iglesias DM, Hueber PA, Chu L, Campbell R, Patenaude AM, Dziarmaga AJ, Quinlan J, Mohamed O, Dufort D, Goodyer PR (2007) Canonical WNT signaling during kidney development. Am J Phys Renal Phys 293(2):F494–F500 (Epub 2007 May 9. PubMed PMID: 17494089)

Jones CV, Williams TM, Walker KA, Dickinson H, Sakkal S, Rumballe BA, Little MH, Jenkin G, Ricardo SD (2013) M2 macrophage polarisation is associated with alveolar formation during postnatal lung development. Respir Res 14:41. doi:10.1186/1465-9921-14-41

Karner CM, Das A, Ma Z, Self M, Chen C, Lum L, Oliver G, Carroll TJ (2011) Canonical Wnt9b signaling balances progenitor cell expansion and differentiation during kidney development. Development 138(7):1247–1257. doi:10.1242/dev.057646 (Epub 2011 Feb 24. PubMed PMID: 21350016; PubMed Central PMCID: PMC3050658)

Kispert A, Vainio S, McMahon AP (1998) Wnt-4 is a mesenchymal signal for epithelial transformation of metanephric mesenchyme in the developing kidney. Development 125 (21):4225–4234 (PubMed PMID: 9753677)

Kobayashi A, Valerius MT, Mugford JW, Carroll TJ, Self M, Oliver G, McMahon AP (2008) Six2 defines and regulates a multipotent self-renewing nephron progenitor population throughout mammalian kidney development. Cell Stem Cell 3(2):169–181. doi:10.1016/j.stem.2008.05.020 (PubMed PMID: 18682239; PubMed Central PMCID: PMC2561900)

Kobayashi A, Mugford JW, Krautzberger AM, Naiman N, Liao J, McMahon AP (2014) Identification of a multipotent self-renewing stromal progenitor population during mammalian kidney organogenesis. Stem Cell Rep 3(4):650–662. doi:10.1016/j.stemcr.2014.08.008 (Epub 2014 Sep 18. PubMed PMID: 25358792; PubMed Central PMCID: PMC4223698)

Kobayashi H, Liu Q, Binns TC, Urrutia AA, Davidoff O, Kapitsinou PP, Pfaff AS, Olauson H, Wernerson A, Fogo AB, Fong GH, Gross KW, Haase VH (2016) Distinct subpopulations of FOXD1 stroma-derived cells regulate renal erythropoietin. J Clin Invest 126(5):1926–1938. doi:10.1172/JCI83551 (Epub 2016 Apr 18. PubMed PMID: 27088801; PubMed Central PMCID: PMC4855934)

Lam AQ, Freedman BS, Morizane R, Lerou PH, Valerius MT, Bonventre JV (2014) Rapid and efficient differentiation of human pluripotent stem cells into intermediate mesoderm that forms tubules expressing kidney proximal tubular markers. J Am Soc Nephrol 25(6):1211–1225. doi:10.1681/ASN.2013080831 (Epub 2013 Dec 19. PubMed PMID: 24357672; PubMed Central PMCID: PMC4033376)

Levéen P, Pekny M, Gebre-Medhin S, Swolin B, Larsson E, Betsholtz C (1994) Mice deficient for PDGF B show renal, cardiovascular, and hematological abnormalities. Genes Dev 8 (16):1875–1887 (PubMed PMID: 7958863)

Lin EE, Sequeira-Lopez ML, Gomez RA (2014) RBP-J in FOXD1+ renal stromal progenitors is crucial for the proper development and assembly of the kidney vasculature and glomerular mesangial cells. Am J Phys Renal Phys 306(2):F249–F258. doi:10.1152/ajprenal.00313.2013 (Epub 2013 Nov 13. PubMed PMID: 24226518; PubMed Central PMCID: PMC3920017)

Lindahl P, Hellström M, Kalén M, Karlsson L, Pekny M, Pekna M, Soriano P, Betsholtz C (1998) Paracrine PDGF-B/PDGF-Rbeta signaling controls mesangial cell development in kidney glomeruli. Development 125(17):3313–3322 PubMed PMID: 9693135

Lindström NO, Carragher NO, Hohenstein P (2015) The PI3K pathway balances self-renewal and differentiation of nephron progenitor cells through β-catenin signaling. Stem Cell Rep 4 (4):551–560. doi:10.1016/j.stemcr.2015.01.021 (Epub 2015 Mar 5. PubMed PMID: 25754203; PubMed Central PMCID: PMC4400645)

Little MH, Brennan J, Georgas K, Davies JA, Davidson DR, Baldock RA, Beverdam A, Bertram JF, Capel B, Chiu HS, Clements D, Cullen-McEwen L, Fleming J, Gilbert T, Herzlinger D, Houghton D, Kaufman MH, Kleymenova E, Koopman PA, Lewis AG, McMahon AP, Mendelsohn CL, Mitchell EK, Rumballe BA, Sweeney DE, Valerius MT, Yamada G, Yang Y, Yu J (2007) A high-resolution anatomical ontology of the developing murine genitourinary tract. Gene Expr Patterns 7(6):680–699 (Epub 2007 Mar 23. Erratum in: Gene

Expr Patterns. 2007 Dec;8(1):47-50. PubMed PMID: 17452023; PubMed Central PMCID: PMC2117077)
Mae S, Shono A, Shiota F, Yasuno T, Kajiwara M, Gotoda-Nishimura N, Arai S, Sato-Otubo A, Toyoda T, Takahashi K, Nakayama N, Cowan CA, Aoi T, Ogawa S, McMahon AP, Yamanaka S, Osafune K (2013) Monitoring and robust induction of nephrogenic intermediate mesoderm from human pluripotent stem cells. Nat Commun 4:1367. doi:10.1038/ncomms2378 (PubMed PMID: 23340407; PubMed Central PMCID: PMC4447148)
Mao Y, Francis-West P, Irvine KD (2015) Fat4/Dchs1 signaling between stromal and cap mesenchyme cells influences nephrogenesis and ureteric bud branching. Development 142 (15):2574–2585. doi:10.1242/dev.122630 (Epub 2015 Jun 26. PubMed PMID: 26116666; PubMed Central PMCID: PMC4529033)
Marks SC Jr, Lane PW (1976) Osteopetrosis, a new recessive skeletal mutation on chromosome 12 of the mouse. J Hered 67(1):11–18
Marose TD, Merkel CE, McMahon AP, Carroll TJ (2008) Beta-catenin is necessary to keep cells of ureteric bud/Wolffian duct epithelium in a precursor state. Dev Biol 314(1):112–126. doi:10.1016/j.ydbio.2007.11.016 (Epub 2007 Nov 28. PubMed PMID: 18177851; PubMed Central PMCID: PMC2699621)
Mendelsohn C, Batourina E, Fung S, Gilbert T, Dodd J (1999) Stromal cells mediate retinoid-dependent functions essential for renal development. Development 126(6):1139–1148 PubMed PMID: 10021334
Metsuyanim S, Harari-Steinberg O, Buzhor E, Omer D, Pode-Shakked N, Ben-Hur H, Halperin R, Schneider D, Dekel B (2009) Expression of stem cell markers in the human fetal kidney. PLoS One 4(8):e6709. doi:10.1371/journal.pone.0006709 (Erratum in: PLoS One. 2011;6(3). doi:10.1371/annotation/5c42ca40-0a01-4e4e-94e3-ccaa2cc54fc6. PubMed PMID: 19696931; PubMed Central PMCID: PMC2725321)
Michaelson MD, Bieri PL, Mehler MF, Xu H, Arezzo JC, Pollard JW, Kessler JA (1996) CSF-1 deficiency in mice results in abnormal brain development. Development 122(9):2661–2672
Michos O, Cebrian C, Hyink D, Grieshammer U, Williams L, D'Agati V, Licht JD, Martin GR, Costantini F (2010) Kidney development in the absence of Gdnf and Spry1 requires Fgf10. PLoS Genet 6(1):e1000809. doi:10.1371/journal.pgen.1000809 (PubMed PMID: 20084103; PubMed Central PMCID: PMC2797609)
Moore MW, Klein RD, Fariñas I, Sauer H, Armanini M, Phillips H, Reichardt LF, Ryan AM, Carver-Moore K, Rosenthal A (1996) Renal and neuronal abnormalities in mice lacking GDNF. Nature 382(6586):76–79 (PubMed PMID: 8657308)
Morizane R, Lam AQ, Freedman BS, Kishi S, Valerius MT, Bonventre JV (2015) Nephron organoids derived from human pluripotent stem cells model kidney development and injury. Nat Biotechnol 33(11):1193–1200 (PubMed PMID: 26458176; PubMed Central PMCID: PMC4747858)
Mugford JW, Sipilä P, McMahon JA, McMahon AP (2008) Osr1 expression demarcates a multi-potent population of intermediate mesoderm that undergoes progressive restriction to an Osr1-dependent nephron progenitor compartment within the mammalian kidney. Dev Biol 324 (1):88–98. doi:10.1016/j.ydbio.2008.09.010 (Epub 2008 Sep 19. PubMed PMID: 18835385; PubMed Central PMCID: PMC2642884)
Mugford JW, Yu J, Kobayashi A, McMahon AP (2009) High-resolution gene expression analysis of the developing mouse kidney defines novel cellular compartments within the nephron progenitor population. Dev Biol 333(2):312–323. doi:10.1016/j.ydbio.2009.06.043 (Epub 2009 Jul 8. PubMed PMID: 19591821; PubMed Central PMCID: PMC2748313)
Muthukrishnan SD, Yang X, Friesel R, Oxburgh L (2015) Concurrent BMP7 and FGF9 signalling governs AP-1 function to promote self-renewal of nephron progenitor cells. Nat Commun 6:10027. doi:10.1038/ncomms10027
Nateri AS, Spencer-Dene B, Behrens A (2005) Interaction of phosphorylated c-Jun with TCF4 regulates intestinal cancer development. Nature 437(7056):281–285. doi:10.1038/nature03914

O'Brien LL, Guo Q, Lee Y, Tran T, Benazet JD, Whitney PH, Valouev A, McMahon AP (2016) Differential regulation of mouse and human nephron progenitors by the Six family of transcriptional regulators. Development 143(4):595–608. doi:10.1242/dev.127175 (PubMed PMID: 26884396; PubMed Central PMCID: PMC4760318)

Ohgushi M, Minaguchi M, Sasai Y (2015) Rho-signaling-directed YAP/TAZ activity underlies the long-term survival and expansion of human embryonic stem cells. Cell Stem Cell 17 (4):448–461. doi:10.1016/j.stem.2015.07.009 (Epub 2015 Aug 27. PubMed PMID: 26321201)

Ohmori T, Tanigawa S, Kaku Y, Fujimura S, Nishinakamura R (2015) Sall1 in renal stromal progenitors non-cell autonomously restricts the excessive expansion of nephron progenitors. Sci Rep 5:15676. doi:10.1038/srep15676 (PubMed PMID: 26511275; PubMed Central PMCID: PMC4625151)

Osafune K, Takasato M, Kispert A, Asashima M, Nishinakamura R (2006) Identification of multipotent progenitors in the embryonic mouse kidney by a novel colony-forming assay. Development 133(1):151–161 (Epub 2005 Nov 30. PubMed PMID: 16319116)

Pachnis V, Mankoo B, Costantini F (1993) Expression of the c-ret proto-oncogene during mouse embryogenesis. Development 119(4):1005–1017 (PubMed PMID: 8306871)

Park JS, Ma W, O'Brien LL, Chung E, Guo JJ, Cheng JG, Valerius MT, McMahon JA, Wong WH, McMahon AP (2012) Six2 and Wnt regulate self-renewal and commitment of nephron progenitors through shared gene regulatory networks. Dev Cell 23(3):637–651. doi:10.1016/j.devcel.2012.07.008 (Epub 2012 Aug 16. PubMed PMID: 22902740; PubMed Central PMCID: PMC3892952)

Pichel JG, Shen L, Sheng HZ, Granholm AC, Drago J, Grinberg A, Lee EJ, Huang SP, Saarma M, Hoffer BJ, Sariola H, Westphal H (1996) Defects in enteric innervation and kidney development in mice lacking GDNF. Nature 382(6586):73–76 (PubMed PMID: 8657307)

Pode-Shakked N, Pleniceanu O, Gershon R, Shukrun R, Kanter I, Bucris E, Pode-Shakked B, Tam G, Tam H, Caspi R, Pri-Chen S, Vax E, Katz G, Omer D, Harari-Steinberg O, Kalisky T, Dekel B (2016) Dissecting stages of human kidney development and tumorigenesis with surface markers affords simple prospective purification of nephron stem cells. Sci Rep 6:23562. doi:10.1038/srep23562 (PubMed PMID: 27020553; PubMed Central PMCID: PMC4810363)

Pollard JW, Hennighausen L (1994) Colony stimulating factor 1 is required for mammary gland development during pregnancy. Proc Natl Acad Sci USA 91(20):9312–9316

Pollard JW, Hunt JS, Wiktor-Jedrzejczak W, Stanley ER (1991) A pregnancy defect in the osteopetrotic (op/op) mouse demonstrates the requirement for CSF-1 in female fertility. Dev Biol 148(1):273–283

Qiao J, Sakurai H, Nigam SK (1999) Branching morphogenesis independent of mesenchymal-epithelial contact in the developing kidney. Proc Natl Acad Sci USA 96(13):7330–7335 (PubMed PMID: 10377414; PubMed Central PMCID: PMC22085)

Rae F, Woods K, Sasmono T, Campanale N, Taylor D, Ovchinnikov DA, Grimmond SM, Hume DA, Ricardo SD, Little MH (2007) Characterisation and trophic functions of murine embryonic macrophages based upon the use of a Csf1r-EGFP transgene reporter. Dev Biol 308 (1):232–246 (Epub 2007 May 25. PubMed PMID: 17597598)

Riccio P, Cebrian C, Zong H, Hippenmeyer S, Costantini F (2016) Ret and Etv4 promote directed movements of progenitor cells during renal branching morphogenesis. PLoS Biol 14(2): e1002382. doi:10.1371/journalpbio1002382 (eCollection 2016 Feb. PubMed PMID: 26894589; PubMed Central PMCID: PMC4760680)

Robert B, St John PL, Hyink DP, Abrahamson DR (1996) Evidence that embryonic kidney cells expressing flk-1 are intrinsic, vasculogenic angioblasts. Am J Phys 271(3 Pt 2):F744–F753 (PubMed PMID: 8853438)

Robert B, Zhao X, Abrahamson DR (2000) Coexpression of neuropilin-1, Flk1, and VEGF(164) in developing and mature mouse kidney glomeruli. Am J Physiol Renal Physiol 279(2):F275–F282 (Erratum in: Am J Physiol Renal Physiol 2001 Sep;281(3):section F following table of contents. PubMed PMID: 10919846)

Rosselot C, Spraggon L, Chia I, Batourina E, Riccio P, Lu B, Niederreither K, Dolle P, Duester G, Chambon P, Costantini F, Gilbert T, Molotkov A, Mendelsohn C (2010) Non-cell-autonomous retinoid signaling is crucial for renal development. Development 137(2):283–292. doi:10.1242/dev.040287 (PubMed PMID: 20040494; PubMed Central PMCID: PMC2799161)

Sainio K, Suvanto P, Davies J, Wartiovaara J, Wartiovaara K, Saarma M, Arumäe U, Meng X, Lindahl M, Pachnis V, Sariola H (1997) Glial-cell-line-derived neurotrophic factor is required for bud initiation from ureteric epithelium. Development 124(20):4077–4087 (PubMed PMID: 9374404)

Saifudeen Z, Liu J, Dipp S, Yao X, Li Y, McLaughlin N, Aboudehen K, El-Dahr SS (2012) A p53-Pax2 pathway in kidney development: implications for nephrogenesis. PLoS One 7(9), e44869. doi:10.1371/journal.pone.0044869

Sánchez MP, Silos-Santiago I, Frisén J, He B, Lira SA, Barbacid M (1996) Renal agenesis and the absence of enteric neurons in mice lacking GDNF. Nature 382(6586):70–73 (PubMed PMID: 8657306)

Saxén L (1987) Organogenesis of the kidney. Cambridge University Press, Cambridge. ISBN-13: 9780521301527

Schmidt-Ott KM, Chen X, Paragas N, Levinson RS, Mendelsohn CL, Barasch J (2006) c-kit delineates a distinct domain of progenitors in the developing kidney. Dev Biol 299(1):238–249 (Epub 2006 Jul 31. PubMed PMID: 16942767)

Schuchardt A, D'Agati V, Larsson-Blomberg L, Costantini F, Pachnis V (1994) Defects in the kidney and enteric nervous system of mice lacking the tyrosine kinase receptor Ret. Nature 367 (6461):380–383 (PubMed PMID: 8114940)

Self M, Lagutin OV, Bowling B, Hendrix J, Cai Y, Dressler GR, Oliver G (2006) Six2 is required for suppression of nephrogenesis and progenitor renewal in the developing kidney. EMBO J 25 (21):5214–5228 (Epub 2006 Oct 12. PubMed PMID: 17036046; PubMed Central PMCID: PMC1630416)

Sequeira Lopez ML, Pentz ES, Robert B, Abrahamson DR, Gomez RA (2001) Embryonic origin and lineage of juxtaglomerular cells. Am J Phys Renal Phys 281(2):F345–F356 (PubMed PMID: 11457727)

Sequeira-Lopez ML, Lin EE, Li M, Hu Y, Sigmund CD, Gomez RA (2015) The earliest metanephric arteriolar progenitors and their role in kidney vascular development. Am J Phys Regul Integr Comp Phys 308(2):R138–R149. doi:10.1152/ajpregu.00428.2014 (Epub 2014 Nov 26. PubMed PMID: 25427768; PubMed Central PMCID: PMC4297861)

Sharmin S, Taguchi A, Kaku Y, Yoshimura Y, Ohmori T, Sakuma T, Mukoyama M, Yamamoto T, Kurihara H, Nishinakamura R (2015) Human induced pluripotent stem cell-derived podocytes mature into vascularized glomeruli upon experimental transplantation. J Am Soc Nephrol. pii: ASN.2015010096 ([Epub ahead of print] PubMed PMID: 26586691)

Shaulian E, Karin M (2001) AP-1 in cell proliferation and survival. Oncogene 20(19):2390–2400. doi:10.1038/sj.onc.1204383

Short KM, Combes AN, Lefevre J, Ju AL, Georgas KM, Lamberton T, Cairncross O, Rumballe BA, McMahon AP, Hamilton NA, Smyth IM, Little MH (2014) Global quantification of tissue dynamics in the developing mouse kidney. Dev Cell 29(2):188–202. doi:10.1016/j.devcel.2014.02.017 (PubMed PMID: 24780737)

Sorenson CM, Rogers SA, Korsmeyer SJ, Hammerman MR (1995) Fulminant metanephric apoptosis and abnormal kidney development in bcl-2-deficient mice. Am J Physiol 268(1 Pt 2):F73–F81

Soriano P (1994) Abnormal kidney development and hematological disorders in PDGF beta-receptor mutant mice. Genes Dev 8(16):1888–1896 (PubMed PMID: 7958864)

Sözen MA, Armstrong JD, Yang M, Kaiser K, Dow JA (1997) Functional domains are specified to single-cell resolution in a *Drosophila epithelium.* Proc Natl Acad Sci USA 94(10):5207–5212 (PubMed PMID: 9144216; PubMed Central PMCID: PMC24657)

Stanley E, Lieschke GJ, Grail D, Metcalf D, Hodgson G, Gall JA, Maher DW, Cebon J, Sinickas V, Dunn AR (1994) Granulocyte/macrophage colonystimulating factor-deficient mice show no major perturbation of hematopoiesis but develop a characteristic pulmonary pathology. Proc Natl Acad Sci USA 91(12):5592–5596

Taguchi A, Kaku Y, Ohmori T, Sharmin S, Ogawa M, Sasaki H, Nishinakamura R (2014) Redefining the in vivo origin of metanephric nephron progenitors enables generation of complex kidney structures from pluripotent stem cells. Cell Stem Cell 14(1):53–67. doi:10.1016/j.stem.2013.11.010 (Epub 2013 Dec 12. PubMed PMID: 24332837)

Takasato M, Er PX, Becroft M, Vanslambrouck JM, Stanley EG, Elefanty AG, Little MH (2014) Directing human embryonic stem cell differentiation towards a renal lineage generates a self-organizing kidney. Nat Cell Biol 16(1):118–126. doi:10.1038/ncb2894 (Epub 2013 Dec 15. PubMed PMID: 24335651)

Takasato M, Er PX, Chiu HS, Maier B, Baillie GJ, Ferguson C, Parton RG, Wolvetang EJ, Roost MS, Chuva de Sousa Lopes SM, Little MH (2015) Kidney organoids from human iPS cells contain multiple lineages and model human nephrogenesis. Nature 526(7574):564–568. doi:10.1038/nature15695 (Epub 2015 Oct 7. PubMed PMID: 26444236)

Tanigawa S, Wang H, Yang Y, Sharma N, Tarasova N, Ajima R, Yamaguchi TP, Rodriguez LG, Perantoni AO (2011) Wnt4 induces nephronic tubules in metanephric mesenchyme by a non-canonical mechanism. Dev Biol 352(1):58–69. doi:10.1016/j.ydbio.2011.01.012 (Epub 2011 Jan 21. PubMed PMID: 21256838; PubMed Central PMCID: PMC3049843)

Tanigawa S, Sharma N, Hall MD, Nishinakamura R, Perantoni AO (2015) Preferential propagation of competent SIX2+ nephronic progenitors by LIF/ROCKi treatment of the metanephric mesenchyme. Stem Cell Rep 5(3):435–447. doi:10.1016/j.stemcr.2015.07.015 (Epub 2015 Aug 28. PubMed PMID: 26321142; PubMed Central PMCID: PMC4618653)

Tomlinson V, Gudmundsdottir K, Luong P, Leung KY, Knebel A, Basu S (2010) JNK phosphorylates Yes-associated protein (YAP) to regulate apoptosis. Cell Death Dis 1, e29. doi:10.1038/cddis.2010.7

Toualbi K, Guller MC, Mauriz JL, Labalette C, Buendia MA, Mauviel A, Bernuau D (2007) Physical and functional cooperation between AP-1 and beta-catenin for the regulation of TCF-dependent genes. Oncogene 26(24):3492–3502. doi:10.1038/sj.onc.1210133

Tufró A (2000) VEGF spatially directs angiogenesis during metanephric development in vitro. Dev Biol 227(2):558–566 (PubMed PMID: 11071774)

Vega QC, Worby CA, Lechner MS, Dixon JE, Dressler GR (1996) Glial cell line-derived neurotrophic factor activates the receptor tyrosine kinase RET and promotes kidney morphogenesis. Proc Natl Acad Sci USA 93(20):10657–10661 (PubMed PMID: 8855235; PubMed Central PMCID: PMC38210)

Vize PD, Smith HW (2004) A homeric view of kidney evolution: a reprint of H.W. Smith's classic essay with a new introduction. Evolution of the kidney. 1943. Anat Rec A Discov Mol Cell Evol Biol 277(2):344–354 (PubMed PMID: 15052662)

Walker KA, Sims-Lucas S, Bates CM (2016) Fibroblast growth factor receptor signaling in kidney and lower urinary tract development. Pediatr Nephrol 31(6):885–895. doi:10.1007/s00467-015-3151-1 (Epub 2015 Aug 21. Review. PubMed PMID: 26293980; PubMed Central PMCID: PMC4761523)

Watanabe K, Ueno M, Kamiya D, Nishiyama A, Matsumura M, Wataya T, Takahashi JB, Nishikawa S, Nishikawa S, Muguruma K, Sasai Y (2007) A ROCK inhibitor permits survival of dissociated human embryonic stem cells. Nat Biotechnol 25(6):681–686 (Epub 2007 May 27. PubMed PMID: 17529971)

Wiktor-Jedrzejczak W, Urbanowska E, Szperl M (1994) Granulocyte-macrophage colony-stimulating factor corrects macrophage deficiencies, but not osteopetrosis, in the colony-stimulating factor-1-deficient op/op mouse. Endocrinology 134(4):1932–1935. doi:10.1210/endo.134.4.8137761

Yao M, Wang Y, Zhang P, Chen H, Xu Z, Jiao J, Yuan Z (2014) BMP2-SMAD signaling represses the proliferation of embryonic neural stem cells through YAP. J Neurosci 34(36):12039–12048. doi:10.1523/JNEUROSCI.0486-14.2014 (PubMed PMID: 25186749)

Yochum GS, Cleland R, Goodman RH (2008) A genome-wide screen for beta-catenin binding sites identifies a downstream enhancer element that controls c-Myc gene expression. Mol Cell Biol 28(24):7368–7379. doi:10.1128/mcb.00744-08

Yu J, Carroll TJ, Rajagopal J, Kobayashi A, Ren Q, McMahon AP (2009) A Wnt7b-dependent pathway regulates the orientation of epithelial cell division and establishes the cortico-medullary axis of the mammalian kidney. Development 136(1):161–171. doi:10.1242/dev.022087 (PubMed PMID: 19060336; PubMed Central PMCID: PMC2685965)

Yuri S, Nishikawa M, Yanagawa N, Jo OD, Yanagawa N (2015) Maintenance of mouse nephron progenitor cells in aggregates with gamma-secretase inhibitor. PLoS One 10(6):e0129242. doi:10.1371/journal.pone.0129242 (eCollection 2015. PubMed PMID: 26075891; PubMed Central PMCID: PMC4468097)

Zanconato F, Forcato M, Battilana G, Azzolin L, Quaranta E, Bodega B, Rosato A, Bicciato S, Cordenonsi M, Piccolo S (2015) Genome-wide association between YAP/TAZ/TEAD and AP-1 at enhancers drives oncogenic growth. Nat Cell Biol 17(9):1218–1227. doi:10.1038/ncb3216

Zubkov VS, Combes AN, Short KM, Lefevre J, Hamilton NA, Smyth IM, Little MH, Byrne HM (2015) A spatially-averaged mathematical model of kidney branching morphogenesis. J Theor Biol 379:24–37. doi:10.1016/j.jtbi.2015.04.015 (Epub 2015 Apr 24. PubMed PMID: 25913880)

Chapter 7
Development and Diseases of the Collecting Duct System

Lihe Chen, Paul J. Higgins, and Wenzheng Zhang

Abstract The collecting duct of the mammalian kidney is important for the regulation of extracellular volume, osmolarity, and pH. There are two major structurally and functionally distinct cell types: principal cells and intercalated cells. The former regulates Na^+ and water homeostasis, while the latter participates in acid–base homeostasis. In vivo lineage tracing using Cre recombinase or its derivatives such as CreGFP and $CreER^{T2}$ is a powerful new technique to identify stem/progenitor cells in their native environment and to decipher the origins of the tissue that they give rise to. Recent studies using this technique in mice have revealed multiple renal progenitor cell populations that differentiate into various nephron segments and collecting duct. In particular, emerging evidence suggests that like principal cells, most of intercalated cells originate from the progenitor cells expressing water channel Aquaporin 2. Mutations or malfunctions of the channels, pumps, and transporters expressed in the collecting duct system cause various human diseases. For example, gain-of-function mutations in ENaC cause Liddle's syndrome, while loss-of-function mutations in ENaC lead to Pseudohypoaldosteronism type 1. Mutations in either AE1 or V-ATPase B1 result in distal renal tubular acidosis. Patients with disrupted AQP2 or AVPR2 develop nephrogenic diabetes insipidus. A better understanding of the function and development of the collecting duct system may facilitate the discovery of new therapeutic strategies for treating kidney disease.

L. Chen
Epithelial Systems Biology Laboratory, Systems Biology Center, NHLBI, Bethesda, MD 20892-1603, USA

P.J. Higgins • W. Zhang (✉)
Department of Regenerative and Cancer Cell Biology, Albany Medical College, MC-165, 47 New Scotland Avenue, Albany, NY 12208, USA
e-mail: zhangw1@mail.amc.edu

R.K. Miller (ed.), *Kidney Development and Disease*, Results and Problems in Cell Differentiation 60, DOI 10.1007/978-3-319-51436-9_7

7.1 Kidney Structure and Function

The adult kidney is a complex filtration system structurally comprised of discrete cell types including glomerulus parietal cells, glomerulus podocytes, proximal tubule brush border cells, loop of Henle thin segment cells, thick ascending limb cells, collecting duct principal cells (PC), collecting duct intercalated cells (IC), and stromal cells. They reside in the cortex, medulla, and papilla regions. In humans, a normal kidney contains 800,000–1.5 million nephrons (Guyton and Hall 2006), which are the basic filtering and excretory renal units. The differentiated segments of each nephron, from proximal to distal, consist of the glomerulus, proximal tubule, loop of Henle, and distal convoluted tubule. A series of tubules and ducts connects the nephrons to the ureter and is referred as the collecting duct system.

The major functions of the kidney are (1) maintenance of electrolyte, water, and acid–base balance; (2) excretion of nitrogenous waste metabolites; (3) hormonal secretion; and (4) regulation of blood pressure. They filter about 180 l of blood per day, with only about 1.5 l finally leaving the body as urine. Urine is formed by nephrons through three precisely regulated processes: filtration, reabsorption, and secretion.

Blood flows into the glomerulus through the afferent arteriole. The filtration barrier prevents the passage of albumin and blood cells. These large substances exit the glomerulus via the efferent arteriole. In contrast, small molecules including ions, water, sugar, and nitrogenous waste are passed through the barrier into Bowman's capsule to produce the glomerular filtrate.

Reabsorption, the next step in urine formation, is the process during which ions and water are reabsorbed back into the bloodstream. This is achieved when the glomerular filtrate passes through the proximal tubule, the loop of Henle, the distal convoluted tubule, and the collecting duct system. The resulting product after reabsorption in the nephron is known as the tubular fluid.

Secretion takes place in the epithelial cells that form the renal tubules and collecting ducts and involves the removal of excessive quantities of certain dissolved substances in the blood to the tubular fluid. Secretion also keeps the blood at a normal pH, which typically ranges from pH 7.35 to pH 7.45.

7.2 Different Origins of the Collecting Duct and Nephron During Development

7.2.1 Ureteric Bud and Metanephric Mesenchyme

In mammals, the pronephros and mesonephros exist transiently (Little and McMahon 2012). Mouse nephrogenesis, which starts at E11, is characterized by an outgrowth of the ureteric bud from the Wolffian duct and an extensive invasion of the ureteric bud into the surrounding metanephric mesenchyme (Little and

McMahon 2012). At this stage, the ureteric bud is evident as a T-shape structure and thereafter undergoes a series of branching morphogenesis. The ureteric bud is able to undergo multiple rounds of branching and creates the whole collecting duct system. Signals secreted from the ureteric bud tip induce the surrounding mesenchymal cells to produce cap mesenchyme, and initiate the mesenchyme-to-epithelial transition of the cap mesenchyme to generate the early nephron structures. The induced cap mesenchyme cells around the ureteric bud tip form the renal vesicles with a small lumen. The comma-shaped body then develops with the appearance of a cleft in the renal vesicles. The s-shaped body becomes evident when a second cleft appears in the comma-shaped body. The distal end of the s-shaped structure attaches to the ureteric bud and forms the continuous tubule through a yet undefined mechanism, while the proximal part of the s-shaped body interacts with endothelial cells to begin glomerulogenesis.

The metanephric mesenchyme is a pool for renal stem and progenitor cells. In addition to giving rise to the entire nephron complement, the stem or progenitor cells within the metanephric mesenchyme also give rise to either vascular progenitors to create the blood vessels (Mugford et al. 2008) or stromal progenitors to generate the stromal tissues (Little and McMahon 2012; Humphreys et al. 2010). With the continuation of ureteric bud branching, new ureteric bud tips form and repeat the same process. Connection of the nephrons with the ureteric bud allows the urine to be voided continuously via the ureters.

7.2.2 *In vivo Genetic Lineage Tracing: A Powerful Tool to Identify the Stem/Progenitor Cells*

Identification and clarifying the biology of the stem and progenitor cell cohorts within the organ hold promise to translate stem cell research to clinical applications. Lineage tracing represents the most powerful tool to identify the stem/progenitor cells in their native environment and decipher the origins of the tissue that they give rise to.

The key components of the genetic lineage tracing technique include Cre recombinase and Cre-dependent reporters. Cre recombinase is an enzyme that catalyzes the recombination of DNA between two loxP target cleavage sites. A loxP site consists of 34 base pairs (bp) with two 13-bp palindromic sequences separated by an 8-bp spacer region. A floxed allele is one that is placed between two loxP sites. Cre recognizes the loxP targets and mediates the deletion or reversion of the floxed DNA sequence, when the two loxP sites are in the same or opposite direction (Nagy 2000; Humphreys and DiRocco 2014), respectively.

In genetic lineage tracing experiments, one mouse line carries a transgene or a knock-in construct that drives Cre expression under control of a tissue- or cell type-specific promoter. This line is often referred as a deletor or Cre driver. The Cre driver is crossed to a second mouse line harboring a Cre-dependent reporter, such as red fluorescent protein (RFP) or LacZ. In case of RFP, RFP is designed to be under the control of a ubiquitous promoter, usually consisting of a *C*MV-IE enhancer/chicken β-*A*ctin/rabbit β-*G*lobin hybrid promoter (CAG) in which the RFP coding sequence and CAG promoter is separated by a loxP-flanked "STOP" cassette, which prevents transcription of the reporter (Madisen et al. 2010). In double transgenic mice harboring both a Cre driver and a RFP reporter, Cre recombinase-mediated recombination deletes the "STOP" cassette resulting in expression of RFP in the Cre-expressing tissue(s) or cells. Because CAG is a ubiquitously active promoter that is silenced by the "STOP" cassette, RFP expression is achieved only in the tissue(s)/cells that express Cre recombinase to remove the "STOP" cassette. Once a cell activates RFP, its offspring also have a permanently deleted "STOP" cassette, which provides for persistent expression of RFP, leading to a population of RFP-traced cells. This technique is extremely useful in stem cell research, especially in organs with complex cellular hierarchies like the kidney (Humphreys and DiRocco 2014).

In addition, several Cre fusion proteins have been described, including CreGFP, $CreER^{T2}$, and $ER^{T2}CreER^{T2}$. CreGFP is a fusion of Cre to green fluorescence protein that enables visualization of Cre expression (Kobayashi et al. 2008). In $CreER^{T2}$ Cre is linked to a mutant form of the human estrogen receptor (ER^{T2}). ER ^{T2} is not able to bind the natural ligand (17β-estradiol) at physiological concentrations but can bind the synthetic ligand 4-hydroxytamoxifen (tamoxifen). After exposure to tamoxifen, $CreER^{T2}$ moves from the cytoplasm to the nucleus, where it catalyzes recombination. This tamoxifen-dependent induction of recombination system allows spatiotemporal control of gene activity. It can be used to pulse label cells and then monitor the labeled cells overtime [reviewed in (Feil et al. 2009)]. However, the high background activity in some cases (Ronzaud et al. 2011) may limit its application. $ER^{T2}CreER^{T2}$, which places Cre between two ER^{T2} elements, is designed to overcome this limitation. It is reported that $ER^{T2}CreER^{T2}$ apparently has a tighter control of induction and thus a lower background activity than CreER ^{T2} (Matsuda and Cepko 2007). Alternatively, lineage tracing can be done using doxycycline-dependent rtTA-inducible system (Gunther et al. 2002).

Obviously, interpretation of in vivo lineage tracing data relies on the faithfulness of the Cre-driven reporters in recapitulating the expression pattern of the endogenous gene, whose promoter drives Cre or Cre derivatives. Addition of ER^{T2} or GFP to Cre may impact the expression pattern of Cre, leading to false positivity. Hence, a direct comparison of Cre or Cre-driven reporters such as RFP and LacZ with the immunofluorescence staining signal of the endogenous gene may help to demonstrate the faithfulness.

7.2.3 *Identification of Six2⁺ Progenitor Cells Responsible for Generation of the Main Body of Nephrons*

Multiple in vivo lineage tracing mouse models relevant to renal biology have been developed including *Six2-Cre* mice (Kobayashi et al. 2008). *Six2-Cre* mice contain a BAC transgene carrying a *Tet-off-eGFPCre (TGC)* cassette; knock-in strains are available with *TGC*, $CreER^{T2}$, and $eGFPCreER^{T2}$ cassettes residing at the *Six2* initiation codon.

As a homeodomain family member, Six2 is very important in the maintenance of the progenitor cell pool in the metanephric mesenchyme. Six2 deficiency leads to early termination of nephrogenesis and loss of progenitor cells. Six2-expressing cells are multipotent, have self-renewal capacity, and directly contribute to the differentiation of the nephron main body, including podocytes, the proximal tubule, the loop of Henle, and the distal convoluted tubule, but not the collecting duct, stromal cells, and interstitial cells (Kobayashi et al. 2008). Like Six2, Cited1 is also uniquely expressed in the cap mesenchyme (Boyle et al. 2008). In $Cited1\text{-}CreER^{T2}$ BAC knock-in mice, $Cited1^{+}$ progenitor cells are also found to contribute to the nephron but not the collecting duct epithelium when traced by LacZ (Boyle et al. 2008). Most recently, lineage tracing with mice carrying an $eGFPCreER^{T2}$ cassette inserted at the *Lgr5* locus shows that Lgr5 is a marker of a more committed group of cells located within the distal region of s-shaped nephrogenic bodies (Barker et al. 2012) that only give rise to the thick ascending limb of Henle's loop and the distal convoluted tubule.

7.3 Molecular Signals Governing Ureteric Bud Growth and Branching

The metanephric mesenchyme can induce ureteric bud growth and branching. Reciprocally, the ureteric bud is capable of inducing nephrogenesis [reviewed in (Krause et al. 2015)]. Many different stimulating or inhibitory signals regulate this mutual induction.

7.3.1 *Gdnf/Ret Signaling*

The principal molecular pathway responsible for the outgrowth and branching of the ureteric bud involves glial cell-derived neurotrophic factor (Gdnf) secreted by the metanephric mesenchyme and its Ret receptor in the ureteric bud (Little and McMahon 2012). Secreted Gdnf activates a Gfrα1/Ret receptor tyrosine kinase complex, initiating a signaling cascade that increases the expression of Ret receptor and induces the outgrowth of Ret positive cells from the Wolffian duct toward the

Gdnf signal (Costantini and Kopan 2010). The signaling cascade also upregulates different ureteric bud tip target genes such as *Etv4*, *Etv5*, *Met*, *Mmp1*, *Spry*, and *Wnt11*. However, only *Etv4* and *Etv5* are essential for the ureteric bud outgrowth (Lu et al. 2009). Wnt11 is a member of the Wnt superfamily of secreted glycoproteins. Disruption of *Wnt11* in mice led to defects in ureteric branching morphogenesis and consequent kidney hypoplasia (Majumdar et al. 2003). Hence, Gdnf acts as a chemotactic cue for the outgrowth of the ureteric bud toward the metanephric mesenchyme. It should be noted that many pathways including ERK MAP kinase, PI3K, and PLCγ also activate Ret signals and contribute to common ureteric bud branching (Krause et al. 2015).

7.3.2 Wnt Signaling Pathway

The secreted lipid-modified signaling Wnt glycoproteins (Wnt) are critical for organogenesis including the development of kidney. Wnt5a, -5b, -6, -9b, and -11 are expressed in the ureteric bud while Wnt2b, -4, and -5a are present in the metanephric mesenchyme (Krause et al. 2015). Wnt4 is synthesized in the condensing mesenchyme and the comma- and s-shaped bodies. It acts in an autocrine manner to maintain the induced metanephric mesenchyme and plays an important role in the epithelialization of condensed mesenchyme and nephrogenesis (Stark et al. 1994; Kispert et al. 1998). Wnt7b in the collecting duct epithelium is crucial for cortico-medullary axis formation (Yu et al. 2009). Wnt9b and Wnt11 released by the ureteric bud induce the metanephric mesenchyme to form cap mesenchyme as well as promote a mesenchyme-to-epithelial transition to generate the renal vesicles. As a ureteric bud-derived inducer of nephrogenesis, Wnt9b contributes to the development of the transient mesonephric and permanent metanephric tubules and the extension of the Müllerian duct. It is required for the initial inductive response in the metanephric mesenchyme, i.e., the formation of the pretubular aggregate (Carroll et al. 2005). Recently, Wnt5a has been added to the list of the members of Wnt family that participate in kidney development (Nishita et al. 2014). Together with its receptor or coreceptor Ror2, a member of the Ror family of receptor tyrosine kinases, Wnt5a activates a noncanonical Wnt signaling pathway and inhibits the canonical Wnt signaling pathway (Oishi et al. 2003). Disruption of Wnt signaling can cause abnormal embryo development. For example, Robinow syndrome is caused by a Wnt5a mutation. The symptoms of this rare skeletal dysplasia syndrome include dysmorphic features resembling a fetal face, renal and vertebral anomalies, short-limb dwarfism, and hypoplastic external genitalia in males (Roifman et al. 2015).

7.3.3 *Six1 and Sall1 Signaling*

Homeobox protein Six1 in the uninduced metanephric mesenchyme is critical for early kidney development. It is required for the expression of Sall1, a multi-zinc finger transcription factor. *Sall1* is critical for the development of the metanehpros. Ablation of *Six1* results in a failure of ureteric bud invasion into the mesenchyme and subsequent apoptosis of the mesenchyme in mice (Xu et al. 2003). Mice with deficiency in *Sall1* are characterized by a failure in ureteric bud outgrowth, renal agenesis or severe dysgenesis, and perinatal death (Nishinakamura et al. 2001).

7.3.4 *FGF Signaling*

Fibroblast growth factors (FGF) trigger a signal cascade through FGF receptors during early metanephric development. The two important receptors Fgfr1 and Fgfr2 are expressed in the metanephric mesenchyme and ureteric bud. Fgfr2 is presented in all parts of the Wolffian duct and is crucial for Wolffian duct maintenance (Okazawa et al. 2015). While single ablation of *Fgfr1* or *Fgfr2* does not cause abnormal kidney development, the lack of both results in kidney agenesis (Poladia et al. 2006).

As a result of the integration of multiple pathways, the cap mesenchyme gives rise to diverse nephron epithelial segments while the ureteric bud differentiates into the collecting duct.

7.4 Collecting Duct Development

The renal collecting duct system is made up of a series of ducts and tubules that interconnect the nephrons to the ureter. It plays an important role in the regulation of fluid and electrolyte metabolism through excretion and reabsorption, processes regulated by the steroid hormone aldosterone and antidiuretic hormone vasopressin. The collecting duct system contains several components including the connecting tubules, cortical collecting ducts, and medullary collecting ducts (Berrout et al. 2014).

7.4.1 *Function and Cellular Composition of the Collecting Duct*

In the collecting duct, two functionally and morphologically distinct cells are present: PC and IC (Bagnis et al. 2001). The homeostatic balance of water, electrolytes, and acid–base is fine-tuned by the expression of many transporters in PC and IC (Roy et al. 2015). PC express apical epithelial Na^+ channel (ENaC) and the water channel Aqp2 as well as the basolaterally located water channels Aqp3 and Aqp4. Collectively, they mediate the water and sodium balance. Aqp2-positive cells are present in the connecting tubule, cortical collecting duct, outer medullary collecting duct, and inner medullary collecting duct. Consistent with their primary role in regulating the acid–base balance, IC express vacuolar-type H^+-ATPase (V-ATPase) and carbonic anhydrase II (CAII). V-ATPases acidify several intracellular organelles and pump protons across the plasma membranes in a wide array of cell types. Proton transport across intracellular and plasma membranes is coupled to ATP hydrolysis through V-ATPases. The carbonic anhydrases belong to a family of enzymes that catalyze the production of bicarbonate and protons from carbon dioxide and water. IC can be further divided into two types according to their immunological differences: α-IC and β-IC. α-IC display V-ATPase at the apical membrane and anion exchanger 1 (AE1) at the basolateral membrane, which is responsible for secreting protons into the urine and reabsorbing bicarbonates, respectively (Roy et al. 2015).

Acidified urine can prevent bacterial growth and other urinary organisms. Nevertheless, acidification of urine rather than a dedicated role in preventing urinary tract infection has been assigned to α-IC. Recent studies have revealed a novel mechanism by which α-IC function in the innate immune defense of urinary tract infection. α-IC produce and release neutrophil gelatinase-associated lipocalin to chelate the siderophore containing host iron to achieve bacteriostasis (Paragas et al. 2014; Chen and Zhang 2014; Becknell et al. 2015).

While β-IC mirror the polarity and function of α-IC, they are characterized by the apical expression of pendrin that secretes bicarbonates and the basolateral expression of V-ATPase that reabsorbs protons (Roy et al. 2015). In addition, Non-A and Non-B IC express both pendrin and V-ATPase at the apical side and likely secrete bicarbonates and protons (Roy et al. 2015).

Unlike the wide distribution of PC throughout the kidney, IC have a more restricted presence. α-IC are found in connecting tubules, cortical collecting ducts, outer medulla collecting ducts, and the initial part of inner medulla collecting ducts. β-IC are abundant only in connecting tubules and cortical collecting ducts (Roy et al. 2015). Besides these major transporters, other important channels are also found in the collecting duct cells like chloride/bicarbonate exchangers SLC26A7 and SLC26A11, AE4, and Na^+/K^+ATPase (Roy et al. 2015).

7.4.2 Molecular Regulators of Collecting Duct Differentiation

7.4.2.1 Circumferential Evidence Suggesting a Common Origin of PC and IC

Based on expression of PC and IC markers as well as functional characteristics, an early study shows that sorted PC in primary culture do not change their PC phenotype and are thus deemed as terminally differentiated cells. Unlike PC, β-IC differentiate into PC, α-IC, and PC/IC "hybrids", a behavior of multipotent cells (Fejes-Toth and Naray-Fejes-Toth 1992). Consistently, metabolic acidosis decreases β-IC and increases α-IC without changing the total IC population, indicating β-IC conversion to α-IC (Schwartz et al. 1985). Disruption of the extracellular matrix protein hensin/DMBT1 results in absence of typical α-IC in mice (Gao et al. 2010). Mice deficient in the forkhead transcription factor Foxi1 contain a single cell type ($Aqp2^+$ $CAII^+$) that has undetectable expression of other IC markers (V-ATPase B1, AE1, AE4, and Pendrin) in the cortex and medulla, arguing that PC and IC arise from progenitor cells characterized by $Foxi1^-$ $Aqp2^+$ $CAII^+$ (Blomqvist et al. 2004). Analyses of IC markers AE1 and H^+-ATPase expression in CAII knockout mice indicate that IC are severely diminished and replaced by PC (Breton et al. 1995).

Loss of Notch signaling results in an increase in the number of IC at the expense of PC. A disintegrin and metalloproteinase domain 10 (Adam10) is a member of the ADAM family of cell membrane-anchored proteins. Adam10 is involved in Notch and other signaling pathways. Inactivation of Adam10 in ureteric bud derivatives with a *HoxB7-Cre* driver leads to reduced PC and increased IC (Guo et al. 2015). A similar decrease of PC/IC ratio in the collecting ducts is observed in *HoxB7-Cre* $Mib1^{f/f}$ mice (Jeong et al. 2009). As an E3 ubiquitin protein ligase, Mib1 plays a key role in the Notch and Wnt/receptor-like tyrosine kinase signaling pathways. These changes in PC to IC ratios in the Notch collecting duct mutants, along with the known conserved role of Notch signaling in binary cell fate choices, are highly suggestive that these PC and IC differentiate from a common precursor cell type. Notch signaling likely plays a conserved role in the "salt-and-pepper" patterning of the collecting duct epithelium with intermingled cell types just as it does in other biological settings including drosophila imaginal discs, frog skin, lung epithelium, and zebrafish pronephros (Quigley et al. 2011; Janicke et al. 2007).

The *HoxB7-Cre* transgene permits Cre expression under the control of a 1.3 kb mouse homeobox B7 (*HoxB7*) enhancer and promoter. Cre-mediated activation of the *ROSA* reporter is visible in the mesonephric duct as early as E9.5, in the ureteric bud by E10.25, and in all ureteric bud epithelial cells by E12.5. Dorsal root ganglia and the spinal cord have low levels of Cre activity (Yu et al. 2002). Given the detection of the *HoxB7-Cre* activity specifically within the mesonephric duct and its derivatives (the Wolffian duct, the collecting duct, and the ureteral epithelium), PC and IC may be differentiated from a pool of progenitor cells marked by HoxB7 expression. However, the faithfulness of the *HoxB7-Cre* transgene and expression of HoxB7 in IC remain obscure. This is because a direct comparison of Cre expression with the endogenous HoxB7 and co-expression of either the endogenous HoxB7 or Cre with IC markers, to our knowledge, have not been reported. Nevertheless, the *HoxB7-Cre* driver has been used to ablate multiple genes including *sonic hedgehog*, *αENaC,* and *Aqp2* as well as *Adam10* and *Mib1* in the collecting duct (Yu et al. 2002; Rubera et al. 2003; Rojek et al. 2006).

7.4.2.2 *Dot1l* as a Primary Epigenetic Regulator of ENaC in Kidney

Disruptor of telomeric silencing (*Dot1*) was originally identified in the yeast *Saccharomyces cerevisiae* as a gene regulating telomeric silencing (Singer et al. 1998). *Dot1* and its mammalian homologs (*Dot1l*) encode a unique family of methyltransferases specific for histone H3 K79 (Feng et al. 2002; van Leeuwen et al. 2002; Zhang et al. 2004). Unlike other methyltransferases, *Dot1 and Dot1l* do not contain a classical enhancer of zeste or trithorax domain (van Leeuwen et al. 2002). Dot1l is widely expressed and highly conserved and catalyzes the addition of one, two, or three methyl groups to lysine 79 at the global domain of histone H3 (Krogan et al. 2003; van Leeuwen et al. 2002; Wu et al. 2013). Dot1l functions in the regulation of transcription, development, erythropoiesis, differentiation, proliferation, and leukemogenesis (Chang et al. 2010; Jo et al. 2011; Okada et al. 2005). The global deletion of *Dot1l* causes embryonic lethality (Jones et al. 2008).

Our previous studies on Dot1l function and regulation in renal physiology lead to (1) cloning of mouse *Dot1l* and identification of five alternative splicing variants (Dot1a-e) (Zhang et al. 2004); (2) demonstration of high Dot1l expression in mouse kidney and strong H3 dimethyl K79 (H3m2K79) staining in PC and IC (Zhang et al. 2006a; Zhang et al. 2004); (3) identification of Dot1a as an integral component of aldosterone signaling that regulates mRNA expression and activity of ENaC in mouse cortical collecting duct M1 cells and in mouse inner medulla collecting duct IMCD3 cells (Zhang et al. 2006a, b, 2007; Chen et al. 2015); and (4) generation and characterization of *Dot1l*$^{f/f}$ *Aqp2Cre* or *Dot1l*^{AC} mice (Chang et al. 2010; Wu et al. 2013; Xiao et al. 2015). In the *Aqp2Cre* transgene, Cre is inserted at the ATG of *Aqp2* in a mouse genomic P1-derived artificial chromosome, which harbors a 125-kb 5′ upstream region and a 31-kb 3′ downstream region of *Aqp2*. The *Aqp2Cre* driver mice carry two copies of the *Aqp2Cre* transgene and have been employed to inactivate *NR3C2* encoding mineralocorticoid receptor (MR) (Ronzaud et al. 2007). As predicted, *Dot1l*^{AC} versus *Dot1*$^{f/f}$ mice had increased αENaC mRNA levels (Zhang et al. 2013).

7.4.2.3 Derivation of IC from Aqp2^{+} Progenitor Cells in *Dot1l*^{AC} Mice

Dot1l^{AC} versus *Dot1l*$^{f/f}$ mice have ~20% less PC and a similar percentage more IC. This change is associated with a significant increase in 24-h urine volume in the

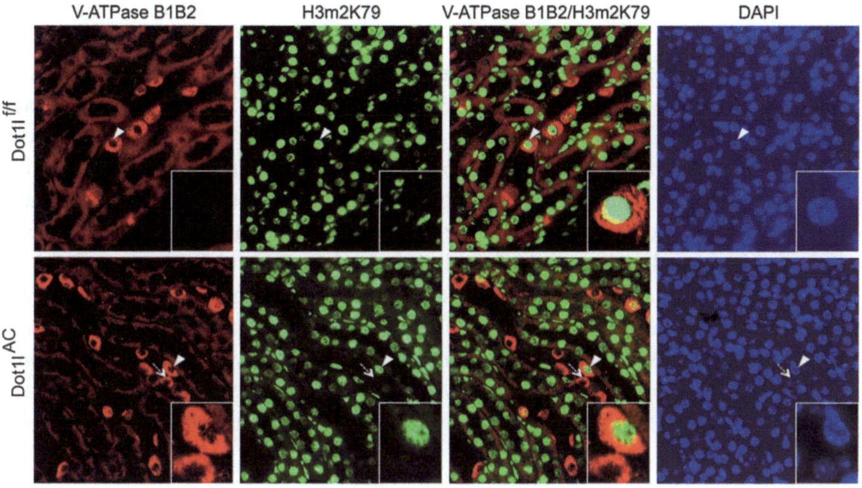

Fig. 7.1 Loss of H3m2K79 is observed in the intercalated cells in *Dot1l*^{AC} mice. Kidney sections from *Dot1l*$^{f/f}$ and *Dot1l*$^{f/f}$ *Aqp2Cre* (*Dot1l*^{AC}) were subjected to double immunofluorescence staining with a rabbit antibody specific for dimethyl histone H3 K79 (H3m2K79) to determine the functional status of Dot1l and a goat antibody recognizing V-ATPase subunits B1 and B2 to identify the intercalated cells. H3m2K79 was robustly detected in all intercalated cells in *Dot1l*$^{f/f}$ mice but undetectable in the most of intercalated cells in *Dot1l*^{AC} mice. Hence, Cre-mediated inactivation of *Dot1l* driven by *Aqp2* promoter in the *Aqp2Cre* transgene occurs in the intercalated cells, indicating that *Aqp2* promoter is activated in the intercalated cells during development. *Arrow* and *arrowhead*: Cells with abolished or intact H3m2K79, respectively. These cells were amplified in the inserts. Scale bar: 50 and 16.3 μM for insert

mutant mice, compared to the control mice. This change suggests a switch in cell fate in the absence of *Dot1l*. Inactivation of *Dot1l* in PC completely abolishes mono-, di-, and tri-methylation at H3 K79. This finding not only indicates that Dot1l is the sole enzyme for all methylation events at H3 K79 but also suggests that loss of H3m2K79 may be used to trace the Aqp2^{+} lineage in *Dot1l*^{AC} mice (Wu et al. 2013; Xiao et al. 2015).

Unexpectedly, *Dot1l*^{AC} mice also display undetectable H3m2K79 in 67–75% of IC identified by staining with an antibody recognizing V-ATPase B1 and B2 subunits (Fig. 7.1) and in 84–90% of cells labeled by other 4 IC markers (V-ATPase A, CAII, AE1, and Pendrin). AE1 and Pendrin mark α-IC and β-IC, respectively (Wu et al. 2013).

Activation of the *Aqp2Cre* transgene may occur before the Aqp2^{+} progenitors switch to IC fate. As a result, Cre-dependent recombination leads to permanent inactivation of *Dot1l* and, thus, loss of H3m2K79 in the Aqp2^{+} lineage. However, Cre expressed in the Aqp2^{+} progenitor cells will be eventually diluted out and the synthesis of new Cre molecules will be dependent on the *Aqp2* promoter, which is active in PC and silenced in IC. As a result, detectable Cre should be observed only in Aqp2^{+} PC cells, but not in the Aqp2^{-} IC^{+} cells, if the *Aqp2Cre* is faithful.

Alternatively, if the *Aqp2Cre* lacks faithfulness, it remains active to direct Cre expression in Aqp2⁻ IC cells.

As predicted, Cre is exclusively expressed and readily detected in PC. However, Cre expression is undetectable in IC [Fig. 7.2 and (Wu et al. 2013)]. Hence, the lack of H3m2K79 in IC results from Cre expression before Aqp2⁺ progenitor cells switch to IC fate, rather than from promiscuous and constitutive Cre expression in IC. The exclusive detection of Cre expression in PC also suggests the faithfulness of the *Aqp2Cre* transgene. Hence, our study provides, for the first time, direct in vivo evidence showing that IC including both α-IC and β-IC, as defined by existing standard markers, can be derived from Aqp2⁺ progenitor cells in *Dot1l*^{AC} mice. Derived IC can be found in the developing and adult kidneys. Inactivation of *Dot1l* promotes this process, leading to increased IC/PC ratio and urine volume (Wu et al. 2013). Given expression of Aqp2 in some connecting tubule cells, it is possible that Aqp2⁺ progenitor cells may contribute to formation of the connecting

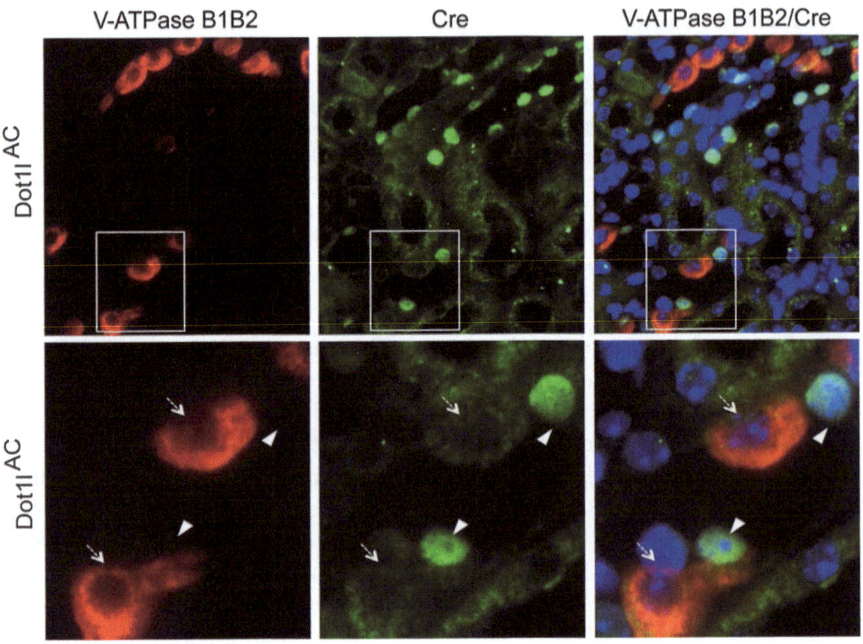

Fig. 7.2 Loss of H3m2K79 in the intercalated cells of *Dot1l*^{AC} mice is not due to promiscuous expression of Cre expression. Kidney sections from *Dot1l*^{AC} were subjected to double immunofluorescence staining with a rabbit antibody specific for Cre recombinase to detect constant and constitutive expression of Cre and the goat antibody recognizing V-ATPase subunits B1 and B2 to identify the intercalated cells. Cre is robustly detected in B1B2⁻ tubular cells, which are principal cells that have constant and constitutive Aqp2 expression. However, Cre is undetectable in the intercalated cells, suggesting that *Aqp2* promoter is silenced after the Aqp2⁺ progenitor cells are differentiated into the intercalated cells during development. *Arrow*: B1B2⁺Cre⁻ (Intercalated cells). *Arrowhead*: B1B2⁻Cre⁺ (Principal cells). *Boxed area* is magnified and shown at the *bottom*. Scale bar: 50 μM (*top*) and 16.3 μM (*bottom*)

tubule. This is consistent with the notion that the connecting tubule originates from the ureteric bud (Howie et al. 1993) or that the connecting tubule can be considered as hybrid epithelia generated by mutual induction from adjoining segments (Schmitt et al. 1999).

Nevertheless, three important questions remain. First, does the *Aqp2Cre* transgene truly recapitulate the spatial and temporal pattern of the endogenous *Aqp2*? Secondly, does IC derivation from $Aqp2^+$ progenitor cells occur in normal physiological condition (i.e., with intact *Dot1l* function)? Thirdly, can the mature PC give rise to IC in adult kidney? These questions may be addressed by extensively characterizing the spatiotemporal faithfulness of the *Aqp2Cre* transgene and by generating an inducible $Aqp2ER^{T2}CreER^{T2}$ knock-in model combined with the Cre-dependent RFP reporter.

7.5 Genetic Diseases of Collecting Duct System

As listed in Table 7.1, mutations or malfunctions of the channels, pumps, and transporters expressed in the collecting duct system have been observed and characterized in different human diseases (Roy et al. 2015; Bruce et al. 2000; Karet et al. 1999; Smith et al. 2000; Rossier and Schild 2008). Gain-of-function mutations in ENaC cause hypertension (Liddle's syndrome), while loss-of-function mutations in ENaC lead to hypotension (Pseudohypoaldosteronism type 1) (Rossier and Schild 2008; Roy et al. 2015). Mutations in either AE1 or V-ATPase cause a failure to acidify urine and decrease plasma bicarbonates, leading to distal renal tubular acidosis (Roy et al. 2015). Patients with mutations in *AQP2* or *AVPR2*, for example, develop nephrogenic diabetes insipidus (NDI), which is characterized by the inability to concentrate urine and excessive urine production. Thus, a better

Table 7.1 Genetic diseases of collecting duct

Disorder	OMIM#	Inheritance	Gene	Gene Product
Liddle's syndrome	177200	AD	*SCNN1B*	βENaC
			SCNN1G	γENaC
PHA1	264350	AR	*SCNN1A*	αENaC
			SCNN1B	βENaC
			SCNN1G	γENaC
	177735	AD	*NR3C2*	Mineralocorticoid receptor
dRTA	179800	AD	*SLC4A1*	AE1
	611590	AR	*SLC4A1*	AE1
	267300		*ATP6V1B1*	V-ATPase B1
	602722		*ATP6V0A4*	V-ATPase a4
NDI	304800	XLR	*AVPR2*	Arginine vasopressin receptor-2
	300539			
	125800	AD, AR	*AQP2*	AQP2

understanding of the function and development of the collecting duct may facilitate the discovery of new therapeutic strategies for kidney disease.

7.5.1 Liddle's Syndrome

7.5.1.1 Clinical Characteristics

The original family described by Liddle et al. has early onset of severe salt-sensitive hypertension in multiple siblings (Liddle et al. 1963). Some of them also display hypokalemia, decreased plasma renin activity, and negligible aldosterone secretion, even on a low-sodium diet (Liddle et al. 1963). These features exclude primary aldosteronism. Other clinical findings such as the high ratios of sodium to potassium in saliva and sweat, the ineffectiveness of spironolactone on electrolyte metabolism and high blood pressure, and normal urinary levels of glucocorticoid metabolites rule out excessive secretion or ingestion of other mineralocorticoids. While the patients on a low sodium diet do not respond to pharmacological challenges with MR inhibitors, prominent amelioration in both hypertension and hypokalemia is achieved through the use of triamterene, a specific ENaC inhibitor.

Renal failure is eventually developed in the proband woman of the original Liddle's kindred. After receiving a cadaveric renal transplant, her abnormalities in both blood pressure and serum potassium levels are corrected, with normal response of the renin–aldosterone axis to salt restriction (Botero-Velez et al. 1994).

Several independent studies (Rodriguez et al. 1981; Wang et al. 1981; Nakada et al. 1987) have confirmed the original Liddle's description (Liddle et al. 1963) and showed that the effective treatments are triamterene and amiloride, but not spironolactone, for hypertension and hypokalemia in Liddle's patients as long as they have restricted dietary Na^+ intake. The disease exists worldwide, without predilection for any race or either sex. Affected pedigrees have been documented in the USA (Shimkets et al. 1994), Japan (Inoue et al. 1998; Uehara et al. 1998), Sweden (Melander et al. 1998), France (Jeunemaitre et al. 1997), the UK (Jackson et al. 1998), China (Gao et al. 2013), Thailand (Sawathiparnich et al. 2009), and Africa (Rayner et al. 2003).

7.5.1.2 Genetic Basis of Liddle's Syndrome

The extended pedigree of the proband in the original Liddle's kindred clearly demonstrates autosomal dominant inheritance. Hypertension associated with hypokalemia indicates that the high blood pressure might result from excessive Na^+ reabsorption in the distal nephron, where Na^+ reabsorption through the amiloride-sensitive, aldosterone-regulated ENaC is indirectly linked to K^+ secretion. Mapping analyses of the original Liddle's kindred demonstrate complete linkage of the disease to *SCNN1B*, the gene encoding βENaC on chromosome 16. Molecular analysis of *SCNN1B* leads to identification of a premature stop codon, which truncates the cytoplasmic C-terminus of βENaC in the affected subjects. Frameshift mutations as well as premature termination in this same C-terminal domain have been found in all affected subjects from four additional kindreds with Liddle's syndrome (Shimkets et al. 1994). Liddle's syndrome can also arise from a mutation ablating the C-terminus of γENaC encoded by *SCNN1G* (Hansson et al. 1995a). Currently, over 20 mutations in *SCNN1B* and *SCNN1G* have been identified to cause Liddle's syndrome [reviewed in (Yang et al. 2014)]. These findings collectively demonstrate gain-of-function mutations in the genes encoding β and γENaC as the molecular defects of Liddle's syndrome and highlight the genetic heterogeneity of the disease.

7.5.1.3 Structure and Function of ENaC

ENaC belongs to epithelial sodium channel/degenerin family of ion channels. It resides at the apical membrane of epithelia in multiple tissues such as lung, exocrine glands, colon, and distal nephron. The channel plays a critical role in the regulation of sodium transport in the collecting duct and hence sodium balance and blood pressure control. The significance of ENaC-mediated Na^+ transport has been dramatically illustrated by the fact that Liddle's syndrome and pseudohypoaldosteronism type 1 (see below) have been linked to gain-of-function and loss-of-function mutations, respectively, in the genes encoding the ENaC subunits.

ENaC consists of three homologous subunits, α, β, and γ (Canessa et al. 1994), which are assembled with a stoichiometry of 2α, 1β, and 1γ (Firsov et al. 1998). Each subunit possesses short intracellular N- and C-termini, a large extracellular loop, and two transmembrane domains (Snyder et al. 1994). Highly conserved proline-rich domains, including the PY motif (xPPxY), appear in the C-terminal region of each subunit (Snyder et al. 1995). The importance of this C-terminal region in the pathogenesis of Liddle's syndrome is well established. As discussed above, all of the original mutations leading to premature stop codons and frameshift identified to spawn Liddle's syndrome result in removal of the C-terminus, with ablation of the PY motifs of β or γENaC (Shimkets et al. 1994; Hansson et al. 1995a). Furthermore, some cases are caused by specific point mutations within the PY motif (Hansson et al. 1995b; Inoue et al. 1998), indicating this region as a critical site for ENaC regulation and as a "hot spot" of the disorder.

7.5.1.4 Molecular Mechanism of Liddle's Syndrome

Studies on the molecular mechanism by which Liddle's mutations of β or γENaC alter the channel function lead to identification of a conserved motif in the C-terminus of all 3 ENaC subunits (Snyder et al. 1995). Loss of the C-terminus of β or γENaC can cause an increase in amiloride-sensitive Na^+ current. The elevated ENaC activity is not a result of an increase in the conductance of Na^+

through each channel or in the open probability (PO), the proportion of time that each channel is in the open state. Instead, the Liddle's truncations increase the number of active channels in the plasma membrane. Moreover, both the truncation of the C terminus and the mutation of the highly conserved C-terminal motif elevate surface abundance of hybrid proteins possessing the C-terminus of the β subunit (Snyder et al. 1995). Therefore, by eliminating a conserved motif, mutations in the Liddle's syndrome increase the apical location of Na^+ channels, leading to an increase in renal Na^+ reabsorption and a predisposition to hypertension. In other words, increase in ENaC activity with increase in renal Na^+ reabsorption in the collecting duct is considered as the primary cause of hypertension in Liddle's syndrome.

The ubiquitin-protein ligase Nedd4 interacts with the PY motifs of either β or γENaC and ubiquitination regulates ENaC (Staub et al. 1996, 1997). Overexpression of wild-type Nedd4, together with ENaC, can reduce the channel activity in Xenopus oocytes. In contrast, overexpression of a catalytically inactive Nedd4 can stimulate the ENaC activity, possibly by serving as a competitive antagonist to the endogenous Nedd4. Nedd4-mediated alternations in the channel activity are not observed in ENaC lacking the PY motifs, indicating the dependency of the Nedd4-mediated inhibition of ENaC on the C-terminal PY motifs. The alteration in the channel activity is attributable entirely to the change in the channel number at the plasma membrane (Abriel et al. 1999). These studies suggest that Nedd4 negatively regulates ENaC and that loss of Nedd4 binding sites in ENaC found in Liddle's syndrome might account for the increased channel number at the cell surface, the increased Na^+ resorption by the collecting duct, and hence hypertension. While the increased channel number at the cell surface probably results from changes in protein–protein interactions regulating the channel degradation through Nedd4 (Staub et al. 1996; Abriel et al. 1999), the hyperactivity of the channel can also be achieved by Liddle's mutations through impairing the clathrin-dependent endocytosis of ENaC (Shimkets et al. 1997). Phosphorylation of Nedd4 by Sgk1 adds another layer of regulation since such modification decreases the affinity of Nedd4 toward ENaC (Bhalla et al. 2005; Debonneville et al. 2001). Consequently, ENaC becomes less ubiquitinated, leading to the accumulation of ENaC channels in the cell surface. In addition, ENaC is tightly regulated at multiple levels by other various mechanisms including Dot1a-Af9-Af17-Sgk1-mediated transcription regulation [reviewed in (Chen et al. 2015; Soundararajan et al. 2012)].

7.5.1.5 Animal Models of Liddle's Syndrome

To generate a mouse model mimicking the human Liddle's syndrome, Cre/loxP-mediated recombination is used as a gene replacement strategy (Pradervand et al. 1999). In this model, a Liddle's allele (L) is generated by introducing a stop codon in the mouse *SCNN1B* locus, imitating human βR566X identified in the original Liddle's kindred. Under normal salt intake, the heterozygotes (L/+) and homozygotes (L/L) at the age of 3 months display no developmental defects. While these mice have increased Na^+ reabsorption in distal colon and low plasma aldosterone, suggesting chronic hypervolemia, their blood pressure is not significantly different from that in WT mice. When challenged with a high salt diet, the Liddle mice develop hypertension, hypokalemia, and metabolic alkalosis associated with cardiac and renal hypertrophy. Hence, the Liddle mouse model replicates to a large extent Liddle's syndrome and establishes βR566X mutation as the cause of the disease.

Analyses of the renal Na^+ transport in vivo, ex vivo with intact perfused tubules, and in vitro using primary culture of the cortical collecting ducts from the Liddle's mice consistently show the constitutive hyperactivity of ENaC (Pradervand et al. 2003). The cortical collecting ducts of L/L mice under a low salt diet have the highest current (ENaC activity), as measured by patch clamp (Dahlmann et al. 2003). Conforming to the proposed pathogenic and molecular mechanism for the Liddle's mutation, the increase in whole cell current is attributed to an increase in the density of conducting channels (Dahlmann et al. 2003).

Despite a primary renal mechanism of Liddle's syndrome, dysregulation of ENaC in other tissues may also be involved. Endothelial cells from the Liddle's mice have enhanced cortical stiffness and ENaC expression (Jeggle et al. 2013). Since the vascular endothelium plays a key role in the control of vascular tone, it is possible that ENaC in the vascular endothelia regulates cellular mechanics and hence plays a role in the control of vascular function (Jeggle et al. 2013). In addition, ENaC-mediated Na^+ transport is enhanced with an increased responsiveness to aldosterone in the colon of Liddle mice. This may be of pathophysiological relevance in subjects with Liddle's syndrome, in particular under a high salt diet, when suppression of aldosterone is not likely to be sufficient to decrease Na^+ absorption to an appropriate level (Nesterov et al. 2012). The pathology of Liddle's syndrome might, hence, relate to ENaC dysfunction not only in the kidney but also in other tissues such as the vascular endothelia and colon.

7.5.1.6 Genetic Testing of Liddle's Syndrome

Traditionally, Liddle's syndrome is diagnosed on the basis of clinical characteristics that may overlap with other Mendelian forms of hypertension. Furthermore, factors including the mild phenotype, atypical cases, and the presence of hypertension without a family history make an accurate diagnosis much more difficult. The variability in the severity of hypertension and hypokalemia in this disease raises the possibility that this disease may be underdiagnosed, particularly among patients with essential hypertension. While several functional tests to determine the response to MR inhibitors vs. ENaC inhibitors or aldosterone response to cosyntropin may facilitate the diagnosis, they are time-consuming and inaccurate. Identification of Liddle's syndrome as a monogenic disorder makes it possible for genetic testing and counseling. To date, over 20 Liddle's mutations have been identified in the exon 13 of *SCNN1B* or *SCNN1G* [reviewed in (Yang et al. 2014)]. Most of these mutations are clustered in a very short region of the C-termini of either β or γENaC. Therefore, direct DNA sequencing of exon 13 of both *SCNN1B* and *SCNN1G* may be exploited to develop a quick, sensitive, and specific genetic screening test for this disorder (Yang et al. 2014).

Family screening is one of the biggest advantages of genetic testing. Through family screening based on the index case, clinicians can identify individuals with potential Liddle's syndrome. Such individuals may be asymptomatic carriers who harbor a known family or de novo mutation. Once the diagnosis of Liddle's syndrome is well established, it offers subjects the possibility to avoid or delay severe complications. Subjects with symptoms can receive tailored therapy containing ENaC antagonists such as amiloride. Early interventions can also be applied to asymptomatic carriers of pathogenic mutations. While the results of a genetic test alone are not deterministic, genetic risk assessment can assist asymptomatic individuals to achieve better prognosis and change their lifestyle, such as decreasing salt intake. Several groups have reported prospective genetic diagnosis of Liddle's syndrome in extended pedigrees, enabling unbiased assessment of the clinical manifestations due to the disease-causing mutations (Findling et al. 1997; Wang et al. 2015).

Genetic testing of exon 13 of both *SCNN1B* and *SCNN1G* led to the identification of 5 pathogenic mutations associated with Liddle's syndrome in 5 index patients and 12 of their relatives from a total of 330 young hypertensive subjects of 14–40 years old (Wang et al. 2015). They are nonsense, missense, and frameshift mutations, altering or truncating the conserved PY motif of β and γENaC. These Liddle's patients have an earlier onset of hypertension, a stronger family history of hypertension, and higher blood pressure than those with essential hypertension. Hence, Liddle's syndrome is a significant etiology of hypertension in this young population (Wang et al. 2015). Genetic testing should be considered in young subjects with early onset of hypertension, particularly those with unknown causes of increased blood pressure and low levels of potassium and renin in blood.

7.5.2 *Pseudohypoaldosteronism Type 1*

7.5.2.1 Clinical Characteristics

The name of pseudohypoaldosteronism type 1 (PHA1) comes from its clinical features, which simulate low levels of aldosterone that regulates sodium reabsorption. However, patients with PHA1 possess high levels of aldosterone.

Based on the severity, the genes involved, and the pattern of inheritance, PHA1 can be divided into two types: autosomal recessive PHA1 (also known as generalized or systemic PHA1, arPHA1) and autosomal dominant PHA1 (also named as renal PHA1, adPHA1). arPHA1 is characterized by Na^+ loss from the kidneys, the sweat glands, the salivary glands, the colon, and other organs. The disease is more severe and does not get better with age. Patients with adPHA1 are presented with excessive Na^+ loss from the kidneys. The condition is relatively mild and often improves in early childhood.

Excessive Na^+ loss leads to hyponatremia and hyperkalemia. Infants with PHA1 can also suffer from metabolic acidosis. Hyponatremia, hyperkalemia, or metabolic acidosis can induce nonspecific symptoms including muscle weakness, vomiting, nausea, and fatigue in infants with PHA1.

The involvement of multiple organs in infants with arPHA1 causes additional signs and symptoms including lesions on the skin, cardiac arrhythmia, and recurrent lung infections. These conditions usually do not develop in adults with arPHA1 (Geller et al. 1998). PHA1 is a rare condition and affects 1 in 80,000 newborns.

7.5.2.2 Molecular Genetics of PHA1

Loss-of-function mutations in one of four different genes involved in Na^+ metabolism cause arPHA1 or adPHA1. Affected subjects in five families with arPHA1 are homozygous for mutations in either *SCNN1A* or *SCNN1B*, which encode α and βENaC, respectively (Chang et al. 1996). A 3′ splice site mutation in *SCNN1G* identified in three families with arPHA1 causes abnormal splicing, generating two splicing variants. One variant arises from skipping the downstream exon and the other from activation of an adjacent cryptic splice site (Strautnieks et al. 1996). Unlike those in Liddle's syndrome, all of these ENaC mutations cause premature termination, missense mutations, or frameshift that lead to loss of the channel activity. Loss-of-function mutations of ENaC are thus causative of arPHA1.

Mutations in *NR3C2*, which encodes MR, result in adPHA1. In six dominant and seven sporadic PHA1 kindreds screened, the affected individuals of four adPHA1 kindreds and one sporadic case have heterozygous *NR3C2* mutations and lack mutations in ENaC genes. These *NR3C2* mutations include two frameshift mutations, two premature termination codons, and one splice donor mutation. They co-segregate with PHA1 and are absent in unaffected subjects (Geller et al. 1998).

Analyses of a Japanese family with adPHA1 lead to identification of a heterozygous point mutation L924P in all affected members. This mutation abolishes aldosterone response in reporter assays in COS-1 cells, suggesting that the missense mutation in *NR3C2* is responsible for the adPHA1 in this family (Tajima et al. 2000). Ins2871C in exon 9 of *NR3C2* of a sporadic case with adPHA1 results in a new frameshift mutation, altering the last 27 residues of the hormone-binding domain (Viemann et al. 2001).

In 14 kindreds with sporadic or adPHA1, there are 6 heterozygous *NR3C2* mutations, which impact MR structure and function differently (Sartorato et al. 2003). The DNA binding domain mutant G633R has decreased maximal transactivation. Q776R and L979P mutations in the ligand-binding domain abolish or reduce aldosterone binding and thus aldosterone-dependent transactivation capacity, respectively. L979P also possesses a transdominant negative effect on WT MR function, while G633R and Q776R might induce haploinsufficiency.

NR3C2 mutations are a common feature of adPHA1, being identified in 70% of the familial cases examined (Sartorato et al. 2003).

Among six heterozygous *NR3C2* mutations detected in seven PHA1 patients from six unrelated families are two previously described frameshift mutations (c.1131dupT and c.2871dupC), two novel nonsense mutations generating a truncated receptor protein, and two novel missense mutations (S818L and E972G) differently affecting MR function (Riepe et al. 2006). Since S818L mutant does not bind aldosterone, it neither translocates into the nucleus nor activates transcription. E972G mutant has a significantly decreased ligand-binding affinity and only 9% of WT transcriptional activity (Riepe et al. 2006). Within 22 *NR3C2* abnormalities in 33 patients with PHA1, 68% of the mutations are dominantly transmitted and 18% de novo mutations (Pujo et al. 2007). These findings collectively demonstrate that heterozygous *NR3C2* mutations cause PHA1, highlighting the critical role of MR function in regulating salt and blood pressure homeostasis in humans.

7.5.2.3 Mouse Models of PHA1

αENaC(-/-) knockout mice are unable to clear fetal lung liquid at birth and die in early neonatal stage, excluding the evaluation of the PHA1 phenotype (Hummler et al. 1997). A transgene expressing αENaC under the control of a cytomegalovirus promoter in αENaC(-/-) knockout mice [αENaC(-/-)Tg] can rescue the perinatal lethality and partially mend Na^+ transport in kidney, colon, and lung. αENaC(-/-)Tg mice develop clinical characteristics of severe PHA1 with metabolic acidosis, urinary salt-wasting, growth retardation, and 50% mortality by days 5–9. The PHA1 phenotype is compensated in adult αENaC(-/-)Tg survivors, with normal electrolyte and acid/base measurements but sixfold higher plasma aldosterone than WT littermates (Hummler et al. 1997).

7.5.2.4 Genetic Testing of PHA1

While genetic testing for Liddle's syndrome can be focused on the C-terminals of *SCNN1B* and *SCNN1G*, genetic testing for PHA1 is more expensive, requiring deletion/duplication analysis, targeted variant analysis, or even sequence analyses of the entire coding regions of four target genes: *SCNN1A*, *SCNN1B*, *SCNN1G*, and *NR3C2*. Such tests have been available in multiple laboratories in Germany, Colombia, Spain, Portugal, and the USA. A complete list of these laboratories can be found at http://www.ncbi.nlm.nih.gov/gtr/tests/?term=C1449842[DISCUI]&test_type=Clinical&display_string=Pseudohypoaldosteronism%20type%201%20autosomal%20dominant&condition=C1449842.

7.5.3 Distal Renal Tubular Acidosis

7.5.3.1 Clinical Characteristics

As a rare genetic disorder, distal renal tubular acidosis (dRTA) arises from the failure of kidneys to acidify the urine normally. This disorder can be acquired or inherited. The clinical features of distal RTA include (1) hypokalemia; (2) normal anion gap metabolic acidosis/acidemia; (3) nephrocalcinosis due to calcium deposition in the kidney; (4) urinary stone formation related to low urinary citrate, hypercalciuria, and alkaline urine; and (5) bone demineralization causing rickets in children and osteomalacia in adults. These symptoms and sequelae of dRTA are inconstant. The entire clinical spectrum of the disease varies from being asymptomatic; to recurrent or persistent flank pain and hematuria from kidney stones, to severe rickets in childhood, growth impairment, progressive nerve deafness, renal failure, and even death (Laing et al. 2005). The disorder is frequently also manifested with sensorineural hearing loss (SNHL). The renal defect of dRTA lies at α-IC. In patients with dRTA, α-IC fails to secrete acid to acidify the urine. dRTA often causes Na^+ loss and volume contraction, leading to a compensatory increase in circulating aldosterone levels. Aldosterone promotes reabsorption of Na^+ and loss of K^+ in the collecting duct. The latter causes hypokalemia, a common feature of dRTA [reviewed in (Batlle and Haque 2012)].

7.5.3.2 Genetics of dRTA

To date, defects in three different genes have been demonstrated to be responsible for dRTA. These genes encode the components of the renal urinary acidification machinery. The first mutations responsible for dRTA are identified in *SLC4A1*, which encodes bicarbonate/chloride exchanger AE1, from patients with autosomal dominant dRTA (Bruce et al. 1997). All affected individuals in four kindreds are heterozygous for one of three *SLC4A1* mutations (R589H, R589C, and S613F). None of these mutations are observed in any of nine normal family members examined. The disease is co-segregated with a genetic marker close to *SLC4A1* (Bruce et al. 1997).

Since no *SLC4A1* mutations are found in the early studies of the families with autosomal recessive dRTA, *SLC4A1* mutations are initially thought to always cause autosomal dominant dRTA (Bruce et al. 1997; Karet et al. 1998). However, this belief is overturned after the discovery of the association of *SLC4A1* mutations with recessive dRTA (Choo et al. 2006; Khositseth et al. 2007). Both affected brother and sister in a Thai family are triply homozygous for two benign mutations M31T and K56E and for the loss-of-function mutation, G701D in AE1 (Choo et al. 2006).

Southeast Asian ovalocytosis (SAO) refers to a deletion of *SLC4A1*, leading to ablation of AE1 residues 400–408 and hence erythrocyte abnormalities. In three Malaysian and six Papua New Guinean families with dRTA examined, SAO occurs in multiple families but does not itself cause dRTA (Bruce et al. 2000). Compound heterozygotes of each of three dRTA mutations (G701D; A858D; and ΔV850) with SAO all result in dRTA associated with abnormal red cell properties and hemolytic anemia. A858D is transmitted in a dominant manner while ΔV850 and G701D show a pseudodominant phenotype when SAO allele is also present (Bruce et al. 2000).

Different compound heterozygous *SLC4A1* mutations have been reported in two Thai families with recessive dRTA (Sritippayawan et al. 2004). The proband of the first family carries compound heterozygous G701D/S773P mutations, which are inherited from his clinically normal mother and father, respectively. In the second family, an affected brother and sister have dRTA and are compound heterozygotes for SAO and R602H mutations. The sister has only mild metabolic acidosis whereas her brother exhibits a severe form of dRTA, suggesting that other modifying factors or genes might regulate the severity of the disorder.

Defects in genes encoding V-ATPase B1 and a4 subunits, *ATP6V1B1* and *ATP6V0A4*, are the genetic cause of recessive dRTA with and without SNHL, respectively (Karet et al. 1999; Smith et al. 2000; Stover et al. 2002; Vargas-Poussou et al. 2006). In three unrelated kindreds with dRTA with SNHL, the affected subjects have a homozygous point mutation that converts codon 31 of *ATP6B1* from CGA to TGA (i.e., R31X) (Karet et al. 1999).

In 26 autosomal recessive dRTA kindreds, *ATP6V1B1* mutations in 10 kindreds and *ATP6V0A4* mutations in 12 kindreds exist. The remaining four families lack linkage to *ATP6V1B1* or *ATP6V0A4*. Several kindreds with mutations in *ATP6V0A4* develop later onset of hearing loss than other dRTA families with progressive SNHL and mutations in *ATP6V1B1* (Stover et al. 2002). These results suggested that dRTA with or without SNHL is the same disorder. In agreement with the associated hearing loss, both *ATP6V1B1* and *ATP6V0A4* are also expressed in the cochlea (Stover et al. 2002; Vargas-Poussou et al. 2006).

A brother and sister from a consanguineous Turkish kindred have an unusual coexistence of two separate recessive disorders: dRTA and osteopetrosis (Borthwick et al. 2003). The dRTA is a result of homozygosity for a G78R substitution in *ATP6V1B1*, while the osteopetrosis is caused by a concurrent homozygous deletion in *TCIRG1*. *TCIRG1* encodes an osteoclast-specific isoform of V-ATPase subunit a (Borthwick et al. 2003). The exceptionality lies in two aspects. Firstly, the coinheritance of the two rare recessive disorders creates a phenocopy of CAII deficiency, offering a novel genetic explanation for the coexistence of osteopetrosis and dRTA; Secondly these disorders affect two different subunits of the V-ATPase that have opposite effects on bone density (Borthwick et al. 2003).

A Japanese family with dRTA associated with early onset of hearing loss possesses two novel *ATP6V1B1* mutations: 15 bp deletion (c.756_770del) in exon 7 and 1 bp insertion (c.1242_1243insC) in exon 12. Both of these mutations are detected in a compound heterozygous state (Yashima et al. 2010). Presumably, mutations in any of genes encoding at least 13 subunits of V-ATPase complex could cause dRTA, highlighting further genetic heterogeneity of the disorder.

7.5.3.3 Molecular Mechanisms of dRTA

The molecular mechanisms of inherited dRTA are complex. The mutations may cause an array of defects ranging from internal sequestration of the mutated proteins in endoplasmic reticulum or Golgi apparatus to their mistargeting to the plasma membrane.

AE1 R589C and S613F mutants have normal chloride transport activity in Xenopus oocytes (Bruce et al. 1997). These mutant proteins are also detected in erythrocytes and oocyte cell membranes, implying that they may be able to normally traffic to the surface of the α-IC of human kidney (Bruce et al. 1997). AE1 R589H mutant has impaired chloride transport activity. It causes a severe trafficking defect of the kidney kAE1 but not of erythrocyte AE1 in human embryonic kidney cells. Although functional, it is apparently trapped in the endoplasmic reticulum, rather than sorted to the basolateral membrane. Hence, bicarbonate could not exit the cell (Bruce et al. 1997; Jarolim et al. 1998). Unlike the dominant R589H mutant, recessive kAE1 G701D and S773P mutants show normal localization to the basolateral membrane when co-expressed with WT kAE1. Glycophorin A is absolutely required for G701D mutant protein movement to the cell surface. Glycophorin A increases G701D surface expression, resulting in 100% return of transport activity to normal. This is suggested as an explanation for the absence of dRTA in heterozygous recessive patients (Bruce et al. 1997).

The ΔV850 and A858D mutant proteins have greatly decreased anion transport when expressed in red blood cells and Xenopus oocytes as compound heterozygotes (ΔV850/A858D, ΔV850/SAO, or A858D/SAO) (Bruce et al. 2000). The surface expression and the transport activity of ΔV850 and A858D mutants are only partially rescued by glycophorin A. The dominant A858D mutant protein may be mistargeted to an inappropriate plasma membrane domain in IC. The decreased anion transport activity may be the cause of dRTA in patients with the recessive ΔV850 mutation (Bruce et al. 2000). The recessive mutation S667P also exhibits a similar glycophorin A rescue pattern. Glycophorin A is expressed in red blood cells but not in kidney cells. This could explain why kAE1 mutations rarely cause a red blood cell defect.

Since polarized MDCK cells are more representative of α-IC in the kidney than non-polarized MDCK cells, they are used to define the underlying defects caused by AE1 mutations leading to dRTA (Cordat et al. 2006). These studies have uncovered several mechanisms by which AE1 mutations lead to reduced chloride-bicarbonate exchange activity and thus dRTA. For example, dominant R589H and S613F mutants are initially sequestered in the endoplasmic reticulum and rapidly become barely detectable, possibly due to degradation. The mutant proteins are also mistargeted to the apical membrane in some cells. Recessive G701D is retained in the Golgi apparatus or sorted to the basolateral membrane to a lesser extent than the WT kAE1 (Cordat et al. 2006). In contrast, recessive AE1 S667P is sequestrated in the endoplasmic reticulum (Toye et al. 2008). Dominant M909T and G609R mutants are targeted to both the apical and basolateral membrane of polarized MDCK cells (Fry et al. 2012).

V-ATPase a4 G820R mutation abolishes phosphofructokinase-1 (PFK-1) binding to the mutant protein without affecting PFK-1 activity and decreases proton transport by 78% and ATPase activity by 36% (Su et al. 2008). These data suggest that a4/PFK-1 binding is important for coupling the ATPase activity to proton transport.

WT V-ATPase B1 subunits, but none of its seven mutants (L81P, R124W, M174R, T275P, G316E, P346R, and G364S) examined, are assembled into the multi-subunit V-ATPase, as demonstrated by co-immunoprecipitation in IMCD cells (Yang et al. 2006). The unassembled mutant subunits are sorted to the apical membrane. Such sorting is increased in response to acidic stimulation and impaired apical membrane abundance of the native V-ATPase. Therefore, B1 mutants may serve as antagonists of V-ATPase function by competing with the endogenous intact V-ATPase for the same apical trafficking pathway (Yang et al. 2006).

7.5.3.4 Mouse Models of dRTA

Genetic ablation of various acid–base transporters leads to generation of mouse models of dRTA. These models reproduce some phenotypic features of human hereditary dRTA. Multiple clinical features observed in patients with *SCL4A1* mutations are similar to those in *Slc4a1*$^{-/-}$ mice (Stehberger et al. 2007). *Slc4a1*$^{-/-}$ mice exhibit dRTA, characterized by hyperchloremic metabolic acidosis associated with inappropriately alkaline urine without bicarbonaturia, low net acid excretion, nephrocalcinosis, hyperphosphaturia, hypercalciuria, and hypocitraturia. Severe urinary concentration defect is also noted, which is evidenced by elevated blood osmolarity and low urine osmolarity, and could be attributed to dysregulated expression and localization of Aqp2 (Stehberger et al. 2007).

In mice deficient in the B1 subunit (*Atp6v1b1*$^{-/-}$), urine pH is more alkaline and oral acid challenge worsens the metabolic acidosis, indicating impaired urinary acidification (Finberg et al. 2005). Phenotypically, the *Atp6v1b1*$^{-/-}$ mice, however, grow normally and do not suffer from systemic acidosis on a normal rodent diet. Nephrocalcinosis, hypercalciuria, and any skeletal defects are not found, possibly due to the absence of metabolic acidosis development (Finberg et al. 2005). While the a4 subunit is also expressed in mice, knockout mice (*Atp6v0a4*$^{-/-}$), to our knowledge, have not been described yet.

7.5.3.5 Genetic Testing of dRTA

SNHL is often seen in autosomal recessive dRTA caused by *ATP6V1B1* and *ATP6V0A4* mutations. By contrast, *SCL4A1* mutations are not usually associated with any hearing loss [reviewed in (Batlle and Haque 2012)]. This clinical distinguishing feature may be useful. Practically, mutations in *ATP6V1B1* and

ATP6V0A4 should be first suspected in dRTA associated with SNHL. However, although the presence of hearing loss and the type are considered as critical clues to the genetic defects underlying inherited dRTA, it is now becoming clearer that they are only helpful as an initial diagnostic impression. Since <50% of the inherited RTA cases exhibit deafness and most of them might develop deafness later in life, genetic testing is left as the solely definitive tool in identifying the defects accurately and early in life.

7.5.4 Congenital Nephrogenic Diabetes Insipidus

7.5.4.1 Description

Diabetes insipidus consists of two major forms: central diabetes insipidus and nephrogenic diabetes insipidus (NDI). The former results from impaired hypothalamic–pituitary axis to produce or release sufficient amounts of vasopressin. The latter is caused by vasopressin insensitivity or hyporesponsiveness due to an intrinsic abnormality of the collecting duct. In both forms of the disorder, the failure of the kidney to concentrate the urine induces polyuria. NDI may be either acquired or congenital in origin. The acquired NDI is much more common than the congenital NDI. Various factors can cause acquired NDI. Chronic administration of the drug lithium is the major cause. Lithium seems to enter into the collecting duct PC through ENaC. ENaC-mediated lithium entry into PC contributes to the pathogenesis of lithium-induced NDI (Christensen et al. 2011). Among less common causes are hypokalemia, hypercalcemia, protein malnutrition, a variety of kidney diseases, and obstruction of the urinary tract (Khanna 2006).

7.5.4.2 Clinical Findings

Patients with congenital NDI usually develop symptoms shortly after birth, with most infants being diagnosed within the first year of life. In addition to the main symptoms (polyuria, polydipsia, and nocturia), the disorder may be associated with a wide range of other symptoms including vomiting, unexplained fevers, retching, irritability, lethargy, diarrhea, constipation, and poor feeding. Consequently, some affected individuals may fail to grow or gain weight at the expected rate. The severity of these symptoms varies from patient to patient. The chronic polyuria may lead to development of hydronephrosis, hydroureter, megacystis, and orthostatic hypotension in adult stage.

7.5.4.3 Genetics of Congenital NDI

The inheritance of congenital NDI in most cases is X-linked recessive, as first described in a Mormon family with three cases of male-to-male transmission of NDI. None of the females traced back to 1813 carry the disease (Cannon 1955). The disorder has now been linked to the defects in antidiuretic hormone arginine vasopressin receptor 2 gene (*AVPR2*) located on the long arm of the X chromosome (Xq28) (van den Ouweland et al. 1992). To date, more than 200 mutations in *AVPR2* have been identified to cause NDI. Most of these mutations are missense mutations (Shida et al. 2013). Other types (i.e., nonsense mutations, insertions, deletions) are less frequent.

In rare cases, the disorder displays an autosomal recessive or dominant mode of inheritance. The recessive and dominant modes account approximately 10% and 1% of cases of hereditary NDI, respectively. In either case, the disease is caused by mutations of *AQP2* on the long arm of chromosome 12 (12q13). Since the identification of a male NDI patient as a compound heterozygote for two *AQP2* missense mutations (R187C and S217P) (Deen et al. 1994), forty-six putative disease-causing *AQP2* mutations have been reported in 52 NDI families [reviewed in (Bichet et al. 2012)]. In brief, inactivation of *AVPR2* and *AQP2* is the molecular basis of congenital NDI.

7.5.4.4 AVPR2 and AQP2

AVPR2 serves as the cognate G protein-coupled receptor for arginine-vasopressin (AVP). It resides in the basolateral plasma membrane of PC. The binding of AVP to AVPR2 triggers a signaling cascade, resulting in the activation of adenylyl cyclases AC3 and AC6 by the stimulatory G protein and thus conversion of ATP to cyclic adenosine monophosphate (cAMP). Protein kinase A (PKA) is one of the cAMP effectors. PKA holoenzyme is a tetramer of two regulatory subunits and two inactive catalytic subunits each bound to one regulatory subunit. There are four regulatory subunits (RIα, RIβ, RIIα, RIIβ) and four catalytic subunits (Cα, Cβ, Cγ, Prkx) identified in mouse and human genomes. cAMP binds to the PKA regulatory subunits (two cAMP molecules per subunit), which results in the release of the catalytic subunits from the tetramer. The free and now active catalytic subunits act as serine-threonine kinases, phosphorylating numerous targets including AQP2.

Apical accumulation of AQP2 in PC increases the membrane permeability, permitting water to move down its osmotic gradient into the interstitium and enter the blood. AQP2 undertakes a constitutive recycling. It moves dynamically between its intracellular storage site and the plasma membrane under resting conditions. In response to AVP in an inadequate fluid status, the equilibrium is shifted toward plasma membrane insertion (Moeller et al. 2013). Subsequently, pro-urinary water reabsorption via coordinated action of AQP2 at the apical side and AQP3 and AQP4 at the basolateral membrane occurs to generate concentrated urine. This process is reversed by a reduction in circulating AVP levels, which reflects the establishment of isotonicity (Kim et al. 2005). AQP2 is under both short-term regulation and long-term regulation by AVP. The short-term regulation occurs in minutes by redistributing AQP2 from cell cytoplasm to cell membrane in response to AVP administration. The long-term regulation involves the transcriptional regulation of AQP2 (Boone and Deen 2008). Hence, redistribution of AQP2 between cell apical membrane and cell cytoplasm is the key for water reabsorption.

In the canonical pathway, regulation of AQP2 trafficking is mediated via AVP-AVPR2-cAMP-PKA signaling as described above. PKA-catalyzed AQP2 phosphorylation on the cytoplasmic C-terminus (notably at S256) increases the rate of exocytosis (Brown 2003). Consistently, changes in cAMP levels and/or the activity of PKA by compounds such as calcitonin or prostaglandin E2 are translated into changes in AQP2 phosphorylation and trafficking (Bouley et al. 2011; Olesen et al. 2011).

S261, S264, and S269 in AQP2 have been identified as additional residues with phosphorylation states modulated by AVP (Hoffert et al. 2006, 2008). Emerging evidence suggests that phosphorylation at these residues can also affect AQP2 trafficking. S261 is dephosphorylated in response to AVP treatment (Hoffert et al. 2007). pS261 is detected mainly in intracellular vesicles after ubiquitination and endocytosis, suggesting a potential role in stabilizing intracellular AQP2 (Hoffert et al. 2007). In contrast, pS269 has been found only on the plasma membrane (Hoffert et al. 2008) and may confer a resistance of AQP2 to endocytosis (Moeller et al. 2010). pS256 is not essential for AQP2 recycling, since AQP2-S256A recycles rapidly and constitutively. However, it is necessary for downstream phosphorylation of other C-terminal serines (Hoffert et al. 2008).

AVP-independent noncanonical pathways for AQP2 membrane accumulation have also been reported. Activation of protein kinase G (PKG) in response to increased cGMP can regulate phosphorylation of the AQP2 C-terminus (Bouley et al. 2005). Depolymerization of actin cytoskeleton can itself cause the membrane translocation of AQP2 in the absence of vasopressin (Klussmann et al. 2001). Modulation of endocytosis and/or exocytosis can impact the membrane accumulation of AQP2 (Brown 2003). Pharmacologically blocking endocytosis with statins (Wade 2011) or methyl-β-cyclodextrin (Russo et al. 2006) causes the redistribution of AQP2 on the plasma membrane independent of phosphorylation.

7.5.4.5 Molecular Mechanism

In vitro functional studies in Xenopus oocytes and in polarized renal tubular cells reveal that mutations in *AVPR2* might lead to (1) impaired binding affinity of the receptor to the AVP, (2) misfolding of the protein that is trapped intracellularly and unable to traffic to the cell membrane to interact with AVP, or (3) ineffectively transcribed *AVPR2* mRNAs (Fujiwara and Bichet 2005). Most of the *AVPR2* mutations result in misfolded proteins (Hermosilla et al. 2004). These misfolded proteins likely accumulate in the Golgi compartment and subsequently degraded by ubiquitin-proteasome system, resulting in loss-of-function phenotypes of AVPR2 unrelated to the actual function of the protein (Romisch 2004). Based on this hypothesis, strategies may be developed to rescue these functionally normal but misfolded proteins. The central idea is to promote the release of the trapped protein in the internal compartment of the cell. For example, the nonpeptide V2 and V1 receptor antagonists could serve as pharmacologic chaperones instead of endocytosis inhibitors to rescue the AVPR2 mutants (Morello et al. 2000). Patients with NDI treated with the vasopressin antagonist SR49059 have significantly decreased urine volume and water intake while other kidney functions like creatinine excretions and electrolyte levels remain constant (Bernier et al. 2006). These studies indicate that pharmacologic chaperones might have therapeutic implications for the treatment of several forms of NDI.

Most of the patients with autosomal recessive NDI have mutations throughout the six transmembrane domains and five connecting loops of AQP2. The AQP2 mutants fail to form tetramers with wild-type AQP2. Similar to most of AVPR2 mutations, these recessive AQP2 mutations also lead to misfolded protein that are trapped in the endoplasmic reticulum and degraded by proteasome machinery (Boone and Deen 2008). Mutations in the autosomal dominant NDI are located in the C-terminal tail of AQP2. The general mechanism of this type of disease is the misrouting of the AQP2 because these mutants tetramerize with wild-type AQP2 and reduce the plasma membrane expression of wild-type AQP2 (Marr et al. 2002).

7.5.4.6 Animal Models

The fortuitous finding of a laboratory rat with DI in Brattleboro, Vermont, USA, in 1961 has led to the development of a strain of rats with the disease. The Brattleboro rat has a deletion of a single base, guanine, in the AVP gene precursor coding region. The deletion causes a frameshift leading to the loss of the normal stop codon. The prolonged C-terminus impairs the folding and configuration necessary for normal processing of the AVP precursor in the endoplasmic reticulum. The mutated precursor is trapped in the endoplasmic reticulum and does not reach the Golgi apparatus. As a result, the rat has no detectable circulating AVP and displays central DI (Kim and Schrier 1998). It has become a very useful model to study the mechanism and treatment strategy of DI.

Mouse models with a collecting duct-specific or global deletion of *Aqp2* are generated using the Cre/loxP technology (Rojek et al. 2006). To generate these NDI models, transgenic mice with loxP sites inserted into intron 2 and 3 of *Aqp2* are bred to the *HoxB7-Cre* driver (to generate collecting duct-specific KO mice) or to mice expressing Cre recombinase under a globally expressed promoter. The mice with collecting duct-specific deletion of *Aqp2* survive with decreased urine osmolality and 10 times more urine, mimicking the severe phenotype of NDI. In contrast, mice with global deletion of *Aqp2* die postnatally due to severe urine concentration defects (Rojek et al. 2006). These two NDI models highlight the importance of Aqp2 in controlling water balance and its unique role in the collecting duct. To mimic the recessive form of NDI, a knock-in mouse model is generated by replacing wild-type Aqp2 with Aqp2-T216M (Yang et al. 2001). The mutants lack obvious abnormalities at 2–3 days after birth but are unable to thrive and generally die by day 6 if supplemental fluid is not offered. A urinary concentrating defect is evidenced by low urine osmolality and serum hyperosmolality. Administration of a V2 vasopressin agonist fails to increase the urine osmolarity. Mutant kidneys show dilated collecting duct, atrophied papillary, and some Aqp2 staining at the plasma membrane. While WT Aqp2 displays complex glycosylation, Aqp2-T126M exhibits mainly endoglycosidase H-sensitive core glycosylation, indicating retention in the endoplasmic reticulum. Since the phosphorylation of Aqp2 by PKA is important for Aqp2 trafficking (Hoffert et al. 2006), a mouse model with S256L mutation is also created. This mutation results in a recessive NDI phenotype due to intracellular trapping of the mutated Aqp2 without apical sorting (McDill et al. 2006). Taken together, these studies suggest a critical role for Aqp2 in renal function in mice and establish multiple mouse models of human autosomal NDI.

References

Abriel H, Loffing J, Rebhun JF, Pratt JH, Schild L, Horisberger JD, Rotin D, Staub O (1999) Defective regulation of the epithelial Na+ channel by Nedd4 in Liddle's syndrome. J Clin Invest 103(5):667–673. doi:10.1172/JCI5713

Bagnis C, Marshansky V, Breton S, Brown D (2001) Remodeling the cellular profile of collecting ducts by chronic carbonic anhydrase inhibition. Am J Physiol Renal Physiol 280(3):F437–F448

Barker N, Rookmaaker MB, Kujala P, Ng A, Leushacke M, Snippert H, van de Wetering M, Tan S, Van Es JH, Huch M, Poulsom R, Verhaar MC, Peters PJ, Clevers H (2012) Lgr5(+ve) stem/progenitor cells contribute to nephron formation during kidney development. Cell Rep 2 (3):540–552. doi:10.1016/j.celrep.2012.08.018

Batlle D, Haque SK (2012) Genetic causes and mechanisms of distal renal tubular acidosis. Nephrol Dial Transplant 27(10):3691–3704. doi:10.1093/ndt/gfs442

Becknell B, Schwaderer A, Hains DS, Spencer JD (2015) Amplifying renal immunity: the role of antimicrobial peptides in pyelonephritis. Nat Rev Nephrol 11(11):642–655. doi:10.1038/nrneph.2015.105

Bernier V, Morello JP, Zarruk A, Debrand N, Salahpour A, Lonergan M, Arthus MF, Laperriere A, Brouard R, Bouvier M, Bichet DG (2006) Pharmacologic chaperones as a potential treatment for X-linked nephrogenic diabetes insipidus. J Am Soc Nephrol 17(1):232–243. doi:10.1681/ASN.2005080854

Berrout J, Mamenko M, Zaika OL, Chen L, Zang W, Pochynyuk O, O'Neil RG (2014) Emerging role of the calcium-activated, small conductance, SK3 K+ channel in distal tubule function: regulation by TRPV4. PLoS One 9(4):e95149. doi:10.1371/journal.pone.0095149

Bhalla V, Daidie D, Li H, Pao AC, LaGrange LP, Wang J, Vandewalle A, Stockand JD, Staub O, Pearce D (2005) Serum- and glucocorticoid-regulated kinase 1 regulates ubiquitin ligase neural precursor cell-expressed, developmentally down-regulated protein 4-2 by inducing interaction with 14-3-3. Mol Endocrinol 19(12):3073–3084

Bichet DG, El Tarazi A, Matar J, Lussier Y, Arthus MF, Lonergan M, Bockenhauer D, Bissonnette P (2012) Aquaporin-2: new mutations responsible for autosomal-recessive nephrogenic diabetes insipidus-update and epidemiology. Clin Kidney J 5(3):195–202. doi:10.1093/ckj/sfs029

Blomqvist SR, Vidarsson H, Fitzgerald S, Johansson BR, Ollerstam A, Brown R, Persson AE, Bergstrom GG, Enerback S (2004) Distal renal tubular acidosis in mice that lack the forkhead transcription factor Foxi1. J Clin Invest 113(11):1560–1570. doi:10.1172/JCI20665

Boone M, Deen PM (2008) Physiology and pathophysiology of the vasopressin-regulated renal water reabsorption. Pflugers Arch Eur J Physiol 456(6):1005–1024. doi:10.1007/s00424-008-0498-1

Borthwick KJ, Kandemir N, Topaloglu R, Kornak U, Bakkaloglu A, Yordam N, Ozen S, Mocan H, Shah GN, Sly WS, Karet FE (2003) A phenocopy of CAII deficiency: a novel genetic explanation for inherited infantile osteopetrosis with distal renal tubular acidosis. J Med Genet 40(2):115–121

Botero-Velez M, Curtis JJ, Warnock DG (1994) Brief report: Liddle's syndrome revisited—a disorder of sodium reabsorption in the distal tubule. N Engl J Med 330(3):178–181. doi:10.1056/NEJM199401203300305

Bouley R, Pastor-Soler N, Cohen O, McLaughlin M, Breton S, Brown D (2005) Stimulation of AQP2 membrane insertion in renal epithelial cells in vitro and in vivo by the cGMP phosphodiesterase inhibitor sildenafil citrate (Viagra). Am J Physiol Renal Physiol 288(6):F1103–F1112. doi:10.1152/ajprenal.00337.2004

Bouley R, Lu HA, Nunes P, Da Silva N, McLaughlin M, Chen Y, Brown D (2011) Calcitonin has a vasopressin-like effect on aquaporin-2 trafficking and urinary concentration. J Am Soc Nephrol 22(1):59–72. doi:10.1681/ASN.2009121267

Boyle S, Misfeldt A, Chandler KJ, Deal KK, Southard-Smith EM, Mortlock DP, Baldwin HS, de Caestecker M (2008) Fate mapping using Cited1-CreERT2 mice demonstrates that the cap mesenchyme contains self-renewing progenitor cells and gives rise exclusively to nephronic epithelia. Dev Biol 313(1):234–245. doi:10.1016/j.ydbio.2007.10.014

Breton S, Alper SL, Gluck SL, Sly WS, Barker JE, Brown D (1995) Depletion of intercalated cells from collecting ducts of carbonic anhydrase II-deficient (CAR2 null) mice. Am J Phys 269(6 Pt 2):F761–F774

Brown D (2003) The ins and outs of aquaporin-2 trafficking. Am J Physiol Renal Physiol 284(5): F893–F901. doi:10.1152/ajprenal.00387.2002

Bruce LJ, Cope DL, Jones GK, Schofield AE, Burley M, Povey S, Unwin RJ, Wrong O, Tanner MJ (1997) Familial distal renal tubular acidosis is associated with mutations in the red cell anion exchanger (Band 3, AE1) gene. J Clin Invest 100(7):1693–1707. doi:10.1172/JCI119694

Bruce LJ, Wrong O, Toye AM, Young MT, Ogle G, Ismail Z, Sinha AK, McMaster P, Hwaihwanje I, Nash GB, Hart S, Lavu E, Palmer R, Othman A, Unwin RJ, Tanner MJ (2000) Band 3 mutations, renal tubular acidosis and South-East Asian ovalocytosis in Malaysia and Papua New Guinea: loss of up to 95% band 3 transport in red cells. Biochem J 350 (Pt 1):41–51

Canessa CM, Schild L, Buell G, Thorens B, Gautschi I, Horisberger JD, Rossier BC (1994) Amiloride-sensitive epithelial Na+ channel is made of three homologous subunits. Nature 367(6462):463–467. doi:10.1038/367463a0

Cannon JF (1955) Diabetes insipidus; clinical and experimental studies with consideration of genetic relationships. AMA Arch Intern Med 96(2):215–272

Carroll TJ, Park JS, Hayashi S, Majumdar A, McMahon AP (2005) Wnt9b plays a central role in the regulation of mesenchymal to epithelial transitions underlying organogenesis of the mammalian urogenital system. Dev Cell 9(2):283–292. doi:10.1016/j.devcel.2005.05.016

Chang SS, Grunder S, Hanukoglu A, Rosler A, Mathew PM, Hanukoglu I, Schild L, Lu Y, Shimkets RA, Nelson-Williams C, Rossier BC, Lifton RP (1996) Mutations in subunits of the epithelial sodium channel cause salt wasting with hyperkalaemic acidosis, pseudohypoaldosteronism type 1. Nat Genet 12(3):248–253

Chang MJ, Wu H, Achille NJ, Reisenauer MR, Chou CW, Zeleznik-Le NJ, Hemenway CS, Zhang W (2010) Histone H3 Lysine 79 methyltransferase Dot1 is required for immortalization by MLL oncogenes. Cancer Res 70 (24):10234-10242. doi:10.1158/0008-5472.CAN-10-3294. 70/24/10234 [pii]

Chen L, Zhang W (2014) Kidney α-intercalated cells, NGAL and urinary tract infection. Austin J Nephrol Hypertens 1(4):1017

Chen L, Zhang X, Zhang W (2015) Regulation of alphaENaC transcription. Vitam Horm 98:101–135. doi:10.1016/bs.vh.2014.12.004

Choo KE, Nicoli TK, Bruce LJ, Tanner MJ, Ruiz-Linares A, Wrong OM (2006) Recessive distal renal tubular acidosis in Sarawak caused by AE1 mutations. Pediatr Nephrol 21(2):212–217. doi:10.1007/s00467-005-2061-z

Christensen BM, Zuber AM, Loffing J, Stehle JC, Deen PM, Rossier BC, Hummler E (2011) alphaENaC-mediated lithium absorption promotes nephrogenic diabetes insipidus. J Am Soc Nephrol 22(2):253–261. doi:10.1681/ASN.2010070734

Cordat E, Kittanakom S, Yenchitsomanus PT, Li J, Du K, Lukacs GL, Reithmeier RA (2006) Dominant and recessive distal renal tubular acidosis mutations of kidney anion exchanger 1 induce distinct trafficking defects in MDCK cells. Traffic 7(2):117–128. doi:10.1111/j.1600-0854.2005.00366.x

Costantini F, Kopan R (2010) Patterning a complex organ: branching morphogenesis and nephron segmentation in kidney development. Dev Cell 18(5):698–712. doi:10.1016/j.devcel.2010.04.008

Dahlmann A, Pradervand S, Hummler E, Rossier BC, Frindt G, Palmer LG (2003) Mineralocorticoid regulation of epithelial Na+ channels is maintained in a mouse model of Liddle's syndrome. Am J Physiol Renal Physiol 285(2):F310–F318

Debonneville C, Flores SY, Kamynina E, Plant PJ, Tauxe C, Thomas MA, Munster C, Chraibi A, Pratt JH, Horisberger JD, Pearce D, Loffing J, Staub O (2001) Phosphorylation of Nedd4-2 by Sgk1 regulates epithelial Na(+) channel cell surface expression. EMBO J 20(24):7052–7059

Deen PM, Verdijk MA, Knoers NV, Wieringa B, Monnens LA, van Os CH, van Oost BA (1994) Requirement of human renal water channel aquaporin-2 for vasopressin-dependent concentration of urine. Science 264(5155):92–95

Feil S, Valtcheva N, Feil R (2009) Inducible Cre mice. Methods Mol Biol 530:343–363. doi:10.1007/978-1-59745-471-1_18

Fejes-Toth G, Naray-Fejes-Toth A (1992) Differentiation of renal beta-intercalated cells to alpha-intercalated and principal cells in culture. Proc Natl Acad Sci USA 89(12):5487–5491

Feng Q, Wang H, Ng HH, Erdjument-Bromage H, Tempst P, Struhl K, Zhang Y (2002) Methylation of H3-lysine 79 is mediated by a new family of HMTases without a SET domain. Curr Biol 12(12):1052–1058

Finberg KE, Wagner CA, Bailey MA, Paunescu TG, Breton S, Brown D, Giebisch G, Geibel JP, Lifton RP (2005) The B1-subunit of the H(+) ATPase is required for maximal urinary acidification. Proc Natl Acad Sci USA 102(38):13616–13621. doi:10.1073/pnas.0506769102

Findling JW, Raff H, Hansson JH, Lifton RP (1997) Liddle's syndrome: prospective genetic screening and suppressed aldosterone secretion in an extended kindred. J Clin Endocrinol Metab 82(4):1071–1074. doi:10.1210/jcem.82.4.3862

Firsov D, Gautschi I, Merillat AM, Rossier BC, Schild L (1998) The heterotetrameric architecture of the epithelial sodium channel (ENaC). EMBO J 17(2):344–352. doi:10.1093/emboj/17.2.344

Fry AC, Su Y, Yiu V, Cuthbert AW, Trachtman H, Karet Frankl FE (2012) Mutation conferring apical-targeting motif on AE1 exchanger causes autosomal dominant distal RTA. J Am Soc Nephrol 23(7):1238–1249. doi:10.1681/ASN.2012020112

Fujiwara TM, Bichet DG (2005) Molecular biology of hereditary diabetes insipidus. J Am Soc Nephrol 16(10):2836–2846. doi:10.1681/ASN.2005040371

Gao L, Wang L, Liu Y, Zhou X, Hui R, Hu A (2013) A family with Liddle syndrome caused by a novel missense mutation in the PY motif of the beta-subunit of the epithelial sodium channel. J Pediatr 162(1):166–170. doi:10.1016/j.jpeds.2012.06.017

Gao X, Eladari D, Leviel F, Tew BY, Miro-Julia C, Cheema F, Miller L, Nelson R, Paunescu TG, McKee M, Brown D, Al-Awqati Q (2010) Deletion of hensin/DMBT1 blocks conversion of {beta}- to {alpha}-intercalated cells and induces distal renal tubular acidosis. Proc Natl Acad Sci USA. doi:10.1073/pnas.1010364107. 1010364107 [pii]

Geller DS, Rodriguez-Soriano J, Vallo Boado A, Schifter S, Bayer M, Chang SS, Lifton RP (1998) Mutations in the mineralocorticoid receptor gene cause autosomal dominant pseudohypoaldosteronism type I. Nat Genet 19(3):279–281. doi:10.1038/966

Gunther EJ, Belka GK, Wertheim GB, Wang J, Hartman JL, Boxer RB, Chodosh LA (2002) A novel doxycycline-inducible system for the transgenic analysis of mammary gland biology. FASEB J 16(3):283–292. doi:10.1096/fj.01-0551com

Guo Q, Wang Y, Tripathi P, Manda KR, Mukherjee M, Chaklader M, Austin PF, Surendran K, Chen F (2015) Adam10 mediates the choice between principal cells and intercalated cells in the kidney. J Am Soc Nephrol 26(1):149–159. doi:10.1681/ASN.2013070764

Guyton AC, Hall JE (2006) Textbook of medical physiology. Elsevier, Philadelphia

Hansson JH, Nelson-Williams C, Suzuki H, Schild L, Shimkets R, Lu Y, Canessa C, Iwasaki T, Rossier B, Lifton RP (1995a) Hypertension caused by a truncated epithelial sodium channel gamma subunit: genetic heterogeneity of Liddle syndrome. Nat Genet 11(1):76–82

Hansson JH, Schild L, Lu Y, Wilson TA, Gautschi I, Shimkets R, Nelson-Williams C, Rossier BC, Lifton RP (1995b) A de novo missense mutation of the beta subunit of the epithelial sodium channel causes hypertension and Liddle syndrome, identifying a proline-rich segment critical for regulation of channel activity. Proc Natl Acad Sci USA 92(25):11495–11499

Hermosilla R, Oueslati M, Donalies U, Schonenberger E, Krause E, Oksche A, Rosenthal W, Schulein R (2004) Disease-causing V(2) vasopressin receptors are retained in different compartments of the early secretory pathway. Traffic 5(12):993–1005. doi:10.1111/j.1600-0854.2004.00239.x

Hoffert JD, Fenton RA, Moeller HB, Simons B, Tchapyjnikov D, McDill BW, Yu MJ, Pisitkun T, Chen F, Knepper MA (2008) Vasopressin-stimulated increase in phosphorylation at Ser269 potentiates plasma membrane retention of aquaporin-2. J Biol Chem 283(36):24617–24627. doi:10.1074/jbc.M803074200 M803074200 [pii]

Hoffert JD, Nielsen J, Yu MJ, Pisitkun T, Schleicher SM, Nielsen S, Knepper MA (2007) Dynamics of aquaporin-2 serine-261 phosphorylation in response to short-term vasopressin treatment in collecting duct. Am J Physiol Renal Physiol 292(2):F691–F700. doi:10.1152/ajprenal.00284.2006 00284.2006 [pii]

Hoffert JD, Pisitkun T, Wang G, Shen RF, Knepper MA (2006) Quantitative phosphoproteomics of vasopressin-sensitive renal cells: regulation of aquaporin-2 phosphorylation at two sites. Proc Natl Acad Sci USA 103(18):7159–7164. doi:10.1073/pnas.0600895103

Howie AJ, Smithson N, Rollason TP (1993) Reconsideration of the development of the distal tubule of the human kidney. J Anat 183(Pt 1):141–147

Hummler E, Barker P, Talbot C, Wang Q, Verdumo C, Grubb B, Gatzy J, Burnier M, Horisberger JD, Beermann F, Boucher R, Rossier BC (1997) A mouse model for the renal salt-wasting syndrome pseudohypoaldosteronism. Proc Natl Acad Sci USA 94(21):11710–11715

Humphreys BD, DiRocco DP (2014) Lineage-tracing methods and the kidney. Kidney Int 86 (3):481–488. doi:10.1038/ki.2013.368

Humphreys BD, Lin SL, Kobayashi A, Hudson TE, Nowlin BT, Bonventre JV, Valerius MT, McMahon AP, Duffield JS (2010) Fate tracing reveals the pericyte and not epithelial origin of myofibroblasts in kidney fibrosis. Am J Pathol 176(1):85–97. doi:10.2353/ajpath.2010.090517

Inoue J, Iwaoka T, Tokunaga H, Takamune K, Naomi S, Araki M, Takahama K, Yamaguchi K, Tomita K (1998) A family with Liddle's syndrome caused by a new missense mutation in the beta subunit of the epithelial sodium channel. J Clin Endocrinol Metab 83(6):2210–2213. doi:10.1210/jcem.83.6.5030

Jackson SN, Williams B, Houtman P, Trembath RC (1998) The diagnosis of Liddle syndrome by identification of a mutation in the beta subunit of the epithelial sodium channel. J Med Genet 35 (6):510–512

Janicke M, Carney TJ, Hammerschmidt M (2007) Foxi3 transcription factors and Notch signaling control the formation of skin ionocytes from epidermal precursors of the zebrafish embryo. Dev Biol 307(2):258–271. doi:10.1016/j.ydbio.2007.04.044

Jarolim P, Shayakul C, Prabakaran D, Jiang L, Stuart-Tilley A, Rubin HL, Simova S, Zavadil J, Herrin JT, Brouillette J, Somers MJ, Seemanova E, Brugnara C, Guay-Woodford LM, Alper SL (1998) Autosomal dominant distal renal tubular acidosis is associated in three families with heterozygosity for the R589H mutation in the AE1 (band 3) Cl-/HCO3- exchanger. J Biol Chem 273(11):6380–6388

Jeggle P, Callies C, Tarjus A, Fassot C, Fels J, Oberleithner H, Jaisser F, Kusche-Vihrog K (2013) Epithelial sodium channel stiffens the vascular endothelium in vitro and in Liddle mice. Hypertension 61(5):1053–1059. doi:10.1161/HYPERTENSIONAHA.111.199455

Jeong HW, Jeon US, Koo BK, Kim WY, Im SK, Shin J, Cho Y, Kim J, Kong YY (2009) Inactivation of Notch signaling in the renal collecting duct causes nephrogenic diabetes insipidus in mice. J Clin Invest 119(11):3290–3300. doi:10.1172/JCI38416

Jeunemaitre X, Bassilana F, Persu A, Dumont C, Champigny G, Lazdunski M, Corvol P, Barbry P (1997) Genotype-phenotype analysis of a newly discovered family with Liddle's syndrome. J Hypertens 15(10):1091–1100

Jo SY, Granowicz EM, Maillard I, Thomas D, Hess JL (2011) Requirement for Dotl1 in murine postnatal hematopoiesis and leukemogenesis by MLL translocation. Blood 117 (18):4759–4768. doi:10.1182/blood-2010-12-327668 blood-2010-12-327668 [pii]

Jones B, Su H, Bhat A, Lei H, Bajko J, Hevi S, Baltus GA, Kadam S, Zhai H, Valdez R, Gonzalo S, Zhang Y, Li E, Chen T (2008) The histone H3K79 methyltransferase Dot1L is essential for mammalian development and heterochromatin structure. PLoS Genet 4(9):1–11

Karet FE, Finberg KE, Nelson RD, Nayir A, Mocan H, Sanjad SA, Rodriguez-Soriano J, Santos F, Cremers CW, Di Pietro A, Hoffbrand BI, Winiarski J, Bakkaloglu A, Ozen S, Dusunsel R, Goodyer P, Hulton SA, Wu DK, Skvorak AB, Morton CC, Cunningham MJ, Jha V, Lifton RP (1999) Mutations in the gene encoding B1 subunit of H+-ATPase cause renal tubular acidosis with sensorineural deafness. Nat Genet 21(1):84–90. doi:10.1038/5022

Karet FE, Gainza FJ, Gyory AZ, Unwin RJ, Wrong O, Tanner MJ, Nayir A, Alpay H, Santos F, Hulton SA, Bakkaloglu A, Ozen S, Cunningham MJ, di Pietro A, Walker WG, Lifton RP (1998) Mutations in the chloride-bicarbonate exchanger gene AE1 cause autosomal dominant but not autosomal recessive distal renal tubular acidosis. Proc Natl Acad Sci USA 95 (11):6337–6342

Khanna A (2006) Acquired nephrogenic diabetes insipidus. Semin Nephrol 26(3):244–248. doi:10.1016/j.semnephrol.2006.03.004

Khositseth S, Sirikanerat A, Wongbenjarat K, Opastirakul S, Khoprasert S, Peuksungnern R, Wattanasirichaigoon D, Thongnoppakhun W, Viprakasit V, Yenchitsomanus PT (2007) Distal renal tubular acidosis associated with anion exchanger 1 mutations in children in Thailand. Am J Kidney Dis 49(6):841–850 . doi:10.1053/j.ajkd.2007.03.002e841

Kim JK, Schrier RW (1998) Vasopressin processing defects in the Brattleboro rat: implications for hereditary central diabetes insipidus in humans? Proc Assoc Am Physicians 110(5):380–386

Kim SW, Gresz V, Rojek A, Wang W, Verkman AS, Frokiaer J, Nielsen S (2005) Decreased expression of AQP2 and AQP4 water channels and Na,K-ATPase in kidney collecting duct in AQP3 null mice. Biol Cell 97(10):765–778. doi:10.1042/BC20040148

Kispert A, Vainio S, McMahon AP (1998) Wnt-4 is a mesenchymal signal for epithelial transformation of metanephric mesenchyme in the developing kidney. Development 125 (21):4225–4234

Klussmann E, Tamma G, Lorenz D, Wiesner B, Maric K, Hofmann F, Aktories K, Valenti G, Rosenthal W (2001) An inhibitory role of Rho in the vasopressin-mediated translocation of aquaporin-2 into cell membranes of renal principal cells. J Biol Chem 276(23):20451–20457. doi:10.1074/jbc.M010270200

Kobayashi A, Valerius MT, Mugford JW, Carroll TJ, Self M, Oliver G, McMahon AP (2008) Six2 defines and regulates a multipotent self-renewing nephron progenitor population throughout mammalian kidney development. Cell Stem Cell 3(2):169–181. doi:10.1016/j.stem.2008.05.020

Krause M, Rak-Raszewska A, Pietila I, Quaggin SE, Vainio S (2015) Signaling during kidney development. Cells 4(2):112–132. doi:10.3390/cells4020112

Krogan NJ, Dover J, Wood A, Schneider J, Heidt J, Boateng MA, Dean K, Ryan OW, Golshani A, Johnston M, Greenblatt JF, Shilatifard A (2003) The Paf1 complex is required for histone H3 methylation by COMPASS and Dot1p: linking transcriptional elongation to histone methylation. Mol Cell 11(3):721–729

Laing CM, Toye AM, Capasso G, Unwin RJ (2005) Renal tubular acidosis: developments in our understanding of the molecular basis. Int J Biochem Cell Biol 37(6):1151–1161. doi:10.1016/j.biocel.2005.01.002

Liddle GW, Bledsoe T, Coppage WS Jr (1963) A familial renal disorder simulating primary aldosteronism but with negligible aldosterone secretion. Trans Assoc Am Phys 76:199–213

Little MH, McMahon AP (2012) Mammalian kidney development: principles, progress, and projections. Cold Spring Harb Perspect Biol 4(5). doi:10.1101/cshperspect.a008300

Lu BC, Cebrian C, Chi X, Kuure S, Kuo R, Bates CM, Arber S, Hassell J, MacNeil L, Hoshi M, Jain S, Asai N, Takahashi M, Schmidt-Ott KM, Barasch J, D'Agati V, Costantini F (2009) Etv4 and Etv5 are required downstream of GDNF and Ret for kidney branching morphogenesis. Nat Genet 41(12):1295–1302. doi:10.1038/ng.476

Madisen L, Zwingman TA, Sunkin SM, Oh SW, Zariwala HA, Gu H, Ng LL, Palmiter RD, Hawrylycz MJ, Jones AR, Lein ES, Zeng H (2010) A robust and high-throughput Cre reporting and characterization system for the whole mouse brain. Nat Neurosci 13(1):133–140. doi:10.1038/nn.2467

Majumdar A, Vainio S, Kispert A, McMahon J, McMahon AP (2003) Wnt11 and Ret/Gdnf pathways cooperate in regulating ureteric branching during metanephric kidney development. Development 130(14):3175–3185

Marr N, Bichet DG, Lonergan M, Arthus MF, Jeck N, Seyberth HW, Rosenthal W, van Os CH, Oksche A, Deen PM (2002) Heteroligomerization of an Aquaporin-2 mutant with wild-type Aquaporin-2 and their misrouting to late endosomes/lysosomes explains dominant nephrogenic diabetes insipidus. Hum Mol Genet 11(7):779–789

Matsuda T, Cepko CL (2007) Controlled expression of transgenes introduced by in vivo electroporation. Proc Natl Acad Sci USA 104(3):1027–1032. doi:10.1073/pnas.0610155104

McDill BW, Li SZ, Kovach PA, Ding L, Chen F (2006) Congenital progressive hydronephrosis (cph) is caused by an S256L mutation in aquaporin-2 that affects its phosphorylation and apical membrane accumulation. Proc Natl Acad Sci USA 103(18):6952–6957. doi:10.1073/pnas.0602087103

Melander O, Orho M, Fagerudd J, Bengtsson K, Groop PH, Mattiasson I, Groop L, Hulthen UL (1998) Mutations and variants of the epithelial sodium channel gene in Liddle's syndrome and primary hypertension. Hypertension 31(5):1118–1124

Moeller HB, Praetorius J, Rutzler MR, Fenton RA (2010) Phosphorylation of aquaporin-2 regulates its endocytosis and protein-protein interactions. Proc Natl Acad Sci USA 107(1):424–429. doi:10.1073/pnas.0910683107

Moeller HB, Rittig S, Fenton RA (2013) Nephrogenic diabetes insipidus: essential insights into the molecular background and potential therapies for treatment. Endocr Rev 34(2):278–301. doi:10.1210/er.2012-1044

Morello JP, Salahpour A, Laperriere A, Bernier V, Arthus MF, Lonergan M, Petaja-Repo U, Angers S, Morin D, Bichet DG, Bouvier M (2000) Pharmacological chaperones rescue cell-surface expression and function of misfolded V2 vasopressin receptor mutants. J Clin Invest 105(7):887–895. doi:10.1172/JCI8688

Mugford JW, Sipila P, McMahon JA, McMahon AP (2008) Osr1 expression demarcates a multi-potent population of intermediate mesoderm that undergoes progressive restriction to an Osr1-dependent nephron progenitor compartment within the mammalian kidney. Dev Biol 324(1):88–98. doi:10.1016/j.ydbio.2008.09.010

Nagy A (2000) Cre recombinase: the universal reagent for genome tailoring. Genesis 26 (2):99–109

Nakada T, Koike H, Akiya T, Katayama T, Kawamata S, Takaya K, Shigematsu H (1987) Liddle's syndrome, an uncommon form of hyporeninemic hypoaldosteronism: functional and histopathological studies. J Urol 137(4):636–640

Nesterov V, Dahlmann A, Krueger B, Bertog M, Loffing J, Korbmacher C (2012) Aldosterone-dependent and -independent regulation of the epithelial sodium channel (ENaC) in mouse distal nephron. Am J Physiol Renal Physiol 303(9):F1289–F1299. doi:10.1152/ajprenal.00247.2012

Nishinakamura R, Matsumoto Y, Nakao K, Nakamura K, Sato A, Copeland NG, Gilbert DJ, Jenkins NA, Scully S, Lacey DL, Katsuki M, Asashima M, Yokota T (2001) Murine homolog of SALL1 is essential for ureteric bud invasion in kidney development. Development 128 (16):3105–3115

Nishita M, Qiao S, Miyamoto M, Okinaka Y, Yamada M, Hashimoto R, Iijima K, Otani H, Hartmann C, Nishinakamura R, Minami Y (2014) Role of Wnt5a-Ror2 signaling in morphogenesis of the metanephric mesenchyme during ureteric budding. Mol Cell Biol 34 (16):3096–3105. doi:10.1128/MCB.00491-14

Oishi I, Suzuki H, Onishi N, Takada R, Kani S, Ohkawara B, Koshida I, Suzuki K, Yamada G, Schwabe GC, Mundlos S, Shibuya H, Takada S, Minami Y (2003) The receptor tyrosine kinase Ror2 is involved in non-canonical Wnt5a/JNK signalling pathway. Genes Cells 8(7):645–654

Okada Y, Feng Q, Lin Y, Jiang Q, Li Y, Coffield VM, Su L, Xu G, Zhang Y (2005) hDOT1L links histone methylation to leukemogenesis. Cell 121(2):167–178

Okazawa M, Murashima A, Harada M, Nakagata N, Noguchi M, Morimoto M, Kimura T, Ornitz DM, Yamada G (2015) Region-specific regulation of cell proliferation by FGF receptor signaling during the Wolffian duct development. Dev Biol 400(1):139–147. doi:10.1016/j.ydbio.2015.01.023

Olesen ET, Rutzler MR, Moeller HB, Praetorius HA, Fenton RA (2011) Vasopressin-independent targeting of aquaporin-2 by selective E-prostanoid receptor agonists alleviates nephrogenic diabetes insipidus. Proc Natl Acad Sci USA 108(31):12949–12954. doi:10.1073/pnas.1104691108

Paragas N, Kulkarni R, Werth M, Schmidt-Ott KM, Forster C, Deng R, Zhang Q, Singer E, Klose AD, Shen TH, Francis KP, Ray S, Vijayakumar S, Seward S, Bovino ME, Xu K, Takabe Y, Amaral FE, Mohan S, Wax R, Corbin K, Sanna-Cherchi S, Mori K, Johnson L, Nickolas T, D'Agati V, Lin CS, Qiu A, Al-Awqati Q, Ratner AJ, Barasch J (2014) alpha-Intercalated cells defend the urinary system from bacterial infection. J Clin Invest 124(7):2963–2976. doi:10.1172/JCI71630

Poladia DP, Kish K, Kutay B, Hains D, Kegg H, Zhao H, Bates CM (2006) Role of fibroblast growth factor receptors 1 and 2 in the metanephric mesenchyme. Dev Biol 291(2):325–339. doi:10.1016/j.ydbio.2005.12.034

Pradervand S, Vandewalle A, Bens M, Gautschi I, Loffing J, Hummler E, Schild L, Rossier BC (2003) Dysfunction of the epithelial sodium channel expressed in the kidney of a mouse model for Liddle syndrome. J Am Soc Nephrol 14(9):2219–2228

Pradervand S, Wang Q, Burnier M, Beermann F, Horisberger JD, Hummler E, Rossier BC (1999) A mouse model for Liddle's syndrome. J Am Soc Nephrol 10(12):2527–2533

Pujo L, Fagart J, Gary F, Papadimitriou DT, Claes A, Jeunemaitre X, Zennaro MC (2007) Mineralocorticoid receptor mutations are the principal cause of renal type 1 pseudohypoaldosteronism. Hum Mutat 28(1):33–40. doi:10.1002/humu.20371

Quigley IK, Stubbs JL, Kintner C (2011) Specification of ion transport cells in the Xenopus larval skin. Development 138(4):705–714. doi:10.1242/dev.055699

Rayner BL, Owen EP, King JA, Soule SG, Vreede H, Opie LH, Marais D, Davidson JS (2003) A new mutation, R563Q, of the beta subunit of the epithelial sodium channel associated with low-renin, low-aldosterone hypertension. J Hypertens 21(5):921–926. doi:10.1097/01.hjh.0000059009.82022.9b

Riepe FG, Finkeldei J, de Sanctis L, Einaudi S, Testa A, Karges B, Peter M, Viemann M, Grotzinger J, Sippell WG, Fejes-Toth G, Krone N (2006) Elucidating the underlying molecular pathogenesis of NR3C2 mutants causing autosomal dominant pseudohypoaldosteronism type 1. J Clin Endocrinol Metab 91(11):4552–4561. doi:10.1210/jc.2006-1161

Rodriguez JA, Biglieri EG, Schambelan M (1981) Pseudohyperaldosteronism with renal tubular resistance to mineralocorticoid hormones. Trans Assoc Am Phys 94:172–182

Roifman M, Marcelis CL, Paton T, Marshall C, Silver R, Lohr JL, Yntema HG, Venselaar H, Kayserili H, van Bon B, Seaward G, Consortium FC, Brunner HG, Chitayat D (2015) De novo WNT5A-associated autosomal dominant Robinow syndrome suggests specificity of genotype and phenotype. Clin Genet 87(1):34–41. doi:10.1111/cge.12401

Rojek A, Fuchtbauer EM, Kwon TH, Frokiaer J, Nielsen S (2006) Severe urinary concentrating defect in renal collecting duct-selective AQP2 conditional-knockout mice. Proc Natl Acad Sci USA 103(15):6037–6042. doi:10.1073/pnas.0511324103 0511324103 [pii]

Romisch K (2004) A cure for traffic jams: small molecule chaperones in the endoplasmic reticulum. Traffic 5(11):815–820. doi:10.1111/j.1600-0854.2004.00231.x

Ronzaud C, Loffing J, Bleich M, Gretz N, Grone HJ, Schutz G, Berger S (2007) Impairment of sodium balance in mice deficient in renal principal cell mineralocorticoid receptor. J Am Soc Nephrol 18(6):1679–1687. doi:10.1681/ASN.2006090975 ASN.2006090975 [pii]

Ronzaud C, Loffing J, Gretz N, Schutz G, Berger S (2011) Inducible renal principal cell-specific mineralocorticoid receptor gene inactivation in mice. Am J Physiol Renal Physiol 300(3): F756–F760. doi:10.1152/ajprenal.00728.2009 300/3/F756 [pii]

Rossier BC, Schild L (2008) Epithelial sodium channel: mendelian versus essential hypertension. Hypertension 52(4):595–600. doi:10.1161/HYPERTENSIONAHA.107.097147

Roy A, Al-bataineh MM, Pastor-Soler NM (2015) Collecting duct intercalated cell function and regulation. Clin J Am Soc Nephrol 10(2):305–324. doi:10.2215/CJN.08880914

Rubera I, Loffing J, Palmer LG, Frindt G, Fowler-Jaeger N, Sauter D, Carroll T, McMahon A, Hummler E, Rossier BC (2003) Collecting duct-specific gene inactivation of alphaENaC in the mouse kidney does not impair sodium and potassium balance. J Clin Invest 112(4):554–565

Russo LM, McKee M, Brown D (2006) Methyl-beta-cyclodextrin induces vasopressin-independent apical accumulation of aquaporin-2 in the isolated, perfused rat kidney. Am J Physiol Renal Physiol 291(1):F246–F253. doi:10.1152/ajprenal.00437.2005

Sartorato P, Lapeyraque AL, Armanini D, Kuhnle U, Khaldi Y, Salomon R, Abadie V, Di Battista E, Naselli A, Racine A, Bosio M, Caprio M, Poulet-Young V, Chabrolle JP, Niaudet P, De Gennes C, Lecornec MH, Poisson E, Fusco AM, Loli P, Lombes M, Zennaro MC (2003) Different inactivating mutations of the mineralocorticoid receptor in fourteen families affected by type I pseudohypoaldosteronism. J Clin Endocrinol Metab 88 (6):2508–2517. doi:10.1210/jc.2002-021932

Sawathiparnich P, Sumboonnanonda A, Weerakulwattana P, Limwongse C (2009) A novel mutation in the beta-subunit of the epithelial sodium channel gene (SCNN1B) in a Thai family with Liddle's syndrome. J Pediatr Endocrinol Metab 22(1):85–89

Schmitt R, Ellison DH, Farman N, Rossier BC, Reilly RF, Reeves WB, Oberbaumer I, Tapp R, Bachmann S (1999) Developmental expression of sodium entry pathways in rat nephron. Am J Phys 276(3 Pt 2):F367–F381

Schwartz GJ, Barasch J, Al-Awqati Q (1985) Plasticity of functional epithelial polarity. Nature 318(6044):368–371

Shida Y, Matsuoka H, Chiga M, Uchida S, Sasaki S, Sugihara S (2013) Characterization of AQP-2 gene mutation (R254Q) in a family with dominant nephrogenic DI. Pediatr Int 55(1):105–107. doi:10.1111/j.1442-200X.2012.03614.x

Shimkets RA, Lifton RP, Canessa CM (1997) The activity of the epithelial sodium channel is regulated by clathrin-mediated endocytosis. J Biol Chem 272(41):25537–25541

Shimkets RA, Warnock DG, Bositis CM, Nelson-Williams C, Hansson JH, Schambelan M, Gill JR Jr, Ulick S, Milora RV, Findling JW, Canessa CM, Rossier B, Lifton RP (1994) Liddle's syndrome: heritable human hypertension caused by mutations in the beta subunit of the epithelial sodium channel. Cell 79(3):407–414

Singer MS, Kahana A, Wolf AJ, Meisinger LL, Peterson SE, Goggin C, Mahowald M, Gottschling DE (1998) Identification of high-copy disruptors of telomeric silencing in *Saccharomyces cerevisiae*. Genetics 150(2):613–632

Smith AN, Skaug J, Choate KA, Nayir A, Bakkaloglu A, Ozen S, Hulton SA, Sanjad SA, Al-Sabban EA, Lifton RP, Scherer SW, Karet FE (2000) Mutations in ATP6N1B, encoding a new kidney vacuolar proton pump 116-kD subunit, cause recessive distal renal tubular acidosis with preserved hearing. Nat Genet 26(1):71–75. doi:10.1038/79208

Snyder PM, McDonald FJ, Stokes JB, Welsh MJ (1994) Membrane topology of the amiloride-sensitive epithelial sodium channel. J Biol Chem 269(39):24379–24383

Snyder PM, Price MP, McDonald FJ, Adams CM, Volk KA, Zeiher BG, Stokes JB, Welsh MJ (1995) Mechanism by which Liddle's syndrome mutations increase activity of a human epithelial Na+ channel. Cell 83(6):969–978

Soundararajan R, Lu M, Pearce D (2012) Organization of the ENaC-regulatory machinery. Crit Rev Biochem Mol Biol 47(4):349–359. doi:10.3109/10409238.2012.678285

Sritippayawan S, Sumboonnanonda A, Vasuvattakul S, Keskanokwong T, Sawasdee N, Paemanee A, Thuwajit P, Wilairat P, Nimmannit S, Malasit P, Yenchitsomanus PT (2004) Novel compound heterozygous SLC4A1 mutations in Thai patients with autosomal recessive distal renal tubular acidosis. Am J Kidney Dis 44(1):64–70

Stark K, Vainio S, Vassileva G, McMahon AP (1994) Epithelial transformation of metanephric mesenchyme in the developing kidney regulated by Wnt-4. Nature 372(6507):679–683. doi:10.1038/372679a0

Staub O, Dho S, Henry P, Correa J, Ishikawa T, McGlade J, Rotin D (1996) WW domains of Nedd4 bind to the proline-rich PY motifs in the epithelial Na+ channel deleted in Liddle's syndrome. EMBO J 15(10):2371–2380

Staub O, Gautschi I, Ishikawa T, Breitschopf K, Ciechanover A, Schild L, Rotin D (1997) Regulation of stability and function of the epithelial Na+ channel (ENaC) by ubiquitination. EMBO J 16(21):6325–6336

Stehberger PA, Shmukler BE, Stuart-Tilley AK, Peters LL, Alper SL, Wagner CA (2007) Distal renal tubular acidosis in mice lacking the AE1 (band3) Cl-/HCO3- exchanger (slc4a1). J Am Soc Nephrol 18(5):1408–1418. doi:10.1681/ASN.2006101072

Stover EH, Borthwick KJ, Bavalia C, Eady N, Fritz DM, Rungroj N, Giersch AB, Morton CC, Axon PR, Akil I, Al-Sabban EA, Baguley DM, Bianca S, Bakkaloglu A, Bircan Z, Chauveau D, Clermont MJ, Guala A, Hulton SA, Kroes H, Li Volti G, Mir S, Mocan H, Nayir A, Ozen S, Rodriguez Soriano J, Sanjad SA, Tasic V, Taylor CM, Topaloglu R, Smith AN, Karet FE (2002) Novel ATP6V1B1 and ATP6V0A4 mutations in autosomal recessive distal renal tubular acidosis with new evidence for hearing loss. J Med Genet 39(11):796–803

Strautnieks SS, Thompson RJ, Gardiner RM, Chung E (1996) A novel splice-site mutation in the gamma subunit of the epithelial sodium channel gene in three pseudohypoaldosteronism type 1 families. Nat Genet 13(2):248–250. doi:10.1038/ng0696-248

Su Y, Blake-Palmer KG, Sorrell S, Javid B, Bowers K, Zhou A, Chang SH, Qamar S, Karet FE (2008) Human H+ATPase a4 subunit mutations causing renal tubular acidosis reveal a role for interaction with phosphofructokinase-1. Am J Physiol Renal Physiol 295(4):F950–F958. doi:10.1152/ajprenal.90258.2008

Tajima T, Kitagawa H, Yokoya S, Tachibana K, Adachi M, Nakae J, Suwa S, Katoh S, Fujieda K (2000) A novel missense mutation of mineralocorticoid receptor gene in one Japanese family with a renal form of pseudohypoaldosteronism type 1. J Clin Endocrinol Metab 85(12):4690–4694. doi:10.1210/jcem.85.12.7078

Toye AM, Williamson RC, Khanfar M, Bader-Meunier B, Cynober T, Thibault M, Tchernia G, Dechaux M, Delaunay J, Bruce LJ (2008) Band 3 Courcouronnes (Ser667Phe): a trafficking mutant differentially rescued by wild-type band 3 and glycophorin A. Blood 111(11):5380–5389. doi:10.1182/blood-2007-07-099473

Uehara Y, Sasaguri M, Kinoshita A, Tsuji E, Kiyose H, Taniguchi H, Noda K, Ideishi M, Inoue J, Tomita K, Arakawa K (1998) Genetic analysis of the epithelial sodium channel in Liddle's syndrome. J Hypertens 16(8):1131–1135

van den Ouweland AM, Dreesen JC, Verdijk M, Knoers NV, Monnens LA, Rocchi M, van Oost BA (1992) Mutations in the vasopressin type 2 receptor gene (AVPR2) associated with nephrogenic diabetes insipidus. Nat Genet 2(2):99–102. doi:10.1038/ng1092-99

van Leeuwen F, Gafken PR, Gottschling DE (2002) Dot1p modulates silencing in yeast by methylation of the nucleosome core. Cell 109(6):745–756

Vargas-Poussou R, Houillier P, Le Pottier N, Strompf L, Loirat C, Baudouin V, Macher MA, Dechaux M, Ulinski T, Nobili F, Eckart P, Novo R, Cailliez M, Salomon R, Nivet H, Cochat P, Tack I, Fargeot A, Bouissou F, Kesler GR, Lorotte S, Godefroid N, Layet V, Morin G, Jeunemaitre X, Blanchard A (2006) Genetic investigation of autosomal recessive distal renal tubular acidosis: evidence for early sensorineural hearing loss associated with mutations in the ATP6V0A4 gene. J Am Soc Nephrol 17(5):1437–1443. doi:10.1681/ASN.2005121305

Viemann M, Peter M, Lopez-Siguero JP, Simic-Schleicher G, Sippell WG (2001) Evidence for genetic heterogeneity of pseudohypoaldosteronism type 1: identification of a novel mutation in the human mineralocorticoid receptor in one sporadic case and no mutations in two autosomal dominant kindreds. J Clin Endocrinol Metab 86(5):2056–2059. doi:10.1210/jcem.86.5.7449

Wade JB (2011) Statins affect AQP2 traffic. Am J Physiol Renal Physiol 301(2):F308. doi:10.1152/ajprenal.00248.2011

Wang C, Chan TK, Yeung RT, Coghlan JP, Scoggins BA, Stockigt JR (1981) The effect of triamterene and sodium intake on renin, aldosterone, and erythrocyte sodium transport in Liddle's syndrome. J Clin Endocrinol Metab 52(5):1027–1032. doi:10.1210/jcem-52-5-1027

Wang LP, Yang KQ, Jiang XJ, Wu HY, Zhang HM, Zou YB, Song L, Bian J, Hui RT, Liu YX, Zhou XL (2015) Prevalence of Liddle syndrome among young hypertension patients of undetermined cause in a Chinese population. J Clin Hypertens 17(11):902–907. doi:10.1111/jch.12598

Wu H, Chen L, Zhou Q, Zhang X, Berger S, Bi J, Lewis DE, Xia Y, Zhang W (2013) Aqp2-expressing cells give rise to renal intercalated cells. J Am Soc Nephrol 24(2):243–252. doi:10.1681/ASN.2012080866 Editorial in the same issue of J Am Soc Nephrol pg 163-165, Selection by Faculty of 1000

Xiao Z, Chen L, Zhou Q, Zhang W (2015) Dot1l deficiency leads to increased intercalated cells and upregulation of V-ATPase B1 in mice. Exp Cell Res. doi:10.1016/j.yexcr.2015.09.014

Xu PX, Zheng W, Huang L, Maire P, Laclef C, Silvius D (2003) Six1 is required for the early organogenesis of mammalian kidney. Development 130(14):3085–3094

Yang B, Gillespie A, Carlson EJ, Epstein CJ, Verkman AS (2001) Neonatal mortality in an aquaporin-2 knock-in mouse model of recessive nephrogenic diabetes insipidus. J Biol Chem 276(4):2775–2779. doi:10.1074/jbc.M008216200 M008216200 [pii]

Yang KQ, Xiao Y, Tian T, Gao LG, Zhou XL (2014) Molecular genetics of Liddle's syndrome. Clin Chim Acta 436:202–206. doi:10.1016/j.cca.2014.05.015

Yang Q, Li G, Singh SK, Alexander EA, Schwartz JH (2006) Vacuolar H+ -ATPase B1 subunit mutations that cause inherited distal renal tubular acidosis affect proton pump assembly and trafficking in inner medullary collecting duct cells. J Am Soc Nephrol 17(7):1858–1866. doi:10.1681/ASN.2005121277

Yashima T, Noguchi Y, Kawashima Y, Rai T, Ito T, Kitamura K (2010) Novel ATP6V1B1 mutations in distal renal tubular acidosis and hearing loss. Acta Otolaryngol 130 (9):1002–1008. doi:10.3109/00016481003631529

Yu J, Carroll TJ, McMahon AP (2002) Sonic hedgehog regulates proliferation and differentiation of mesenchymal cells in the mouse metanephric kidney. Development 129(22):5301–5312

Yu J, Carroll TJ, Rajagopal J, Kobayashi A, Ren Q, McMahon AP (2009) A Wnt7b-dependent pathway regulates the orientation of epithelial cell division and establishes the cortico-medullary axis of the mammalian kidney. Development 136(1):161–171. doi:10.1242/dev.022087

Zhang W, Hayashizaki Y, Kone BC (2004) Structure and regulation of the mDot1 gene, a mouse histone H3 methyltransferase. Biochem J 377(Pt 3):641–651

Zhang W, Xia X, Jalal DI, Kuncewicz T, Xu W, Lesage GD, Kone BC (2006a) Aldosterone-sensitive repression of ENaCalpha transcription by a histone H3 lysine-79 methyltransferase. Am J Phys Cell Physiol 290(3):C936–C946

Zhang W, Xia X, Reisenauer MR, Hemenway CS, Kone BC (2006b) Dot1a-AF9 complex mediates histone H3 Lys-79 hypermethylation and repression of ENaC{alpha} in an aldosterone-sensitive manner. J Biol Chem 281(26):18059–18068

Zhang W, Xia X, Reisenauer MR, Rieg T, Lang F, Kuhl D, Vallon V, Kone BC (2007) Aldosterone-induced Sgk1 relieves Dot1a-Af9-mediated transcriptional repression of epithelial Na+ channel alpha. J Clin Invest 117(3):773–783 (Comment in the same issue of J Clin Invest pg 592–595, Selection by Faculty of 1000)

Zhang W, Yu Z, Wu H, Chen L, Kong Q, Kone BC (2013) An Af9 cis-element directly targets Dot1a to mediate transcriptional repression of the alphaENaC gene. Am J Physiol Renal Physiol 304(4):F367–F375. doi:10.1152/ajprenal.00537.2011

Chapter 8
Origin and Function of the Renal Stroma in Health and Disease

Christopher J. Rowan, Sepideh Sheybani-Deloui, and Norman D. Rosenblum

Abstract The renal stroma is defined as a heterogeneous population of cells that serve both as a supportive framework and as a source of specialized cells in the renal capsule, glomerulus, vasculature, and interstitium. In this chapter, we review published evidence defining what, where, and why stromal cells are important. We describe the functions of the renal stroma andhow stromal derivatives are crucial for normal kidney function. Next, we review the specification of stromal cells from the *Osr1*+ intermediate mesoderm and *T*+ presomitic mesoderm during embryogenesis and stromal cell differentiation. We focus on stromal signaling mechanisms that act in both a cell and non-cell autonomous manner in communication with the nephron

Christopher J. Rowan and Sepideh Sheybani-Deloui: These authors contributed equally to this work.

C.J. Rowan
Program in Developmental and Stem Cell Biology, The Hospital for Sick Children, Toronto, ON, Canada

Department of Laboratory Medicine and Pathobiology, University of Toronto, Toronto, ON, Canada

S. Sheybani-Deloui
Program in Developmental and Stem Cell Biology, The Hospital for Sick Children, Toronto, ON, Canada

Department of Physiology, University of Toronto, Toronto, ON, Canada

N.D. Rosenblum (✉)
Program in Developmental and Stem Cell Biology, The Hospital for Sick Children, Toronto, ON, Canada

Department of Laboratory Medicine and Pathobiology, University of Toronto, Toronto, ON, Canada

Department of Physiology, University of Toronto, Toronto, ON, Canada

Division of Nephrology, Department of Paediatrics, University of Toronto, Toronto, ON, Canada

Peter Gilgan Centre for Research and Learning, Hospital for Sick Children, 686 Bay St., Rm 16-9-706, Toronto, ON, M5G 0A4, Canada
e-mail: norman.rosenblum@sickkids.ca

R.K. Miller (ed.), *Kidney Development and Disease*, Results and Problems in Cell Differentiation 60, DOI 10.1007/978-3-319-51436-9_8

progenitor and ureteric lineages. To conclude, stromal cells and the contribution of stromal cells to renal fibrosis and chronic kidney disease are described.

8.1 Introduction

The stroma has been defined classically as a connective framework of support cells and associated extracellular matrix. In the mammalian adult kidney, the stroma was historically defined as the renal interstitial matrix and resident fibroblasts and any elements that are neither nephrogenic nor derived from ureteric branches (Hatini et al. 1996). More recently, this definition has been expanded to include vascular smooth muscle, pericytes, mesangial cells, renal capsular cells, and resident fibroblasts (Fig. 8.1) (Humphreys et al. 2010; Kobayashi et al. 2014).

Stromal elements reside within both the cortical and medullary zones of the kidney. Surrounding the cortical portion of the kidney lies the renal capsule (Fig. 8.1a), which is thought to be involved in detachment of the kidney from the body wall during embryonic development (Levinson et al. 2005; Kobayashi et al. 2014; Li et al. 2014). Vascular smooth muscle cells, pericytes, and mesangial cells are stromal components associated with major renal vessels, peritubular, and glomerular capillaries, respectively (Fig. 8.1b). Pericytes and mesangial cells are collectively involved in the glomerular vasculature response to various physical stimuli and in regulation of renal hemodynamics. Finally, resident fibroblasts are

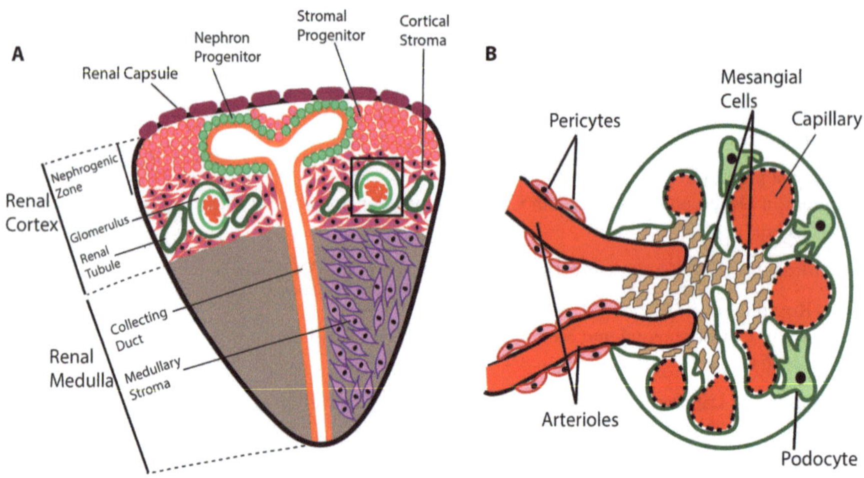

Fig. 8.1 Structural organization of the prenatal embryonic kidney and development of the renal stroma. Stromal cells are present in various locations in the prenatal kidney (E18.5). (**a**) A schematic showing the outer nephrogenic zone and inner medullary zone containing stromal elements including cortical stroma (*pink*), renal capsule (*dark purple*), and medullary stroma (*light purple*). A glomerulus is outlined in the black box. (**b**) A higher resolution illustration of the outlined area in (**a**). The glomerulus contains stromal derivatives including pericytes (*pink*), mesangial cells (*brown*), and endothelial cells of the renal vasculature (*black*). Abbreviations: *NP* nephron progenitors; *NZ* nephrogenic zone

located in the outer nephrogenic zone and inner medullary region interspersed between tubules and nephrons and contain supportive fibroblast cells and ECM components (Fig. 8.1a).

A large body of work has recently provided critical insights into stromal signaling mechanisms during development. Key studies have demonstrated that the stroma plays an important role in controlling branching morphogenesis, nephrogenesis, and vascular development [for review, see: (Li et al. 2014)]. In addition to physiological roles in the developing and adult kidney, stromal cells are also thought to play a role in disease states, especially renal fibrosis (Falke et al. 2015). Despite existing knowledge of stromal cell types, and their normal function, the embryonic origin of these cells and their role in renal diseases are not fully elucidated. In this chapter, we describe the origin, differentiation, and function of embryonic stromal cell types in the developing kidney

8.2 Embryological Origin of the Stromal Cell Lineage

8.2.1 Specification and Lineage Tracing

Stromal cells are one of the earliest lineages specified in the metanephric mesenchyme (MM) and are marked by expression of the transcription factor *Foxd1* as early as Embryonic day (E) 10.5–E11.5 (Kobayashi et al. 2014). *Foxd1*+ cells are initially found anteriorly to nephron progenitors (NP), condensing around ureteric bud (UB) tips (Yallowitz et al. 2011; Kobayashi et al. 2014). As development proceeds, *Foxd1*+ cells occupy a more posterior position under the regulation of *Hox10* and integrate into the nephrogenic zone (NZ) and intercalate between nephrogenic elements (Fig. 8.1a) (Yallowitz et al. 2011). A comprehensive analysis of the fate of lineage-tagged *Foxd1*+ cells demonstrated the self-renewing capacity and multipotent potential of these cells to give rise to all the different types of stromal cells including interstitial fibroblasts, pericytes, and mesangial cells (Kobayashi et al. 2014). In contrast, the potential of *Foxd1*+ cells to give rise to endothelial cells is still debated. *Foxd1*+ derivatives, marked by *β-gal* expression, were not found to contribute to *Flk1*+ endothelial cells (Kobayashi et al. 2014). However, in a separate study in which *Foxd1*+ cells and their derivatives were marked with Red Fluorescence Protein (RFP), PECAM and FLK1, the two markers of endothelial cells, co-localized with RFP in peritubular capillaries, suggesting the potential of *Foxd1*+ cells to give rise to a subset of endothelial cells in the kidney (Sims-Lucas et al. 2013).

The origin of *Foxd1*+ cells has been investigated in relation to several key regulators of the intermediate mesoderm (IM) and the MM. Odd skipped regulator 1, *Osr1*, is a zinc finger transcription factor broadly expressed in the early IM at E7.5. While *Osr1* expression is excluded from the ureteric epithelial lineage at E9.5, it remains strongly expressed in the differentiating MM from E9.5 to E15.5

(James et al. 2006). *Osr1* is essential for the initial formation of the MM. Expression of several key regulators of MM development, including *Six2*, *Pax2*, and *Sall*, is disrupted in the absence of *Osr1*, and the newly formed MM undergoes apoptosis (James et al. 2006). Mugford et al. (2008) demonstrated that *Osr1*+ cells pulse labeled between E8.5 and E10.5 contribute to both *Six2*+ NP and *Foxd1*+ stromal progenitors, while *Osr1*+ cells labeled at E11.5 onwards are restricted to the NP lineage (Mugford et al. 2008). Thus, it is clear that both the NP and stromal lineages of the kidney emerge within the multipotent progenitor pool of *Osr1*+ IM prior to the onset of kidney development (E11.5) (Fig. 8.2b).

Despite advances made in understanding the signaling mechanisms establishing the NP compartment, regulation of stromal lineage specification from the *Osr1*+ progenitor pool remains poorly understood (Taguchi et al. 2014). Analysis of both OSR1 and FOXD1 deficient mice showed that expression of neither *Osr1* nor

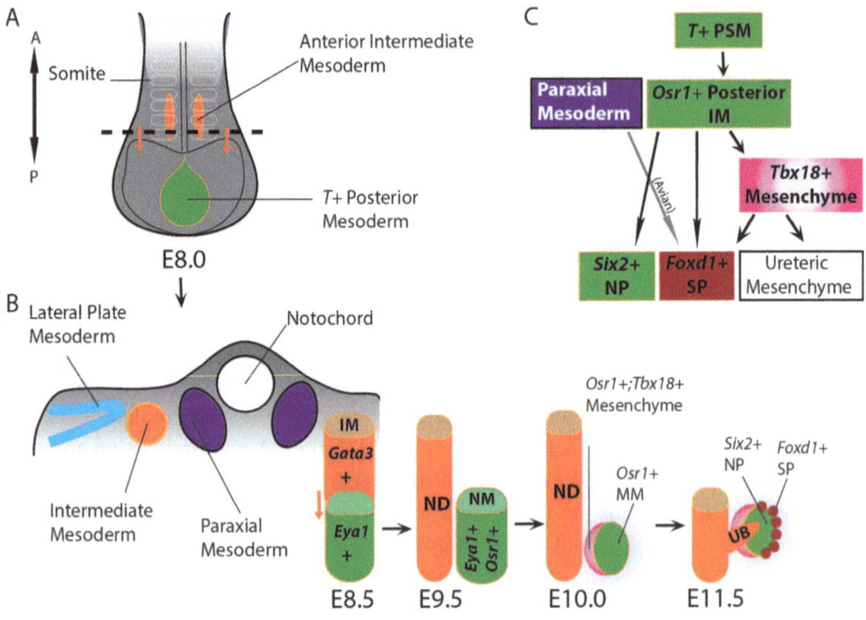

Fig. 8.2 Development of the rostral and caudal intermediate mesoderm. (**a**) A diagram illustrating the elongating anterior IM (*orange*) and the nascent *T*+ PM (*green*) in an E8.0 embryo. (**b**) Cross section of an E8.5 embryo at the position marked with a *dashed line* in (**a**). *Gata3*+ anterior IM and its derivative, the ND, are shown in *orange*. *T*+ PSM and its derivative *Osr1*+*Eya1*+ posterior IM are shown in *green*. *Tbx18*+ cells, located adjacent to the ND and the MM, is shown in *pink*. At E11.5 *Six2*+ nephrogenic (*green*) and *Foxd1*+ stromal progenitors (*red*) (NP and SP, respectively) are specified from the *Osr1*+ progenitor pool. (**c**) Summary of the proposed model of the origin of the renal stroma. *Foxd1*+ stromal progenitors arise from the *Osr1*+ posterior IM. This precursor population develops independently from the anterior IM and specifically originates from the T+ PSM. Uncommitted *Tbx18*+ cells located in between the MM and the ND lose *Tbx18* expression and give rise to *Foxd1*+ cells. Paraxial mesoderm is the principle origin of the *Foxd1*+ cells in chicken. *NP* nephron progenitors; *SP* stromal progenitors; *UB* ureteric bud; *MM* metanephric mesenchyme; *NM* nephrogenic mesenchyme

Foxdl is required for specification of stromal progenitors (Yallowitz et al. 2011; Levinson et al. 2005). A detailed understanding of temporal fate restriction in common precursors of the stromal and nephron progenitors will provide us with a deeper insight into the origin of stromal cells and could thereby potentially lead to identification of genes involved in stromal lineage specification.

Until recently, all the different lineages involved in kidney development, including the nephrogenic and stromal progenitors, as well as the ureteric epithelium, were thought to originate from the elongated rostral (anterior) IM (Dressler 2006). However, it is now appreciated that the *Osr1+* metanephric field and the ureteric epithelium develop from the caudal (posterior) IM and the rostral (anterior) IM, respectively (Grote et al. 2006). Several types of evidence demonstrated the distinct developmental roles of the rostral and caudal IM. Abrogation of rostral IM elongation and UB formation in mice with deficiency of *Gata3* does not adversely affect formation of the *Pax2*-expressing MM (Grote et al. 2006). Inducible labeling of caudal IM cells expressing *Eya1*, a transcription factor essential for the initial formation of the MM, demonstrated the nephron-restricted fate of these cells, and lack of any contribution to ND-derived structures (Fig. 8.2b) (Gong et al. 2007; Xu et al. 2014a). *Brachyury* (*T*) is a marker of immature caudal pre-somitic mesoderm (PSM) that persists in the posterior end of the embryo (Fig. 8.2). Investigation of the fate of *Brachyury* (*T*)+ cells labeled at E8.0 in the mouse demonstrated the contribution of these cells to the *Osr1+* MM, while labeled cells were not detected in the ND (Taguchi et al. 2014). It is now believed that caudal trunk progenitors residing within the presomitic mesoderm (PSM) that lose *Brachyury* (*T*) expression turn on *Eya1* expression at around E8.5 to form the posterior IM, which subsequently gives rise to *Osr1+* renal precursors (Takasato and Little 2015).

Studies defining T-box transcription factor 18 (*Tbx18*) expression in the urogenital region have provided further insight into the spatiotemporal pattern of stromal lineage specification. *Tbx18* was originally demonstrated to mark undifferentiated ureteric mesenchyme. Subsequent analysis of *Tbx18* expression at E11.5 uncovered a population of *Tbx18+Osr1+* cells present in a narrow band in between the mesenchyme surrounding the ND and the nephrogenic mesenchyme (Fig. 8.2) (Airik et al. 2006; Bohnenpoll et al. 2013). Lineage tracing of *Tbx18+* cells has demonstrated the contribution of this spatially restricted population of *Tbx18+* cells to *Foxdl+* derivative cells (Bohnenpoll et al. 2013). Hence, it was postulated that at E10.5, in addition to the already specified *Foxdl+* and *Six2+* lineages, there are uncommitted *Tbx18+* cells within the *Osr1+* cell population that give rise to either stromal or ureteric mesenchymal lineages (Bohnenpoll et al. 2013). The mechanism whereby *Tbx18+* cells that do not differentiate into the mesenchymal cell types of the ureter lose *Tbx18* expression between E10.5–E11.5 and contribute to the renal stroma remains to be defined.

Studies in the chick and mouse embryo suggest the existence of a distinct progenitor of renal stromal cells residing outside the IM. First, direct fate-mapping studies in the chick embryo identified the adjacent paraxial mesoderm (PM) as the principle source of the renal stroma. This was evident by localization of lineage-tagged cells derived from the PM to the *Foxdl+* stromal zone of the kidney

(Guillaume et al. 2009). Moreover, a recent high-throughput gene expression analysis in mice lacking *Pax2* provided indirect evidence that lower *Pax2* expression in the mammalian PM might govern the contribution of the PM to the renal stroma (Ranghini and Dressler 2015). When *Pax2* is deficient in the IM, IM cells are characterized by a pattern of gene expression that is more consistent with the PM, and genes normally expressed in the common derivatives of *Foxd1*+ progenitors are upregulated (Ranghini and Dressler 2015). Therefore, it was hypothesized that after *Osr1* expression is restricted to NP at E11.5, new stromal progenitors likely arise from the PM and migrate to the induced MM. Yet, to date, there has been no direct *in vivo* evidence demonstrating the contribution of PM cells to the renal stroma.

In summary, the published body of work strongly suggests that *Osr1*+ progenitor pool derived from the *T*+ PSM has the potential to give rise to both stromal and nephron progenitors. Further studies have shown the presence of a subpopulation of non-nephrogenic *Osr1*+ cells expressing *Tbx18* that contributes to both ureteric mesenchyme and the renal stromal lineage. It is also likely that mammalian stromal progenitors are a heterogeneous population of cells, a subset of which are derived from the PM. Despite these advances in defining the developmental origin of the stromal progenitors, the molecular mechanisms regulating the separation of the stromal lineage from either nephrogenic or ureteric mesenchymal lineages remains incompletely defined.

8.2.2 Differentiated Stromal Cell Types

8.2.2.1 Renal Capsule

The outer layer of the kidney contains the *Foxd1*+-derived renal capsule, which consists of a flattened layer of stromal cells defined by expression of the molecular markers *Foxd1*, *Raldh2*, and *Sfrp1* (Hatini et al. 1996; Levinson et al. 2005). *Foxd1* plays a crucial role in proper renal capsule development. In the absence of *Foxd1*, the capsule is present but consists of a thicker layer that ectopically expresses markers specific to the endothelial linage of the kidney, *Flk1* and PECAM (Levinson et al. 2005). Using *Foxd1*$^{GFP/GFP}$ mice, it was shown that the endothelial cells present in the capsule of mutant mice are not derived from *Foxd1*+ progenitors but are ectopically located in the capsule. These observations suggest that FOXD1 regulates the establishment of the correct population of cells in the capsule (Levinson et al. 2005). *Tbx18* and *Hox10* are strongly expressed in the renal capsule from E12.5 onwards and are required for normal development of the capsule, possibly through acting in a similar regulatory pathway as *Foxd1* (Xu et al. 2014b; Yallowitz et al. 2011). The mechanisms by which each of these factors regulate the differentiation and integration of the capsular stroma is undefined.

8.2.2.2 Vascular Pericytes and Mesangial Cells

Mature renal vessels contain a special contractile cells surrounding the endothelial cells and share some characteristics, such as α-SMA expression, with smooth muscle cells resident in major vessels (Am et al. 2013). The polymorphic α-SMA + cells surrounding the peritubular capillaries are called pericytes while the specialized cells residing within the glomerular space are called mesangial cells (Fig. 8.1). Collectively, these cells regulate medullary blood flow and contribute to kidney fibrosis by differentiation into collagen-generating myofibroblasts (Quaggin and Kreidberg 2008; Am et al. 2013; Humphreys et al. 2010). Fate-mapping experiments have demonstrated that *Foxd1+* cells undergo distinct differentiation pathways to give rise to both pericytes and mesangial cells (discussed below) (Quaggin and Kreidberg 2008; Kobayashi et al. 2014; Humphreys et al. 2010).

The study of pericyte differentiation has been limited by the lack of specific markers for various intermediate stages of cellular development. Renin, an enzyme synthesized and secreted in the kidney and non-renal tissues, catalyzes formation of the angiotensin peptides that regulate blood pressure, renal hemodynamics, and electrolyte balance (Sequeira-Lopez et al. 2015). Renin-expressing cells arise from *Foxd1+* cells at E12–E14 at the juxtaglomerular location where they mainly regulate elongation and branching of the renal vasculature. A subset of renin+ cells differentiate to become α-SMA+ pericytes (Sequeira Lopez et al. 2001, 2015). The manner by which renin+ cells differentiate to α-SMA+ pericyte fate is not well defined. Studies in *Foxd1*-DTA "stroma-less" mice showed that FOXD1 activity does not regulate differentiation of α-SMA+ pericytes (Hum et al. 2014). *Wnt4* deficiency in the medullary stroma has been shown to cause a loss of α-SMA+ pericytes which is accompanied by a complete lack of *Bmp4* expression in kidneys (Itäranta et al. 2006). Bmp4 is crucial for smooth muscle development around the ureteric epithelium. Thus, it is likely that *Bmp4* regulation is also involved in the medullary stroma to determine the fate of the pericytic smooth muscles via Wnt4 signaling (Miyazaki et al. 2003; Itäranta et al. 2006). The role of other signaling pathways in pericyte differentiation remains to be determined.

Like pericytes, glomerular mesangial cells also express *Pdgfrβ* and desmin during early stages of renal development and α-SMA+ later in development (Lindahl et al. 1998). However, mesangial cells are functionally specialized in that they contribute to the proper formation of the glomerular capillary loop and the filtration function of the glomeruli (Quaggin and Kreidberg 2008). Mesangial cell development is dependent on normal endothelial migration into the developing glomerulus, secretion of platelet-derived growth factor (PDGFβ) by the glomerular endothelium, and initial specification of mesangial cells outside the glomerulus followed by migration of mesangial cells into the glomerular cleft (Chen et al. 2011; Lindahl et al. 1998). Mutations in either *Pdgfβ* or *Pdgfrβ* result in failure of migration of mouse mesangial precursors to the glomerular tuft. In these mice, desmin+ mesangial precursors are still found in close proximity to but not inside the glomeruli (Gaengel et al. 2009).

It is not clear how mesangial cell precursors are specified and fated to respond to PDGFβ signaling such that only these cells migrate to the glomerular tuft. Analysis of *Pdgfrβ*-null mice has shown that intrarenal arterial and arteriolar smooth muscle cells are also critical targets for PDGFβ signaling (Lindahl et al. 1998). Also, *Pdgfrβ* expression is itself regulated by PBX1, a transcription factor which is highly expressed in the interstitial components of the developing kidney, and is not specific to a particular cell type (Hurtado et al. 2015). However, studies investigating the role of Notch signaling in the formation of the mesangium and kidney vasculature uncovered a Notch-dependent signal that enables only mesangial cells to respond to PDGFβ signals from endothelial cells within the developing glomerular cleft (Shawber and Kitajewski 2004). Utilizing mice in which cells responding to Notch activation and all their descendants are marked with a reporter, it was first shown that mesangial cells respond to NOTCH1 and NOTCH2 signaling during development (Boyle et al. 2014). Specific removal of the obligate binding partner of NOTCH intracellular domains from *Foxd1*+ stromal progenitors resulted in a complete loss of PDGFRβ+ mesangial cells. Loss of these cells is likely due to failure of the stromal progenitors to produce desmin+ cells capable of assuming a mesangial cell fate (Boyle et al. 2014). The identity and location of the Notch ligand(s) required for mesangial cell fate determination remain to be further elucidated.

8.2.2.3 Interstitial Fibroblasts

Interstitial fibroblasts, characterized as widely spaced cells with long cellular projections, also arise from *Foxd1*+ stromal progenitors (Fig. 8.1) (Cui et al. 2003; Kobayashi et al. 2014). Distinct cortical and medullary interstitial cells, with marked differences in gene expressions and physiological functions, are established during embryonic development. For example, cortical peritubular fibroblasts display the membrane-bound enzyme ecto-5′-nucleotidase (5′NT), produce extracellular adenosine, and are postnatally the main source of renal erythropoietin (Marxer-Meier et al. 1998; Dawson et al. 1989; Le Hir and Kaissling 1989, 1993). In fact, adult interstitial cells are critical for erythropoiesis, and recent work done by Kobayashi et al. (2016) has demonstrated that distinct subpopulations of the stroma regulate erythropoietin via the HIF2-PHD signalling axis (Kobayashi et al. 2016). In addition, a specific subset of the medullary interstitial cells are marked by CDKN1C (P57KIP2) expression, which is a cyclin-dependent kinase (Cdk) inhibitor essential for proper development of the medullary region (Caspary et al. 1999; Kobayashi et al. 2014). Although the embryonic mechanisms underlying differentiation of the cortical and medullary interstitium from the *Foxd1*+ progenitor population are not fully understood, it has been shown that P57KIP2 may be a direct target of Wnt7b signaling during renal medulla morphogenesis and establishment of the cortico-medullary axis of the kidney (discussed further below) (Yu et al. 2009). Therefore, it is likely that similar Wnt7b-dependent pathways could be involved in stromal progenitors to control their cortical versus medullary interstitial fate.

8.3 Signaling Mechanisms and the Control of Renal Stromal Progenitors

Normal renal development is highly dependent on proper cell–cell communication between the ureteric, nephrogenic, and stromal lineages in the nephrogenic zone. Indeed, stromal cells play an active role in regulating the two main processes in renal development, branching morphogenesis and nephrogenesis. Various signaling pathways that have been implicated in mediating stromal-NP and stromal-UB crosstalk are discussed below (Table 8.1).

Table 8.1 Genes involved in stromal cell autonomous and non-autonomous signaling mechanisms and stromal cell differentiation

Gene name	Localization	Genetic deletion phenotype	Reference
Foxd1	CS, podocytes	Hypoplastic, fused kidneys; reduced branching, expanded NP	Hatini et al. (1996), Levinson et al. (2005)
Decorin	CS, MS	Partial rescue of *Foxd1* null phenotype	Fetting et al. (2014)
Rarα/β	*Rarα*: ubiquitous; *Rarβ*: CS, MS	*Rarαβ* double KO: expanded CS domain, reduced nephrons, inhibited branching	Mendelsohn et al. (1999), Batourina et al. (2001)
Raldh2	CS	Similar to *Rarαβ* double KO	Batourina et al. (2005), Rosselot et al. (2010)
Wnt4	MS, differentiating NP, PTA	Failure of epithelial RV formation	Stark et al. (1994)
β-catenin	CS, MS, NP, UB	CKO using *Foxd1Cre*: lack of medullary zone, improper oriented cell divisions; decreased *Wnt9b*, depleted NP	Yu et al. (2009), Boivin et al. (2015)
Sfrp1	Capsule, CS (<E14.5), LOH (>E15.5)	Reduced medullary zone, increased glomeruli (in some adult mutants)	Trevant et al. (2008)
Fat4	CS, MS, NP (weakly)	Small, cystic kidneys; CKO using *Foxd1Cre*: expanded NP, disrupted cell organization	Saburi et al. (2008), Das et al. (2013), Bagherie-Lachidan et al. (2015), Mao et al. (2015)
Pod1	CS, MS, NP, podocytes	Hypoplasia, reduced UB branching, expanded CS	Cui et al. (2003), Quaggin et al. (1999)

(continued)

Table 8.1 (continued)

Gene name	Localization	Genetic deletion phenotype	Reference
Pbx1	CS, MS, NP (weakly)	Renal hypoplasia/aplasia, expanded NP, ectopic *Ret* expression	Schnabel et al. (2003), Hurtado et al. (2015)
		CKO using *Foxd1Cre*: diminished *Foxd1*, increased *Pdgfrb*, and perturbed arterial patterning	
Sall1	CS, NP	CKO using *Foxd1Cre*: expansion of NP, dilated tubules, and glomeruli	Ohmori et al. (2015)
Dicer	CS, NP, UB, podocytes, renin cells, proximal tubules, juxtaglomerular cells	CKO using *Foxd1Cre*: defects in NP compartment, vasculture, and glomeruli	Nakagawa et al. (2015), Phua et al. (2015)
Notch1/2	*Notch1:* VSMC, MGP; *Notch2:* EC, Interstitial fibroblasts	Loss of desmin+ cells close to glomerular cup, glomerular aneurysm, perinatal death	Boyle et al. (2014)
Hox10	CS, NP	Failure of CS differentiation and integration into the NZ; similar to *Foxd1* null	Yallowitz et al. (2011)
Pdgfrβ	MGP, EC	Failure of desmin+ cells to migrate to the glomeruli, microaneurysms similar to capillary ballooning	Lindahl et al. (1998)

CS cortical stroma; *MS* medullary stroma; *NP* nephron progenitors; *PTA* pretubular aggregate; *RV* renal vesicle; *UB* ureteric bud; *LOH* loop of Henle; *CKO* conditional knockout; *MGP* mesangial precursor; *VSMC* vascular smooth muscle cell; *EC* endothelial cell

8.3.1 FOXD1

The spatial arrangement of stromal cells in the developing NZ is ideal for and communicating with both NP and UB tip cells (Fig. 8.3). Loss of *Foxd1*, a key stromal marker, results in kidneys that have abnormal vasculature, less nephrons, and abnormal capsular morphology (Hatini et al. 1996; Levinson et al. 2005). Using *Foxd1Cre;DTA* (diphtheria toxin) mice, Hum et al. (2014) showed that ablation of the renal stroma phenocopies *Foxd1*-KO mice, implicating *Foxd1* in maintaining the functional integrity of the stroma (Fig. 8.3a) (Das et al. 2013; Hum et al. 2014). Further analysis by Levinson et al. (2005) showed that the *Foxd1*-KO mice are characterized by numerous cell autonomous defects, including decreased capsular expression of *Sfrp1*, a secreted WNT inhibitor involved in endothelial migration

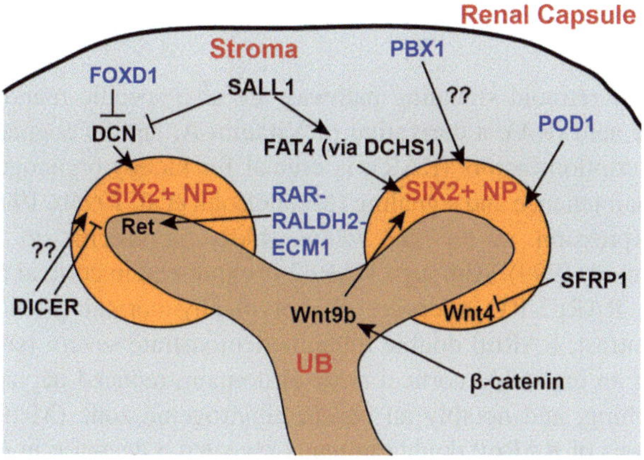

Fig. 8.3 Stromal cell autonomous and non-cell autonomous signaling mechanisms. A summary of stromal signaling in the nephrogenic zone, with major compartments labeled in *red text* (*gray*, stroma; *orange*, NP; *brown*, UB). Many different signaling pathways contribute to the interactions between the UB, NP, and stromal compartments. *Arrows* indicate genetic interactions and/or signaling directions. Genes in *blue text* are known to play cell-autonomous roles in the stromal compartment (or stromal-derived compartments). Stromal FOXD1 represses the expression of BMP-inhibitor Decorin (DCN), which in turn signals to NP by inhibiting BMP-mediated nephron differentiation. SALL1 (possibly via DCN and FAT4), FAT4, PBX1, and POD1 act to restrict NP expansion and promote differentiation in contrast to inductive signals from the UB. Stromal β-catenin modulates *Wnt9b* expression in the UB, which is required for induction of NP. The RAR-RALDH2-ECM1 signaling axis drives *Ret* expression in the UB tip that controls branching morphogenesis. Stromal SFRP1 inhibits WNT4 activity and inhibits epithelial conversion. DICER may have a role in controlling NP maintenance via a variety of miRNAs

and patterning, decreased *Raldh2*, involved in retinoid signalling, and ectopic Bmp4-expressing cells in the renal capsule (Dufourcq et al. 2002; Levinson et al. 2005). *Foxd1*-null kidneys are also characterized by expanded NP and perturbed nephrogenesis, suggesting an additional non-cell autonomous role for *Foxd1* in nephron formation (Hatini et al. 1996; Levinson et al. 2005). Analysis of NP cells in the *Foxd1*-KO mouse demonstrated that the expanded NP cells remain in a self-renewing state and are not sensitive to the effect of WNT-induced differentiation (Fetting et al. 2014). It was hypothesized that ectopic Bmp4 expression in the capsule is the cause of the abnormal NP differentiation (Levinson et al. 2005). These data suggest that *Foxd1* plays an important role in signaling between the stroma and NP domain to restrict NP expansion.

To understand how stromal-BMP signaling might be affecting NPs in *Foxd1*-KO mice, transcriptome analysis of *Foxd1*-KO kidneys was performed. Decorin, a BMP inhibitor, was found to be upregulated in the *Foxd1*-null kidney (Fetting et al. 2014) (Fig. 8.3). FOXD1 directly represses DCN such that deletion of *Dcn* in the *Foxd1*-null background rescues NP differentiation by restoring SMAD-mediated BMP signaling in NP (Brown et al. 2013; Fetting et al. 2014). These data suggest that FOXD1 acts to repress DCN activity in mediating BMP-induced differentiation of NP (Fig. 8.3).

8.3.2 Retinoid Signaling

Members of the retinoid signaling pathway are also specific markers of stromal cells. Retinoic acid (RA), a derivative of Vitamin A, and its cognate nuclear RA receptor transcription factors (RAR), is crucial for kidney organogenesis and for stromal cell maintenance and signaling (Mendelsohn et al. 1994b, 1999). RARα has ubiquitous expression in the NZ while RARβ is specific to stromal cells (Mendelsohn et al. 1999). Although RARβ2 is expressed in cortical and medullary stromal cells, RARβ2-KO mice are phenotypically normal (Mendelsohn et al. 1994a). In contrast, RARαβ double mutants demonstrate severe renal abnormalities, including an expanded cortical stromal domain, reduced nephrons, inhibited ureteric branching, and notably an absent nephrogenic zone (Mendelsohn et al. 1994b). Analysis of RARαβ double mutants revealed a decrease in *Ret*, a receptor tyrosine kinase that is normally expressed in the UB and is required for UB outgrowth and branching (Schuchardt et al. 1996; Mendelsohn et al. 1999). The downregulation of *Ret* expression in RARαβ double mutants suggests that stromal-retinoid signals control ureteric branching through regulating the expression of *Ret*, thus indicating a novel non-cell autonomous role of stromal signaling (Fig. 8.3). This finding is in contrast to *Foxd1*-KO mice, in which *Ret* expression is inappropriately maintained in the UB trunk and branches, and suggests that *Foxd1* and retinoids act via separate signaling mechanisms in the stroma (Hatini et al. 1996). Furthermore, constitutive expression of *Ret* rescues stromal cell patterning and branching in RARαβ double mutants, further demonstrating that *Ret* acts downstream of stromal-retinoid signaling (Batourina et al. 2001).

Enzymes responsible for RA synthesis, members of the retinaldehyde dehydrogenase (RALDH) family, are also important for renal development (Batourina et al. 2005). *Raldh2* is strongly expressed in stromal cells, and deletion of *Raldh2* results in a less severe but similar phenotype to RARαβ double mutant mice (Rosselot et al. 2010; Batourina et al. 2005). To understand the mechanism to how RAR-RALDH2 signaling influences branching, Paroly et al. (2013) used microarray analysis to analyze the effect of RA treatment on stromal cell gene expression (Paroly et al. 2013). Analysis of kidneys treated with RA showed an increased expression of *Ecm1*, which is secreted from cortical stromal cells (Paroly et al. 2013). Neutralization of ECM1 results in expanded *Ret* expression and decreased branching, suggesting that RA-dependent *Ret* acts via stromal ECM1 to control branching (Paroly et al. 2013). Taken together, these data suggest a key paracrine role for RA signaling between stromal cells in controlling *Ret* expression in the ureteric bud (Fig. 8.3).

8.3.3 WNT Signaling

Both canonical and noncanonical WNT signaling pathways play critical roles during branching morphogenesis and NP cell differentiation (Karner et al. 2009, 2011; Park et al. 2007, 2012; Carroll et al. 2005). In contrast, comparatively little is known about WNT signaling mechanisms in the stroma. Evidence of WNT signaling in the stroma was first demonstrated in the medullary stroma. Deletion of *Wnt4* results in defects in the medullary stroma, specifically a decrease in expression of α-SMA and a lack in smooth muscle differentiation (Itäranta et al. 2006). Itäranta et al. (2006) also showed that a lack of *Wnt4* results in a loss of *Bmp4* expression, suggesting a mechanism whereby Wnt4-dependent *Bmp4* expression controls the fate of smooth muscle cells in the medullary stroma (Itäranta et al. 2006). A role for *Wnt4* in the adult renal intersitium is suggested by strong *Wnt4* expression in tubule cells of the renal papilla and in proliferating medullary myofibroblasts during renal fibrosis (DiRocco et al. 2013).

Wnt7b is expressed in the medullary collecting duct network. *Wnt7b* deficiency results in kidneys that lack a medullary zone and tubular epithelial cells that are biased to divide along the radial (versus the longitudinal) axis (Yu et al. 2009). Yu et al. (2009) demonstrated that *Wnt7b* signals to the interstitium via a canonical mechanism. Evidence of canonical WNT signaling to adjacent medullary stromal cells was demonstrated by examination of canonical WNT targets *Lef1* and *Axin2*, which showed *Wnt7b*-dependent expression (Yu et al. 2009). In addition, the adjacent medullary stromal cells co-expresssed LEF1 with a canonical WNT reporter transgene demonstrating that Wnt7b acts through β-catenin (Yu et al. 2009). To confirm the paracrine function of Wnt7b, Yu et al. (2009) deleted *β-catenin* in adjacent medullary stromal cells using *Foxd1Cre*, which phenocopied *Wnt7b* mutant mice, suggesting that Wnt7b signals in a paracrine manner to the medullary stroma that controls cortico-medullary axis establishment (Yu et al. 2009). Deletion of *β-catenin* using *Foxd1Cre* also results in abnormalities in the cortical NZ including depletion of NP cells and decreased *Wnt9b*, a gene crucial for inducing NP differentiation (Boivin et al. 2015). These data suggest that canonical WNT signaling is important in both cortical and medullary stromal cells in controlling differentiation of adjacent cell populations (Fig. 8.3). Interestingly, overexpression of *β-catenin* in stromal progenitors also results in defects characteristic of human dysplastic kidneys (Boivin et al. 2016). Analysis of *β-catenin* gain of function (GOF) mutants revealed that dysplastic kidneys had abnormal stromal differentiation characterized by an expansion of stromal-like cells ectopically expressing *Wnt4* and *Bmp4* (Boivin et al. 2016). The ectopic expression of these two genes was suggested to be the mechanism by which endothelial migration and formation is abnormal in *β-catenin* overexpression mutant mice. These findings are similar to those in human renal dysplatic tissue and together suggest that disruption of WNT signaling in the stromal population) results in abnormal stromal expansion and ectopic gene expression (Boivin et al. 2015, 2016).

The secreted WNT inhibitor *Sfrp1* is expressed in the cortical and capsular stroma early in development and in the loop of Henle after E15.5 (Leimeister et al. 1998; Levinson et al. 2005). SFRP1 binds and inhibits Wnt4, which is responsible for aggregation of NP cells into pretubular aggregates (PTA) and their subsequent differentiation into nephrons (Yoshino et al. 2001). Deletion of *Sfrp1* in the murine embryo results in no gross abnormalities in the kidney, aside from reduced size of the renal papilla and a small increase in the total number of glomeruli (Trevant et al. 2008). However, *Sfrp1* is downregulated in *Foxd1*-KO mice, which are characterized by an expanded NP domain (Levinson et al. 2005). These data suggest that stromal SFRP1 may act to restrict Wnt4-induced NP differentiation to generate the appropriate number of nephrons during development. It remains to be elucidated how SFRP1 directly influences nephrogenesis; targeted deletion of *Sfrp1* will allow for a more complete understanding of the mechanism. In addition, there is evidence of regulation of WNT signaling by other signaling pathways, notably Hedgehog (HH) signaling, which has been shown to be upstream of *Sfrp1* (He et al. 2006). Analysis of HH signaling in the stroma may provide a more thorough understanding of how SFRP1 regulates NP differentiation.

8.3.4 Fat-Hippo Signaling

In the kidney, normal tissue organization is regulated by proteins involved in planar cell polarity (PCP), of which the protocadherin *Fat* has a major role (McNeill 2009). The atypical cadherin *Fat4* is expressed in stromal cells and weakly in NP cells (Saburi et al. 2008; Rock et al. 2005; Bagherie-Lachidan et al. 2015). Deletion of *Fat4* results in kidney abnormalities, including disoriented cell divisions and tubule elongation resulting in grossly cystic kidneys (Saburi et al. 2008). Analyses of *Fat4*-KO kidneys by Das et al. (2013) demonstrated that stromal-FAT4 signaling acts via downstream YAP/TAZ (Hippo) signaling effectors (Das et al. 2013). A model was suggested whereby *Fat4* from the stroma acts on adjacent NP cells to phosphorylate YAP/TAZ that stimulates WNT-induced differentiation genes (Das et al. 2013). Conversely, NP cells that were not in direct contact with stromal-FAT4 signals have decreased phosphorylation of YAP/TAZ, which results in WNT-induced renewal genes. (Das et al. 2013). Das et al. (2013) suggest that *Fat4* signals from the stroma promote NP differentiation and antagonize NP renewal by modulating the response to UB-derived Wnt9b (Das et al. 2013). In contrast to reported decreases of pYAP in NP by Das et al. (2013), both Bagherie-Lachidan et al. (2015) and Mao et al. (2015) show that FAT4 acts through *Dchs1* and *Dchs2* and not via YAP/TAZ to regulate NP differentiation (Mao et al. 2015; Bagherie-Lachidan et al. 2015). Interestingly, deletion of YAP alone in the *Fat4*-null background fails to rescue the expanded NP, and deletion of TAZ alone in the *Fat4*-null background not only fails to rescue but results in grossly cystic kidneys (Bagherie-Lachidan et al. 2015; Mao et al. 2015). It is not immediately clear why the work done by Das et al. (2013) differs from both Mao et al. (2015) and

Bagherie-Lachidan et al. (2015); however, one possibility is that YAP and TAZ share functional redundancy. Based on these new data, it is suggested that FAT4 in the stroma binds to DCHS1/2 to restrict renewal of NP and provides a novel non-cell autonomous role for *Fat4* in the stroma (Fig. 8.3).

8.3.5 POD1 and PBX1

Pod1 is a transcription factor expressed in the renal stroma and thought to be expressed at sites of epithelial–mesenchymal interactions (Cui et al. 2003). Cui et al. (2003) investigated the function of *Pod1* using a chimeric mouse model as *Pod1*-KO mice suffer from gross renal abnormalities, limiting the ability to define cell-specific functional roles (Quaggin et al. 1999). Like many other stromal disruption models, loss of *Pod1* results in an expansion of NP cells (Cui et al. 2003). Interestingly, *Pod1* is not required in NP cells as nephrogenesis is disrupted only when *Pod1* is absent in stromal cells (Cui et al. 2003). These data suggest a non-cell autonomous role for *Pod1* in the stroma (Fig. 8.3). Further analyses using conditional knockout mouse models is needed for a more thorough understanding of signaling mechanisms regulated by this transcription factor in development.

Pbx1 is a transcription factor widely expressed in a variety of mesenchymal tissue during embryonic development (Schnabel et al. 2001). Disruption of *Pbx1* results in embryonic lethality at E15.5, with renal defects ranging from hypoplasia to aplasia (Schnabel et al. 2003). *Pbx1* is expressed in the cortical and medullary stroma and weakly in the NP cell population (Schnabel et al. 2001, 2003). *Pbx1*-null embryonic kidneys are characterized by an expanded and proliferating NP domain, decreased branching morphogenesis, and ectopic *Ret* expression (Schnabel et al. 2003). Expression of *Foxd1* and *Pod1* are normal in *Pbx1*-null kidneys suggesting stromal cells are both specified and unaltered. Similarities between the *Foxd1*-KO mouse and the *Pbx1*-KO mouse suggest a similar signaling mechanism to *Foxd1* in the stromal compartment (Fig. 8.3) (Schnabel et al. 2003).

Recently, Hurtado et al. (2015) demonstrated a functional role for PBX1 in stromal progenitors (Hurtado et al. 2015). Interestingly, conditional deletion of *Pbx1* in stromal progenitors results in cell autonomous defects in the stroma and its derivatives, including diminished *Foxd1*, increased *Pdgfrβ* and perturbed arterial patterning (Hurtado et al. 2015). Hurtado et al. (2015) further demonstrate that PBX1 binds to regulatory elements in the *Pdgfrβ* gene and acts to repress its transcription (Hurtado et al. 2015). Deletion of one copy of *Pdgfrβ* in the stromal-*Pbx1* mutant rescues lethality and arterial patterning, suggesting that *Pdgfrβ* lies downstream of *Pbx1* (Hurtado et al. 2015). These data suggest a cell autonomous role for PBX1 in the stroma in controlling the expression of *Pdgfrβ* (Lindahl et al. 1998; Hurtado et al. 2015). As Hurtado et al. (2015) report no expansion in NP or defective nephrogenesis in contrast to observations reported in the *Pbx1*-null mouse, *Pbx1* may also play a cell autonomous role within the NP population (Fig. 8.3) (Schnabel et al. 2003; Hurtado et al. 2015). Conditional deletion of

Pbx1 in the NP cell lineage using *Six2Cre* will likely provide a more detailed understanding of the functional significance of this transcription factor in kidney development.

8.3.6 Other Signaling Pathways Active in Stromal Cells

Aside from the common signaling pathways discussed above, there are other factors implicated to play crucial roles in the stromal compartment. *Sall1* is a transcription factor expressed in both stromal and NP cells and is required for NP cell self-renewal (Kanda et al. 2014; Basta et al. 2014). Ohmori et al. (2015) showed that *Sall1* also has a functional role in the stroma. Specifically, conditional deletion of *Sall1* using *Foxd1Cre* results in an expansion of NP and dilated tubules and glomeruli (Ohmori et al. 2015). Perhaps more importantly, Ohmori and colleagues showed that SALL1 binds to genetic elements in both *Decorin* and *Fat4*, and that a loss of stromal *Sall1* results in less *Fat4* and increased *Decorin* (Ohmori et al. 2015). These data suggest a non-cell autonomous role for *Sall1* in controlling NP expansion and self-renewal (Fig. 8.3). These findings also implicate a connection between different signaling pathways in the stroma and how they might work in tandem to control adjacent NP cells.

Emerging studies in renal stromal signalling have focused on posttranscriptional regulation. Recently, two groups have published findings describing the role of microRNA (miRNA) posttranscriptional regulation in the stroma. Specifically, conditional deletion of miRNA processing enzyme *Dicer* results in defects in the NP compartment, vasculature, and glomeruli (Nakagawa et al. 2015; Phua et al. 2015). Phua et al. (2015) show that loss of *Dicer* results in increased NP (a finding that is consistent with many stromal disruption models), decreased nephrons, and mesangial defects (Phua et al. 2015). The authors attributed these defects in part to dysregulated cellular apoptosis (Phua et al. 2015). In contrast, Nakagawa et al. (2015) show that loss of *Dicer* in stromal cells results in depletion of NP cells and decreased WNT signaling (Nakagawa et al. 2015). Abnormalities associated with glomerular development, mesangial cells, and vasculature pattering were reported by both groups (Phua et al. 2015; Nakagawa et al. 2015). The opposing phenotypes in NP expansion and WNT signaling was hypothesized to be due to background mouse strain differences and mosaicism in the Cre (Phua et al. 2015). Notwithstanding the observed differences in phenotype, these studies suggest a mechanism where miRNA functions in the stroma to signal to NP, in addition to autonomous control of mesangial cell development (Fig. 8.3).

Although many stromal signaling pathways have been investigated, there is still a lack of understanding how stromal signaling mechanisms work in tandem. Many models of stromal disruption have overlapping phenotypes (i.e., expanded NP domain) yet little has been done to investigate how the signaling pathways might be interconnected or cooperate. It is imperative that future functional studies of the stroma attempt to understand the complex mechanisms at play. A current summary of stromal signaling mechanisms in the nephrogenic zone is illustrated in Fig. 8.3. Interestingly, many stromal disruption models, including those for *Fat4*, *Sall1*, *Pbx1*, and *Pod1* have demonstrated phenotypes similar to the *Foxd1*-KO mouse; however, *Foxd1* expression is normal in these mutants (Bagherie-Lachidan et al. 2015; Ohmori et al. 2015; Mao et al. 2015; Schnabel et al. 2003; Cui et al. 2003). These data suggest that *Foxd1* may function upstream of some of these molecular pathways to control nephron progenitor expansion and renewal.

8.3.7 Signaling Mechanisms in Stromal Derivatives

Many of the studies described in this chapter have focused primarily on signaling in stromal cell progenitors, that is the *Foxd1*+ cells that give rise to the cortical and medullary stroma of the embryonic kidney (Fig. 8.1). In contrast, a greater understanding of signaling pathways and cell–cell interactions in differentiated stromal elements is required. Signaling mechanisms that control differentiated stromal cells such as the renal capsule, interstitial fibroblasts, glomerular mesangial cells, and the renal vasculature are only beginning to be elucidated (Sequeira Lopez and Gomez 2011; Sequeira-Lopez et al. 2015; Quaggin and Kreidberg 2008). Many of the studies published to date focus on signaling in differentiated stromal cell types in models of renal fibrosis and have led to the discovery that pathways such as TGFβ, Hedgehog, and PDGF are active in differentiated stromal cells. Below, we review evidence concerning these signaling pathways and the contribution of the stroma in renal disease.

8.4 Stroma in Renal Diseases

8.4.1 Horseshoe Kidney

In normal renal development, kidneys flank the midline of the embryo before separating from the body wall, rotating and then separating from each other (Fotter 2008). However, kidneys maintain their connection to the dorsal body wall through a thin layer of connective tissue (Levinson et al. 2005; Kobayashi et al. 2014). Given the strong expression of *Foxd1* in the connective tissue surrounding the kidney and the renal capsule, it is plausible that FOXD1 functions to detach the kidneys from the dorsal body wall through proper capsule formation (Levinson et al. 2005). In *Foxd1*-null mutants where the capsule does not form properly, kidneys remain attached to the body wall and do not ascend to the lumbar region (Hatini et al. 1996; Levinson et al. 2005). A similar phenomenon occurs in human patients where failure of proper kidney ascension results in horseshoe kidneys with an incidence rate of one in 400–600 live births. Horseshoe kidneys are fused at the lower pole, forming a U or horseshoe shape. Failure of ascension also results in the "pelvic kidney" in which the kidney fails to ascend from the pelvis (Nemes et al. 2015; Weizer et al. 2003). In some cases, the horseshoe kidneys can present with ureteropelvic obstruction leading to urinary tract infections, abdominal mass, and hematuria (Rodriguez 2014).

8.4.2 Renal Fibrosis in Chronic Kidney Disease

Chronic kidney diseases (CKD) that cause chronic kidney injury and loss of renal function are characterized by fibrosis where functional tissue is replaced by permanent interstitial fibrotic tissue (Boor and Floege 2012). Fibrotic scar tissue disrupts normal organ structure and hinders regeneration and normal function. The key characteristics of renal fibrosis are the extensive deposition of ECM and expansion of a distinct α-SMA+ population of fibroblasts called myofibroblasts (Grgic et al. 2014; Falke et al. 2015). By definition, myofibroblasts are contractile cells that are the principle source of interstitial collagens, including fibrillar collagens I and III (Grgic et al. 2014).

Studying the exact origin of myofibroblasts during CKD can have implications in anti-fibrotic therapies, but is also challenging due to the lack of markers specific to myofibroblasts. *In vitro* studies initially suggested that myofibroblasts form via transition of epithelial cells to a mesenchymal phenotype, a process termed epithelial to mesenchymal transition (EMT) (Cheng et al. 2010; Iwano et al. 2002). Expression of α-SMA has been detected in tubular and glomerular epithelia in association with disease progression, in both a remnant kidney model and during experimental glomerulonephritis (Iwano et al. 2002). However, lineage-tracing studies using epithelial cell-specific *γ-glutamyl transpeptidase (GT)*-driven GFP and *Cadherin* (*Cdh*)*16* promoter in a model of obstructive nephropathy failed to detect the contribution of epithelial cells to the myofibroblast pool, raising doubt as to the epithelial origin of the myofibroblasts (Li et al. 2010; Iwano et al. 2002). More recently, studies by Humphreys et al. (2010) reported a marked expansion of pulse labeled interstitial cells that expressed α-SMA, following either complete unilateral ureteric obstruction (UUO) or ischemia-reperfusion (IR) injury (Grgic et al. 2012; Humphreys et al. 2010). In fact, an elegant lineage-tracing analysis by Humphreys et al. (2010) revealed that majority of myofibroblasts originate specifically from *Foxd1*+-derived percicytes (Humphreys et al. 2010). Interestingly, lineage tracing of both NP- and UB-derived elements using *Six2Cre* and *Hoxb7Cre* demonstrated no contribution to myofibroblasts in both UUO and IR model of fibrosis, contradicting the initially thought contribution of EMT process to the myofibroblast pool (Humphreys et al. 2010). These data suggest that *Foxd1*+ stromal progenitors are important potential contributors to the fibrotic response.

Factors regulating the differentiation of pericytes and interstitial fibroblasts to myofibroblasts in response to injury are not clear. There are new functional studies suggesting that activated PDGFR signaling in response to injury leads to proliferation and differentiation of the pericytes to myofibroblasts, and that TGFβ1 may be an important contributing cytokine in transducing activated PDGFR signaling (Chen et al. 2011). Moreover, shown for the first time in an in vivo model of fibrosis, administration of anti-PDGFR antibody markedly decreases the number of α-SMA+ myofibroblasts (Chen et al. 2011). Contrary to the aforementioned findings, other studies using PDGFRβ- and NG2-promoter-driven mice to functionally deplete pericytes in models of kidney injury did not result in any significant reduction in fibrosis (LeBleu et al. 2013). Similarly, lineage tracing of NG2+ and PDGFRβ+ pericytes did not show a predominant contribution of labeled cells to the myofibroblast population in the fibrotic kidney (LeBleu et al. 2013).

New evidence has implicated Hedgehog signaling as a major contributor to the fibrotic response. It was shown that pericytes express Hedgehog effectors *Gli1* and *Gli2*, and that *Gli1*+ pericyte proliferation is greatly enhanced (11-fold greater) during fibrosis (Fabian et al. 2012). Further evidence to support Hedgehog signaling as a key contributor to fibrosis was shown using lineage tracing of *Gli1*+ cells in the kidney after acute injury (Kramann et al. 2015a). Genetic ablation of tissue-resident *Gli1*+ mesenchymal stem cells protected against injury-induced renal fibrosis (Kramann et al. 2015a). It was further shown that GLI2 inhibition prevented myofibroblast progression and limited the fibrotic response (Kramann et al. 2015b). These data suggest a key regulatory role for GLI2 in inducing myofibroblast formation from pericytes in response to injury. Further studies are needed to fully delineate the mechanism and interactions of Hedgehog signaling in the fibrotic response.

Wnt signaling has also been implicated in contributing to the fibrotic response in the kidney. Wnt4 is expressed in medullary stromal cells and NP and is important for the mesenchymal-to-epithelial transition of NP to form epithelial renal vesicles (Stark et al. 1994). Recently, DiRocco et al. (2013) studied *Wnt4* function in medullary myofibroblasts. Interestingly, during fibrosis, medullary myofibroblasts strongly express *Wnt4* whereas tubule epithelial cells do not (DiRocco et al. 2013). Although exogenous Wnt4 stimulated myofibroblast differentiation in an *in vitro* system, deletion of *Wnt4* in interstitial cells did not reduce myofibroblast proliferation or gene expression during fibrosis (DiRocco et al. 2013). DiRocco et al. (2013) also demonstrated that constitutively active Wnt/β-catenin signaling was sufficient to drive myofibroblast differentiation *in vivo*. These data suggest that Wnt4 is not required for myofibroblast differentiation but *Wnt* signaling through β-catenin is sufficient to stimulate myofibroblast differentiation during fibrosis (DiRocco et al. 2013).

Overall, there are discrepancies concerning the origin of myofibroblasts in CKD most likely due to limitations in the available studies, including the use of overlapping markers, different mouse strains, choice of injury model, and the duration and severity of the injury affecting the outcome (Falke et al. 2015). A thorough and detailed analysis of the lineage and signaling mechanisms of the stroma will allow for a greater understanding of the functional significance of this population. These new insights will allow for better treatment options and outcomes of fibrosis in patients.

8.5 Conclusions

Our understanding of the renal stroma has greatly changed over the past decade. With new advances in the field, the stroma is now understood to encompass different kinds of cells in the adult kidney that function in various roles. Further, it is now understood that stromal progenitors play a crucial role in signaling to nephron progenitors and ureteric cells during development. Despite a large advancement in understanding, the mechanism and function of the stroma is still not fully understood. The use of genomic tools and conditional mouse models will pave the avenue for future studies on stromal origins and signaling mechanisms.

Acknowledgements This work was supported by grants from the Canadian Institute of Health Research, the Kidney Foundation of Canada and the Canada Research Chairs Program (to NDR), the Natural Sciences and Engineering Research Council of Canada (to SS), and the Research Training Centre at The Hospital for Sick Children (to CR).

References

Airik R, Bussen M, Singh MK, Petry M, Kispert A (2006) Tbx18 regulates the development of the ureteral mesenchyme. J Clin Investig 116(3):663–674

Am S, Péault P, Jj M (2013) Renal pericytes: multifunctional cells of the kidneys. Pflugers Arch Eur J Physiol 465(6):767–773

Bagherie-Lachidan M, Reginensi A, Pan Q, Zaveri HP, Scott DA, Blencowe BJ, Helmbacher F, McNeill H (2015) Stromal Fat4 acts non-autonomously with Dchs1/2 to restrict the nephron progenitor pool. Development 142(15):2564–2573

Basta JM, Robbins L, Kiefer SM, Dorsett D, Rauchman M (2014) Sall1 balances self-renewal and differentiation of renal progenitor cells. Development 141(5):1047–1058

Batourina E, Gim S, Bello N, Shy M, Clagett-Dame M, Srinivas S, Costantini F, Mendelsohn C (2001) Vitamin A controls epithelial/mesenchymal interactions through Ret expression. Nat Genet 27(1):74–78

Batourina E, Tsai S, Lambert S, Sprenkle P, Viana R, Dutta S, Hensle T, Wang F, Niederreither K, McMahon AP et al (2005) Apoptosis induced by vitamin A signaling is crucial for connecting the ureters to the bladder. Nat Genet 37(10):1082–1089

Bohnenpoll T, Bettenhausen E, Weiss A-C, Foik AB, Trowe M-O, Blank P, Airik R, Kispert A (2013) Tbx18 expression demarcates multipotent precursor populations in the developing urogenital system but is exclusively required within the ureteric mesenchymal lineage to suppress a renal stromal fate. Dev Biol 380(1):25–36

Boivin FJ, Sarin S, Dabas P, Karolak M, Oxburgh L, Bridgewater D (2016) Stromal β-catenin overexpression contributes to the pathogenesis of renal dysplasia. J Pathol 239(2):174–185

Boivin FJ, Sarin S, Lim J, Javidan A, Svajger B, Khalili H, Bridgewater D (2015) Stromally expressed β-catenin modulates Wnt9b signaling in the ureteric epithelium. PLoS One 10(3): e0120347

Boor P, Floege J (2012) The renal (myo-)fibroblast: a heterogeneous group of cells. Nephrol Dial Transplant 27(8):3027–3036

Boyle SC, Liu Z, Kopan R (2014) Notch signaling is required for the formation of mesangial cells from a stromal mesenchyme precursor during kidney development. Development 141 (2):346–354

Brown AC, Muthukrishnan SD, Guay JA, Adams DC, Schafer DA, Fetting JL, Oxburgh L (2013) Role for compartmentalization in nephron progenitor differentiation. PNAS 110 (12):4640–4645

Carroll TJ, Park J-S, Hayashi S, Majumdar A, McMahon AP (2005) Wnt9b plays a central role in the regulation of mesenchymal to epithelial transitions underlying organogenesis of the mammalian urogenital system. Dev Cell 9(2):283–292

Caspary T, Cleary MA, Perlman EJ, Zhang P, Elledge SJ, Tilghman SM (1999) Oppositely imprinted genes p57Kip2 and Igf2 interact in a mouse model for Beckwith–Wiedemann syndrome. Genes Dev 13(23):3115–3124

Chen Y-T, Chang F, Wu C-F, Chou Y, Hsu H, Chiang W, Shen J, Chen Y, Wu K, Tsai T et al (2011) Platelet-derived growth factor receptor signaling activates pericyte–myofibroblast transition in obstructive and post-ischemic kidney fibrosis. Kidney Int 80(11):1170–1181

Cheng J, Truong LD, Wu X, Kuhl D, Lang F, Du J (2010) Serum- and glucocorticoid-regulated kinase 1 is upregulated following unilateral ureteral obstruction causing epithelial-mesenchymal transition. Kidney Int 78(7):668–678

Cui S, Schwartz L, Quaggin SE (2003) Pod1 isuri required in stromal cells for glomerulogenesis. Dev Dyn 226(3):512–522

Das A, Tanigawa S, Karner CM, Xin M, Lum L, Chen C, Olson EN, Perantoni AO, Carroll TJ (2013) Stromal-epithelial crosstalk regulates kidney progenitor cell differentation. Nat Cell Biol 15(9):1035–1044

Dawson TP, Gandhi R, Le Hir M, Kaissling B (1989) Ecto-5′-nucleotidase: localization in rat kidney by light microscopic histochemical and immunohistochemical methods. J Histochem Cytochem 37(1):39–47

DiRocco DP, Kobayashi A, Taketo MM, McMahon AP, Humphreys BD (2013) Wnt4/β-catenin signaling in medullary kidney myofibroblasts. J Am Soc Nephrol 24(9):1399–1412

Dressler GR (2006) The cellular basis of kidney development. Annu Rev Cell Dev Biol 22:509–529

Dufourcq P, Couffinhal T, Ezan J, Barandon L, Moreau C, Daret D, Duplàa C (2002) FrzA, a secreted frizzled related protein, induced angiogenic response. Circulation 106(24):3097–3103

Fabian SL, Penchev RR, St-Jacques B, Rao AN, Sipilä P, West KA, McMahon AP, Humphreys BD (2012) Hedgehog-Gli pathway activation during kidney fibrosis. Am J Pathol 180 (4):1441–1453

Falke LL, Gholizadeh S, Goldschmeding R, Kok RJ, Nguyen TQ (2015) Diverse origins of the myofibroblast—implications for kidney fibrosis. Nat Rev Nephrol 11(4):233–244

Fetting JL, Guay JA, Karolak MJ, Iozzo RV, Adams DC, Maridas DE, Brown AC, Oxburgh L (2014) FOXD1 promotes nephron progenitor differentiation by repressing decorin in the embryonic kidney. Development 141(1):17–27

Fotter R (ed) (2008) Pediatric uroradiology. Springer, Berlin, Heidelberg

Gaengel K, Genové G, Armulik A, Betsholtz C (2009) Endothelial-mural cell signaling in vascular development and angiogenesis. Arterioscler Thromb Vasc Biol 29(5):630–638

Gong K-Q, Yallowitz AR, Sun H, Dressler GR, Wellik DM (2007) A Hox-Eya-Pax complex regulates early kidney developmental gene expression. Mol Cell Biol 27(21):7661–7668

Grgic I, Duffield JS, Humphreys BD (2012) The origin of interstitial myofibroblasts in chronic kidney disease. Pediatr Nephrol 27(2):183–193

Grgic I, Krautzberger AM, Hofmeister A, Lalli M, DiRocco DP, Fleig SV, Liu J, Duffield JS, McMahon AP, Aronow B et al (2014) Translational profiles of medullary myofibroblasts during kidney fibrosis. J Am Soc Nephrol 25(9):1979–1990

Grote D, Souabni A, Busslinger M, Bouchard M (2006) Pax 2/8-regulated Gata 3 expression is necessary for morphogenesis and guidance of the nephric duct in the developing kidney. Development 133(1):53–61

Guillaume R, Bressan M, Herzlinger D (2009) Paraxial mesoderm contributes stromal cells to the developing kidney. Dev Biol 329(2):169–175

Hatini V, Huh SO, Herzlinger D, Soares VC, Lai E (1996) Essential role of stromal mesenchyme in kidney morphogenesis revealed by targeted disruption of Winged Helix transcription factor BF-2. Genes Dev 10(12):1467–1478

He J, Sheng T, Stelter AA, Li C, Zhang X, Sinha M, Luxon BA, Xie J (2006) Suppressing Wnt signaling by the hedgehog pathway through sFRP-1. J Biol Chem 281(47):35598–35602

Hum S, Rymer C, Schaefer C, Bushnell D, Sims-Lucas S (2014) Ablation of the renal stroma defines its critical role in nephron progenitor and vasculature patterning. PLoS One 9(2): e88400

Humphreys BD, Lin S, Kobayashi A, Hudson TE, Nowlin BT, Bonventre JV, Valerius MT, Mcmahon AP, Duffield JS (2010) Fate tracing reveals the pericyte and not epithelial origin of myofibroblasts in kidney fibrosis. Am J Pathol 176(1):85–97

Hurtado R, Zewdu R, Mtui J, Liang C, Aho R, Kurylo C, Selleri L, Herzlinger D (2015) Pbx1-dependent control of VMC differentiation kinetics underlies gross renal vascular patterning. Development 142(15):2653–2664

Itäranta P, Chi L, Seppänen T, Niku M, Tuukkanen J, Peltoketo H, Vainio S (2006) Wnt-4 signaling is involved in the control of smooth muscle cell fate via Bmp-4 in the medullary stroma of the developing kidney. Dev Biol 293(2):473–483

Iwano M, Plieth D, Danoff TM, Xue C, Okada H, Neilson EG (2002) Evidence that fibroblasts derive from epithelium during tissue fibrosis. J Clin Investig 110(3):341–350

James RG, Kamei CN, Wang Q, Jiang R, Schultheiss TM (2006) Odd-skipped related 1 is required for development of the metanephric kidney and regulates formation and differentiation of kidney precursor cells. Development 133(15):2995–3004

Kanda S, Tanigawa S, Ohmori T, Taguchi A, Kudo K, Suzuki Y, Sato Y, Hino S, Sander M, Perantoni AO et al (2014) Sall1 maintains nephron progenitors and nascent nephrons by acting as both an activator and a repressor. J Am Soc Nephrol 25(11):2584–2595

Karner CM, Chirumamilla R, Aoki S, Igarashi P, Wallingford JB, Carroll TJ (2009) Wnt9b signaling regulates planar cell polarity and kidney tubule morphogenesis. Nat Genet 41 (7):793–799

Karner CM, Das A, Ma Z, Self M, Chen C, Lum L, Oliver G, Carroll TJ (2011) Canonical Wnt9b signaling balances progenitor cell expansion and differentiation during kidney development. Development 138(7):1247–1257

Kobayashi H, Liu Q, Binns TC, Urrutia AA, Davidoff O, Kapitsinou PP, Pfaff AS, Olauson H, Wernerson A, Fogo AB et al (2016) Distinct subpopulations of FOXD1 stroma-derived cells regulate renal erythropoietin. J Clin Invest 126(5):1926–1938

Kobayashi A, Mugford JW, Krautzberger AM, Naiman N, Liao J, McMahon AP (2014) Identification of a multipotent self-renewing stromal progenitor population during mammalian kidney organogenesis. Stem Cell Rep 3(4):650–662

Kramann R, Fleig SV, Schneider RK, Fabian SL, DiRocco DP, Maarouf O, Wongboonsin J, Ikeda Y, Heckl D, Chang SL et al (2015a) Pharmacological GLI2 inhibition prevents myofibroblast cell-cycle progression and reduces kidney fibrosis. J Clin Invest 125 (8):2935–2951

Kramann R, Schneider RK, DiRocco DP, Machado F, Fleig S, Bondzie PA, Henderson JM, Ebert BL, Humphreys BD (2015b) Perivascular Gli1+ progenitors are key contributors to injury-induced organ fibrosis. Cell Stem Cell 16(1):51–66

Le Hir M, Kaissling B (1989) Distribution of 5′-nucleotidase in the renal interstitium of the rat. Cell Tissue Res 258(1):177–182

Le Hir M, Kaissling B (1993) Distribution and regulation of renal ecto-5′-nucleotidase: implications for physiological functions of adenosine. Am J Physiol 264(3 Pt 2):F377–F387

LeBleu VS, Taduri G, O'Connell J, Teng Y, Cooke VG, Woda C, Sugimoto H, Kalluri R (2013) Origin and function of myofibroblasts in kidney fibrosis. Nat Med 19(8):1047–1053

Leimeister C, Bach A, Gessler M (1998) Developmental expression patterns of mouse sFRP genes encoding members of the secreted frizzled related protein family. Mech Dev 75(1–2):29–42

Levinson RS, Batourina E, Choi C, Vorontchikhina M, Kitajewski J, Mendelsohn CL (2005) Foxd1-dependent signals control cellularity in the renal capsule, a structure required for normal renal development. Development 132(3):529–539

Li W, Hartwig S, Rosenblum ND (2014) Developmental origins and functions of stromal cells in the normal and diseased mammalian kidney. Dev Dyn 243(7):853–863

Li L, Zepeda-Orozco D, Black R, Lin F (2010) Autophagy is a component of epithelial cell fate in obstructive uropathy. Am J Pathol 176(4):1767–1778

Lindahl P, Hellström M, Kalén M, Karlsson L, Pekny M, Pekna M, Soriano P, Betsholtz C (1998) Paracrine PDGF-B/PDGF-Rbeta signaling controls mesangial cell development in kidney glomeruli. Development 125(17):3313–3322

Mao Y, Francis-West P, Irvine KD (2015) Fat4/Dchs1 signaling between stromal and cap mesenchyme cells influences nephrogenesis and ureteric bud branching. Development 142 (15):2574–2585

Mao Y, Mulvaney J, Zakaria S, Yu T, Morgan KM, Allen S, Basson MA, Francis-West P, Irvine KD (2011) Characterization of a Dchs1 mutant mouse reveals requirements for Dchs1-Fat4 signaling during mammalian development. Development 138(5):947–957

Marxer-Meier A, Hegyi I, Loffing J, Kaissling B (1998) Postnatal maturation of renal cortical peritubular fibroblasts in the rat. Anat Embryol 197:143–153

McNeill H (2009) Planar cell polarity and the kidney. J Am Soc Nephrol 20(10):2104–2111

Mendelsohn C, Mark M, Dollé P, Dierich A, Gaub MP, Krust A, Lampron C, Chambon P (1994a) Retinoic acid receptor beta 2 (RAR beta 2) null mutant mice appear normal. Dev Biol 166 (1):246–258

Mendelsohn C, Lohnes D, Décimo D, Lufkin T, LeMeur M, Chambon P, Mark M (1994b) Function of the retinoic acid receptors (RARs) during development (II). Multiple abnormalities at various stages of organogenesis in RAR double mutants. Development 120(10):2749–2771

Mendelsohn C, Batourina E, Fung S, Gilbert T, Dodd J (1999) Stromal cells mediate retinoid-dependent functions essential for renal development. Development 1148:1139–1148

Miyazaki Y, Oshima K, Fogo A, Ichikawa I (2003) Evidence that bone morphogenetic protein 4 has multiple biological functions during kidney and urinary tract development. Kidney Int 63 (3):835–844

Mugford JW, Sipilä P, McMahon JA, McMahon AP (2008) Osr1 expression demarcates a multi-potent population of intermediate mesoderm that undergoes progressive restriction to an Osr1-dependent nephron progenitor compartment within the mammalian kidney. Dev Biol 324 (1):88–98

Nakagawa N, Xin C, Roach AM, Naiman N, Shankland SJ, Ligresti G, Ren S, Szak S, Gomez IG, Duffield JS (2015) Dicer1 activity in the stromal compartment regulates nephron differentiation and vascular patterning during mammalian kidney organogenesis. Kidney Int 87 (6):1125–1140

Nemes B, Kanyári Z, Zádori G, Zsom L, Berhés M, Hamar M, Kóbor K, Péter A (2015) Horseshoe kidney transplantation. Intervent Med Appl Sci 7(2):85–89

Ohmori T, Tanigawa S, Kaku Y, Fujimura S (2015) Sall1 in renal stromal progenitors non-cell autonomously restricts the excessive expansion of nephron progenitors. Sci Rep 5:1–11

Oxburgh L, Brown AC, Muthukrishnan SD, Fetting JL (2014) Bone morphogenetic protein signaling in nephron progenitor cells. Pediatr Nephrol 29(4):531–536

Park J-S, Ma W, O'Brien LL, Chung E, Guo J-J, Cheng J-G, Valerius MT, McMahon JA, Wong WH, McMahon AP (2012) Six2 and Wnt regulate self-renewal and commitment of nephron progenitors through shared gene regulatory networks. Dev Cell 23(3):637–651

Park J-S, Valerius MT, McMahon AP (2007) Wnt/beta-catenin signaling regulates nephron induction during mouse kidney development. Development 134(13):2533–2539

Paroly SS, Wang F, Spraggon L, Merregaert J, Batourina E, Tycko B, Schmidt-Ott KM, Grimmond S, Little M, Mendelsohn C (2013) Stromal protein Ecm1 regulates ureteric bud patterning and branching. PLoS One 8(12):e84155

Phua YL, Chu JYS, Marrone AK, Bodnar AJ, Sims-Lucas S, Ho J (2015) Renal stromal miRNAs are required for normal nephrogenesis and glomerular mesangial survival. Physiol Rep 3(10): e12537

Quaggin SE, Kreidberg JA (2008) Development of the renal glomerulus: good neighbors and good fences. Development 135(4):609–620

Quaggin SE, Schwartz L, Cui S, Igarashi P, Deimling J, Post M, Rossant J (1999) The basic-helix-loop-helix protein pod1 is critically important for kidney and lung organogenesis. Development 126(24):5771–5783

Ranghini EJ, Dressler GR (2015) Evidence for intermediate mesoderm and kidney progenitor cell specification by Pax2 and PTIP dependent mechanisms. Dev Biol 399(2):296–305

Rock R, Schrauth S, Gessler M (2005) Expression of mouse dchs1, fjx1, and fat-j suggests conservation of the planar cell polarity pathway identified in Drosophila. Dev Dyn 234 (3):747–755

Rodriguez MM (2014) Congenital anomalies of the kidney and the urinary tract (CAKUT). Fetal Pediatr Pathol 33(5–6):293–320

Rosselot C, Spraggon L, Chia I, Batourina E, Riccio P, Lu B, Niederreither K, Dolle P, Duester G, Chambon P et al (2010) Non-cell-autonomous retinoid signaling is crucial for renal development. Development 137(2):283–292

Saburi S, Hester I, Fischer E, Pontoglio M, Eremina V, Gessler M, Quaggin SE, Harrison R, Mount R, McNeill H (2008) Loss of Fat4 disrupts PCP signaling and oriented cell division and leads to cystic kidney disease. Nat Genet 40(8):1010–1015

Schnabel CA, Godin RE, Cleary ML (2003) Pbx1 regulates nephrogenesis and ureteric branching in the developing kidney. Dev Biol 254(2):262–276

Schnabel CA, Selleri L, Jacobs Y, Warnke R, Cleary ML (2001) Expression of Pbx1b during mammalian organogenesis. Mech Dev 100(1):131–135

Schuchardt A, D'Agati V, Pachnis V, Costantini F (1996) Renal agenesis and hypodysplasia in ret-k-mutant mice result from defects in ureteric bud development. Development 122 (6):1919–1929

Sequeira Lopez MLS, Gomez RA (2011) Development of the renal arterioles. J Am Soc Nephrol 22(12):2156–2165

Sequeira Lopez ML, Pentz ES, Robert B, Abrahamson DR, Gomez RA (2001) Embryonic origin and lineage of juxtaglomerular cells. Am J Physiol Renal Physiol 281(2):F345–F356

Sequeira-Lopez MLS, Lin EE, Li M, Hu Y, Sigmund CD, Gomez RA (2015) The earliest metanephric arteriolar progenitors and their role in kidney vascular development. Am J Physiol Regul Integr Comp Physiol 308(2):R138–R149

Shawber CJ, Kitajewski J (2004) Notch function in the vasculature: insights from zebrafish, mouse and man. BioEssays 26(3):225–234

Sims-Lucas S, Schaefer C, Bushnell D, Ho J, Logar A, Prochownik E, Gittes G, Bates CM (2013) Endothelial progenitors exist within the kidney and lung mesenchyme. PLoS One 8(6):e65993

Stark K, Vainio S, Vassileva G, McMahon AP (1994) Epithelial transformation of metanephric mesenchyme in the developing kidney regulated by Wnt-4. Nature 372(6507):679–683

Taguchi A, Kaku Y, Ohmori T, Sharmin S, Ogawa M, Sasaki H, Nishinakamura R (2014) Redefining the in vivo origin of metanephric nephron progenitors enables generation of complex kidney structures from pluripotent stem cells. Cell Stem Cell 14(1):53–67

Takasato M, Little MH (2015) The origin of the mammalian kidney: implications for recreating the kidney in vitro. Development 142(11):1937–1947

Trevant B, Gaur T, Hussain S, Symons J, Komm BS, Bodine PVN, Stein GS, Lian JB (2008) Expression of secreted frizzled related protein 1, a Wnt antagonist, in brain, kidney, and skeleton is dispensable for normal embryonic development. J Cell Physiol 217(1):113–126

Weizer AZ, Silverstein AD, Auge BK, Delvecchio FC, Raj G, Albala DM, Leder R, Preminger GM (2003) Determining the incidence of horseshoe kidney from radiographic data at a single institution. J Urol 170(5):1722–1726

Xu J, Wong EYM, Cheng C, Li J, Sharkar MTK, Xu CY, Chen B, Sun J, Jing D, Xu P-X (2014a) Eya1 interacts with Six2 and Myc to regulate expansion of the nephron progenitor pool during nephrogenesis. Dev Cell 31(4):434–447

Xu J, Nie X, Cai X, Cai CL, Xu PX (2014b) Tbx18 is essential for normal development of vasculature network and glomerular mesangium in the mammalian kidney. Dev Biol 391 (1):17–31

Yallowitz AR, Hrycaj SM, Short KM, Smyth IM, Wellik DM (2011) Hox10 genes function in kidney development in the differentiation and integration of the cortical stroma. PLoS One 6 (8):e23410

Yoshino K, Rubin JS, Higinbotham KG, Üren A, Anest V, Plisov SY, Perantoni AO (2001) Secreted frizzled-related proteins can regulate metanephric development. Mech Dev 102 (1–2):45–55

Yu J, Carroll TJ, Rajagopal J, Kobayashi A, Ren Q, McMahon AP (2009) A Wnt7b-dependent pathway regulates the orientation of epithelial cell division and establishes the cortico-medullary axis of the mammalian kidney. Development 136(1):161–171

Part III
Nephron Morphogenesis and Maintenance

Chapter 9
Imaging, Analysing and Interpreting Branching Morphogenesis in the Developing Kidney

Kieran M. Short and Ian M. Smyth

Abstract The kidney develops as an outgrowth of the epithelial nephric duct known as the ureteric bud, in a position specified by a range of rostral and caudal factors which serve to ensure two kidneys form in the appropriate positions in the body. At its simplest level, kidney development can be viewed as the process by which this single bud then undergoes a process of arborisation to form a complex connected network of ducts which will serve to drain urine from the nephrons in the adult organ. The process of bud elaboration is dictated by factors expressed by both the bud itself and by surrounding cells of the metanephric mesenchyme which control cell division and bifurcation. These cells play two critical roles. Firstly, they potentiate the ongoing elaboration of the ureteric tree: remove them and branching ceases. Secondly, they harbour progenitor cells which are fated to undergo their own process of tubulogenesis to form the nephrons of the adult organ. In this chapter, we will discuss how the ureteric bud arises in the developing embryo, how it undergoes branching, how we can measure and study this process and finally the likely relevance that this process has for our understanding of congenital and acquired kidney disease.

K.M. Short
Department of Biochemistry and Molecular Biology, Monash Biomedicine Discovery Institute, Monash University, 19 Innovation Walk, Clayton, VIC 3800, Australia

I.M. Smyth (✉)
Department of Biochemistry and Molecular Biology, Monash Biomedicine Discovery Institute, Monash University, 19 Innovation Walk, Clayton, VIC 3800, Australia

Department of Anatomy and Developmental Biology, Monash Biomedicine Discovery Institute, Monash University, 19 Innovation Walk, Clayton, VIC 3800, Australia
e-mail: ian.smyth@monash.edu

R.K. Miller (ed.), *Kidney Development and Disease*, Results and Problems in Cell Differentiation 60, DOI 10.1007/978-3-319-51436-9_9

9.1 Normal Kidney Development

The metanephric kidney is the third kidney to form in higher vertebrates, after the largely vestigial pronephros and the more elaborate mesonephros, which contributes cells to other parts of the urogenital system. All three of these organs are derived from the intermediate mesoderm (IM), a field of cells specified by factors produced from the surrounding paraxial mesoderm, lateral plate mesoderm and epidermis [reviewed in (Takasato and Little 2015)]. At the rostral end of the intermediate mesoderm, a population of cells undergoes a mesenchymal to epithelial transition to form the nephric duct which begins a rostro-caudal elongation and migration which will eventually result in its connection to the cloaca. A second IM derivative is the metanephric mesenchyme, a teardrop-shaped collection of cells whose gene expression is distinct from other IM components and which is essential for specifying the position and outgrowth of the ureteric bud from the nephric duct. The factors involved in IM differentiation and temporal specification have been detailed elsewhere (Takasato and Little 2015). The *Hox* gene family, particularly those of the *Hox11* class, are critical for activating a range of genes in the metanephric mesenchyme which specify the position of metanephric kidney development (Wellik et al. 2002), one of the most important of which is the glial cell line-derived neurotrophic factor (*Gdnf*) (Fig. 9.1). A range of other genes serve to reinforce *Gdnf* expression including *Pax2*, *Eya1*, *Wt1* and *Sall1* (Brophy et al. 2001; Xu et al. 1999; Gross et al. 2003; Nishinakamura et al. 2001) while others like *Foxc1*, *Foxc2* and the SLIT-ROBO receptor ligand pathways ensure that expression of *Gdnf* or reception of Gdnf signalling is suppressed at positionally inappropriate locations along the IM (Kume et al. 2000; Grieshammer et al. 2004) (Fig. 9.1). SLIT/ROBO signalling has recently been proposed to be necessary for detachment of the nephrogenic mesenchyme from the duct, therefore mechanically

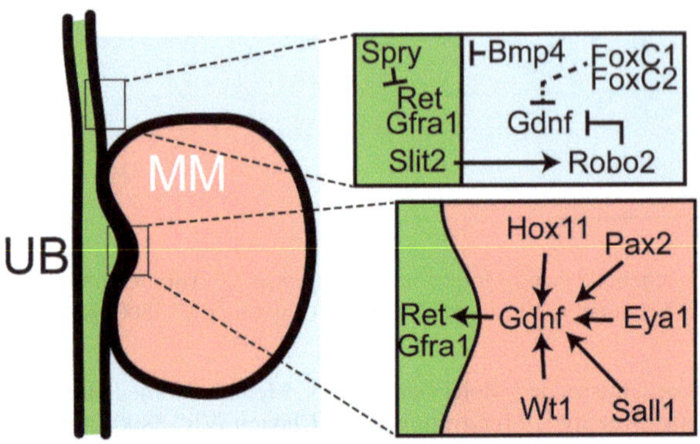

Fig. 9.1 Simplified genetic pathways regulating ureteric budding. The location of the ureteric bud (UB, *green*) is defined by the suppression of factors such as Ret in the nephric duct and Gdnf in the uninduced mesenchyme (*blue*). Metanephric mesenchyme (MM, *pink*) is actively producing Gdnf under the control of multiple factors which induces Ret signalling and UB growth

limiting the range of GDNF activity rather than directly controlling its expression (Wainwright et al. 2015). GDNF interacts with its co-receptors RET and GFRα1 which are expressed in the nephric duct, and it is this interaction which initiates UB outgrowth (Durbec et al. 1996; Schuchardt et al. 1996; Vega et al. 1996). This process begins with a pseudo stratification of the epithelia in the duct (Chi et al. 2009) and then progresses to form the ureteric bud (UB), which invades the neighbouring receptive metanephric mesenchyme. As we discuss later in this chapter, the UB then undergoes a process of (principally) bifurcation to form a complex arborised tree which is fated to form the collecting duct system of the adult kidney.

The formation of the collecting duct system from the UB is intimately linked with the generation of nephrons which drain urine into it. This association is driven by the intimate developmental link between these two tissues. As the UB elaborates within the metanephric mesenchyme, a proportion of the mesenchyme coalesces around the UB tips and changes gene expression through the induction of genes like *Six1*, *Six2* and *Cited1* (Self et al. 2006; Boyle et al. 2007). Cell lineage tracing based on the expression of the two latter genes shows that these cells are fated to form all of the tubular components of the adult nephrons (Kobayashi et al. 2008; Boyle et al. 2008). Following the induction of pre-tubular aggregates and their development through renal vesicle, S- and comma-shaped bodies, the nascent nephrons re-integrate into the collecting duct system and are carried with the tip as it grows and branches (Georgas et al. 2009; Kao et al. 2012). The specification of nephron number can therefore be viewed as a product of (a) the extent of UB branching (thereby forming the "environment" for nephron formation), (b) the staged commitment of progenitor cell populations in the mesenchymal caps to form nephrons and (c) the retention and proliferation of nephron progenitor cells over the course of developmental time. The emerging links between nephron number (which varies dramatically between individuals) and adult onset disease mean that studying branching morphogenesis is of considerable value (Luyckx et al. 2013).

9.2 Branching in Other Organs

Branching morphogenesis is a commonly employed developmental mechanism evident in the foetal kidney, lung and in many other organs such as the salivary and mammary glands. In these and other contexts, it serves as a way to break up large tissue masses permitting secretion, excretion and exchange of nutrients and gases. Its relative importance in facilitating such exchange is evident in its evolutionary conservation, forming structures as diverse as the Drosophila trachea and the planarian gut. The commonalities and differences between these different organs are enlightening. All of these systems involve, to one extent or another, interactions between the elaborating epithelia and an associated population of mesenchymal cells, and the nature of these interactions is often instructive and essential. Typically, each tissue component provides signals to the other to direct cell proliferation and differentiation. The inductive nature of these interactions, while mooted by Hans Spemann in the 1920s, was expounded to great effect in a series of pioneering experiments by Clifford Grobstein in the 1950s and 1960s. This work showed that, in isolation, epithelia and mesenchyme of developing branching organs failed to differentiate (Grobstein 1953a) but when placed in adjacent culture, separated by an intervening membrane, the two cell populations supported each other (Grobstein 1953b). In particular, his work in the kidney served to highlight the interactions between UB and cap mesenchyme (Grobstein 1955) and provided the first insights into the nature of the molecules mediating the communication between these cells, proposing the important role of the extracellular matrix in this interaction (Grobstein 1956, 1961).

Until recently, the best understood model of developmental branching was the foetal mouse lung. Lungs are structurally different to kidney in that they possess long and linear main bronchii from which smaller airways arise. This type of branching has been termed domain branching, and it is important for establishing the initial lobar arrangement of the organ and the arrangement of the main airways. Metzger and colleagues (2008) performed an exhaustive manual annotation survey of foetal mouse lung development using organs stained with antibodies to epithelial cells. They proposed a highly stereotyped model of branching employing different branch modalities in addition to domain branching. Described were two specific conformations of bifurcation which differed in their rotation between successive bifurcations. "Planar bifurcations" describe branching events with no rotation between branches which typically occur towards the margins of the organ whereas "orthogonal bifurcations" undergo a 90° rotation and are often towards the centre of the tissue. It was proposed that domain, planar and orthogonal branching were combined to execute a highly spatially stereotyped programme of branching that establishes organ structure.

9.3 Quantifying Branching Morphogenesis

The branching epithelium of the developing kidney, lung, glands and other organs form complex structures that do not lend themselves to assessment by individual histological sections. Fully appreciating 3D context of the branches and measuring the architecture of this structure is essential for understanding normal development and also the impact of genetic or environmental changes to the developmental programme. For this reason, culture of organs at air–liquid interfaces has historically played an important role in establishing the basic mechanisms shaping elaboration of this structure. This is no better illustrated by the work of Grobstein we have previously discussed. Readers are directed to recent reviews for broader assessments of the value of organ culture (Rak-Raszewska et al. 2015; Nigam 2013). However, these systems are hampered by a number of features which make them less than ideal for assessing how the kidney develops. Firstly, the cultured organs are unusually flattened as a consequence of surface tension, resulting in abnormal juxtaposition of tissue elements. Secondly, they develop much more slowly compared to their in vivo counterparts, possibly as a consequence of the physical constrictions imposed by culture but also likely because of a lack of a connected vasculature and the associated lack of exposure to the growth factors present in the foetal milieu. Because of these drawbacks, a number of studies have employed exhausting section-based reconstruction to "rebuild" foetal organs from individually stained and scanned slices of the foetal organ (Cebrian et al. 2004; Sims-Lucas et al. 2009; Sampogna et al. 2015). Such studies have made important contributions to our understanding of both normal kidney development and to the analysis of mutant models. However, the laborious nature of the sectioning, staining and reconstruction combined with the inevitable tissue distortion associated with sectioning do not lend such approaches to high throughput or extensive analysis.

The ability to quantitatively analyse these small structures at a global scale, in three dimensions and without recourse to sectioning, has long been hampered by a lack of sufficient imaging and quantification tools. A fundamental advance in imaging whole organs in model organisms was the development of Optical Projection Tomography (OPT) (Sharpe et al. 2002). OPT is a microscopy method very close in practice to medical X-ray computed tomography (CT). The primary difference is that with OPT, visible light is used rather than X-rays, and the sample image is treated chemically so light can pass through it (X-rays pass through solid opaque tissue, light does not). When OPT was first released, it showed great promise for the analysis of model organism development *qualitatively* (Kerwin et al. 2004), but a major issue was that the imaging technology was introduced to a world bereft of software tools to *quantitatively* analyse the complex medical CT-like data it produces.

With new imaging technologies, one difficulty in translating large bodies of information from qualitative to quantitative representations of the samples imaged, is a change in the type/scale/form of data that are generated. This leads to existing quantitative methods to either fail or be inadequate due to a deficit in computational algorithms "in the wild" which accurately convert this data into a form that can be measured. This was the case for much of the embryonic kidney. To address this issue, tools have been recently developed which enable the quantitative analysis of stained ureteric trees (the primordial collecting duct system) in embryonic mouse kidneys from Optical Projection Tomography datasets (Short et al. 2010, 2013).

9.4 Visualising the Ureteric Tree and Computationally Deciphering Branched Structures

From the earliest stages of budding through to an adult collecting duct, the UB maintains a tubular form. In order to visualise the epithelial UB by OPT, as would be performed for confocal microscopy, the epithelia must first be stained with fluorescent markers. Antibody staining is used to do this, and the antibodies used typically recognise components of cellular architecture unique to epithelia such as anti-pan-cytokeratin and anti-Trop2 antisera which are more specific to the UB and others such as e-cadherin which label both the UB and more differentiated nephron epithelia (Combes et al. 2014). OPT, like all tomographic imaging modalities, generates cross-section data that is isotropic (i.e. the same resolution in X, Y and Z dimensions). This is a primary feature of OPT that sets it apart from confocal microscopy because confocal microscopy has excellent X and Y resolution (far better than OPT) but comparatively poor Z resolution when compared with OPT. This makes OPT more amenable to 3D analysis of complex features such as the ureteric tree. So when a kidney stained for epithelial markers is imaged by OPT, the fluorescence captured shows the collecting ducts in cross-section, appearing as either filled or open shapes (Fig. 9.2a). Therefore, one of the primary hurdles for the analysis of embryonic UB data is the appearance of the ducts in three dimensions.

Computationally, the study of the UB becomes a question of analysing a connected series of ducts or tubes. This is also true of the vascular system when analysing arterial trees, which have been the subject of study for well over 100 years (Stiles 1897). In the vascular field, as imaging technologies have advanced and permitted the imaging of vascular trees in three dimensions by Magnetic Resonance Imaging (MRI) and X-ray CT, the analysis of these datasets has laid the groundwork for much of the development of methods for analysing other hollow branching networks. Two major modes of computational analysis have emerged: binary image processing and grey-level image processing. Binary images are those where any duct signal is turned on (white), and any non-duct signal is turned off (black). Greyscale images represent the original data acquired with the microscopy/imaging hardware. In the vascular field, image processing using binary data (containing two colours) has dominated, in part because it has generally been computationally easier and therefore is the basis for many more measurement tools. However, direct measurement of greyscale source data has also been demonstrated (Aylward et al. 1996).

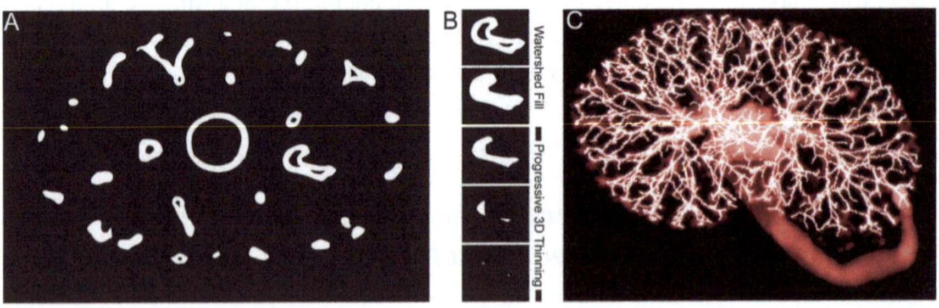

Fig. 9.2 Optical projection tomography-reconstructed ureteric tree data appears as many different shapes in cross section (**a**) which can be a challenge to quantify. Watershed filling of hollow branch lumens generates solid structures and many methods erode these filled structures to determine a skeleton (**b**) which is represented in 3D (**c**)

To date, binary image processing has been the basis for most embryonic kidney branching analysis (Cullen-McEwen et al. 2002; Short et al. 2010; Watanabe and Costantini 2004). In binary image processing, greyscale source images are turned into binary images using thresholding (where a certain signal intensity is set as the "cut off"; any lower intensity in the image is turned off and everything above is turned on). Various implementations using this approach have been used in previous quantitative studies of embryonic kidney branching (Cullen-McEwen et al. 2002; Short et al. 2010; Watanabe and Costantini 2004). Typically, global thresholding is used which applies a manual automatically calculated threshold (Otsu 1979) to the data. Alternative methods exist which account for variation in signal intensity throughout the imaged sample, such as adaptive thresholding, which applies thresholds locally in small volumes within in the dataset (Short et al. 2010). Depending on the imaging technology and magnification used, UBs can present (in cross-section) as either solid or hollow circles/ovals. Some approaches such as low magnification confocal microscopy produce data that is filled (Cullen-McEwen et al. 2002; Watanabe and Costantini 2004) and require no further processing prior to further computation. However, high resolution confocal or OPT imaging generates images which can resolve the tubular lumen.

To measure the structure of branched structures such as the renal ureteric tree (UT), image processing approaches often apply thinning which "skeletonises" the data for subsequent quantification. A skeletal representation of an object is relatively simple to quantify computationally. However, in order to perform digital thinning on the binary data, the hollow components need to first be filled, and a Watershed algorithm is often used to do this (Vincent and Soille 1991). This method floods enclosed areas (like circles and ovals) in two dimensions with pixels to produce solid filled binary objects and in 3D, solid filled binary ducts. The simplest computational method of converting this binary data into a form that is quantifiable is to progressively erode the outermost layers of the ducts either in two dimensions (Cullen-McEwen et al. 2002; Watanabe and Costantini 2004) or three dimensions (Short et al. 2010), to find their midpoint (Fig. 9.2a, b). The resulting data represents the skeleton of a UT in either 2D or 3D (Fig. 9.2c). The skeleton is then computationally "traversed" from a single end point to all the remaining tips, with the number of voxels (in three dimensional space terms, a $1 \times 1 \times 1$ pixel is a voxel) counted between branch points. Quantification of tip number is a simple sum of all terminal points, and branch length is calculated by multiplying the number of voxels by the real-world voxel dimensions. This particular skeletonisation method has been used as the basis for the analysis of branching in organs for a long time from multiple imaging modalities (Malkusch et al. 1995), particularly in kidneys (Cullen-McEwen et al. 2002; Watanabe and Costantini 2004) and was first adapted for kidney OPT data in 2010 (Short et al. 2010).

The problem with this technique when applied in three dimensions is that the developing UT is rarely round in cross-section (Fig. 9.2a), particularly at branch points and oblique sections through the UT. In addition, while OPT data has relatively high signal-to-noise ratio, often background staining can occur and erosion-based thinning is very sensitive to noise. Noise is a problem not restricted to OPT, and attempts to circumvent this problem vary from the use of additional smoothing filters (Short et al. 2010; Cullen-McEwen et al. 2002). However, the effectiveness of smoothing is limited and can lead to a skeletal structure being produced that, while approximating the true centre of the branched object, contains many flaws in the process that do not represent the nature of the branching organ.

9.5 Improved Image Processing Approaches for 3D Branching Analysis

In order to improve analysis of tubular branching systems, different approaches have now been established that operate on both binary and greyscale formats of the source data (Short et al. 2013; Wörz et al. 2009). These new techniques utilise mathematically descriptive methods and stem from initial approaches to skeletonise vasculature captured by MRI imaging (Aylward et al. 1996). Two such approaches have emerged as useful for the definition of branching systems. One uses particle filtering to detect the walls of the tubular objects in 3D space (Wörz et al. 2009) and the other utilises vector magnitude to identify the centre of the tube (Short et al. 2013). While particle filtering methods are excellent at describing continuous tubes such as major arteries, the UT is more varied in shape. The kidney has a thick ureter and at latter stages of development, a large luminar pelvis, which is not a tube per se. This type of morphology is better computationally described by vector magnitude approaches which are applicable to more than simple tubes. To this end, software called Tree Surveyor was developed using this method with the particular focus of the measurement of OPT images of embryonic mouse UTs as one of its primary goals (Short et al. 2013).

Tree Surveyor was the first software application that applied vector magnitude in three dimensions and takes cues from two dimensional Optical Character Recognition algorithms (Goh and Chan 2007). Rather than relying on the processing of pure binary data and derivation of a skeleton and measurements from that data, this method derives the skeleton of the branching object mathematically, using vectors (Fig. 9.3). The process begins with isotropic OPT or confocal source data being temporarily turned into a binary format and watershed filled (Fig. 9.3a). This initial stage is similar to rudimentary binarisation approaches previously described in this chapter, but the analysis is dramatically different from this point onward. After watershed filling, the UT data is then processed in "linear scale space" in three dimensions. Scale space analysis starts with Gaussian blurring of the filled UT, which (visually) makes it blurry with the highest intensity in the centre and the lowest intensity at the furthermost edges of the branches and tips (Fig. 9.3b). The gradient from low to high intensity can be analysed in linear scale space by looking at the magnitude and direction of the gradient, which describes the shape of the UT in three dimensions. The magnitude and direction of the gradient are defined by vectors, and vectors are used to generate a "flow field" in linear scale space which

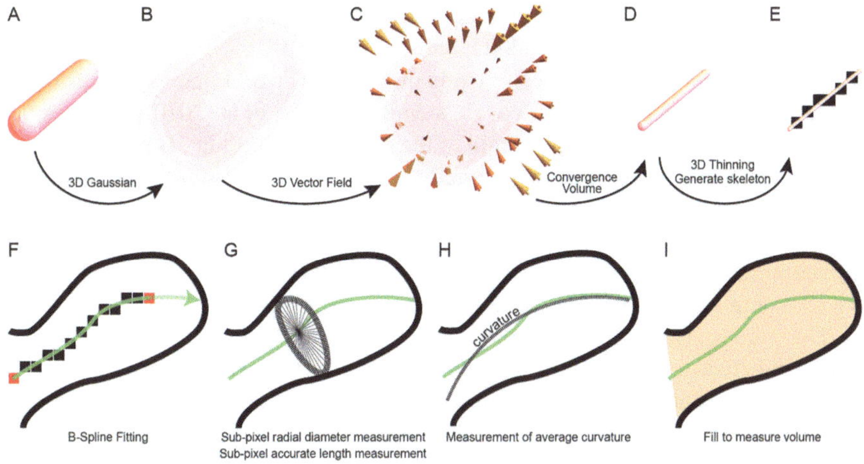

Fig. 9.3 Tree surveyor determines the skeleton of a filled ureteric tree branch (**a**) by Gaussian blurring the dataset (**b**), before generating a 3D vector field that is based on the blurred branch (**c**). The direction and magnitude of the vector field are used to determine the centre of mass (**d**), before a thinning process is used to generate a skeleton that is made of voxels (3D pixels) (**e**). Control points for B-splines are mapped to the voxel skeleton and are used to generate mathematical 3D curves which fit the branches (**f**). Quantification of spline parameters is used to determine diameter (**g**), curvature (**h**) and volume (**i**)

indicate the direction of the gradient, from darkest to brightest intensity signal (see arrows in Fig. 9.3c). The vectors which define the flow field are generated from "seed points", and using more seed points per unit volume means that more vectors are generated which form a more complex flow field and (generally) a more detailed description of the UT. In Tree Surveyor, the amount of Gaussian blurring and the number of vector seeds used to assess an object can be modified. This allows refinement of the UT detection process because it permits adaptation of the gradient and flow field for the detection of smaller or larger features depending on the settings. This is helpful when analysing the UTs from younger kidneys (with more blurring, less seeding; defining larger tips), and older embryonic ureteric trees (with less blurring, more seeding; defining finer, thinner tips).

Once a flow field of vectors is generated, the direction and magnitude of the vectors are used to define a region of convergence in linear scale space. This region is converted to a volume, and it represents the highest intensity signal of the UT dataset (i.e. the centre of the branches and tips, Fig. 9.3c, d). As the convergence volume defined by the vectors is a volume and not necessarily a thin structure, thinning approaches are then applied to the convergence volume to produce a single voxel thickness skeleton (Fig. 9.3e). At this point, the skeleton appears visually similar to a skeleton produced by binary erosion, but Tree Surveyor again takes a different approach to skeleton definition which provides significantly greater precision and offers other benefits for quantification of the UT. Prior to the release of Tree Surveyor, other software designed to measure UT development directly counted the number voxels in order to measure the length and other metrics of branches (Cullen-McEwen et al. 2002; Short et al. 2010). This is inaccurate because simply multiplying the number of voxels by their real-world pixel sizes results in overestimation of length due to voxel "stepping" (Short et al. 2010) and does not always accurately represent the length of the biological object in 3D space. For this reason, some approximations are required to help reduce this (Short et al. 2010). To avoid this problem completely, Tree Surveyor uses the voxel skeleton as the source for generating b-splines (Fig. 9.3f) which are smooth mathematically generated curved lines in 3D space which pass through multiple points (known in mathematics as "control points") in 3D space with sub-pixel accuracy.

The splines start and end at branch points (or tips), and the number of control points in-between that are used for generating the curves depends on the length of the branch/tip segment (coarsely estimated by a voxel count). Longer segments are marked with more control points for the spline curve to pass through, and shorter segments with less. In a special case for UT tips, Tree Surveyor projects tip splines in a direction based on the incident curve into the terminal tip voxel, through to the surface of the original UT dataset. The splines that describe the branched ureteric tree are used for highly accurate measurements of length, curvature, diameter and angle (Fig. 9.3f–i) and provide a new degree of precision for the measurement of tip morphology. The advantage of splines is that they are described by maths, so measurement units can be smaller than the voxel size that Tree Surveyor is analysing. An example of this advantage is that diameter is measured at sub-micron level by analysing the length of a line projected at a normal angle to the spline curve at 50 points per step along the branches and tips. In this way, it samples many diameters per branch, so uses a median of all samples per branch as a measurement of the segment.

9.6 What Happens in the Mouse?

The advent of tomographic and section-based imaging and reconstruction techniques, as well as programmes like Tree Surveyor, has facilitated the examination of the process of elaboration of the UB over time in significant detail. As we have previously discussed, the mouse UB grows from the nephric duct and undergoes its first branching event at embryonic day 11.5. Over the course of the next 5 days, it undergoes an extremely rapid process of elaboration such that by E16.5 the organ primordia has ~1300 tips (Short et al. 2014) (Fig. 9.4a). Given that the final tip number of the kidney is estimated to be ~3000, this early period of rapid branching comprises the vast majority of branching events needed to form the organ. While the "finished" kidney is composed entirely of bifurcating branches, it is notable that during these early stages tri-foil tip morphologies are noted, although their occurrence decreases rapidly over the first 72 h of development such that they are only rarely observed after E15.5 (Short et al. 2014). The significance of these different morphologies is unclear. They may serve to increase the branch complexity/depth

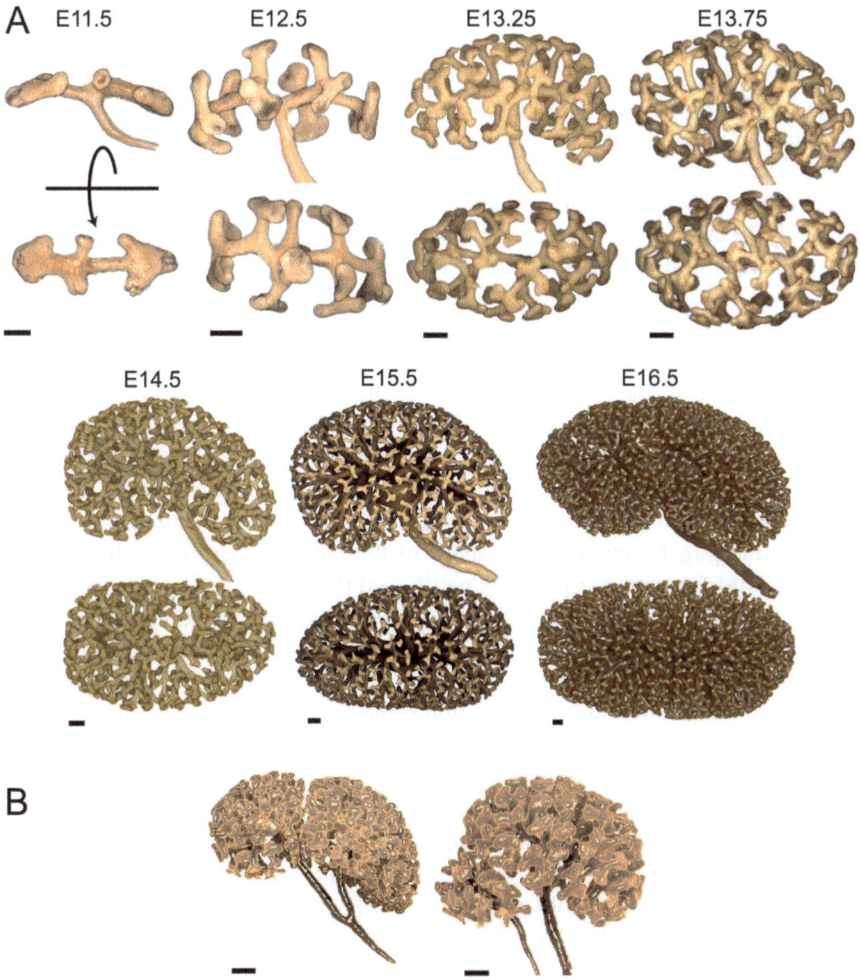

Fig. 9.4 Elaboration of the ureteric bud in the mouse visualised by optical projection tomography (OPT) of organs stained with a marker to the ureteric epithelium (pan-cytokeratin). The period depicted is between embryonic day 11.5 and E16.5 (**a**). Note the swelling in the earliest branches marking the formation of the renal pelvis into which the single papillae of the mouse organ will eventually project. The shape of the kidney is dictated to a considerable extent by the shape of the field of metanephric mesenchyme. In mouse mutants in which ectopic ureteric buds form, such as *Asciz*$^{-/-}$ (**b**, unpublished data), the shape of the organ remains largely unchanged, despite the presence of multiple or bifid ureters

in different parts of the organ or they may represent the product of a relaxation of the processes which normally restrain branching to "ordinary" bifid conformations. This relaxation might be driven by unusually rapid division of cells in the UB tips early in development (Short et al. 2014) or by the generation of fields on the surface of the organ in which branch suppressing signals, such as Sprouty 1 (Basson et al. 2006), are reduced. The latter scenario could be facilitated by rapid increases in the surface area of the organ associated with organ or somatic growth. Interestingly, branch morphologies suggesting de novo formation of new "buds" from established branches are never observed—a finding which stands in contrast to observations of branching in cultured organs (Watanabe and Costantini 2004). It seems likely that events of this type are actively suppressed by the renal interstitium and that such repression is not operant or is ineffective in culture, wherein the branches are unusually arranged and tissue elements are abnormally juxtaposed.

Another interesting observation is that the extent of branching is not uniform across the organ. Kidneys are longer in their rostral–caudal axis, and this elongation is characterised (and to some extent possibly driven) by more branching along this axis (Short et al. 2014). This raises the broader question as to the factors which affect the shape of the kidney. This is something we will return to later in our discussion of kidneys in other organisms, but it is likely that two major factors are at play. Firstly, the shape of the organ is considerably influenced by the shape of the field of induced metanephric mesenchyme into which the ureteric bud grows. As an example, kidneys in which ectopic ureteric buds have contacted and grown into the mesenchyme largely retain overall shape of the kidney, even though the ureteric tree within this volume is effectively partitioned (Fig. 9.4b). Such findings suggest that the shape of the MM field dictates, to a considerable extent, the shape of the tree/organ. The rostro-caudal extension of branching might therefore reflect the "exploration" of this field by the UB, associated with increases in branch complexity. A second and important consideration is the size of the embryo which changes very rapidly during this period which would conceivably offer a larger volume in which the kidney could develop.

The computational advances provided by programmes like Tree Surveyor allows for the analysis of individual generations of branching over time. Examination of this type highlights the highly dynamic nature of the ureteric tree. Individual branches change quite dramatically in their shape, size and organisation over this time (Short et al. 2014). For the most part, branch generations become longer over developmental time, and this is most true for branch generations 4–8. Branch volumes also change, although less dramatically than length. Perhaps the most striking alteration in branch form occurs during the formation of the renal pelvis. This vault within the kidney serves as a collecting point for urine delivered from the renal papilla, a finger-like structure which projects into it. The first evidence for the formation of the renal pelvis comes in the form of a swelling of the ureter at the base of the first bifurcation of the ureteric tree which over time becomes considerably larger. Analysis of tip number versus branch number reveals a disconnect from E15.5 to E16.5, suggesting that at least two and possibly more of the branch generations nearest the ureter swell and fuse, forming a void which will become the renal pelvis (Short et al. 2014). The renal papillae which fills this space typically has 4–6 collecting ducts at its terminus (Phua et al. 2016), supporting the notion that branch resorption/fusion drives the formation of this structures. The molecular basis for this remodelling and the changes in tissue architecture which accompany it remain largely unknown.

The spatial relationship between different branches also changes over time and the branch points themselves alter considerably in form. Initial branching events are relatively large, often as high as 180° (essentially forming a "T" shape) but as development progresses and as extra branches are added, these angles reduce (Sampogna et al. 2015; Short et al. 2014; Cebrian et al. 2004). For branches which have already formed, developmental progression sees a lessening of these angles. Notably these changes are evident when considering the angle formed between a branch point and its immediate offspring. However, when considering the architecture of the branch itself (i.e. the angle at the branch point formed during bifurcation, without reference to future branches), the angle stays notably fixed. A second measure of a given branch is it's rotation in 3D space. As we have discussed in the context of lung development, there is some support for the existence of branching programmes with fixed dihedral architecture—so-called planar or orthogonal branching events in which the rotation between the planes of one branching relative to its offspring is either 0° or 90°, respectively. In the kidney, there is little to no evidence of such a rigid arrangement of branches, with most dihedral rotation being around 65° (Short et al. 2014). Interestingly, despite the changes in branch shape and size and the differentiation of nephrons, the dihedral distribution of branches in the kidney remains relatively static across the developmental period. This suggests that once the gross organisation of branches is fixed in 3D space, they remain in that confirmation for much of embryonic development.

Because of the links between the cap and tip cell niches during renal development, the generation of nephrons is intimately associated with the elaboration of the ureteric tree. The full endowment of nephrons in the mouse kidney is ~14,000 (Merlet-Benichou et al. 1999), and parallel cellular studies have found that the vast majority of these nephrons are formed either in the later stages of organ development or after birth (Short et al. 2014). Renal vesicles that are formed from the cap mesenchyme reintegrate into the tip from which they are derived, but they are also largely co-transported with that tip as the kidney grows over development. During the period of most rapid branching from E11.5 to E15.5, the number of nephrons per tip increases to two and then remains relatively constant until 2 days after birth, at which point there is a rapid burst of new nephron formation (Short et al. 2014). The emergence of these latter nephrons is distinct from earlier in development. Instead of forming in the "armpits" of the branching tree, they arise from around the un-dividing terminal UB ampullae, including on the distal side closest to the renal capsule (Rumballe et al. 2011). The triggers for this change are unclear, but may relate to an active signalling event triggered at birth and/or by a change in ratio of cap and tip cells. The process of tubulogenesis associated with nephron differentiation occurs as the ureteric bud continues to grow and divide. The impact of this growth and elaboration is currently unclear, though we would envisage that this process would reinforce the separation between different branches of the tree and may in fact shape the capacity of a given tip to divide.

9.7 What Happens in Humans?

Kidney development in humans appears to be driven by a similar process to mice (outgrowth and ramification of the ureteric bud). There have been relatively few studies of human kidney development and very few in the last 20–30 years. Superficially at least, human kidney development looks different to mouse development, principally because the human kidney has between 4 and 18 papillae (where mice have one) which empty to minor and major calyces (Cullen McEwan and Black 2015). As a consequence (and unlike the mouse), its surface is somewhat lobulated. The renal calyces are thought to form through a process of branch resorption, swelling and/or rearrangement which would appear to be analogous to the remodelling which forms the renal pelvis in mice (Short et al. 2014). The ureteric bud grows from the nephric duct at 4–5 weeks of gestation (E10.5–11.0 in mice) and, like the mouse kidney, it undergoes a period of very rapid branching over the next 10 weeks. Nephron formation is thought to begin at around 8 weeks of gestation, after the initial rounds of branching morphogenesis. Prior to 15 weeks of gestation, nephron formation is associated with the bifurcating tips of the UB, and nephrons remain connected to the tips from which they arose. However, following this initial burst of nephron formation, a second round of generation (~15–21 weeks) occurs, during which nephrons are generated and often connected through a common collecting duct in an arrangement known as an "arcade". Osathanondh and Potter contend that nephron formation in arcades occurs after branching ceases (Osathanondh and Potter 1963), but the dissections of Oliver (1968) indicate that this is not necessarily the case. In any event, the proportion of nephrons in this conformation has been assessed as contributing ~1/6 of the total endowment of the human kidney, and most of these elaborate structures reside at the cortico-medullary boundary (Oliver 1968). Arcade formation has not been reported in mice, although there are strong indications that it is a feature of organ development in the rat and rabbit (Steer et al. 2004; Zent et al. 2001). The final period of kidney development and nephron formation in humans is characterised by the formation of nephrons with very little branching of the ureteric tree. This period spans from ~21 weeks until the cessation of renal development (at 36 weeks, although this has not been well studied). It is entirely unclear as to how this process (nephron formation without branching) occurs and in particular whether the same inductive interactions with the cap dictate the formation of renal vesicles and the geometry/architecture of these events.

9.8 Cell Movement and Reorganisation in the UB

It is clear that the process of branching morphogenesis is a dynamic one, but what is known of the cellular events which drive this developmental mechanism? Considerable experimental evidence indicates that branching per se does not require the presence of mesenchyme, but rather a small number of growth factors provide the "inductive mechanism" identified decades ago by Grobstein (1953a). For example, dissected UBs branch extensively in ECM-based culture systems supplemented with factors like GDNF (Qiao et al. 1999). However, it is clear that the mesenchyme is critically important in "tuning" the process of branching morphogenesis. This is evident in the observation that UBs cultured in matrix do not form structures that are particularly reminiscent of kidneys; however, UBs grown in complex with lung mesenchyme (or kidney mesenchyme) form structures similar to lungs (or kidneys) (Lin et al. 2003). Real-time studies of kidneys in culture have provided us with considerable insights into how the process of branching occurs during foetal development (Watanabe and Costantini 2004). It is clear based on morphological and gene expression studies that different cell populations inhabit the branching ureteric tree. In particular, there exists a population of Wnt11 expressing cells at the most distal end of the branching tree (Majumdar et al. 2003) which represents a persistent population of cells driving the branching process. Cells which exit this progenitor cell pool contribute to the forming trunks, but cells do not appear to move in the other direction (from trunk to tip) (Shakya et al. 2005). Studies in culture suggest that trunk cells still have the capacity to initiate budding, but this requires induction by factors produced by a transplanted mesenchymal source (Sweeney et al. 2008). The decision as to whether a cell remains in the tip or moves to the trunk is not an arbitrary one. Elegant cell mosaic studies in mice show that cells lacking Ret, or its downstream effector Etv4, are less likely to remain in the tip cell niche and move to the trunk, a pattern which importantly does not reflect changes in cell proliferation (Riccio et al. 2016). These studies also recapitulate similar work examining UB initiation and outgrowth in which cells lacking Ret are evenly distributed in the nephric duct during early development but which segregate away from the tips of the outgrowing UB at the expense of wild type cells upon induction of metanephric kidney formation (Chi et al. 2009).

Cell proliferation in the branching ureteric tree is highest in the tips, though there is no discernible pattern in the distribution of these cells (Fisher et al. 2001; Short et al. 2014). Moreover, there is little evidence to suggest that alterations in proliferation specifically drive the formation of new branches or initiate branching events. Indeed, there is little experimental evidence to suggest that this is the case in any branching system [reviewed in (Varner and Nelson 2014)]. However, cell division in the tips is proportional to the rates of branching, with highest levels seen during the period between E12.5 and E15.5 in mice when ramification of the tree is accelerating fastest (Short et al. 2014). The cells of the UB tips surround a lumen which features in one of the more remarkable cellular events in branching morphogenesis (Packard et al. 2013). Ordinarily, a cell dividing in an epithelial sheet produces daughters which integrate into that sheet in adjacent positions. Instead in the kidney (tips, not trunk), a dividing cell extends into the lumen (while maintaining a basal attachment) and then divides to produce one "tethered" daughter which returns to its original position and a second daughter which reintegrates into the sheet, apparently at random, up to 3 cells diameters away. While fascinating and unrecognised over decades of research, the relevance of this "mitosis associated cell dispersal" is currently unclear.

Our understanding of the processes which drive the decision of ampulla to divide or extend are also somewhat limited. It seems likely that changes in the actin cytoskeleton of the UB are important, as mice carrying mutations in actin depolymerising molecules like COFILIN1 and DESTRIN develop dysmorphic tip morphologies with abnormal branch patterns (Kuure et al. 2010). In other organs like the lung and submandibular glands, tip bifurcation is associated with (and/or mediated by) the presence of cables of ECM or smooth muscle (Kim et al. 2015) which act somewhat like a cheese cutter to divide an ampullae (Larsen et al. 2006). However, little evidence exists for the presence of a similar mechanism in the kidney. It is clear that the ECM plays a central role in regulating kidney development and branching, as loss of many members of this protein family result in renal agenesis [reviewed in (Kim and Nelson 2012)].

9.9 Stereotypy and Patterns in Branching

How similar is the pattern of branching between any two kidneys? Does the programme of branching in the organ rigorously define the timing and position of branching? These questions are largely prompted by the previously discussed studies of lung branching (Metzger et al. 2008). This proposes a spatially and architecturally defined programme which underlies the development of the organ, such that individual branching events could be classified and were of defined geometry. While some recent analysis has called this programme into question (Short et al. 2013), it is worth considering whether such a programme operates in the kidney. Attempts to manually or computationally overlay ureteric tree volumes from different organs in 3D space have largely proven unsuccessful (our unpublished data), but analysis of branching from several groups has established a repeatable pattern of major branch lobes within the organ (Sampogna et al. 2015; Short et al. 2014). Surprisingly, when these lobes are specified, it is possible to almost perfectly superimpose the branch patterns of one kidney within another, particularly if comparing a kidney with one incrementally younger in developmental age (Short et al. 2014; Lamberton et al. 2015). This finding indicates that although elaboration within the 3D space occupied by the organ is not spatially constrained or stereotypic (i.e. the same branching event does not occur in the same 3D space in different organs), it does indicate that the extent to which a clade of branches can elaborate within the organ is much more strictly programmed. The mechanisms by which this control are exerted remain unclear, but given that branching in the organ occurs on its surface it seems likely that it is shaped to a considerable extent by the interactions between tips. Tips do not collide or fuse, suggesting strongly that physical interference or the actions of secreted factors which mediate tip avoidance are likely at work. On the basis of organ culture experiments, Bmp7 has been proposed as a tip avoidance cue in the kidney, as knockdown or interference with its actions results in tip collision and fusion (Davies et al. 2014). The kidney stroma may also play a central role, in the same way in which it appears to suppress branching of tips in the trunks of the ureteric tree.

In considering kidney shape and form the morphological diversity that characterises the organ in different species of animals is striking. Broadly speaking, kidneys come in three types. The "unipapillary" kidney found in mice and in a range of other organisms including dogs, horses, bats and rabbits (Sperber 1944). This form of the organ also encompasses so-called "crest" kidneys. As the name suggests, this type of kidney has a single papillae draining urine to the renal pelvis and bladder. The other two renal forms have multiple papillae (multi-renculated). These organs can assume the form of a single, "compound" organ with several calyces (such as in humans, in which the multiple papillae contribute to the organs uneven, lobular surface) (Oliver 1968). Alternately, the multiple calyces are "discrete" and look much like a bunch of grapes (as in the whales or seals) (Sperber 1944). If such different organ forms reflected different developmental mechanisms, one might expect them to appear in organisms related by phylogeny. That is to say a "novel" organ form might have evolved in one branch of the mammalian lineage and was then employed by other organs of related descent. However, this is not the case. Instead, organ architecture is more or less randomly distributed amongst the mammals (Sperber 1944). These observations suggest strongly that a single unifying mode of development is employed in kidney development and that other changes, perhaps in body size or the shape and size of the metanephric mesenchyme, dictate the overall structure of the organ. This concept is perhaps best embodied by the whale's kidney—a discrete multi-renculated organ (Cave and Aumonier 1964)—in which many hundreds of individual "kidney-lets" each strikingly similar to the uni-papillary organ or the mouse—are networked to form the adult organ.

9.10 Role of Branching Morphogenesis in Kidney Disease

Given its central role in organ development, it is perhaps not surprising the defects in the programme of branching morphogenesis can lead to kidney disease. Malformations of this type contribute significantly to the disease spectrum known as CAKUT—congenital malformations of the kidney and urogenital tract—which affects ~1/500 births and is the leading cause of renal failure in infants (Wiesel et al. 2005; Nakanishi and Yoshikawa 2003). Defects particularly relevant to branching include renal agenesis and ectopic kidney development which likely result from defects in UB formation and/or growth. Renal hypoplasia is also commonly reported and might be associated with factors involved in ongoing elaboration of the ureteric tree. Indeed, mutations in a number of genes which drive both UB outgrowth and elaboration have been found in CAKUT patients. These include genes such as *RET* (Skinner et al. 2008), *GATA3* (Van Esch et al. 2000), *EYA1*

(Abdelhak et al. 1997), *ITGA8* (Humbert et al. 2014), *FRAS1* (Vrontou et al. 2003), *FREM1* (Alazami et al. 2009), *FREM2* (Jadeja et al. 2005), *GRIP1* (Vogel et al. 2012), *PAX2* (Porteous et al. 2000), *FGF20* (Barak et al. 2012), *UPK3A* (Jenkins et al. 2005), *WNT4* (Vivante et al. 2013), *SIX1* (Ruf et al. 2004) and *SIX2* and *BMP4* (Weber et al. 2008). In addition to the striking developmental abnormalities associated with CAKUT, nephron number is also known to vary dramatically between individuals (Hughson et al. 2003). Although these individuals may be otherwise healthy, there is an emerging consensus that low nephron endowment is associated with predisposition to hypertension and chronic kidney disease (Luyckx et al. 2013). Although these links are relatively poorly understood and are largely correlative, a number of studies, principally in animal models, have shown that nephron number is considerably reduced in offspring of pregnant mothers with compromised gestation. Maternal insults such as dietary protein restriction (Harrison and Langley-Evans 2009), alcohol consumption (Gray et al. 2010), maternal hypoglycaemia (Hokke et al. 2013), placental insufficiency (Wlodek et al. 2007) and vitamin A deficiency (Lelievre-Pegorier et al. 1998) (amongst others) have all been linked to reduced nephron endowment, in many cases as a consequence of reduced or abnormal branching morphogenesis.

9.11 Conclusions

Branching morphogenesis is a driving developmental mechanism which establishes the basic architecture of the mammalian kidney and other organs such as the mammary gland and the lungs. Its developmental association with the formation of nephrons and the maintenance of nephron progenitor cells further highlight its importance for the establishment of normal adult organ function. Recent advances in imaging technologies and the ability to interpret these image datasets places us in a unique position to analyse how the organ grows and changes in form. These insights will be important for understanding the role of different genes or genetic perturbations to kidney development. These advances will in turn be important for better understanding the processes which lead to disease in the organ.

References

Abdelhak S, Kalatzis V, Heilig R, Compain S, Samson D, Vincent C, Weil D, Cruaud C, Sahly I, Leibovici M, Bitner-Glindzicz M, Francis M, Lacombe D, Vigneron J, Charachon R, Boven K, Bedbeder P, Van Regemorter N, Weissenbach J, Petit C (1997) A human homologue of the Drosophila eyes absent gene underlies branchio-oto-renal (BOR) syndrome and identifies a novel gene family. Nat Genet 15(2):157–164. doi:10.1038/ng0297-157

Alazami AM, Shaheen R, Alzahrani F, Snape K, Saggar A, Brinkmann B, Bavi P, Al-Gazali LI, Alkuraya FS (2009) FREM1 mutations cause bifid nose, renal agenesis, and anorectal malformations syndrome. Am J Hum Genet 85(3):414–418. doi:10.1016/j.ajhg.2009.08.010

Aylward S, Pizer S, Eberly D, Bullitt E (1996) Intensity ridge and widths for tubular object segmentation and description. In: Paper presented at the Proceedings of the 1996 Workshop on mathematical methods in biomedical image analysis (MMBIA '96)

Barak H, Huh SH, Chen S, Jeanpierre C, Martinovic J, Parisot M, Bole-Feysot C, Nitschke P, Salomon R, Antignac C, Ornitz DM, Kopan R (2012) FGF9 and FGF20 maintain the stemness of nephron progenitors in mice and man. Dev Cell 22(6):1191–1207. doi:10.1016/j.devcel.2012.04.018

Basson MA, Watson-Johnson J, Shakya R, Akbulut S, Hyink D, Costantini FD, Wilson PD, Mason IJ, Licht JD (2006) Branching morphogenesis of the ureteric epithelium during kidney development is coordinated by the opposing functions of GDNF and Sprouty1. Dev Biol 299 (2):466–477. doi:10.1016/j.ydbio.2006.08.051

Boyle S, Misfeldt A, Chandler KJ, Deal KK, Southard-Smith EM, Mortlock DP, Baldwin HS, de Caestecker M (2008) Fate mapping using Cited1-CreERT2 mice demonstrates that the cap mesenchyme contains self-renewing progenitor cells and gives rise exclusively to nephronic epithelia. Dev Biol 313(1):234–245. doi:10.1016/j.ydbio.2007.10.014

Boyle S, Shioda T, Perantoni AO, de Caestecker M (2007) Cited1 and Cited2 are differentially expressed in the developing kidney but are not required for nephrogenesis. Dev Dyn 236 (8):2321–2330. doi:10.1002/dvdy.21242

Brophy PD, Ostrom L, Lang KM, Dressler GR (2001) Regulation of ureteric bud outgrowth by Pax2-dependent activation of the glial derived neurotrophic factor gene. Development 128 (23):4747–4756

Cave AJ, Aumonier FJ (1964) The reniculus in certain balaenopterids. J R Microsc Soc 83:255–264

Cebrian C, Borodo K, Charles N, Herzlinger DA (2004) Morphometric index of the developing murine kidney. Dev Dyn 231(3):601–608. doi:10.1002/dvdy.20143

Chi X, Michos O, Shakya R, Riccio P, Enomoto H, Licht JD, Asai N, Takahashi M, Ohgami N, Kato M, Mendelsohn C, Costantini F (2009) Ret-dependent cell rearrangements in the Wolffian duct epithelium initiate ureteric bud morphogenesis. Dev Cell 17(2):199–209. doi:10.1016/j.devcel.2009.07.013

Combes AN, Short KM, Lefevre J, Hamilton NA, Little MH, Smyth IM (2014) An integrated pipeline for the multidimensional analysis of branching morphogenesis. Nat Protoc 9 (12):2859–2879. doi:10.1038/nprot.2014.193

Cullen McEwan L, Black MJ (2015) The human kidney: parallels in structure, spatial development, and timing of nephrogenesis. In: Little M (ed) Kidney development, disease, repair and regeneration. Academic Press, Boston, MA

Cullen-McEwen LA, Fricout G, Harper IS, Jeulin D, Bertram JF (2002) Quantitation of 3D ureteric branching morphogenesis in cultured embryonic mouse kidney. Int J Dev Biol 46 (8):1049–1055

Davies JA, Hohenstein P, Chang CH, Berry R (2014) A self-avoidance mechanism in patterning of the urinary collecting duct tree. BMC Dev Biol 14:35. doi:10.1186/s12861-014-0035-8

Durbec P, Marcos-Gutierrez CV, Kilkenny C, Grigoriou M, Wartiowaara K, Suvanto P, Smith D, Ponder B, Costantini F, Saarma M et al (1996) GDNF signalling through the Ret receptor tyrosine kinase. Nature 381(6585):789–793. doi:10.1038/381789a0

Fisher CE, Michael L, Barnett MW, Davies JA (2001) Erk MAP kinase regulates branching morphogenesis in the developing mouse kidney. Development 128(21):4329–4338

Georgas K, Rumballe B, Valerius MT, Chiu HS, Thiagarajan RD, Lesieur E, Aronow BJ, Brunskill EW, Combes AN, Tang D, Taylor D, Grimmond SM, Potter SS, McMahon AP, Little MH (2009) Analysis of early nephron patterning reveals a role for distal RV proliferation in fusion to the ureteric tip via a cap mesenchyme-derived connecting segment. Dev Biol 332 (2):273–286. doi:10.1016/j.ydbio.2009.05.578

Goh W, Chan K (2007) The multiresolution gradient vector field skeleton. Pattern Recogn 40 (4):1255–1269. doi:10.1016/j.patcog.2006.08.007

Gray SP, Denton KM, Cullen-McEwen L, Bertram JF, Moritz KM (2010) Prenatal exposure to alcohol reduces nephron number and raises blood pressure in progeny. J Am Soc Nephrol JASN 21(11):1891–1902. doi:10.1681/ASN.2010040368
Grieshammer U, Le M, Plump AS, Wang F, Tessier-Lavigne M, Martin GR (2004) SLIT2-mediated ROBO2 signaling restricts kidney induction to a single site. Dev Cell 6(5):709–717
Grobstein C (1953a) Inductive epitheliomesenchymal interaction in cultured organ rudiments of the mouse. Science 118(3054):52–55
Grobstein C (1953b) Morphogenetic interaction between embryonic mouse tissues separated by a membrane filter. Nature 172(4384):869–870
Grobstein C (1955) Inductive interaction in the development of the mouse metanephros. J Exp Zool 130(2):319–339. doi:10.1002/jez.1401300207
Grobstein C (1956) Trans-filter induction of tubules in mouse metanephrogenic mesenchyme. Exp Cell Res 10(2):424–440
Grobstein C (1961) Cell contact in relation to embryonic induction. Exp Cell Res Suppl 8:234–245
Gross I, Morrison DJ, Hyink DP, Georgas K, English MA, Mericskay M, Hosono S, Sassoon D, Wilson PD, Little M, Licht JD (2003) The receptor tyrosine kinase regulator Sprouty1 is a target of the tumor suppressor WT1 and important for kidney development. J Biol Chem 278 (42):41420–41430. doi:10.1074/jbc.M306425200
Harrison M, Langley-Evans SC (2009) Intergenerational programming of impaired nephrogenesis and hypertension in rats following maternal protein restriction during pregnancy. Br J Nutr 101 (7):1020–1030. doi:10.1017/S0007114508057607
Hokke SN, Armitage JA, Puelles VG, Short KM, Jones L, Smyth IM, Bertram JF, Cullen-McEwen LA (2013) Altered ureteric branching morphogenesis and nephron endowment in offspring of diabetic and insulin-treated pregnancy. PLoS One 8(3):e58243. doi:10.1371/journal.pone.0058243
Hughson M, Farris AB 3rd, Douglas-Denton R, Hoy WE, Bertram JF (2003) Glomerular number and size in autopsy kidneys: the relationship to birth weight. Kidney Int 63(6):2113–2122. doi:10.1046/j.1523-1755.2003.00018.x
Humbert C, Silbermann F, Morar B, Parisot M, Zarhrate M, Masson C, Tores F, Blanchet P, Perez MJ, Petrov Y, Khau Van Kien P, Roume J, Leroy B, Gribouval O, Kalaydjieva L, Heidet L, Salomon R, Antignac C, Benmerah A, Saunier S, Jeanpierre C (2014) Integrin alpha 8 recessive mutations are responsible for bilateral renal agenesis in humans. Am J Hum Genet 94 (2):288–294. doi:10.1016/j.ajhg.2013.12.017
Jadeja S, Smyth I, Pitera JE, Taylor MS, van Haelst M, Bentley E, McGregor L, Hopkins J, Chalepakis G, Philip N, Perez Aytes A, Watt FM, Darling SM, Jackson I, Woolf AS, Scambler PJ (2005) Identification of a new gene mutated in Fraser syndrome and mouse myelencephalic blebs. Nat Genet 37(5):520–525. doi:10.1038/ng1549
Jenkins D, Bitner-Glindzicz M, Malcolm S, Hu CC, Allison J, Winyard PJ, Gullett AM, Thomas DF, Belk RA, Feather SA, Sun TT, Woolf AS (2005) De novo Uroplakin IIIa heterozygous mutations cause human renal adysplasia leading to severe kidney failure. J Am Soc Nephrol JASN 16(7):2141–2149. doi:10.1681/ASN.2004090776
Kao RM, Vasilyev A, Miyawaki A, Drummond IA, McMahon AP (2012) Invasion of distal nephron precursors associates with tubular interconnection during nephrogenesis. J Am Soc Nephrol JASN 23(10):1682–1690. doi:10.1681/ASN.2012030283
Kerwin J, Scott M, Sharpe J, Puelles L, Robson SC, Martinez-de-la-Torre M, Ferran JL, Feng G, Baldock R, Strachan T, Davidson D, Lindsay S (2004) 3 dimensional modelling of early human brain development using optical projection tomography. BMC Neurosci 5:27. doi:10.1186/1471-2202-5-27
Kim HY, Nelson CM (2012) Extracellular matrix and cytoskeletal dynamics during branching morphogenesis. Organogenesis 8(2):56–64. doi:10.4161/org.19813
Kim HY, Pang MF, Varner VD, Kojima L, Miller E, Radisky DC, Nelson CM (2015) Localized smooth muscle differentiation is essential for epithelial bifurcation during branching morphogenesis of the mammalian lung. Dev Cell 34(6):719–726. doi:10.1016/j.devcel.2015.08.012

Kobayashi A, Valerius MT, Mugford JW, Carroll TJ, Self M, Oliver G, McMahon AP (2008) Six2 defines and regulates a multipotent self-renewing nephron progenitor population throughout mammalian kidney development. Cell Stem Cell 3(2):169–181. doi:10.1016/j.stem.2008.05.020

Kume T, Deng K, Hogan BL (2000) Murine forkhead/winged helix genes Foxc1 (Mf1) and Foxc2 (Mfh1) are required for the early organogenesis of the kidney and urinary tract. Development 127(7):1387–1395

Kuure S, Cebrian C, Machingo Q, Lu BC, Chi X, Hyink D, D'Agati V, Gurniak C, Witke W, Costantini F (2010) Actin depolymerizing factors cofilin1 and destrin are required for ureteric bud branching morphogenesis. PLoS Genet 6(10):e1001176. doi:10.1371/journal.pgen.1001176

Lamberton TO, Lefevre J, Short KM, Smyth IM, Hamilton NA (2015) Comparing and distinguishing the structure of biological branching. J Theor Biol 365:226–237. doi:10.1016/j.jtbi.2014.10.001

Larsen M, Wei C, Yamada KM (2006) Cell and fibronectin dynamics during branching morphogenesis. J Cell Sci 119(Pt 16):3376–3384. doi:10.1242/jcs.03079

Lelievre-Pegorier M, Vilar J, Ferrier ML, Moreau E, Freund N, Gilbert T, Merlet-Benichou C (1998) Mild vitamin A deficiency leads to inborn nephron deficit in the rat. Kidney Int 54 (5):1455–1462. doi:10.1046/j.1523-1755.1998.00151.x

Lin Y, Zhang S, Tuukkanen J, Peltoketo H, Pihlajaniemi T, Vainio S (2003) Patterning parameters associated with the branching of the ureteric bud regulated by epithelial-mesenchymal interactions. Int J Dev Biol 47(1):3–13

Luyckx VA, Bertram JF, Brenner BM, Fall C, Hoy WE, Ozanne SE, Vikse BE (2013) Effect of fetal and child health on kidney development and long-term risk of hypertension and kidney disease. Lancet 382(9888):273–283. doi:10.1016/S0140-6736(13)60311-6

Majumdar A, Vainio S, Kispert A, McMahon J, McMahon AP (2003) Wnt11 and Ret/Gdnf pathways cooperate in regulating ureteric branching during metanephric kidney development. Development 130(14):3175–3185

Malkusch W, Hellinger A, Konerding M, Bruch J, Obertacke U (1995) Morphometry of experimental lung contusion: an improved quantitative method. Anal Cell Pathol J Eur Soc Anal Cell Pathol 8(4):279–286

Merlet-Benichou C, Gilbert T, Vilar J, Moreau E, Freund N, Lelievre-Pegorier M (1999) Nephron number: variability is the rule. Causes and consequences. Lab Invest J Tech Methods Pathol 79 (5):515–527

Metzger RJ, Klein OD, Martin GR, Krasnow MA (2008) The branching programme of mouse lung development. Nature 453(7196):745–750. doi:10.1038/nature07005

Nakanishi K, Yoshikawa N (2003) Genetic disorders of human congenital anomalies of the kidney and urinary tract (CAKUT). Pediatr Int Off J Jpn Pediatr Soc 45(5):610–616

Nigam SK (2013) Concise review: can the intrinsic power of branching morphogenesis be used for engineering epithelial tissues and organs? Stem Cells Transl Med 2(12):993–1000. doi:10.5966/sctm.2013-0076

Nishinakamura R, Matsumoto Y, Nakao K, Nakamura K, Sato A, Copeland NG, Gilbert DJ, Jenkins NA, Scully S, Lacey DL, Katsuki M, Asashima M, Yokota T (2001) Murine homolog of SALL1 is essential for ureteric bud invasion in kidney development. Development 128 (16):3105–3115

Oliver J (1968) Nephrons and kidneys: a quantitative study of developmental and evolutionary mammalian renal archtectonics. Harper & Row, New York, NY

Osathanondh V, Potter EL (1963) Development of human kidney as shown by microdissection. III. formation and interrelationship of collecting tubules and nephrons. Arch Pathol 76:290–302

Otsu N (1979) A threshold selection method from gray-level histograms. Syst Man Cybern IEEE Trans 9(1):62–66. doi:10.1109/TSMC.1979.4310076

Packard A, Georgas K, Michos O, Riccio P, Cebrian C, Combes AN, Ju A, Ferrer-Vaquer A, Hadjantonakis AK, Zong H, Little MH, Costantini F (2013) Luminal mitosis drives epithelial cell dispersal within the branching ureteric bud. Dev Cell 27(3):319–330. doi:10.1016/j.devcel.2013.09.001

Phua YL, Gilbert T, Combes A, Wilkinson L, Little MH (2016) Neonatal vascularization and oxygen tension regulate appropriate perinatal renal medulla/papilla maturation. J Pathol 238 (5):665–676. doi:10.1002/path.4690

Porteous S, Torban E, Cho NP, Cunliffe H, Chua L, McNoe L, Ward T, Souza C, Gus P, Giugliani R, Sato T, Yun K, Favor J, Sicotte M, Goodyer P, Eccles M (2000) Primary renal hypoplasia in humans and mice with PAX2 mutations: evidence of increased apoptosis in fetal kidneys of Pax2(1Neu) +/- mutant mice. Hum Mol Genet 9(1):1–11

Qiao J, Sakurai H, Nigam SK (1999) Branching morphogenesis independent of mesenchymal-epithelial contact in the developing kidney. Proc Natl Acad Sci USA 96(13):7330–7335

Rak-Raszewska A, Hauser PV, Vainio S (2015) Organ in vitro culture: what have we learned about early kidney development? Stem Cells Int 2015:959807. doi:10.1155/2015/959807

Riccio P, Cebrian C, Zong H, Hippenmeyer S, Costantini F (2016) Ret and Etv4 promote directed movements of progenitor cells during renal branching morphogenesis. PLoS Biol 14(2): e1002382. doi:10.1371/journal.pbio.1002382

Ruf RG, Xu PX, Silvius D, Otto EA, Beekmann F, Muerb UT, Kumar S, Neuhaus TJ, Kemper MJ, Raymond RM Jr, Brophy PD, Berkman J, Gattas M, Hyland V, Ruf EM, Schwartz C, Chang EH, Smith RJ, Stratakis CA, Weil D, Petit C, Hildebrandt F (2004) SIX1 mutations cause branchio-oto-renal syndrome by disruption of EYA1-SIX1-DNA complexes. Proc Natl Acad Sci USA 101(21):8090–8095. doi:10.1073/pnas.0308475101

Rumballe BA, Georgas KM, Combes AN, Ju AL, Gilbert T, Little MH (2011) Nephron formation adopts a novel spatial topology at cessation of nephrogenesis. Dev Biol 360(1):110–122. doi:10.1016/j.ydbio.2011.09.011

Sampogna RV, Schneider L, Al-Awqati Q (2015) Developmental programming of branching morphogenesis in the kidney. J Am Soc Nephrol JASN 26(10):2414–2422. doi:10.1681/ASN.2014090886

Schuchardt A, D'Agati V, Pachnis V, Costantini F (1996) Renal agenesis and hypodysplasia in ret-k- mutant mice result from defects in ureteric bud development. Development 122 (6):1919–1929

Self M, Lagutin OV, Bowling B, Hendrix J, Cai Y, Dressler GR, Oliver G (2006) Six2 is required for suppression of nephrogenesis and progenitor renewal in the developing kidney. EMBO J 25 (21):5214–5228. doi:10.1038/sj.emboj.7601381

Shakya R, Watanabe T, Costantini F (2005) The role of GDNF/Ret signaling in ureteric bud cell fate and branching morphogenesis. Dev Cell 8(1):65–74. doi:10.1016/j.devcel.2004.11.008

Sharpe J, Ahlgren U, Perry P, Hill B, Ross A, Hecksher-Sorensen J, Baldock R, Davidson D (2002) Optical projection tomography as a tool for 3D microscopy and gene expression studies. Science 296(5567):541–545. doi:10.1126/science.1068206 296/5567/541 [pii]

Short KM, Combes AN, Lefevre J, Ju AL, Georgas KM, Lamberton T, Cairncross O, Rumballe BA, McMahon AP, Hamilton NA, Smyth IM, Little MH (2014) Global quantification of tissue dynamics in the developing mouse kidney. Dev Cell 29(2):188–202. doi:10.1016/j.devcel.2014.02.017

Short KM, Hodson MJ, Smyth IM (2010) Tomographic quantification of branching morphogenesis and renal development. Kidney Int 77(12):1132–1139. doi:10.1038/ki.2010.42

Short K, Hodson M, Smyth I (2013) Spatial mapping and quantification of developmental branching morphogenesis. Development 140(2):471–478. doi:10.1242/dev.088500 dev.088500 [pii]

Sims-Lucas S, Argyropoulos C, Kish K, McHugh K, Bertram JF, Quigley R, Bates CM (2009) Three-dimensional imaging reveals ureteric and mesenchymal defects in Fgfr2-mutant kidneys. J Am Soc Nephrol JASN 20(12):2525–2533. doi:10.1681/ASN.2009050532

Skinner MA, Safford SD, Reeves JG, Jackson ME, Freemerman AJ (2008) Renal aplasia in humans is associated with RET mutations. Am J Hum Genet 82(2):344–351. doi:10.1016/j.ajhg.2007.10.008

Sperber I (1944) Studies on the mammalian kidney. Zool Bidr Upps 22:249–432

Steer DL, Shah MM, Bush KT, Stuart RO, Sampogna RV, Meyer TN, Schwesinger C, Bai X, Esko JD, Nigam SK (2004) Regulation of ureteric bud branching morphogenesis by sulfated proteoglycans in the developing kidney. Dev Biol 272(2):310–327. doi:10.1016/j.ydbio.2004.04.029

Stiles HJ (1897) Skiagraphy after injection of the blood-vessels with mercury. J Anat Physiol 32 (Pt 1):83–91

Sweeney D, Lindstrom N, Davies JA (2008) Developmental plasticity and regenerative capacity in the renal ureteric bud/collecting duct system. Development 135(15):2505–2510. doi:10.1242/dev.022145

Takasato M, Little MH (2015) The origin of the mammalian kidney: implications for recreating the kidney in vitro. Development 142(11):1937–1947. doi:10.1242/dev.104802

Van Esch H, Groenen P, Nesbit MA, Schuffenhauer S, Lichtner P, Vanderlinden G, Harding B, Beetz R, Bilous RW, Holdaway I, Shaw NJ, Fryns JP, Van de Ven W, Thakker RV, Devriendt K (2000) GATA3 haplo-insufficiency causes human HDR syndrome. Nature 406 (6794):419–422. doi:10.1038/35019088

Varner VD, Nelson CM (2014) Cellular and physical mechanisms of branching morphogenesis. Development 141(14):2750–2759. doi:10.1242/dev.104794

Vega QC, Worby CA, Lechner MS, Dixon JE, Dressler GR (1996) Glial cell line-derived neurotrophic factor activates the receptor tyrosine kinase RET and promotes kidney morphogenesis. Proc Natl Acad Sci USA 93(20):10657–10661

Vincent L, Soille P (1991) Watersheds in digital spaces: an efficient algorithm based on immersion simulations. Pattern Anal Mach Intell IEEE Trans 13(6):583–598. doi:10.1109/34.87344

Vivante A, Mark-Danieli M, Davidovits M, Harari-Steinberg O, Omer D, Gnatek Y, Cleper R, Landau D, Kovalski Y, Weissman I, Eisenstein I, Soudack M, Wolf HR, Issler N, Lotan D, Anikster Y, Dekel B (2013) Renal hypodysplasia associates with a WNT4 variant that causes aberrant canonical WNT signaling. J Am Soc Nephrol JASN 24(4):550–558. doi:10.1681/ASN.2012010097

Vogel MJ, van Zon P, Brueton L, Gijzen M, van Tuil MC, Cox P, Schanze D, Kariminejad A, Ghaderi-Sohi S, Blair E, Zenker M, Scambler PJ, Ploos van Amstel HK, van Haelst MM (2012) Mutations in GRIP1 cause Fraser syndrome. J Med Genet 49(5):303–306. doi:10.1136/jmedgenet-2011-100590

Vrontou S, Petrou P, Meyer BI, Galanopoulos VK, Imai K, Yanagi M, Chowdhury K, Scambler PJ, Chalepakis G (2003) Fras1 deficiency results in cryptophthalmos, renal agenesis and blebbed phenotype in mice. Nat Genet 34(2):209–214. doi:10.1038/ng1168

Wainwright EN, Wilhelm D, Combes AN, Little MH, Koopman P (2015) ROBO2 restricts the nephrogenic field and regulates Wolffian duct-nephrogenic cord separation. Dev Biol 404 (2):88–102. doi:10.1016/j.ydbio.2015.05.023

Watanabe T, Costantini F (2004) Real-time analysis of ureteric bud branching morphogenesis in vitro. Dev Biol 271(1):98–108. doi:10.1016/j.ydbio.2004.03.025

Weber S, Taylor JC, Winyard P, Baker KF, Sullivan-Brown J, Schild R, Knuppel T, Zurowska AM, Caldas-Alfonso A, Litwin M, Emre S, Ghiggeri GM, Bakkaloglu A, Mehls O, Antignac C, Network E, Schaefer F, Burdine RD (2008) SIX2 and BMP4 mutations associate with anomalous kidney development. J Am Soc Nephrol JASN 19(5):891–903. doi:10.1681/ASN.2006111282

Wellik DM, Hawkes PJ, Capecchi MR (2002) Hox11 paralogous genes are essential for metanephric kidney induction. Genes Dev 16(11):1423–1432. doi:10.1101/gad.993302

Wiesel A, Queisser-Luft A, Clementi M, Bianca S, Stoll C, Group ES (2005) Prenatal detection of congenital renal malformations by fetal ultrasonographic examination: an analysis of 709,030 births in 12 European countries. Eur J Med Genet 48(2):131–144. doi:10.1016/j.ejmg.2005.02.003

Wlodek ME, Mibus A, Tan A, Siebel AL, Owens JA, Moritz KM (2007) Normal lactational environment restores nephron endowment and prevents hypertension after placental restriction in the rat. J Am Soc Nephrol JASN 18(6):1688–1696. doi:10.1681/ASN.2007010015

Wörz S, Godinez WJ, Rohr K (2009) Probabilistic tracking and model-based segmentation of 3D tubular structures. In: Meinzer H-P et al (eds) Bildverarbeitung für die Medizin 2009: Algorithmen—Systeme—Anwendungen Proceedings des Workshops vom 22. bis 25. März 2009 in Heidelberg. Springer, Berlin, Heidelberg, pp 41–45

Xu PX, Adams J, Peters H, Brown MC, Heaney S, Maas R (1999) Eya1-deficient mice lack ears and kidneys and show abnormal apoptosis of organ primordia. Nat Genet 23(1):113–117. doi:10.1038/12722

Zent R, Bush KT, Pohl ML, Quaranta V, Koshikawa N, Wang Z, Kreidberg JA, Sakurai H, Stuart RO, Nigam SK (2001) Involvement of laminin binding integrins and laminin-5 in branching morphogenesis of the ureteric bud during kidney development. Dev Biol 238(2):289–302. doi:10.1006/dbio.2001.0391

Chapter 10
Genetic Syndromes Affecting Kidney Development

Abhijeet Pal and Kimberly J Reidy

Abstract Renal anomalies are common birth defects that may manifest as a wide spectrum of anomalies from hydronephrosis (dilation of the renal pelvis and calyces) to renal aplasia (complete absence of the kidney(s)). Aneuploidies and mosaicisms are the most common syndromes associated with CAKUT. Syndromes with single gene and renal developmental defects are less common but have facilitated insight into the mechanism of renal and other organ development. Analysis of underlying genetic mutations with transgenic and mutant mice has also led to advances in our understanding of mechanisms of renal development.

10.1 Introduction

Renal anomalies are among the most common congenital abnormalities. Although the prevalence varies in different populations, congenital abnormalities of the kidney and urinary tract (CAKUT) complicate on average approximately 1 in 500 pregnancies (Song and Yosypiv 2011; Stoll et al. 2014). CAKUT encompass a broad spectrum of anomalies from hydronephrosis (dilation of the renal pelvis and calyces) to renal aplasia [complete absence of the kidney(s)]. CAKUT also include abnormalities of the ureters, bladder, and urethra, including ureteral obstruction at the renal pelvis (UPJ) or entry into the bladder [uretero-vesicle junction (UVJ)], vesicoureteral reflux (backup of urine in the ureters from bladder toward the kidneys), and obstruction of the urethra (e.g., posterior urethral valves). Up to 30% of children with end-stage renal disease (ESRD) in the USA have underlying

A. Pal, M.D.
Children's Hospital at Montefiore/Albert Einstein College of Medicine, Bronx, NY, USA

K.J. Reidy, M.D. (✉)
Children's Hospital at Montefiore/Albert Einstein College of Medicine, Bronx, NY, USA

Pediatric Nephrology, Children's Hospital at Montefiore/Albert Einstein College of Medicine, 3415 Bainbridge Avenue, Bronx, NY 10467, USA
e-mail: kreidy@montefiore.org

R.K. Miller (ed.), *Kidney Development and Disease*, Results and Problems in Cell Differentiation 60, DOI 10.1007/978-3-319-51436-9_10

CAKUT, making it the leading cause of pediatric ESRD (North American Pediatric Renal Trials and Collaborative Studies 2011).

Up to one-third of children with CAKUT may have associated extrarenal manifestations (Stoll et al. 2014). Extrarenal manifestations appear to be most frequent with the severest forms of CAKUT (e.g., renal agenesis) (Stoll et al. 2014). In one large European population-based study (Stoll et al. 2014), over half of the extrarenal manifestations were part of known syndromes or chromosomal abnormalities. The three most common syndromes were VACTERL (which is comprised of vertebral defects, anal atresia, cardiac defects, tracheoesophageal fistula, renal anomalies, and limb abnormalities), Meckel–Gruber syndrome [which is comprised (Sadowski et al. 2015) of renal cystic dysplasia, occipital encephalocele/holoprosencephaly, and postaxial polydactyly], and prune belly syndrome (which is comprised of hydroureteronephrosis, defects in abdominal wall musculature, and cryptorchidism) (Stoll et al. 2014). Chromosomal aneuploidies and mosaicism, including Trisomy 18, 21, and 13, were the second most common likely to be associated with renal anomalies. Collectively, genetic syndromes associated with chromosomal microdeletions/duplications (e.g., DiGeorge) or single gene defects (e.g., Simpson–Golabie–Behmel, Pallister–Hall, Beckwith–Wiedemann) and environmental toxin-induced syndromes (e.g., fetal alcohol syndrome) accounted for the rest of the renal anomalies with identifiable syndromes (Stoll et al. 2014).

The availability of gene sequencing and microarray studies has led to significant advances in the genetics of renal anomalies (Song and Yosypiv 2011). They have rapidly expanded our knowledge of the genes that affect renal and ureter development. In addition, the ability to manipulate genes using transgenic and knockout mice has revealed that a few key signaling pathways regulate multiple stages of kidney development (Reidy and Rosenblum 2009). In particular, gene deletion with conditional knockout mice has revealed cell- and time-specific effects of developmental signaling pathways (Reidy and Rosenblum 2009). Interestingly, several of these key signaling pathways play developmental roles in other organs. Thus, genetic defects lead to both abnormal kidney development and extrarenal manifestations.

Here we will focus on a few key signaling pathways that regulate development in multiple organs and their role in renal epithelial differentiation. In particular, we will emphasize on how the identification of genetic mutations in syndromic renal malformations has advanced our insight into mechanisms of renal epithelial differentiation.

10.2 Manifestations of CAKUT Reflect the Impaired Stage of Renal Development

First, it is important to briefly review the stages of renal development (see Chap. x for a more detailed discussion). The mammalian kidney develops from the intermediate mesoderm. There are three primitive kidneys that form sequentially during human kidney development: the pronephros, mesonephros, and metanephros (Fig. 10.1). The pronephros consists of simple tubules that form and involute.

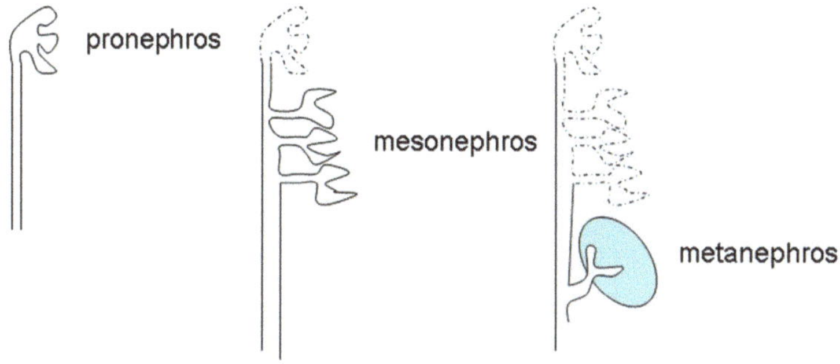

Fig. 10.1 Depiction of the three primitive kidneys

Table 10.1 Single gene defects, impaired stages of renal development, and human syndrome associated with these defects

Impaired stage of renal development	Genes affected	Human syndromes
Defects in UB formation, branching, and elongation	*FGFR2*, *FGFR1*	Craniosynostosis syndrome—Apert's, Antley–Bixler, Pfeiffer, and Beare–Stevenson syndromes
	TRAP1, *FGF8*	VACTERL
Defects in mesenchymal-to-epithelial transformation	*WNT4*	SERKAL syndrome
	RET, *GDNF*	Hirschsprung's disease with unilateral and bilateral renal agenesis
	HNF1B	Renal cysts and diabetes syndrome
	PAX2	Renal coloboma syndrome
	EYA1, *SIX1*, *SIX5*	Branchio-oto-renal syndrome
Defects in terminal differentiation	*JAG1*, *NOTCH2*	Alagille syndrome
	WT1	Denys–Drash and Frasier syndromes
	TSC1, *TSC2*	Tuberous sclerosis

The mesonephros forms caudal to the pronephros and consists of glomeruli and simple tubules that are able to form a filtrate. The mesonephric tubules then form the mesonephric or Wolffian duct. The final adult kidney forms when the ureteric bud grows from the mesonephric duct and invades the surrounding metanephric mesenchyme (Table 10.1).

The reciprocal interactions between metanephric mesenchyme and ureteric bud are required for ureteric bud branching and conversion of the metanephric mesenchyme to renal epithelial cells. The branching ureteric bud forms the collecting ducts. The metanephric mesenchyme condenses and converts to epithelia of the renal vesicle. The renal vesicle becomes a comma and then an S-shape body. The portion of the S-shape closest to the ureteric bud tip forms the distal tubule, the mid

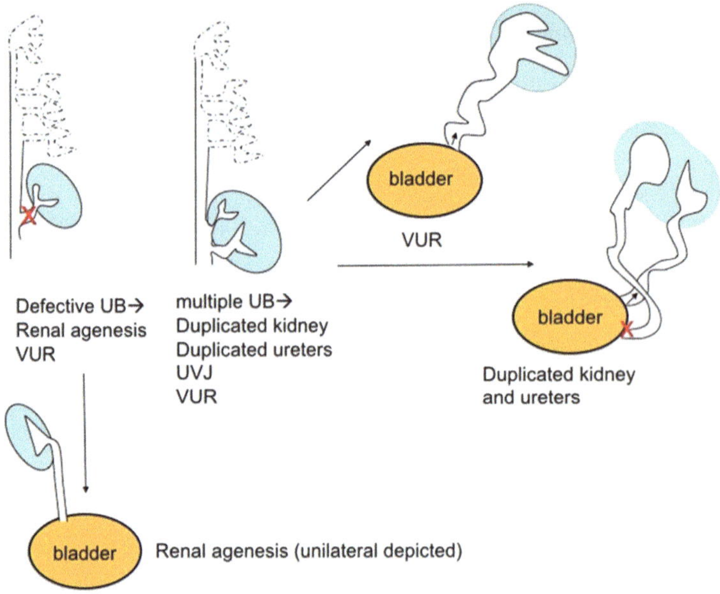

Fig. 10.2 Defects in the ureteric bud (UB) outgrowth lead to renal and ureteral abnormalities. Complete failure of the ureteric bud to grow or to contact the surrounding metanephric mesenchyme leads to apoptosis of the metanephric mesenchyme and renal agenesis. Defects in ureteric bud outgrowth can also result in vesicoureteral reflux (VUR). Outgrowth of multiple ureteric buds is associated with duplication of kidney and/or ureters, which are often associated with obstruction or VUR. With fully duplicated ureters, the upper ureter frequently inserts ectopically into the bladder and may be associated with an obstructing ureterocele (an out-pouching of the ureter). Obstruction at the level of the ureter entering the bladder is known as uretero-vesicular (UVJ) junction obstruction. Duplicated kidney depicted with typical clinical scenario observed (an upper pole with an ectopic, obstructed ureter, and lower pole ureter with VUR)

portion forms the proximal tubule, and the portion furthest from the ureteric bud tip will form the podocytes.

Genetic defects that impair any stage of the development from the outgrowth of the ureteric bud until terminal differentiation into specialized cells of the nephron segments can manifest as developmental renal disease. For example, both defects in ureteric bud outgrowth or extra ureteric bud outgrowth are associated with CAKUT (Fig. 10.2).

Defects in later stages of differentiation can result in renal dysplasia or hypoplasia. Renal dysplasia is the most severe and is classically characterized by primitive ducts with a fibromuscular collar (Kakkar et al. 2006). Non-renal tissues such as cartilage or hematopoietic cells may be present (Kakkar et al. 2006). Mouse models with defective WNT4 signaling between the ureteric bud and metanephric mesenchyme lead to disruptions in the mesenchymal-to-epithelial transition (MET) (Stark et al. 1994). Histologically this appears as primitive ducts surrounded by undifferentiated mesenchyme, analogous to human renal dysplasia (Stark et al. 1994). Supportive of this concept is the finding of dysplasia in a few families with

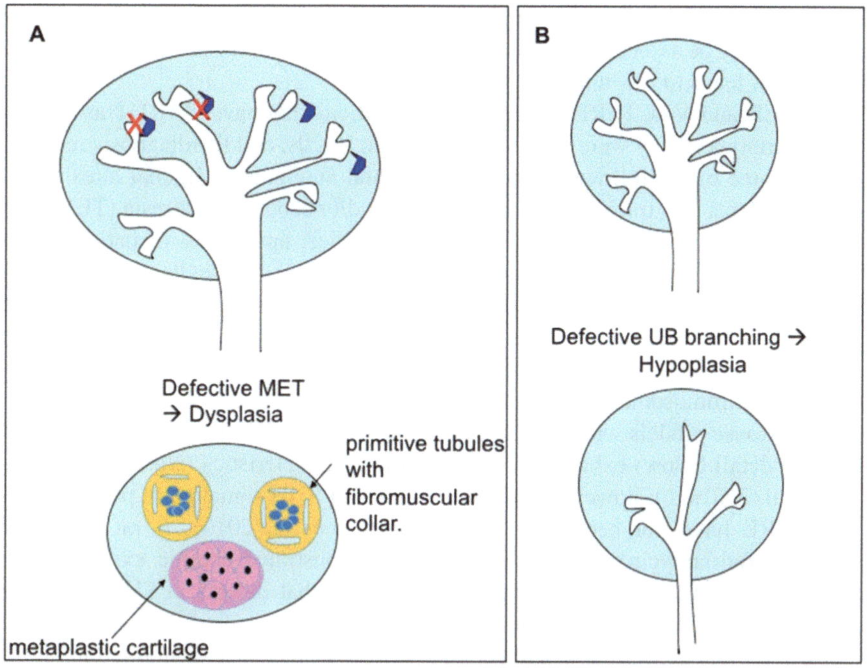

Fig. 10.3 Defects in later differentiation result in renal dysplasia and hypoplasia. (**a**) Defects in the mesenchymal-to-epithelial transition lead to renal dysplasia with primitive tubules surrounded by a fibromuscular collar and metaplastic cartilage. (**b**) Defects in ureteric bud branching result in renal hypoplasia

mutations in *WNT4*, although the frequencies of these mutations have not been assessed in larger studies (Vivante et al. 2013; Biason-Lauber et al. 2004, 2007). Small kidneys with decreased nephron number and histologically normal nephron structures are typical findings in renal hypoplasia (Blake and Rosenblum 2014). In mouse models, defects in ureteric bud branching or premature loss of renal epithelial progenitors lead to renal hypoplasia (Zubkov et al. 2015). Dysplasia and hypoplasia may both be present in human CAKUT (known as hypo-dysplasia) (Fig. 10.3).

10.3 Common Syndromes Associated with CAKUT

In genetics, syndromes are characterized by a set of associated signs and symptoms where a genetic cause is known. In some cases, there is a suspected genetic cause of associated characteristics, but the cause is unknown. These are referred to as genetic associations (e.g., VATER and VACTERL). In contrast, sequences are anomalies associated with a specific cause. For example, the Potter sequence (micrognathia, clubbed feet, and pulmonary hypoplasia) are the result of low amniotic fluid. Potter sequence may be associated with severe CAKUT, as decreased urine production for any reason leads to low amniotic fluid.

VATER and VACTERL are genetic associations that have CAKUT as a common phenotype. They were first described in the 1970s, and the diagnosis requires at least three of the following findings: vertebral anomalies (V), anal atresia (A), cardiac defects (C), tracheoesophageal fistula and/or esophageal atresia (TE), renal anomalies (R), or limb defects (L). In some studies, up to 80% of patients with VACTERL have renal anomalies (Botto et al. 1997). Some have argued that VACTERL results from defects in midline cell fate determination and morphogen gradients, with modulation by genetic-environmental interactions (Lubinsky 2015; Gillick et al. 2003). Genetic studies of patients with VACTERL have revealed a variety of chromosomal micro-deletions and duplications, although none are consistent. Mouse models with defects in sonic hedgehog (SHH) signaling (discussed in more detail below) exhibit many VACTERL characteristics (Ngan et al. 2013; Kim et al. 2001a, b); however, human mutations of SHH signaling in patients with VACTERL have not been identified (Aguinaga et al. 2010). A rat model of teratogen (adriamycin)-induced VACTERL demonstrated ectopic expression of SHH (Gillick et al. 2003), supporting the concept that defects in SHH signaling likely contribute to the pathogenesis of VACTERL.

Rarely, monogenic autosomal recessive mutations have been identified in familial VACTERL, including mutations in *TRAP1* (a heat-shock protein 90-related mitochondrial chaperone) and *FGF8* (fibroblast growth factor 8) (Saisawat et al. 2014; Zeidler et al. 2014). FGFs will be discussed in further detail below.

10.4 Chromosomal Abnormalities in CAKUT

Chromosomal aneuploidies and mosaicism, including Trisomy 13, 18, and 21 as mentioned before, are common causes of syndromic CAKUT. Of these, Trisomy 21 is the most common (1.5 in 1000 births) (Altug-Teber et al. 2007). A study of over 700 cases of Trisomy 21 in Europe revealed 5% had CAKUT, including 2% with obstructive defects of the urinary pelvis (Stoll et al. 2015). Trisomy 13 and 18 are less common (0.05 and 0.15 per 1000 live births, respectively). The exact mechanisms by which trisomies alter renal development are unclear. An analysis of transcripts from amniocytes and chorionic villous cells from trisomy infants supports a model in which there is increased transcription (increased gene dosage) from the additional chromosome (Altug-Teber et al. 2007). The increased expression of genes may itself be deleterious, but it also appears to disrupt expression of downstream genes (not on the trisomy chromosome), known as developmental instability (Altug-Teber et al. 2007). Thus, contributions of both increased gene dosage and developmental instability likely contribute to the observed phenotypes (Altug-Teber et al. 2007).

Smaller chromosomal abnormalities are also common causes of CAKUT. Recently, it has been shown that copy number variants (CNVs) are present in 10–16% of isolated CAKUT (Caruana et al. 2015; Sanna-Cherchi et al. 2012; Xi et al. 2016). Some of these CNVs affected known developmental genes, including *HNF1B*. Mutations in *HNF1B* can manifest as syndromic or isolated CAKUT (discussed in more detail below) (Caruana et al. 2015; Verbitsky et al. 2015). CNVs also are rarely associated with VACTERL (see above) and other less common syndromes (Brosens et al. 2013; Hilger et al. 2013; Materna-Kiryluk et al. 2014; Handrigan et al. 2013). These CNVs affected multiple genes, likely explaining the multiple organ systems affected.

10.5 Challenges of Genotype–Phenotype Correlation

It should be emphasized that both in the setting of CNVs and single nucleotide variations (SNVs), environmental factors likely modify the phenotypic outcome. In several studies, the identified CNVs were also present in siblings and parents of affected children with CAKUT who were either asymptomatic or had mild manifestations (Xi et al. 2016; Handrigan et al. 2013). A large study of 75 patients harboring *HNF1B* mutations revealed a broad spectrum of manifestations, from echogenic kidneys to multi-cystic kidneys to renal hypoplasia/agenesis (Heidet et al. 2010).

10.6 Genetics in *HNF1B* are Commonly Associated with CAKUT and Extrarenal Manifestations

Genetic studies of large cohorts of children with CAKUT have revealed that the most common mutations (identified in up to 15% of European or US Caucasian patients) are in *HNF1B* (associated with renal cysts and diabetes syndrome) and *PAX2* (associated with renal-coloboma syndrome) (Hwang et al. 2014; Madariaga et al. 2013; Weber et al. 2006). Less common but also identified in large cohorts of patients with CAKUT were mutations in *SALL1* (associated with Townes–Brocks syndrome), *EYA1*, and *SIX5* (associated with Branchio-oto-renal syndrome).

Hepatocyte nuclear factor beta (*HNF1B*) encodes a POU (PIT-1, OCT-1, UNC-86) homeodomain containing transcription factor. Heterozygous mutations in this gene are an important cause of developmental diseases of the kidney (Madariaga et al. 2013; Thomas et al. 2011) and renal cysts (Heidet et al. 2010). Also, patients with mutations in this gene are at an increased likelihood of developing maturity onset diabetes mellitus (MODY), secondary to pancreatic hypoplasia, which has led to the description of renal cyst and diabetes (RCAD) syndrome (Bingham et al. 2001). Since initial reporting of the early cases of *HNF1B*-associated disease in 1997 (Iwasaki et al. 1998; Horikawa et al. 1997), it has become evident that additional clinical features also include pancreatic hypoplasia, genital tract malformations, abnormal liver function, hypomagnesemia, hyperuricemia, and early-onset gout (Clissold et al. 2014). *HNF1B*-associated disease is, therefore, considered to be a multisystem disorder. It is a perfect example of a gene that is involved in the development of more than one organ, with its mutations being a cause of multiple phenotypes.

HNF1B shares >80% sequence homology with HNF1A, and they both bind to the same DNA sequence (Mendel et al. 1991). *HNF1A* gene mutations are the most common cause of MODY. More than 50 different *HNF1B* mutations have been reported, including missense, nonsense, frameshift, and splicing mutations. No correlation seems to exist between the type or position of the mutation and particular clinical features (Edghill et al. 2006), and up to 50% are spontaneous mutations not present in parents. *HNF1A* is most highly expressed in the liver. *HNF1B* is predominantly expressed in the kidneys but is also expressed in the gonads, thymus, and lungs. In situ hybridization studies using human tissue samples have shown that HNF1B mRNA is highly expressed in the fetal collecting ducts, with lower levels of expression in the metanephric mesenchyme (Kolatsi-Joannou et al. 2001).

Renal-specific inactivation of *Hnf1b* in mice results in polycystic kidneys. This phenotype is associated with a marked reduction in the transcriptional activation of the cystic disease genes *Umod, Pkhd1,* and *Pkd2*, which contain DNA elements to which HNF1B normally binds (Gresh et al. 2004). Hnf1b appears to act upstream of WNT9B and alters WNT signaling and thereby impairs early renal development (Carroll et al. 2005; Lokmane et al. 2010).

10.7 Identification of Syndromic Genes Reveals Mechanisms of Renal Epithelial Differentiation

It is challenging to categorize genes as affecting only one aspect of kidney development. However, many syndromic genes have helped to elucidate mechanisms of renal epithelial differentiation.

10.7.1 Genetic Syndromes Affecting Early Renal Epithelial Differentiation

Nephron progenitor cells are a self-renewing population of cells in the metanephric mesenchyme that form the nephron (O'Brien et al. 2016). The progenitor cell population is regulated by a complex network of transcription factors (Basta et al. 2014). The proper balance of these transcription factors is necessary to both induce the transcriptional program of renal differentiation and maintain the progenitor population during renal development. Studies of animal models indicate that some of the most common mutations in human CAKUT (*PAX2*, *SALL1*, *EYA1*, and *SIX2*) likely affect nephron progenitor survival and/or self-renewal.

As indicated above, *EYA1, SIX1, and SIX5* mutations are associated with branchio-oto-renal syndrome (Weber et al. 2008; Hoskins et al. 2007; Ruf et al. 2004). *SIX2* is a gene which belongs to the same family. Although *SIX2* is not found to be associated with branchio-oto-renal syndrome, its mutations are associated with hypodysplasia of the kidney (Weber et al. 2008). The EYA1 protein consists of an N-terminal transactivation domain and a C-terminal protein interaction domain. In the kidney, the EYA1 protein interaction domain binds Six proteins, facilitating EYA1 translocation from cytoplasm to the nucleus and DNA binding. Binding of EYA1 to SIX proteins converts the SIX proteins from weak to strong transcriptional activators (Xu et al. 1997; Chen et al. 1997; Pignoni et al. 1997) to specify nephron progenitor population from the intermediate mesoderm. EYA1 coordinates with SIX1 and PAX2 to induce GDNF expression in the metanephric mesenchyme during ureteric bud branching (Sajithlal et al. 2005). EYA1 is also required to induce SIX2 expression (Xu et al. 2014). EYA1 and SIX2 lead to phosphorylation of MYC to prevent apoptosis and maintain the proliferative capacity of nephron progenitors. Genetic deletion of *Six2* leads to premature differentiation, depletion of the nephron progenitor population, and hypoplasia (Park et al. 2012). Genetic variants in *EYA1* and *SIX2* are associated with branchio-oto-renal syndrome (hearing loss, branchial fistulae, preauricular pits or tags, and renal anomalies). Similar to its actions in the nephron progenitor populations, EYA1 stimulates proliferation and inhibits apoptosis in the ear (Li et al. 2010; Zou et al. 2006; Xu et al. 1999). EYA1 regulates the growth of otic epithelium and interacts with PAX2 during the development of all sensory areas in the inner ear (Zou et al. 2006). During craniofacial development in mice, EYA1 is expressed in the cranial placodes. In these placodes, EYA1 interaction domain binds and acts as a transcriptional coactivator with two key proteins in craniofacial development, SIPL1 (Shank-interacting protein-like 1) and RBCK1 (RBCC protein interacting with PKC1) (Landgraf et al. 2010). EYA proteins are required for retinal development in *Drosophila*, but due to functional redundancy, individual EYA mutant mice do not exhibit a phenotype (Tadjuidje and Hegde 2013).

SALL1 autosomal dominant mutations result in Townes–Brock syndrome, which includes imperforate anus, ear malformations and hearing loss, thumb malformations (most often triphalangeal thumbs or preaxial polydactyly), and renal anomalies (Kohlhase 1993). Some patients also have congenital heart disease and genitourinary malformations. The renal phenotype includes abnormal rotation, ectopic kidneys, horseshoe kidneys, polycystic kidneys, and vesicoureteral reflux. Complete deletion of *Sall1* in mice leads to failure of ureteric bud invasion, resulting in aplasia or severe dysplasia (Nishinakamura et al. 2001). Conditional knockout of *Sall1* in the metanephric mesenchyme leads to ureteric bud branching defects with dysregulation of WNT signaling (Kiefer et al. 2010). SALL1 has also been shown to regulate the balance between differentiation and self-renewal in the metanephric mesenchyme by activating progenitor genes in SIX2-positive cells and repressing progenitor genes in SIX2-negative cells (Basta et al. 2014; Kanda et al. 2014). SALL1 was also recently shown to maintain cardiac progenitor cells and promote neurogenesis in mice (Harrison et al. 2008; Morita et al. 2016). Although SALL1 is expressed in the limb bud, ear, anus, heart, and kidney, *Sall1* knockout mice only exhibited a renal phenotype. It is thought that *SALL1* mutations may lead to truncated proteins that exert a dominant-negative effect by suppressing other SALL proteins (Kiefer et al. 2003).

The conversion of metanephric mesenchyme to renal vesicle epithelia is another key step in early nephrogenesis. Landmark work by Stark et al. identified WNT4 as a major factor regulating mesenchyme-to-epithelial transition (Stark et al. 1994). Human mutations in *WNT4* are rare, but a consanguineous kindred was identified with homozygous *WNT4* mutations and a phenotype of female sex reversal and dysgenesis of kidneys, adrenals, and lungs (SERKAL syndrome) (Mandel et al. 2008). Multiple WNTs are present in the developing kidney, and WNT signaling functions at multiple stages of renal development (Vainio 2003). WNT signaling is discussed in detail in Chap. X; thus, we will highlight here only its role in SERKAL syndrome. Canonical WNT signaling occurs by binding of WNTs to Frizzled receptors. This leads to an intracellular signaling cascade that ultimately results in the stabilization of beta-catenin, which translocates to the nucleus to stimulate downstream effectors. Studies suggest that both canonical and noncanonical WNT/planar cell polarity contribute to kidney development (Carroll et al. 2005; Bridgewater et al. 2008; Iglesias et al. 2007; Nishita et al. 2014). WNT9B is secreted from the collecting duct and stimulates expression of *Wnt4*, as well as FGF8, PAX8, and LHX1, thereby inducing nephron differentiation (Carroll et al. 2005). The gradient of WNT signaling also establishes proximal-distal patterning of the nascent nephron (Schneider et al. 2015). As in the kidney, WNT signaling plays a key role in regulation lung branching, proximal-distal patterning, and mesenchyme development (Volckaert and De Langhe 2015). Studies of conditional knockout of beta-catenin in mouse adrenal glands indicate that canonical WNT signaling is required for adrenal cortical cell renewal (Berthon et al. 2012). In the ovary, WNT4/beta-catenin activation is required for oogonial differentiation and entry into meiosis (Bernard and Harley 2007; Bernard et al. 2008; Chassot et al. 2011; Munger et al. 2013), likely explaining the sex-reversal phenotype of SERKAL.

10.7.2 Genetic Syndromes Affecting UB Branching

UB branching establishes the radial architecture, and the number of branches determines nephron number. This is clinically relevant because, in addition to severe defects that result in hypoplasia, the low number of nephrons (even in absence of overt renal disease in childhood) is associated with increased risk for adult-onset chronic kidney disease.

UB branching is regulated by multiple growth factors, many of which also regulate lung branching and cardiac development. Many of the growth factors that can affect branching are membrane-bound tyrosine kinase receptors. RET-GDNF is one key pathway (Schuchardt et al. 1996). RET tyrosine kinase receptors are concentrated at the tips of the ureteric buds (Riccio et al. 2016). GDNF (stimulated by the transcription factors discussed above) is secreted by the metanephric mesenchyme and binds to the RET receptor to activate intracellular signaling pathways such as PI3 kinase and MAPK/ERK proliferation signaling pathways (Shakya et al. 2005; Costantini and Shakya 2006; Costantini 2010). This leads to ureteric bud cell division and branching (Costantini and Shakya 2006; Costantini 2010). A high concentration of RET at the tip of the ureteric bud directs cell movements to regulate ureteric bud outgrowth (Riccio et al. 2016). It is important to have a single ureteric bud in the proper location to induce metanephric mesenchyme. As indicated above, extra ureteric buds or a ureteric bud in the wrong location results in duplication of ureters that may not properly drain in the bladder and become obstructed or have vesicoureteral reflux. A tyrosine kinase inhibitor, Sprouty, opposes tyrosine kinase activity to specify outgrowth of a single ureteric bud (Basson et al. 2006).

RET mutations have been associated with unilateral and bilateral renal agenesis, vesicoureteral reflux, renal hypodysplasia, and ureteropelvic junction obstruction (Hwang et al. 2014; Chatterjee et al. 2012; Jeanpierre et al. 2011). Loss-of-function *GDNF* and *RET* mutations have also been associated with Hirschsprung's disease (congenital megacolon), which is characterized by an absence of enteric nervous system (ENS) ganglion cells in the distal intestine (Martucciello et al. 2000). In mice, GDNF and RET are required for enteric neural crest cell survival and differentiation (Natarajan et al. 2002; Taraviras et al. 1999). In addition, GDNF expressed in the developing gut acts as a chemo-attractant for ENS progenitors, and thus, mutations in *GDNF* or *RET* result in defective migration of ENS progenitors from the neural crest into the intestine (Natarajan et al. 2002). As in the kidney, PI3K and MAPK/ERK pathways appear to mediate GDNF/RET's effects on ENS neural crest cell differentiation, survival, and migration (Natarajan et al. 2002). *Ret*-deficient mice exhibit Hirschspung's disease and renal agenesis. Interestingly, studies indicate between 15 and 25% of Hirschsprung's patients may have associated CAKUT (often hydronephrosis or hypoplasia) (Hofmann et al. 2014; Pini Prato et al. 2009).

Other tyrosine kinases may contribute to ureteric bud outgrowth. In particular, in mice, fibroblast growth factors (FGF) can stimulate ureteric bud formation in the absence of GDNF and Sprouty (Michos et al. 2010). Elegant studies of ureteric bud and metanephric mesenchyme-specific knockout mice of FGF receptors 1 and 2 have revealed time and spatial specific roles for the FGF receptors (Bates 2011). Knockout of both *Fgfr1* and *Fgfr2* in the metanephric mesenchyme led to renal dysgenesis due to the failure of the ureteric bud to elongate and branch (Poladia et al. 2006). In the ureteric bud, FGFR2 is the key receptor, and deletion in the ureteric bud leads to branching defects and hypoplasia (Bates 2011; Zhao et al. 2004). FGFs are also key factors in bone growth and remodeling (Chen and

Deng 2005; Su et al. 2014). Mutations in *FGFR1* and *FGFR2* are associated with craniosynostosis syndromes such as Apert's syndrome, Antley–Bixler, Pfeiffer, Beare–Stevenson syndromes (Marie et al. 2005; Passos-Bueno et al. 1999). *FGFR 1* or *2* mutations may cause craniosynostosis via effects of FGFs on osteoblast differentiation and apoptosis, although differences between humans and mouse models have made understanding the pathophysiology challenging. CAKUT has been described in association with craniosynostosis syndromes, including solitary kidneys, hydroureter, and vesicoureteral reflux (Seyedzadeh et al. 2008; Hains et al. 2010).

Closely related to GDNF/RET and the FGFs are Fraser syndrome *FRAS1* and FRAS1-related extracellular matrix (ECM) gene 2 (*FREM2*) (Pavlakis et al. 2011). Fraser syndrome is characterized by CAKUT (often uni- or bilateral renal agenesis), eye anomalies (Cryptophthalmos), embryonic skin blistering, and cutaneous syndactyly and results from autosomal recessive mutations in *FRAS1*, *FREM2*, or *GRIP1* (a scaffolding protein that links to FRAS1 and FREM2). Autosomal recessive mutations in Fraser syndrome genes are also associated with isolated CAKUT (Kohl et al. 2014). FRAS1 and FREM2 are extracellular matrix proteins that line the ureteric bud epithelia, and mouse models of mutant FRAS1 develop renal agenesis due to the failure of ureteric bud outgrowth (Pitera et al. 2008). Impaired *GDNF* expression likely contributes to the failure of ureteric bud outgrowth. *Fras1* mutant mice can be rescued by decreased Sprouty activity, likely via FGF-mediated ureteric bud induction (Pitera et al. 2012). *Fras1/Frem* contribute to epithelial integrity and mutations lead to disrupted dermal-epidermal attachment in the plane of sublamina densa likely leading to the embryonic skin blistering (Pavlakis et al. 2011). Syndactyly likely results from downregulation of Msx2, a transcription factor that stimulates apoptosis and interdigital cell death (Hines et al. 2016).

10.7.3 Genetic Defects Affecting Renal Patterning

Several developmental patterning signaling pathways including Hox and Sonic Hedgehog (SHH) contribute to renal development (Reidy and Rosenblum 2009). As indicated above, defects in Sonic Hedgehog (SHH) signaling may contribute to VACTERL associated with CAKUT. Genetic mutations in SHH signaling component *GLI3* are associated with Pallister–Hall syndrome, which is characterized by polydactyly, imperforate anus, hypothalamic hamartoma, and CAKUT (Hill et al. 2007). SHH is a secreted morphogen that binds the Patched receptor. The Patched receptor constitutively inhibits Smoothened, and binding of SHH allows Smoothened to stimulate translocation of GLI1 and 2 effectors to the nucleus. In the kidney, GLI1 and 2 induce transcription of cell cycle proliferative proteins MYC and CYCLIND1 as well as nephron differentiation factors PAX2 and SALL1. The *GLI3* mutations associated with Pallister–Hall syndrome lead to the expression of a truncated GLI3 that inhibits GLI1 and GLI2. In the hand, GLI3 restricts regulators of cell cycle entry and also promotes differentiation into chondrocytes (Lopez-Rios et al. 2012). A gradient of SHH also establishes the anteroposterior limb axis and digit identities, and defects in GLI3 lead to polydactyly in mice (Hill et al. 2007).

10.7.4 Genetic Defects Affecting Nephron Segmentation

NOTCH signaling is a highly conserved evolutionary pathway that is important for cell–cell signaling and determines cell fate and differentiation during renal development. To date, five NOTCH ligands (delta-like protein [DLL]-1, DLL-3, DLL-4, Jagged1, and Jagged2) and four NOTCH receptors (NOTCH 1–4) have been identified in mammals. Both the receptors and their ligands are single-pass type I membrane proteins. NOTCH ligands consist of an extracellular N-terminal region called as the delta/serrate/LAG2 (DSL) domain which is required for their binding to the receptors (Kamath et al. 2013). NOTCH receptors consist of an extracellular segment, formed of multiple EGF-like repeats, a transmembrane domain, and an intracellular domain. Once the receptor–ligand interaction has occurred, the intracellular domain of the NOTCH receptor undergoes a series of cleavages that allows the intracellular domain to translocate to the nucleus, where it regulates the transcription of downstream genes, such as *HES1* and *HEY2* (transcription factors that are NOTCH effectors) (Kopan 2012).

In developing kidneys, receptors NOTCH2 and NOTCH1, along with ligands, Jagged1 and DLL-1, are expressed. Both receptors and ligands are expressed in pretubular aggregates. In later stages of glomerular development, the NOTCH receptors are expressed in podocytes and Bowman's capsule, whereas the ligands are expressed in endothelial cells (McCright 2003). This change in expression pattern over development suggests that NOTCH signaling is operating via inductive signaling and lateral inhibition to control cell-fate determination. In inductive signaling, the ligand-expressing cell induces differentiation in receptor-expressing cell. In lateral inhibition, the ligand-expressing cell differentiates and interacts with receptor expressing cell and inhibits it from adopting a differentiated state. The developing kidney may use both types of NOTCH signaling (McCright 2003). This process is comparable to the commitment of hepatoblasts to a biliary cell fate that occurs within the ductal plate of the developing liver, which is also dependent on NOTCH signaling (Si-Tayeb et al. 2010).

Various models of *Jag1* and *Notch2* mutations in mice have been developed to study the role of NOTCH signaling during renal development. Mice homozygous for a hypomorphic allele of *Notch2* die shortly after birth owing to a lack of functional nephrons (McCright et al. 2001). The early stages of kidney development (ureteric bud migration and mesenchymal aggregate formation) occur normally in these mice; however, podocytes and proximal tubules do not form, and the developing glomerulus has vascularization defects. This phenotype is consistent with the known role of NOTCH signaling in tissue specification and the formation of major arteries and veins (Krebs et al. 2000; Uyttendaele et al. 2001; Xue et al.

1999). Mouse metanephrons cultured in the presence of an inhibitor of NOTCH intracellular signaling have a severe deficiency of proximal tubules and glomerular podocytes (Cheng et al. 2003). In the collecting duct, NOTCH acts by lateral inhibition to determine cell fate (Guo et al. 2015). Collecting duct-specific deletion of the ADAM10 metalloproteinase that cleaves the intracellular domain of NOTCH led to a reduction of principal cells and increase of intercalated cells in the collecting duct (Guo et al. 2015).

In humans, abnormalities in NOTCH signaling present as Alagille syndrome, an autosomal dominant disorder with variable multisystem organ involvement that is caused by mutations in one of two genes in the NOTCH signaling pathway, *JAG1* or *NOTCH2*. The diagnosis is based on the presence of intrahepatic bile duct paucity in liver biopsy samples in association with at least three of the following major clinical features: chronic cholestasis, cardiac disease (most often peripheral pulmonary stenosis), skeletal abnormalities (typically butterfly vertebrae), ocular abnormalities (primarily posterior embryotoxon), and characteristic facial features. *JAG1* mutations have been identified in 94% of patients with clinically defined Alagille syndrome (Warthen et al. 2006). Renal anomalies, such as impaired function and structural abnormalities, have been documented in almost 40% of patients with Alagille syndrome who carry *JAG1* mutations. Renal dysplasia (58.9%) is the most common renal abnormality. Other common renal anomalies include renal tubular acidosis (9.5%), ureteropelvic or vesicoureteral urinary obstruction (8.2%), and VUR (also 8.2%). Since renal involvement is common in Alagille syndrome, it has been suggested to be included as the sixth diagnostic criteria for this syndrome (Kamath et al. 2012, 2013).

10.7.5 Genetic Syndromes Affecting Terminal Epithelial Differentiation of Podocytes

Over 40 genes have now been associated with nephrotic syndrome (Sadowski et al. 2015). Some of these gene defects affect podocyte differentiation and are also associated with extrarenal manifestations (Sadowski et al. 2015). Genetic defects in the nuclear transcription factor *WT1* associated with Denys–Drash and Frasier syndrome are perhaps some of the best-understood examples (Morrison et al. 2008). Denys–Drash is associated with autosomal dominant missense mutations in the zinc finger region of exons 8 or 9 of the *WT1* gene (Patek et al. 2003). These mutations lead to decreased nuclear expression of WT1, which allows for abnormal persistent expression of PAX2, likely leading to diffuse mesangial sclerotic lesion. The *WT1* zinc finger mutations alter DNA binding and may act as a dominant-negative fashion during genitourinary development. In contrast, mutations in a WT1 splice site lead to Frasier syndrome (Lefebvre et al. 2015). The splice site mutations lead to loss of the most common WT1 isoform that includes three amino acids (known as +KTS). The WT1 isoform that lacks those three amino acids

(-KTS) has a different DNA binding and nuclear localization patterns (Lefebvre et al. 2015). Thus, loss of this +KTS isoform likely leads to gonadal dysgenesis, male–female sex reversal, progressive podocyte damage, and focal glomerulosclerosis (Lahiri et al. 2007).

10.7.6 Genetic Syndromes with Progressive Defects in Tubular Epithelial Differentiation

Tuberous sclerosis complex (TSC) is an autosomal dominant genetic disorder characterized by the growth of dysgenic lesions in multiple organs including the brain, skin, kidney, heart, lungs, and retina. TSC arises from inactivating mutations of either *TSC1* (chromosome locus 9q34.3) or *TSC2* (16p13.3), which encode hamartin and tuberin, respectively. These proteins are believed to function as tumor suppressors by forming a complex that regulates cellular proliferation (Plank et al. 1998; van Slegtenhorst et al. 1998).

Hamartin and tuberin, together with a third protein, TBC1D7 (Dibble et al. 2012), form the TSC protein complex, which regulates multiple cellular processes and importantly acts as a critical negative regulator of the mechanistic target of rapamycin (mTOR) complex 1 (mTORC1) (DiMario et al. 2015; Inoki et al. 2002; Tee et al. 2002), a serine/threonine kinase that is central to many cell functions including cell growth and proliferation. RHEB (RAS homolog enriched in brain) is a specific GTPase downstream of the TSC protein complex that functionally links TSC1/TSC2 to mTORC1. The TSC1/TSC2 complex functions as a GTPase-activating protein (GAP) for Rheb and stimulates the conversion of RHEB-GTP to a GDP-bound state, thereby inactivating RHEB signaling and thus removing its stimulatory effect on mTORC1. Conversely, loss-of-function mutations in either TSC1 or TSC2 lead to enhanced RHEB-GTP signaling and mTORC1 activation. Constitutively, active mTORC1 signaling thus constitutes the molecular basis of TSC and responsible for the dysgenic lesions. When in its active state, mTORC1 phosphorylates the translational regulators 4E-BP1 (eukaryotic translation initiation factor 4E-binding protein 1) and S6K1 (S6 kinase 1), which, in turn, stimulate protein synthesis. Active mTORC1 stimulates the biosynthesis of ribosomes, lipid biogenesis, glucose metabolism, nucleotide synthesis, mitochondrial and lysosomal biogenesis, ATP, and amino acid production. mTORC1 also functions as a negative regulator of autophagy, another key cellular pathway that has been implicated in TSC. mTORC1 inhibitors have proven efficacy in TSC and are now the first line of therapy in growing angiomyolipomas of the kidney and in surgically unresectable symptomatic giant cell tumors of the brain (Kingswood et al. 2016; Krueger et al. 2010).

TSC most often presents with neurologic symptoms, and approximately 90% of affected individuals experience seizures and about half of patients experience cognitive impairment, autism, or other behavioral disorders. Renal manifestations are the second most common findings associated with TSC, with angiomyolipomas occurring in up to 55–75% and renal cystic disease in 30–45% of patients (Crino et al. 2006; Rakowski et al. 2006).

Angiomyolipomas are the prototype of a family of lesions called perivascular epithelioid cell tumors or PEComas which exhibit immunoreactivity for both melanocytic markers (as detected by the HMB-45 and melanin-A antibodies) and smooth-muscle markers (actin and desmin). All components of angiomyolipomas, including the vascular cells, immature smooth-muscle-like spindle cells, epithelioid cells, and fat cells, contain somatic mutations that, combined with their germline mutation, render the cells deficient in either tuberin or hamartin. Presumably, this deficiency disrupts the integrated control of cell growth leading to the angiomyolipoma (Siroky et al. 2011; Dixon et al. 2011).

Renal cysts are also a common complication of TSC. Fifty percent of the patients with TSC develop renal cysts. New insight has been gained into the different mechanisms by which TSC pathway interacts with polycystin-1 (PC-1). PC-1 is a protein encoded by *PKD1* which is the gene involved in adult-onset polycystic kidney disease (ADPKD). Two percent of patients with TSC have a contiguous germline deletion of TSC2 and PKD1 (chromosome 16p13.3) and develop severe infantile polycystic kidney disease (Brook-Carter et al. 1994), suggesting a functional cooperation between their gene products. ADPKD is discussed in more detail in another chapter and is caused secondary to defects in *PKD1* (81% of the cases) and PKD2 (19% of the cases) (Mao et al. 2016). It has been recently shown that mTORC-1 upregulation in TSC-1 mutant mice causes decreased protein neo-synthesis of PC-1 which cystogenesis in TSC (Pema et al. 2016). There are other studies in *Tsc1*+/- *Tsc2*+/- mice that have shown the presence of early cysts with no evidence of mTOR activation measured by phosphorylated S6K staining, suggesting that many TSC-associated renal lesions initially develop via mTOR-independent pathways. Primary cilia are anchored to the cell via the basal body. Hamartin has been localized to the basal body (Hartman et al. 2009), and tuberin interacts with PC-1 (Shillingford et al. 2006), which has been localized to the primary cilium (Yoder et al. 2002). Similarities between TSC and polycystic kidney disease have been demonstrated by showing how hamartin, tuberin, and polycystin-1 defects affect primary cilia length (Bonnet et al. 2009). Primary cilia is also an organelle important for cell polarity and is thought to underlie numerous disorders associated with cystic kidney disease, as discussed in detail elsewhere in the volume (Hartman et al. 2009). Further defects in apicobasal cell polarity and misoriented dividing tubular cells have been seen, pointing out another possible mechanism for cystogenesis (Bonnet et al. 2009).

10.8 Conclusions

CAKUT is a common feature of many syndromes. The clinical manifestations of CAKUT depend on the stage at which renal development is impaired. Aneuploidies and mosaicisms are the most common syndromes associated with CAKUT. Single gene defects are less common but have facilitated insight into the mechanism of renal and other organ development. Analysis of underlying genetic mutations with transgenic and mutant mice has also led to advances in our understanding of mechanisms of renal development.

References

Aguinaga M, Zenteno JC, Perez-Cano H, Moran V (2010) Sonic hedgehog mutation analysis in patients with VACTERL association. Am J Med Genet Part A 152a:781–783. doi:10.1002/ajmg.a.33293

Altug-Teber O et al (2007) Specific transcriptional changes in human fetuses with autosomal trisomies. Cytogenet Genome Res 119:171–184. doi:10.1159/000112058

Basson MA et al (2006) Branching morphogenesis of the ureteric epithelium during kidney development is coordinated by the opposing functions of GDNF and Sprouty1. Dev Biol 299:466–477. doi:10.1016/j.ydbio.2006.08.051

Basta JM, Robbins L, Kiefer SM, Dorsett D, Rauchman M (2014) Sall1 balances self-renewal and differentiation of renal progenitor cells. Development 141:1047–1058. doi:10.1242/dev.095851

Bates CM (2011) Role of fibroblast growth factor receptor signaling in kidney development. Am J Physiol Renal Physiol 301:F245–F251. doi:10.1152/ajprenal.00186.2011

Bernard P, Harley VR (2007) Wnt4 action in gonadal development and sex determination. Int J Biochem Cell Biol 39:31–43. doi:10.1016/j.biocel.2006.06.007

Bernard P, Sim H, Knower K, Vilain E, Harley V (2008) Human SRY inhibits beta-catenin-mediated transcription. Int J Biochem Cell Biol 40:2889–2900. doi:10.1016/j.biocel.2008.06.006

Berthon A, Martinez A, Bertherat J, Val P (2012) Wnt/beta-catenin signalling in adrenal physiology and tumour development. Mol Cell Endocrinol 351:87–95. doi:10.1016/j.mce.2011.09.009

Biason-Lauber A, Konrad D, Navratil F, Schoenle EJ (2004) A WNT4 mutation associated with Mullerian-duct regression and virilization in a 46,XX woman. N Engl J Med 351:792–798. doi:10.1056/NEJMoa040533

Biason-Lauber A et al (2007) WNT4 deficiency—a clinical phenotype distinct from the classic Mayer–Rokitansky–Kuster–Hauser syndrome: a case report. Hum Reprod 22:224–229. doi:10.1093/humrep/del360

Bingham C et al (2001) Mutations in the hepatocyte nuclear factor-1beta gene are associated with familial hypoplastic glomerulocystic kidney disease. Am J Hum Genet 68:219–224. doi:10.1086/316945

Blake J, Rosenblum ND (2014) Renal branching morphogenesis: morphogenetic and signaling mechanisms. Semin Cell Dev Biol 36:2–12. doi:10.1016/j.semcdb.2014.07.011

Bonnet CS et al (2009) Defects in cell polarity underlie TSC and ADPKD-associated cystogenesis. Hum Mol Genet 18:2166–2176. doi:10.1093/hmg/ddp149

Botto LD et al (1997) The spectrum of congenital anomalies of the VATER association: an international study. Am J Med Genet 71:8–15

Bridgewater D et al (2008) Canonical WNT/beta-catenin signaling is required for ureteric branching. Dev Biol 317:83–94. doi:10.1016/j.ydbio.2008.02.010

Brook-Carter PT et al (1994) Deletion of the TSC2 and PKD1 genes associated with severe infantile polycystic kidney disease—a contiguous gene syndrome. Nat Genet 8:328–332. doi:10.1038/ng1294-328

Brosens E et al (2013) VACTERL association etiology: the impact of de novo and rare copy number variations. Mol Syndromol 4:20–26. doi:10.1159/000345577

Carroll TJ, Park JS, Hayashi S, Majumdar A, McMahon AP (2005) Wnt9b plays a central role in the regulation of mesenchymal to epithelial transitions underlying organogenesis of the mammalian urogenital system. Dev Cell 9:283–292. doi:10.1016/j.devcel.2005.05.016

Caruana G et al (2015) Copy-number variation associated with congenital anomalies of the kidney and urinary tract. Pediatr Nephrol 30:487–495. doi:10.1007/s00467-014-2962-9

Chassot AA et al (2011) RSPO1/β-catenin signaling pathway regulates oogonia differentiation and entry into meiosis in the mouse fetal ovary. PLoS One 6. doi:10.1371/journal.pone.0025641

Chatterjee R et al (2012) Traditional and targeted exome sequencing reveals common, rare and novel functional deleterious variants in RET-signaling complex in a cohort of living US patients with urinary tract malformations. Hum Genet 131:1725–1738. doi:10.1007/s00439-012-1181-3

Chen L, Deng CX (2005) Roles of FGF signaling in skeletal development and human genetic diseases. Front Biosci J Virtual Libr 10:1961–1976

Chen R, Amoui M, Zhang Z, Mardon G (1997) Dachshund and eyes absent proteins form a complex and function synergistically to induce ectopic eye development in Drosophila. Cell 91:893–903

Cheng HT et al (2003) Gamma-secretase activity is dispensable for mesenchyme-to-epithelium transition but required for podocyte and proximal tubule formation in developing mouse kidney. Development 130:5031–5042. doi:10.1242/dev.00697

Clissold RL, Hamilton AJ, Hattersley AT, Ellard S, Bingham C (2014) HNF1B-associated renal and extra-renal disease—an expanding clinical spectrum. Nat Rev Nephrol 11:102–112. doi:10.1038/nrneph.2014.232

Costantini F (2010) GDNF/Ret signaling and renal branching morphogenesis: from mesenchymal signals to epithelial cell behaviors. Organogenesis 6:252–262. doi:10.4161/org.6.4.12680

Costantini F, Shakya R (2006) GDNF/Ret signaling and the development of the kidney. Bioessays 28:117–127. doi:10.1002/bies.20357

Crino PB, Nathanson KL, Henske EP (2006) The tuberous sclerosis complex. N Engl J Med 355:1345–1356. doi:10.1056/NEJMra055323

Dibble CC et al (2012) TBC1D7 is a third subunit of the TSC1-TSC2 complex upstream of mTORC1. Mol Cell 47:535–546. doi:10.1016/j.molcel.2012.06.009

DiMario FJ, Sahin M, Ebrahimi-Fakhari D (2015) Tuberous sclerosis complex. Pediatr Clin N Am 62:633–648. doi:10.1016/j.pcl.2015.03.005

Dixon BP, Hulbert JC, Bissler JJ (2011) Tuberous sclerosis complex renal disease. Nephron Exp Nephrol 118:e15–e20. doi:10.1159/000320891

Edghill EL, Bingham C, Ellard S, Hattersley AT (2006) Mutations in hepatocyte nuclear factor-1beta and their related phenotypes. J Med Genet 43:84–90. doi:10.1136/jmg.2005.032854

Gillick J, Mooney E, Giles S, Bannigan J, Puri P (2003) Notochord anomalies in the adriamycin rat model: a morphologic and molecular basis for the VACTERL association. J Pediatr Surg 38, 469–473; discussion 469–473. doi:10.1053/jpsu.2003.50081

Gresh L et al (2004) A transcriptional network in polycystic kidney disease. EMBO J 23:1657–1668. doi:10.1038/sj.emboj.7600160

Guo Q et al (2015) Adam10 mediates the choice between principal cells and intercalated cells in the kidney. J Am Soc Nephrol JASN 26:149–159. doi:10.1681/asn.2013070764

Hains DS et al (2010) High incidence of vesicoureteral reflux in mice with Fgfr2 deletion in kidney mesenchyma. J Urol 183:2077–2084. doi:10.1016/j.juro.2009.12.095

Handrigan GR et al (2013) Deletions in 16q24.2 are associated with autism spectrum disorder, intellectual disability and congenital renal malformation. J Med Genet 50:163–173. doi:10.1136/jmedgenet-2012-101288

Harrison SJ, Nishinakamura R, Monaghan AP (2008) Sall1 regulates mitral cell development and olfactory nerve extension in the developing olfactory bulb. Cereb Cortex (New York, N.Y.: 1991) 18:1604–1617. doi:10.1093/cercor/bhm191

Hartman TR et al (2009) The tuberous sclerosis proteins regulate formation of the primary cilium via a rapamycin-insensitive and polycystin 1-independent pathway. Hum Mol Genet 18:151–163. doi:10.1093/hmg/ddn325

Heidet L et al (2010) Spectrum of HNF1B mutations in a large cohort of patients who harbor renal diseases. Clin J Am Soc Nephrol CJASN 5:1079–1090. doi:10.2215/cjn.06810909

Hilger A et al (2013) De novo microduplications at 1q41, 2q37.3, and 8q24.3 in patients with VATER/VACTERL association. Eur J Hum Genet EJHG 21:1377–1382. doi:10.1038/ejhg.2013.58

Hill P, Wang B, Ruther U (2007) The molecular basis of Pallister Hall associated polydactyly. Hum Mol Genet 16:2089–2096. doi:10.1093/hmg/ddm156

Hines EA et al (2016) Syndactyly in a novel Fras1(rdf) mutant results from interruption of signals for interdigital apoptosis. Dev Dyn Off Publ Am Assoc Anat 245:497–507. doi:10.1002/dvdy.24389

Hofmann AD, Duess JW, Puri P (2014) Congenital anomalies of the kidney and urinary tract (CAKUT) associated with Hirschsprung's disease: a systematic review. Pediatr Surg Int 30:757–761. doi:10.1007/s00383-014-3529-3

Horikawa Y et al (1997) Mutation in hepatocyte nuclear factor-1 beta gene (TCF2) associated with MODY. Nat Genet 17:384–385. doi:10.1038/ng1297-384

Hoskins BE et al (2007) Transcription factor SIX5 is mutated in patients with branchio-oto-renal syndrome. Am J Hum Genet 80:800–804. doi:10.1086/513322

Hwang DY et al (2014) Mutations in 12 known dominant disease-causing genes clarify many congenital anomalies of the kidney and urinary tract. Kidney Int 85:1429–1433. doi:10.1038/ki.2013.508

Iglesias DM et al (2007) Canonical WNT signaling during kidney development. Am J Physiol Renal Physiol 293:F494–F500. doi:10.1152/ajprenal.00416.2006

Inoki K, Li Y, Zhu T, Wu J, Guan K-L (2002) TSC2 is phosphorylated and inhibited by Akt and suppresses mTOR signalling. Nat Cell Biol 4:648–657. doi:10.1038/ncb839

Iwasaki N et al (1998) Liver and kidney function in Japanese patients with maturity-onset diabetes of the young. Diabetes Care 21:2144–2148

Jeanpierre C et al (2011) RET and GDNF mutations are rare in fetuses with renal agenesis or other severe kidney development defects. J Med Genet 48:497–504. doi:10.1136/jmg.2010.088526

Kakkar N, Menon S, Radotra BD (2006) Histomorphology of renal dysplasia—an autopsy study. Fetal Pediatr Pathol 25:73–86. doi:10.1080/15513810600788764

Kamath BM, Spinner NB, Rosenblum ND (2013) Renal involvement and the role of Notch signalling in Alagille syndrome. Nat Rev Nephrol 9:409–418. doi:10.1038/nrneph.2013.102

Kamath BM et al (2012) Renal anomalies in Alagille syndrome: a disease-defining feature. Am J Med Genet A 158A:85–89. doi:10.1002/ajmg.a.34369

Kanda S et al (2014) Sall1 maintains nephron progenitors and nascent nephrons by acting as both an activator and a repressor. J Am Soc Nephrol JASN 25:2584–2595. doi:10.1681/asn.2013080896

Kiefer SM et al (2003) Expression of a truncated Sall1 transcriptional repressor is responsible for Townes–Brocks syndrome birth defects. Hum Mol Genet 12:2221–2227. doi:10.1093/hmg/ddg233

Kiefer SM et al (2010) Sall1-dependent signals affect Wnt signaling and ureter tip fate to initiate kidney development. Development 137:3099–3106. doi:10.1242/dev.037812

Kim J, Kim P, Hui CC (2001a) The VACTERL association: lessons from the Sonic hedgehog pathway. Clin Genet 59:306–315

Kim PC, Mo R, Hui CC (2001b) Murine models of VACTERL syndrome: role of sonic hedgehog signaling pathway. J Pediatr Surg 36:381–384

Kingswood JC et al (2016) Review of the tuberous sclerosis renal guidelines from the 2012 consensus conference: current data and future study. Nephron 133. doi:10.1159/000448293

Kohl S et al (2014) Mild recessive mutations in six Fraser syndrome-related genes cause isolated congenital anomalies of the kidney and urinary tract. J Am Soc Nephrol JASN 25:1917–1922. doi:10.1681/asn.2013101103

Kohlhase J (1993) In: Pagon RA et al (eds) GeneReviews(R). University of Washington, Seattle University of Washington, Seattle

Kolatsi-Joannou M et al (2001) Hepatocyte nuclear factor-1beta: a new kindred with renal cysts and diabetes and gene expression in normal human development. J Am Soc Nephrol JASN 12:2175–2180

Kopan R (2012) Notch signaling. Cold Spring Harb Perspect Biol 4:a011213. doi:10.1101/cshperspect.a011213

Krebs LT et al (2000) Notch signaling is essential for vascular morphogenesis in mice. Genes Dev 14:1343–1352

Krueger DA et al (2010) Everolimus for subependymal giant-cell astrocytomas in tuberous sclerosis. N Engl J Med 363:1801–1811. doi:10.1056/NEJMoa1001671

Lahiri D et al (2007) Nephropathy and defective spermatogenesis in mice transgenic for a single isoform of the Wilms' tumour suppressor protein, WT1-KTS, together with one disrupted Wt1 allele. Mol Reprod Dev 74:300–311. doi:10.1002/mrd.20491

Landgraf K et al (2010) Sipl1 and Rbck1 are novel Eya1-binding proteins with a role in craniofacial development. Mol Cell Biol 30:5764–5775. doi:10.1128/mcb.01645-09

Lefebvre J et al (2015) Alternatively spliced isoforms of WT1 control podocyte-specific gene expression. Kidney Int 88:321–331. doi:10.1038/ki.2015.140

Li Y, Manaligod J, Weeks D (2010) EYA1 mutations associated with the branchio-oto-renal syndrome result in defective otic development in *Xenopus laevis*. Biol Cell 102:277–292. doi:10.1042/bc20090098

Lokmane L, Heliot C, Garcia-Villalba P, Fabre M, Cereghini S (2010) vHNF1 functions in distinct regulatory circuits to control ureteric bud branching and early nephrogenesis. Development 137:347–357. doi:10.1242/dev.042226

Lopez-Rios J et al (2012) GLI3 constrains digit number by controlling both progenitor proliferation and BMP-dependent exit to chondrogenesis. Dev Cell 22:837–848. doi:10.1016/j.devcel.2012.01.006

Lubinsky M (2015) The VACTERL Association as a disturbance of cell fate determination. Am J Med Genet Part A 167a:2582–2588. doi:10.1002/ajmg.a.37238

Madariaga L et al (2013) Severe prenatal renal anomalies associated with mutations in HNF1B or PAX2 genes. Clin J Am Soc Nephrol CJASN 8:1179–1187. doi:10.2215/cjn.10221012

Mandel H et al (2008) SERKAL syndrome: an autosomal-recessive disorder caused by a loss-of-function mutation in WNT4. Am J Hum Genet 82:39–47. doi:10.1016/j.ajhg.2007.08.005

Mao Z, Chong J, Ong ACM (2016) Autosomal dominant polycystic kidney disease: recent advances in clinical management. F1000Research 5. doi:10.12688/f1000research.9045.1

Marie PJ, Coffin JD, Hurley MM (2005) FGF and FGFR signaling in chondrodysplasias and craniosynostosis. J Cell Biochem 96:888–896. doi:10.1002/jcb.20582

Martucciello G, Ceccherini I, Lerone M, Jasonni V (2000) Special basic science review: pathogenesis of Hirschsprung's disease. J Pediatr Surg 35:1017–1025. doi:10.1053/jpsu.2000.7763

Materna-Kiryluk A et al (2014) The emerging role of genomics in the diagnosis and workup of congenital urinary tract defects: a novel deletion syndrome on chromosome 3q13.31-22.1. Pediatr Nephrol 29:257–267. doi:10.1007/s00467-013-2625-2

McCright B (2003) Notch signaling in kidney development. Curr Opin Nephrol Hypertens 12:5–10. doi:10.1097/01.mnh.0000049802.69874.c0

McCright B et al (2001) Defects in development of the kidney, heart and eye vasculature in mice homozygous for a hypomorphic Notch2 mutation. Development 128:491–502

Mendel DB, Hansen LP, Graves MK, Conley PB, Crabtree GR (1991) HNF-1 alpha and HNF-1 beta (vHNF-1) share dimerization and homeo domains, but not activation domains, and form heterodimers in vitro. Genes Dev 5:1042–1056

Michos O et al (2010) Kidney development in the absence of Gdnf and Spry1 requires Fgf10. PLoS Genet 6:e1000809. doi:10.1371/journal.pgen.1000809

Morita Y et al (2016) Sall1 transiently marks undifferentiated heart precursors and regulates their fate. J Mol Cell Cardiol 92:158–162. doi:10.1016/j.yjmcc.2016.02.008

Morrison AA, Viney RL, Saleem MA, Ladomery MR (2008) New insights into the function of the Wilms tumor suppressor gene WT1 in podocytes. Am J Physiol Renal Physiol 295:F12–F17. doi:10.1152/ajprenal.00597.2007

Munger SC, Natarajan A, Looger LL, Ohler U, Capel B (2013) Fine time course expression analysis identifies cascades of activation and repression and maps a putative regulator of mammalian sex determination. PLoS Genet 9. doi:10.1371/journal.pgen.1003630

North American Pediatric Renal Trials and Collaborative Studies (2011) NAPRTCS 2011 annual dialysis report. The National Institute of Diabetes and Digestive and Kidney Diseases, pp 1–26

Natarajan D, Marcos-Gutierrez C, Pachnis V, de Graaff E (2002) Requirement of signalling by receptor tyrosine kinase RET for the directed migration of enteric nervous system progenitor cells during mammalian embryogenesis. Development 129:5151–5160

Ngan ES, Kim KH, Hui CC (2013) Sonic hedgehog signaling and VACTERL association. Mol Syndromol 4:32–45. doi:10.1159/000345725

Nishinakamura R et al (2001) Murine homolog of SALL1 is essential for ureteric bud invasion in kidney development. Development 128:3105–3115

Nishita M et al (2014) Role of Wnt5a-Ror2 signaling in morphogenesis of the metanephric mesenchyme during ureteric budding. Mol Cell Biol 34:3096–3105. doi:10.1128/mcb.00491-14

O'Brien LL et al (2016) Differential regulation of mouse and human nephron progenitors by the Six family of transcriptional regulators. Development 143:595–608. doi:10.1242/dev.127175

Park JS et al (2012) Six2 and Wnt regulate self-renewal and commitment of nephron progenitors through shared gene regulatory networks. Dev Cell 23:637–651. doi:10.1016/j.devcel.2012.07.008

Passos-Bueno MR et al (1999) Clinical spectrum of fibroblast growth factor receptor mutations. Hum Mutat 14:115–125. doi:10.1002/(sici)1098-1004(1999)14:2<115::aid-humu3>3.0.co;2-2

Patek CE et al (2003) Murine Denys–Drash syndrome: evidence of podocyte de-differentiation and systemic mediation of glomerulosclerosis. Hum Mol Genet 12:2379–2394. doi:10.1093/hmg/ddg240

Pavlakis E, Chiotaki R, Chalepakis G (2011) The role of Fras1/Frem proteins in the structure and function of basement membrane. Int J Biochem Cell Biol 43:487–495. doi:10.1016/j.biocel.2010.12.016

Pema M et al (2016) mTORC1-mediated inhibition of polycystin-1 expression drives renal cyst formation in tuberous sclerosis complex. Nat Commun 7:10786. doi:10.1038/ncomms10786

Pignoni F et al (1997) The eye-specification proteins So and Eya form a complex and regulate multiple steps in Drosophila eye development. Cell 91:881–891

Pini Prato A et al (2009) Hirschsprung disease and congenital anomalies of the kidney and urinary tract (CAKUT): a novel syndromic association. Medicine 88:83–90. doi:10.1097/MD.0b013e31819cf5da

Pitera JE, Scambler PJ, Woolf AS (2008) Fras1, a basement membrane-associated protein mutated in Fraser syndrome, mediates both the initiation of the mammalian kidney and the integrity of renal glomeruli. Hum Mol Genet 17:3953–3964. doi:10.1093/hmg/ddn297

Pitera JE, Woolf AS, Basson MA, Scambler PJ (2012) Sprouty1 haploinsufficiency prevents renal agenesis in a model of Fraser syndrome. J Am Soc Nephrol JASN 23:1790–1796. doi:10.1681/asn.2012020146

Plank TL, Yeung RS, Henske EP (1998) Hamartin, the product of the tuberous sclerosis 1 (TSC1) gene, interacts with tuberin and appears to be localized to cytoplasmic vesicles. Cancer Res 58:4766–4770

Poladia DP et al (2006) Role of fibroblast growth factor receptors 1 and 2 in the metanephric mesenchyme. Dev Biol 291:325–339. doi:10.1016/j.ydbio.2005.12.034

Rakowski SK et al (2006) Renal manifestations of tuberous sclerosis complex: incidence, prognosis, and predictive factors. Kidney Int 70:1777–1782. doi:10.1038/sj.ki.5001853
Reidy KJ, Rosenblum ND (2009) Cell and molecular biology of kidney development. Semin Nephrol 29:321–337. doi:10.1016/j.semnephrol.2009.03.009
Riccio P, Cebrian C, Zong H, Hippenmeyer S, Costantini F (2016) Ret and Etv4 promote directed movements of progenitor cells during renal branching morphogenesis. PLoS Biol 14: e1002382. doi:10.1371/journal.pbio.1002382
Ruf RG et al (2004) SIX1 mutations cause branchio-oto-renal syndrome by disruption of EYA1–SIX1–DNA complexes. Proc Natl Acad Sci USA 101:8090–8095. doi:10.1073/pnas.0308475101
Sadowski CE et al (2015) A single-gene cause in 29.5% of cases of steroid-resistant nephrotic syndrome. J Am Soc Nephrol JASN 26:1279–1289. doi:10.1681/asn.2014050489
Saisawat P et al (2014) Whole-exome resequencing reveals recessive mutations in TRAP1 in individuals with CAKUT and VACTERL association. Kidney Int 85:1310–1317. doi:10.1038/ki.2013.417
Sajithlal G, Zou D, Silvius D, Xu PX (2005) Eya 1 acts as a critical regulator for specifying the metanephric mesenchyme. Dev Biol 284:323–336. doi:10.1016/j.ydbio.2005.05.029
Sanna-Cherchi S et al (2012) Copy-number disorders are a common cause of congenital kidney malformations. Am J Hum Genet 91:987–997. doi:10.1016/j.ajhg.2012.10.007
Schneider J, Arraf AA, Grinstein M, Yelin R, Schultheiss TM (2015) Wnt signaling orients the proximal-distal axis of chick kidney nephrons. Development 142:2686–2695. doi:10.1242/dev.123968
Schuchardt A, D'Agati V, Pachnis V, Costantini F (1996) Renal agenesis and hypodysplasia in ret-k-mutant mice result from defects in ureteric bud development. Development 122:1919–1929
Seyedzadeh A, Kompani F, Esmailie E, Samadzadeh S, Farshchi B (2008) High-grade vesicoureteral reflux in Pfeiffer syndrome. Urol J 5:200–202
Shakya R, Watanabe T, Costantini F (2005) The role of GDNF/Ret signaling in ureteric bud cell fate and branching morphogenesis. Dev Cell 8:65–74. doi:10.1016/j.devcel.2004.11.008
Shillingford JM et al (2006) The mTOR pathway is regulated by polycystin-1, and its inhibition reverses renal cystogenesis in polycystic kidney disease. Proc Natl Acad Sci USA 103:5466–5471. doi:10.1073/pnas.0509694103
Siroky BJ, Yin H, Bissler JJ (2011) Clinical and molecular insights into tuberous sclerosis complex renal disease. Pediatr Nephrol 26:839–852. doi:10.1007/s00467-010-1689-5
Si-Tayeb K, Lemaigre FP, Duncan SA (2010) Organogenesis and development of the liver. Dev Cell 18:175–189. doi:10.1016/j.devcel.2010.01.011
Song R, Yosypiv IV (2011) Genetics of congenital anomalies of the kidney and urinary tract. Pediatr Nephrol 26:353–364. doi:10.1007/s00467-010-1629-4
Stark K, Vainio S, Vassileva G, McMahon AP (1994) Epithelial transformation of metanephric mesenchyme in the developing kidney regulated by Wnt-4. Nature 372:679–683. doi:10.1038/372679a0
Stoll C, Dott B, Alembik Y, Roth MP (2014) Associated nonurinary congenital anomalies among infants with congenital anomalies of kidney and urinary tract (CAKUT). Eur J Med Genet 57:322–328. doi:10.1016/j.ejmg.2014.04.014
Stoll C, Dott B, Alembik Y, Roth MP (2015) Associated congenital anomalies among cases with Down syndrome. Eur J Med Genet 58:674–680. doi:10.1016/j.ejmg.2015.11.003
Su N, Jin M, Chen L (2014) Role of FGF/FGFR signaling in skeletal development and homeostasis: learning from mouse models. Bone Res 2:14003. doi:10.1038/boneres.2014.3
Tadjuidje E, Hegde RS (2013) The eyes absent proteins in development and disease. Cell Mol Life Sci 70:1897–1913. doi:10.1007/s00018-012-1144-9
Taraviras S et al (1999) Signalling by the RET receptor tyrosine kinase and its role in the development of the mammalian enteric nervous system. Development 126:2785–2797

Tee AR et al (2002) Tuberous sclerosis complex-1 and -2 gene products function together to inhibit mammalian target of rapamycin (mTOR)-mediated downstream signaling. Proc Natl Acad Sci USA 99:13571–13576. doi:10.1073/pnas.202476899

Thomas R et al (2011) HNF1B and PAX2 mutations are a common cause of renal hypodysplasia in the CKiD cohort. Pediatr Nephrol 26:897–903. doi:10.1007/s00467-011-1826-9

Uyttendaele H, Ho J, Rossant J, Kitajewski J (2001) Vascular patterning defects associated with expression of activated Notch4 in embryonic endothelium. Proc Natl Acad Sci USA 98:5643–5648. doi:10.1073/pnas.091584598

Vainio SJ (2003) Nephrogenesis regulated by Wnt signaling. J Nephrol 16:279–285

van Slegtenhorst M et al (1998) Interaction between hamartin and tuberin, the TSC1 and TSC2 gene products. Hum Mol Genet 7:1053–1057

Verbitsky M et al (2015) Genomic imbalances in pediatric patients with chronic kidney disease. J Clin Invest 125:2171–2178. doi:10.1172/jci80877

Vivante A et al (2013) Renal hypodysplasia associates with a WNT4 variant that causes aberrant canonical WNT signaling. J Am Soc Nephrol JASN 24:550–558. doi:10.1681/asn.2012010097

Volckaert T, De Langhe SP (2015) Wnt and FGF mediated epithelial-mesenchymal crosstalk during lung development. Dev Dyn Off Publ Am Assoc Anat 244:342–366. doi:10.1002/dvdy.24234

Warthen DM et al (2006) Jagged1 (JAG1) mutations in Alagille syndrome: increasing the mutation detection rate. Hum Mutat 27:436–443. doi:10.1002/humu.20310

Weber S et al (2006) Prevalence of mutations in renal developmental genes in children with renal hypodysplasia: results of the ESCAPE study. J Am Soc Nephrol JASN 17:2864–2870. doi:10.1681/asn.2006030277

Weber S et al (2008) SIX2 and BMP4 mutations associate with anomalous kidney development. J Am Soc Nephrol JASN 19:891–903. doi:10.1681/asn.2006111282

Xi Q et al (2016) Copy number variations in multicystic dysplastic kidney: update for prenatal diagnosis and genetic counseling. Prenat Diagn. doi:10.1002/pd.4807

Xu PX, Cheng J, Epstein JA, Maas RL (1997) Mouse Eya genes are expressed during limb tendon development and encode a transcriptional activation function. Proc Natl Acad Sci USA 94:11974–11979

Xu PX et al (1999) Eya1-deficient mice lack ears and kidneys and show abnormal apoptosis of organ primordia. Nat Genet 23:113–117. doi:10.1038/12722

Xu J et al (2014) Eya1 interacts with Six2 and Myc to regulate expansion of the nephron progenitor pool during nephrogenesis. Dev Cell 31:434–447. doi:10.1016/j.devcel.2014.10.015

Xue Y et al (1999) Embryonic lethality and vascular defects in mice lacking the Notch ligand Jagged1. Hum Mol Genet 8:723–730

Yoder BK, Hou X, Guay-Woodford LM (2002) The polycystic kidney disease proteins, polycystin-1, polycystin-2, polaris, and cystin, are co-localized in renal cilia. J Am Soc Nephrol JASN 13:2508–2516

Zeidler C et al (2014) Heterozygous FGF8 mutations in patients presenting cryptorchidism and multiple VATER/VACTERL features without limb anomalies. Birth Defects Res A Clin Mol Teratol 100:750–759. doi:10.1002/bdra.23278

Zhao H et al (2004) Role of fibroblast growth factor receptors 1 and 2 in the ureteric bud. Dev Biol 276:403–415. doi:10.1016/j.ydbio.2004.09.002

Zou D, Silvius D, Rodrigo-Blomqvist S, Enerbäck S, Xu PX (2006) Eya1 regulates the growth of otic epithelium and interacts with Pax2 during the development of all sensory areas in the inner ear. Dev Biol 298. doi:10.1016/j.ydbio.2006.06.049

Zubkov VS et al (2015) A spatially-averaged mathematical model of kidney branching morphogenesis. J Theor Biol 379:24–37. doi:10.1016/j.jtbi.2015.04.015

Chapter 11
Primary Cilia in Cystic Kidney Disease

Prachee Avasthi, Robin L. Maser, and Pamela V. Tran

Abstract Primary cilia are small, antenna-like structures that detect mechanical and chemical cues and transduce extracellular signals. While mammalian primary cilia were first reported in the late 1800s, scientific interest in these sensory organelles has burgeoned since the beginning of the twenty-first century with recognition that primary cilia are essential to human health. Among the most common clinical manifestations of ciliary dysfunction are renal cysts. The molecular mechanisms underlying renal cystogenesis are complex, involving multiple aberrant cellular processes and signaling pathways, while initiating molecular events remain undefined. Autosomal Dominant Polycystic Kidney Disease is the most common renal cystic disease, caused by disruption of polycystin-1 and polycystin-2 transmembrane proteins, which evidence suggests must localize to primary cilia for proper function. To understand how the absence of these proteins in primary cilia may be remediated, we review intracellular trafficking of polycystins to the primary cilium. We also examine the controversial mechanisms by which primary cilia transduce flow-mediated mechanical stress into intracellular calcium. Further, to better understand ciliary function in the kidney, we highlight the LKB1/AMPK, Wnt, and Hedgehog developmental signaling pathways mediated by primary cilia and misregulated in renal cystic disease.

P. Avasthi, Ph.D.
Department of Anatomy and Cell Biology, University of Kansas Medical Center, Kansas City, KS, USA
e-mail: pavasthi@kumc.edu

R.L. Maser, Ph.D.
Department of Clinical Laboratory Sciences, University of Kansas Medical Center, Kansas City, KS, USA

Kidney Institute, University of Kansas Medical Center, 3901 Rainbow Blvd., MS #3038, Kansas City 66160, KS, USA
e-mail: rmaser@kumc.edu

P.V. Tran, Ph.D. (✉)
Department of Anatomy and Cell Biology, University of Kansas Medical Center, Kansas City, KS, USA

Kidney Institute, University of Kansas Medical Center, 3901 Rainbow Blvd., MS #3038, Kansas City 66160, KS, USA
e-mail: ptran@kumc.edu

R.K. Miller (ed.), *Kidney Development and Disease*, Results and Problems in Cell Differentiation 60, DOI 10.1007/978-3-319-51436-9_11

11.1 Functions and Features of Primary Cilia

Almost all mammalian cells have an apical protrusion that is used to sense the extracellular environment. This protrusion, the cilium (also known as the flagellum), is an ancient organelle thought to be present on the last eukaryotic common ancestor (Mitchell 2007). Mammalian primary cilia were first reported in 1898 on various iron hematoxylin-stained epithelia, and at that time, primary cilia extending into renal tubules were speculated to sense fluid flow (Zimmermann 1898). Following detailed electron microscopic analyses of embryonic and adult cell types, cilia were further proposed to serve as conserved structures that receive and transduce extracellular cues (Poole et al. 1985). Indeed, the primary cilium functions as an antenna to receive light and mechanical and chemical signals. The ciliary membrane, while topologically continuous with the plasma membrane, is distinct in its composition of membrane and protein components. The increase in cell surface area provided by the ciliary membrane protrusion allows for a concentration of receptors that can transduce extracellular signals. Cilia can also shed or lose membrane from their distal tips in order to release materials to the extracellular environment (Hogan et al. 2009; Wood et al. 2013), signal to adjacent organisms (Wang et al. 2014), or turn over its spent protein, including molecular and membrane components (Young 1971; Young and Bok 1969), demonstrating their capacity for two-way communication.

11.1.1 Cilium Assembly

The cilium is composed of a cylindrical arrangement of nine microtubule doublets (A and B tubules) termed the axoneme, which is ensheathed by the cellular plasma membrane (Fig. 11.1). While motile cilia also contain a central pair of microtubules along with accessory structures such as dynein arms and radial spokes for force generation, nonmotile primary cilia lack these structures. Cilia extend and are maintained via intraflagellar transport (IFT), the bidirectional movement of motor and cargo proteins along the microtubular axoneme. IFT was first discovered in the green alga, *Chlamydomonas*, by differential interference contrast (DIC) visualization of particle movement along flagella (Kozminski et al. 1993). Trafficking of proteins from the base to the tip of the cilium in anterograde IFT is powered by motor proteins in the kinesin-2 family (Cole et al. 1998). This anterograde motor sedimented with a large group of proteins could be biochemically purified as two complexes, complex B and complex A (Cole et al. 1998). Complex B proteins are associated with the anterograde kinesin motor and complex A with retrograde transport from the tip to the ciliary base mediated by the cytoplasmic dynein motor (Pazour et al. 1998, 1999; Porter et al. 1999; Signor et al. 1999). Mutations in the complex B IFT proteins or IFT motors result in stunted cilia and flagella, suggesting their requirement for ciliary assembly (Brazelton et al. 2001; Deane

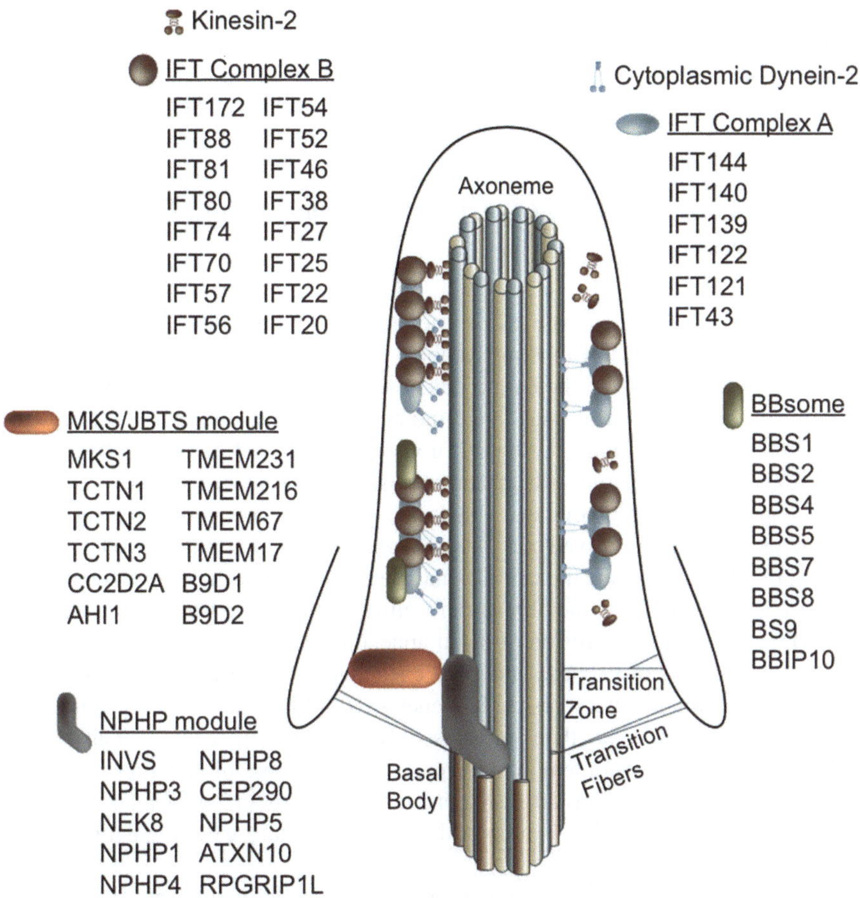

Fig. 11.1 Structure of primary cilium. Multi-protein complexes at the transition zone, basal body, and along the axoneme are required for building and maintaining the primary cilium

et al. 2001; Follit et al. 2006; Fujiwara et al. 1999; Haycraft et al. 2003; Huangfu et al. 2003; Ishikawa et al. 2014; Pazour et al. 2000; Qin et al. 2007; Sun et al. 2004). Accumulation of complex B proteins in swollen ciliary distal tips and impaired retrograde IFT in complex A mutants indicate that complex A proteins are required for retrograde IFT (Blacque et al. 2006; Iomini et al. 2001, 2009; Perkins et al. 1986; Qin et al. 2011; Tran et al. 2008). Additionally, complex A proteins have also been shown to be required for ciliary entry of G-protein coupled receptors in mammalian cells (Mukhopadhyay et al. 2010).

While each type of cilium is specialized to receive appropriate signals within a certain tissue, proteins common to most cilia include tubulins, comprising the structural units of the axoneme, microtubule motors and IFT complexes, the Bardet–Biedl Syndrome-related protein complex (BBsome), which traffics cargo in and out of the cilium, and vesicular trafficking-related small GTPases (ARFs,

ARFs, Rabs). As we will discuss in more detail below, mutant proteins common to several types of cilia result in pleiotropic phenotypes due to ciliary dysfunction in multiple tissues.

11.1.2 Basal Bodies, Early Cilium Formation, and Cilium Disassembly

Cilia are anchored at their base by centriole-derived basal bodies (Fig. 11.1). The basal body is composed of nine microtubule triplets (A, B, and C tubules). The A and B tubules extend from the basal body during cilium formation and become the ciliary axoneme. The basal body is the mother centriole of the pair that forms the centrosome at spindle poles during cell division. During G1 or G0, the basal body acquires accessory structures (distal and subdistal appendages), migrates, and docks to the apical plasma membrane. Docking occurs by attachment of the distal appendages to Golgi-derived vesicles (Sorokin 1962) that fuse to one another and, subsequently, to the apical plasma membrane for cilium extension. This process is mediated by membrane-deforming proteins, EH-domain containing 1 (EHD1) and EHD3 for fusion of distal appendage vesicles into a ciliary vesicle, and by recruitment of GTPase and GTP exchange factors Rab11 and Rabin8 for activation of Rab8 in ciliary vesicle biogenesis (Knodler et al. 2010; Westlake et al. 2011). The basal body then matures through loss of centrosomal protein 110 (CP110) and recruitment of IFT proteins and those of the transition zone (Deane et al. 2001; Lu et al. 2015; Rosenbaum and Witman 2002), a compartment distal to the basal body that is proposed to act as a ciliary gate, regulating ciliary entry of proteins.

Just as there is a mechanism for maturation of centrioles to nucleate cilia, there is also a mechanism to disassemble cilia and release sequestered basal bodies for mitotic progression. Several proteins have been implicated in this process, most notably the Aurora A kinase (Pugacheva et al. 2007), which activates mitotic spindle assembling CDK1-Cyclin-B. Growth factors activate a pathway that stimulates Aurora A kinase and, subsequently, the histone deacetylase, HDAC6, responsible for tubulin deacetylation and cilium destabilization. Another group of kinases, the NIMA-related kinases (NEK), are also involved in ciliary resorption (Hilton et al. 2013; Wloga et al. 2006). Nek2, involved in centrosome separation, is basal body localized and activates a microtubule depolymerizing kinesin, kif24, for ciliary resorption in the G2/M transition (Kim et al. 2015; Spalluto et al. 2012). In *Chlamydomonas*, another microtubule depolymerizing kinesin, kinesin-13, is also required for flagellar disassembly (Piao et al. 2009).

11.1.3 Cilium Length Control

Another fundamental discovery made in *Chlamydomonas* that expanded our understanding of how cilia are maintained was that tubulin continuously turns over at flagellar tips (Marshall and Rosenbaum 2001; Song and Dentler 2001; Stephens 1997). This, along with evidence that flagella exhibit a rapid initial growth that slows as they approach steady state (Rosenbaum and Child 1967) and that flagella disassembly occurs at a constant rate independent of length (Kozminski et al. 1995; Marshall et al. 2005; Marshall and Rosenbaum 2001; Parker and Quarmby 2003), forms the basis of the balance point model of flagellar maintenance (Marshall et al. 2005; Marshall and Rosenbaum 2001). In this model, flagella reach their steady-state length when the assembly and disassembly rates are equal. The length dependence of flagellar assembly is based upon the variable quantity of IFT material entering the flagellum at different lengths (Craft et al. 2015; Engel et al. 2009; Ludington et al. 2013) as well as the variable quantity of IFT material accumulating at basal bodies prior to trafficking into flagella (Ludington et al. 2013). Cilia in different cell types have a characteristic length, which presumably are well suited for their function. A few examples of characteristic ciliary lengths are shown in Table 11.1.

Chemical and genetic screens (Avasthi et al. 2012; Kim et al. 2010, 2016a) along with single gene perturbation have identified many regulators of ciliary length. In addition to IFT motors, proteins, and cargo that are required for proper cilium assembly, proteins involved in tubulin modification (Pugacheva et al. 2007; Sanchez de Diego et al. 2014) and actin dynamics (Abdul-Majeed et al. 2012; Avasthi et al. 2014; Bershteyn et al. 2010; Kim et al. 2010; Oishi et al. 2006; Sharma et al. 2011) also regulate cilium length. A variety of cell surface receptors have also been identified as modulators of cilium length such as dopamine receptors D1, D5, and other G-protein coupled receptors (Abdul-Majeed et al. 2012; Abdul-Majeed and Nauli 2011; Avasthi et al. 2012). Regulators of protein phosphorylation

Table 11.1 Mammalian ciliary lengths

Cell/tissue type	Cilium length (μm)	Citation
Embryonic neural tube	0.97 ± 0.17	He et al. (2014)
Rat tail tendon fascicles	1.35 ± 0.11	Lavagnino et al. (2013)
Mouse embryonic node	2–2.5	Alten et al. (2012)
Mouse olfactory bulb	22	Ying et al. (2014)
Rat cholangiocytes (large bile ducts)	7.35±1.32	Masyuk et al. (2008)
Rat cholangiocytes (small bile ducts)	3.58±1.12	Masyuk et al. (2008)
Pancreatic duct	2	Cano et al. (2004)
Pancreatic islet	2.5	Ait-Lounis et al. (2007)
Kidney	3–3.5	Pazour et al. (2000)
N1 hypothalamic neurons	2–2.5	Han et al. (2014)
Hypothalamic arcuate nucleus	5.5 ± 0.44	Han et al. (2014)
Ventromedial hypothalamus	5.3 ± 0.39	Han et al. (2014)

such as GSK3β, MAP kinases (LF4, Dlk-1, DYF-5), cell cycle-related kinases (Nek1, Nek4, Nek8, Nrk2, Nrk17, Nrk30, Cdc42, Cnk2, LF2, LF5, AurA), and phosphatases (cdc14b) also modulate cilia length (Berman et al. 2003; Burghoorn et al. 2007; Coene et al. 2011; Pugacheva et al. 2007; Tam et al. 2013; Tam et al. 2007; Thiel et al. 2011; van der Vaart et al. 2015; White and Quarmby 2008; Wloga et al. 2006; Zuo et al. 2011). In addition to all of these, cilia respond to second messengers such as calcium, cAMP, inositol 1,4,5-trisphosphate, and associated enzymes protein kinase C, adenylate cyclase III, and inositol polyphosphate 5-phosphatase E, respectively (Abdul-Majeed et al. 2012; Besschetnova et al. 2010; Hatayama et al. 2011; Luo et al. 2012; Ou et al. 2009). Clearly, many different genes are involved in the response of cilia to external signals as well as in coordination with the cell cycle. Additional genes regulating cilium structure and length have been previously reviewed (Avasthi and Marshall 2012; Broekhuis et al. 2013; Keeling et al. 2016), and much is still unknown about how these pathways work together or, indeed, which of them are utilized concurrently. Along with axoneme structure, ciliary membrane must also be regulated for proper assembly, maintenance, and protein composition (Bloodgood 2012). Ciliary formation requires trafficking of Golgi-derived vesicles regulated by small GTPases including Rab8, Rab11, Arf4, and Arl3 (Kim et al. 2014; Knodler et al. 2010; Mazelova et al. 2009; Schwarz et al. 2012; Westlake et al. 2011) as well as membrane regulating proteins EHD1/3 (Lu et al. 2015). In mammalian photoreceptors and *Chlamydomonas* flagella, ciliary membrane is continuously renewed (Dentler 2013; Young 1971; Young and Bok 1969) and membrane/proteins can be shed via ectosome release (Avasthi and Marshall 2013; Hogan et al. 2009; Wang et al. 2014; Wood et al. 2013; Wood and Rosenbaum 2015). While these studies highlight mechanisms of membrane regulation, how ciliary membrane and axoneme dynamics are coordinated remains largely a mystery.

Within the kidney, cilium length increases during development, being shortest in renal vesicles and longest in mature fetal nephron segments (Saraga-Babic et al. 2012). Renal ciliary length is also modulated in response to various endogenous and extracellular factors, including hormones and changes in urinary shear stress, which are thought to alter the sensitivity of the ciliated cell to external cues (Besschetnova et al. 2010; Upadhyay et al. 2014). Alterations in length of primary cilia are also observed in response to renal injury and during regeneration/repair processes (Verghese et al. 2008, 2009; Han et al. 2016).

11.2 Ciliopathies: Linking Ciliary Dysfunction to Renal Cystogenesis

The discovery that the Oak Ridge polycystic kidney disease (*orpk*) mouse model resulted from a defect in *Tg737/IFT88* marked a major breakthrough linking dysfunction of primary cilia to human disease (Pazour et al. 2000). Soon afterward, an explosion of discoveries followed demonstrating that mysterious disorders affecting multiple organ systems including the retina, kidney, skeletal system, reproductive systems, and others could be traced to defects in a single ciliary gene. These syndromic diseases, termed ciliopathies, include Meckel–Gruber Syndrome (MKS), Bardet–Biedl Syndrome (BBS), Joubert Syndrome, Jeune Syndrome, Nephronophthisis (NPHP), and others (Table 11.2). These pediatric diseases range in severity, causing death in infancy of individuals with MKS or during late teenage years in some individuals with NPHP, and have overlapping but varying degrees of clinical manifestations, including kidney cysts, intellectual disabilities, craniofacial and skeletal defects, polydactyly, retinal degeneration, hypogonadism, and obesity. Many of these diseases, such as MKS and NPHP,

Table 11.2 Ciliopathies

Ciliopathy	Organ systems affected	References
Alström Syndrome	Visual, auditory, cardiovascular, endocrine, hepatic, renal, skeletal, neural	Hearn et al. (2005)
Bardet–Biedl Syndrome	Skeletal, neural, visual, renal, reproductive	Ansley et al. (2003)
Birt–Hogg–Dubé Syndrome	Renal (carcinoma and cysts), lung, integumentary	Luijten et al. (2013)
Ellis van Creveld Syndrome	Skeletal, cardiac	Ruiz-Perez and Goodship (2009)
Jeune Syndrome	Skeletal, renal	Beales et al. (2007)
Joubert Syndrome	Renal, visual, neural	Louie and Gleeson (2005)
Juvenile cystic kidney disease	Renal	Smith et al. (2006)
Kartagener Syndrome/Primary ciliary dyskinesia	Respiratory, neural, laterality, reproductive	Afzelius (1976), Camner et al. (1975)
Leber congenital amaurosis	Visual	den Hollander et al. (2007)
Meckel–Gruber Syndrome	Renal, neural, hepatic	Dawe et al. (2007)
Nephronophthisis	Renal	Otto et al. (2003)
Orofaciodigital Syndrome 1	Skeletal	Romio et al. (2004)
Polycystic kidney disease	Renal, hepatic, cardiovascular	Pazour et al. (2000)
Retinitis pigmentosa	Visual	Hong et al. (2001)
Senior–Loken Syndrome	Renal, visual	Otto et al. (2005)
Sensenbrenner Syndrome	Skeletal	Walczak-Sztulpa et al. (2010)
Tuberous sclerosis	Neural, integumentary, cardiovascular, renal, respiratory	Hartman et al. (2009)
Usher Syndrome	Visual, auditory	Bonneau et al. (1993)
Von Hippel–Lindau	Tumors in blood vessels and adrenal glands, renal (carcinoma and cysts)	Esteban et al. (2006)

result from disruption of proteins at the transition zone, a proposed hotspot for disease gene networks (Chih et al. 2012; Garcia-Gonzalo et al. 2011; Williams et al. 2011). Disruption of these proteins alters cilia structure, resulting in cilia that are abnormally short, absent, or, less commonly, elongated (Williams et al. 2011). BBS is caused by mutations in one of the 21 genes (Bujakowska et al. 2015; Heon et al. 2016; Lindstrand et al. 2014; Schaefer et al. 2016; reviewed in Khan et al. 2016), most of which encode proteins comprising or facilitating formation or trafficking of the BBSome complex, which has been demonstrated to be a cargo of IFT in *Chlamydomonas* flagella (Lechtreck et al. 2009) and to transport signaling molecules to membrane compartments and within the ciliary membrane (Guo et al. 2016; Guo and Rahmouni 2011). The BBS core is an 8-subunit complex composed of BBS1, BBS2, BBS4, BBS5, BBS7, BBS8, BBS9, and BBS18/BBIP1 (Nachury et al. 2007; Scheidecker et al. 2014). Additionally, BBS3/ARL6 and BBS17/LZTFL1 regulate ciliary trafficking of the BBSome (Liew et al. 2014; Marion et al. 2012; Seo et al. 2011), while BBS6, BBS10, and BBS12 form a BBS-chaperonin complex together with BBS7 to promote BBSome assembly (Seo et al. 2010; Zhang et al. 2012). Causative mutations in *IFT* genes have also been identified in BBS patients. IFT27/BBS19 (Aldahmesh et al. 2014) facilitates ciliary exit of the BBSome (Eguether et al. 2014; Liew et al. 2014), and while IFT172/BBS20 (Bujakowska et al. 2015; Schaefer et al. 2016) has been shown to regulate IFT particle turnaround at the ciliary distal tip in *Chlamydomonas* (Pedersen et al. 2005), its connection to the BBSome remains to be explored. While disruption of some BBS proteins impairs ciliogenesis (Marion et al. 2009), mutations of other BBS proteins do not overtly affect cilia structure, but impede ciliary entry or exit of signaling molecules (Berbari et al. 2008; Zhang et al. 2011).

Of the many clinical features of ciliopathies, a major hallmark is renal cystic disease (Quinlan et al. 2008). Renal cystic diseases are mostly inherited and can be classified as Polycystic Kidney Disease (PKD) or as non-PKD, depending on the mutated gene. While the non-PKD diseases comprise the ciliopathies mentioned above, PKD results from mutations in *Polycystic Kidney Disease 1* (*PKD1*), *PKD2*, or *Polycystic Kidney and Hepatic Disease 1* (*PKHD1*) genes. Mutations of *PKD1* or *PKD2* cause approximately 90% of cases of Autosomal Dominant Polycystic Kidney Disease (ADPKD), the most common fatal genetic disease, affecting 1:400–1:1000 adults worldwide. Renal cysts can initiate in the fetus and progressively grow, compressing and compromising surrounding parenchyma and causing end-stage renal disease in the 6th decade of life. Mutations in *PKHD1* cause Autosomal Recessive PKD, which affects 1:20,000 children. The *PKD1*, *PKD2*, and *PKHD1* gene products, polycystin 1 (PC1), PC2, and fibrocystin/polyductin (FPC), localize to various subcellular compartments, including the primary cilium (Barr and Sternberg 1999; Yoder et al. 2002). Ciliary localization of polycystin-GFP fusion proteins in *C. elegans* provided the first demonstration of a link between renal cystic disease and primary cilia (Barr and Sternberg 1999). The ciliary localization of mammalian *PKD* gene products together with evidence that ciliary defects cause renal cysts led to the proposal that ciliary dysfunction may present a unifying etiological mechanism for renal cystic diseases. Recently, mutations in

Glucosidase Alpha; Neutral AB (*GANAB*), which catalyzes a step in the *N*-glycosylation pathway in the endoplasmic reticulum (ER), have been reported in ADPKD patients, reflecting a critical role for the ER protein maturation process for functional PC1 and its localization to the primary cilium (Porath et al. 2016), discussed in more detail in Sect. 3.

Multiple cellular processes are disrupted in renal cystogenesis that result in cell dedifferentiation, increased cell proliferation, increased fluid secretion of renal epithelial cells, and increased tubular cell apoptosis (reviewed in Calvet and Grantham 2001; Torres and Harris 2006; Zhou 2009). Studies using ADPKD primary renal cells revealed that the combination of high intracellular cAMP levels and low intracellular calcium (Ca^{2+}) drives cell proliferation and fluid secretion of cyst-lining epithelial cells (Calvet 2008; Wallace 2011). In human ADPKD cells, low homeostatic intracellular Ca^{2+} enables aberrant cAMP-mediated activation of Protein Kinase A (PKA) and ERK, which, in turn, activates both cystic fibrosis transmembrane conductance regulator (CFTR) and ERK signaling pathways, causing fluid secretion by epithelial cells into the cyst lumen and increased proliferation of cyst-lining epithelial cells, respectively (Yamaguchi et al. 2004). Addition of a Ca^{2+} ionophore rescued these aberrant responses, and conversely, addition of a Ca^{2+} channel blocker to normal human kidney (NHK) cells caused ADPKD cellular responses, demonstrating the pivotal role of intracellular Ca^{2+} in influencing cellular phenotype (Yamaguchi et al. 2006). Elevated intracellular cAMP is common to cystic kidneys of both orthologous *Pkd* and ciliary mouse models: the former including *Pkd1* (Hopp et al. 2012) and *Pkd2* (Torres et al. 2004) mutant mice and the *Pkhd1*-mutant Polycystic Kidney (PCK) rat (Wang et al. 2008), while the latter group including *pcy* (Yamaguchi et al. 1997), *cpk* (Tao et al. 2015), *jck* (Smith et al. 2006), *Kif3a* (Choi et al. 2011), and *Thm1*/*Ttc21b* (Tran et al. 2014b). Such observations suggest that increased cAMP may be a universal and fundamental mechanism originating from ciliary dysfunction. Another unifying phenomenon is that loss of PKD or ciliary proteins during kidney development leads to severe, rapidly progressing renal cystic disease, while loss of these proteins after kidney maturation (P12–P14) causes a mild, slowly progressing disease, demonstrating that the developmental state of the kidney strongly influences renal cystogenesis (Piontek et al. 2007). A role for the primary cilium and PC1 in sensing tubular injury and in regulating the repair–regeneration–redifferentiation response has been proposed, and dysfunction of this adaptive mechanism in mutant kidneys may provide a trigger for cyst formation in the adult kidney (Weimbs 2007).

Yet, differences are evident between PKD and non-PKD renal cystic diseases. In PKD, growth of renal cysts greatly enlarges kidney size, while in non-PKD diseases, such as NPHP, kidney size tends to decrease due to abundant fibrosis. Another contrast between PKD and non-PKD diseases is that while most non-PKD proteins have a fundamental role in ciliogenesis, PC1, PC2, and FPC do not. This is also reflected by the majority of studies that do not report altered ciliary length in *Pkd1*/*Pkd2* mutants, with the exception of two. In renal epithelia of the $PC1^{RC/RC}$ mouse model, which carries an ADPKD mutation (Hopp et al. 2012) and in cultured *Pkd1−/−* and *Pkd2−/−* derived mouse renal epithelial cells (Jin et al. 2014b), cilia were lengthened. This may relate to the role of PC1/PC2 in modulating intracellular cAMP and Ca^{2+}, since pharmacological manipulation of intracellular cAMP and Ca^{2+} in mouse inner medullary collecting duct (IMCD) cells showed that low Ca^{2+} lengthens cilia, while high Ca^{2+} shortens cilia (Besschetnova et al. 2010).

Although loss of proteins required for ciliary structure usually causes NPHP in the human population, IFT mouse models and the *Nek8* mutant, *jck*, manifest PKD-like disease with enlarged kidneys. Moreover, heterozygosity for both *Pkd1* and the $Nek8^{jck/+}$ mutation, which in the homozygous form causes longer cilia and an ADPKD phenotype (Smith et al. 2006), caused renal cysts in mice, while single heterozygosity did not, suggesting that molecular mechanisms stemming from deficiency of PC1 and *jck* mutant NEK8 protein may converge (Natoli et al. 2008). Intriguingly, despite that ablation of *Kif3a*, and in turn primary cilia, results in renal cysts, combined loss of primary cilia and PC1/2 ameliorated PC-mediated renal cystogenesis and decreased renal epithelial cell proliferation in mice, suggesting the presence of an as yet unidentified ciliary-mediated signaling pathway that is crucial for PC1/2 cyst development (Ma et al. 2013). Corroborating and extending these data, reduced ciliary length through inhibition of Cyclin Dependent Kinase 5 (CKD5) also ameliorated the renal cystic disease in *jck* mice (Husson et al. 2016). These data underscore the need to define the role of ciliary proteins individually and in PKD to understand mechanisms that exacerbate or attenuate PKD. Disease variability observed among ADPKD patients may be due to mutations/polymorphisms in genes affecting ciliary function.

11.3 Ciliary Localization of Polycystins Is Interdependent and Important for Their Function

Analysis of several ADPKD missense mutations in either *PKD1* or *PKD2* showed that they resulted in the absence of polycystins in cilia (Cai et al. 2014). One of these mutations, $PC2^{W414G}$, present in the first extracellular loop, retained channel activity, suggesting that functional PC2 in the endoplasmic reticulum (ER) is insufficient to prevent cystic disease and that ciliary localization of polycystins is essential for this function. Thus, understanding ciliary-targeting mechanisms of polycystins is important for understanding PKD biology and finding therapeutic targets.

Several ciliary targeting sites have been uncovered in the PC1 C-terminal cystosolic tail, including a KVHPSST sequence (Ward et al. 2011) and sites within and upstream of the coiled-coiled domain required for interaction with PC2 (Su et al. 2015). An RVxP ciliary targeting sequence in the PC2 N-terminus has also been identified (Geng et al. 2006). Accumulating evidence suggests that ciliary localization of PC1 and PC2 is interdependent (Freedman et al. 2013; Gainullin et al. 2015; Kim et al. 2014; Nauli et al. 2003; Su et al. 2015). In ADPKD-derived, *PKD1*-mutant induced pluripotent stem cells and in *Pkd1* −/− MEFs, PC2 ciliary localization was reduced or undetectable, respectively, while in *Pkd2−/−* MEFs, ciliary PC1 was lacking (Freedman et al. 2013; Gainullin et al. 2015; Kim et al. 2014). Conversely, overexpression of PC2 enhanced both WT and mutant PC1 ciliary targeting, while increasing levels of WT PC1, but not of a PC1 construct lacking its C-terminus, promoted PC2 ciliary localization in mouse IMCD3 cells (Freedman et al. 2013; Su et al. 2015), consistent with a requirement for the PC1 C-terminus in ciliary localization.

To reach the primary cilium, PC1 protein must undergo maturation, which has also been shown to depend on PC2 (Gainullin et al. 2015; Kim et al. 2014). PC1 maturation involves N-linked core glycan addition and cleavage at its G-protein coupled proteolysis site (GPS), which generates extracellular N-terminal (NTF) and membrane-embedded C-terminal (CTF) fragments that remain non-covalently associated, and occurs within the ER prior to subsequent trafficking and N-glycan modification in the Golgi (Wei et al. 2007). *PKD1* missense mutations that interfere with GPS cleavage prevent PC1 maturation and ciliary localization (Cai et al. 2014; Chapin et al. 2010; Kurbegovic et al. 2014). PC1–PC2 association occurs before GPS cleavage of PC1 (Gainullin et al. 2015) and is required for the PC complex to reach the trans-Golgi (Kim et al. 2014). In *Pkd2−/−* mouse embryonic fibroblasts (MEF), cleaved PC1 remained endoH sensitive, suggesting that PC1 must complex with PC2 to traffic through the Golgi. Further, cleavage of PC1 is essential for transport through the Golgi, since a non-cleavable PC1 mutant co-precipitated with PC2 remained endoH sensitive (Kim et al. 2014). $Pkd1^{RC/RC}$ hypomorphic mutant mice carrying various *Pkd2*-mutant alleles of differing deficiency revealed that renal cystic disease severity was determined by levels of mature PC1, which in turn correlated with PC2 levels, demonstrating the dependence of PC1 maturation on PC2 in vivo (Gainullin et al. 2015).

11.3.1 Polycystin Transport from the Golgi Apparatus to the Cilium

Golgi-derived vesicles convey proteins to the base of the primary cilium, a process termed polarized vesicle trafficking (reviewed by (Hsiao et al. 2012). This transport is essential for ciliogenesis and is mediated by the Rab and Arf families of small GTPases, which recruit vesicle coating complexes during vesicle budding, docking, and fusion. At the centrosome, Rab11 is essential in vesicular targeting of Rabin8 to recruit and activate Rab8 for ciliary targeting of cargo proteins (Knodler et al. 2010). Two 8-subunit protein complexes, the exocyst and the BBSome, are required for post-Golgi vesicular membrane trafficking and for ciliary entry and exit of signaling proteins, respectively. Within cilia, trafficking of the BBSome and its cargo is mediated by IFT.

Evidence suggests that multi-protein complexes involving the Rab and Arf GTPases, the exocyst, and the BBSome mediate post-Golgi intracellular and ciliary trafficking of polycystins (Fig. 11.2). In photoreceptors, a protein complex consisting of Arf4 and the ArfGAP with SH3 domain, ankyrin repeat, and PH domain 1 GTPase-activating protein (ASAP1) traffics membrane proteins from the Golgi to the base of the modified cilium (Mazelova et al. 2009). Hypothesizing a role for such a complex in trafficking polycystins, Ward et al. (2011) found that via its KVHPSST ciliary targeting sequence, PC1 interacts with Arf4 and ASAP1 in a complex with Rab6 and Rab11 in the Golgi membranes of primary renal cortical epithelial (RCTE) cells (Ward et al. 2011). The RVxP ciliary targeting sequence of PC2 also allowed binding of PC2 to Arf4, which is localized in the trans-Golgi, suggesting that PC2 also passes through the Golgi network. In another study, a yeast two-hybrid screen identified RAB GTPase Binding Effector Protein 1 (Rabep1) to interact with PC1 in a region containing the coiled-coiled domain within the C-terminal tail (Kim et al. 2014). This interaction was confirmed in primary collecting duct (CD) cells of mouse postnatal kidneys and also occurred in *Pkd2−/−-* MEF. Since this study showed that PC1 exits the ER in a complex with PC2, the presence of PC1-Rabep1 complex in *Pkd2−/−* MEF indicates this interaction occurs in a pre-Golgi compartment. Subsequently, Rabep1 was found to bind to Golgi-localized, gamma adaptin ear-containing, ARF-binding protein 1 (GGA1) in CD cells, an interaction that had been shown previously in trans-Golgi vesicles (Mattera et al. 2003). GGA1 was subsequently found to bind to Arf-like 3 (Arl3), which facilitates ciliary localization of cargo proteins (Wright et al. 2011). Taken together, binding of Rabep1 to PC1 in a pre-Golgi compartment links the PC1/PC2 complex to GGA1 at the trans-Golgi, where GGA1 then interacts with Arl3. In contrast to findings of Ward et al. (2011), PC1 did not co-immunoprecipitate with Arf4 in CD cells (Kim et al. 2014), which may point to differences between cell types and requires further investigation.

In contrast to these two studies, glycosylation analysis of PC2 in whole cell lysates of porcine renal proximal tubular epithelial LLC-PK_1 cells, together with live-imaging and electron microscopy, suggested that PC2 destined for primary cilia is trafficked to the cis-Golgi and, subsequently to the cilium, bypassing the mid- and trans-Golgi membranes (Hoffmeister et al. 2011). Supporting this, ciliary protein fractions of RCTE cells contained endoH-resistant PC1 that co-precipitated with endoH-sensitive PC2 (Gainullin et al. 2015). These data imply that PC2 vesicular trafficking is independent of that of PC1. Thus, two possible routes for trafficking of PC2 to the primary cilium are suggested by these studies.

PC1 also interacted with Rab8 (Ward et al. 2011), which is directed to the ciliary base by Abelson Helper Integration Site 1 (AHI1) and targets cargo to the primary cilium. Expression of a dominant-negative Rab8 prevented ciliary localization of PC1 (Ward et al. 2011) and also of PC2 (Hoffmeister et al. 2011). Rab8 is activated by Rab11, which is stimulated by Class II phosphoinositide 3-kinase-C2α (PI3K-C2α), which localizes to pericentriolar recycling endosomes (PRE) near the ciliary base (Franco et al. 2014). shRNA knockdown of PI3K-C2α in IMCD3 cells shortened cilia and resulted in a marked reduction of cells with PC2 ciliary localization, which was rescued upon expression of a constitutively active Rab8 (Franco et al. 2016). *Pi3k-C2α+/−* mice were sensitized to cyst formation on a

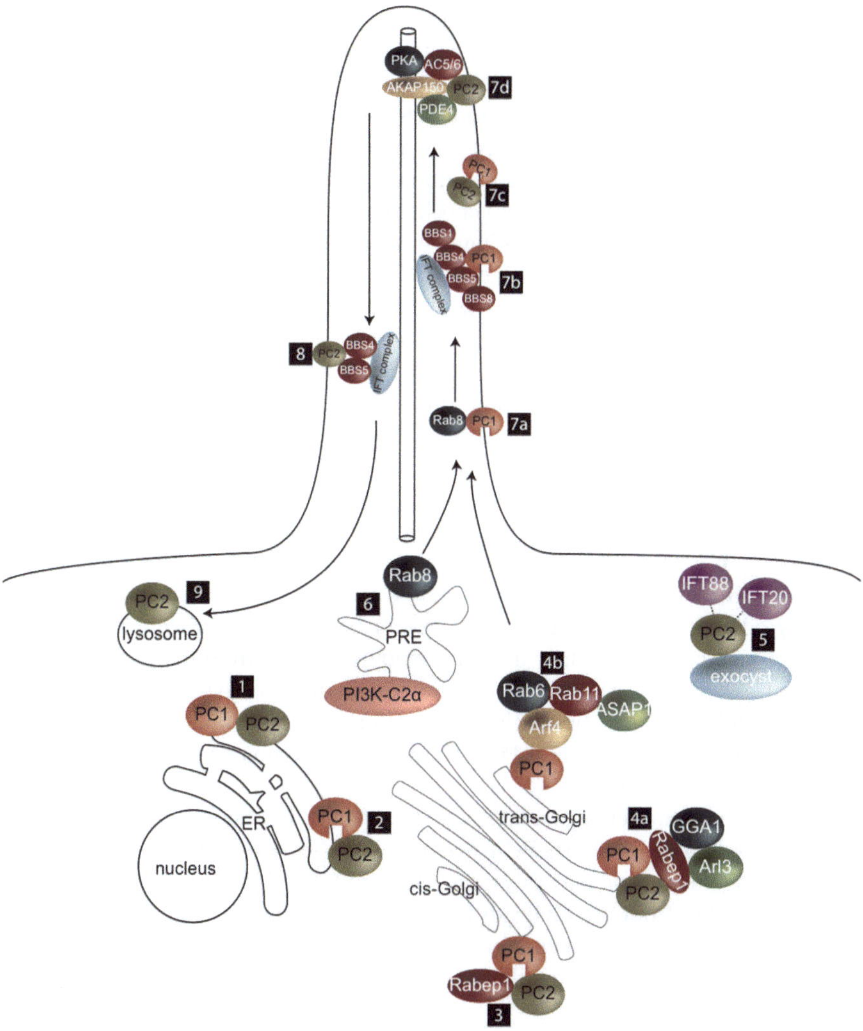

Fig. 11.2 Intracellular trafficking of polycystins to primary cilium. Intracellular and ciliary trafficking of polycystins involves the following steps: (1) PC1 and PC2 interact in the ER (Kim et al. 2014; Gainullin et al. 2015); (2) PC1 is cleaved (Kim et al. 2014; Gainullin et al. 2015); (3) PC1/PC2 complex interacts with Rabep1 in pre-Golgi complex (Kim et al. 2014); (4a) Rabep1 bridges PC1/PC2 complex to GGA1, which binds to Arl3 at the trans-Golgi (Kim et al. 2014); (4b) PC1 in a complex with Arf4, ASAP1, Rab6, and Rab11 in Golgi (Ward et al. 2011); (5) PC2 interacts with exocyst, in a complex with IFT88 and IFT20, which facilitates ciliary entry of PC2 (Fogelgren et al. 2011); (6) PI3K-C2α in the pericentriolar recycling endosome (PRE) activates Rab8 (Franco et al. 2014), which interacts with PC1 and also PC2 and (7a) facilitates ciliary entry of PC1 and also of PC2 (Ward et al. 2011; Hoffmeister et al. 2011; Franco et al. 2016); (7b) Binding of PC1 to BBS1 and the BBSome facilitates ciliary entry of PC1 (Su et al. 2014); (7c) PC1 and PC2 complex in the cilium (Nauli et al. 2003); (7d) PC2 is in a ciliary complex with AKAP150, AC5/6, PDE4, and PKA (Choi et al. 2011); (8) BBS4 and BBS5 mediate ciliary removal of PC2 (Xu et al. 2015); (9) PC2 is degraded in the lysosome (Hu et al. 2007; Xu et al. 2015)

Pkd1+/− or *Pkd2*+/− genetic background or when subjected to ischemia/reperfusion-induced renal injury, substantiating the importance of PC ciliary localization for PC function in vivo.

Both the exocyst and BBSome appear to mediate ciliary entry of polycystins. Deficiency of Sec10, an exocyst subunit, which localizes to the cilium and controls ciliogenesis (Zuo et al. 2009), resulted in undetectable ciliary localization of PC2 (Fogelgren et al. 2011). Morpholino knockdown of Sec10 in zebrafish phenocopied the PC2 morpholino mutant, resulting in a tail curled upwards, left–right laterality defects, and glomerular abnormalities. Sec10 was also shown to co-precipitate with PC2, IFT88, and IFT52, suggesting that the exocyst together in a complex with IFT proteins mediates ciliary entry of PC2. Using a yeast two-hybrid screen and GST pull-down experiments, BBS1, BBS4, BBS5, and BBS8 were identified as interacting with the PC1 C-terminus, although only loss of BBS1 and expression of a dominant-negative form of BBS3/Arl6 in IMCD cells shortened cilia and prevented PC1 ciliary localization (Su et al. 2014). Since Rab11 interacts with the exocyst and also promotes association of BBS1 with Rabin8 (Knodler et al. 2010), PC-BBSome and PC-exocyst complexes may form at the base of the cilium in connection with Rab11. The BBSome is also implicated in ciliary removal of PC2. In RPE cells deficient for both BBS4 and BBS5, PC2 accumulated in the primary cilium, although overall PC2 protein levels were unaffected, suggesting that BBS4 and BBS5 together mediate removal of PC2 from the primary cilium (Xu et al. 2015). PC2 further undergoes ubiquitination and lysosomal degradation (Hu et al. 2007). Treatment of RPE cells with a lysosomal inhibitor, chloroquine, increased ciliary localization of PC2, similar to the combined deficiency of BBS4 and BBS5 (Xu et al. 2015).

Most of these trafficking studies examined PC1 or PC2 alone. Further examination of whether PC-Rab/Arf, PC-BBSome, or PC-exocyst complexes consist of PC1/PC2 together would help ascertain whether PC1 and PC2 traffic independently or as a complex from the Golgi to the primary cilium. If the polycystins traffic independently, such experiments would further determine at what point of trafficking, e.g., prior to or following ciliary entry, the PC1/PC2 complex is formed.

11.4 Primary Cilium is a Mechanosensory Organelle

After more than 100 years, Zimmermann's hypothesis that primary cilia lining the renal tubules act as flow sensors (Zimmermann 1898) was demonstrated. Fluid flow causes primary cilia of cultured renal epithelial cells to deflect and intracellular Ca^{2+} levels to increase via a mechanism that requires extracellular Ca^{2+} influx and ensuing release of intracellular Ca^{2+} stores (Praetorius and Spring 2001). In cells deciliated by chloral hydrate, Ca^{2+} response was abrogated, demonstrating that primary cilia are essential in flow-mediated elevation of cystosolic Ca^{2+} (Praetorius and Spring 2003). A role for cilia was further made evident by the absence of a luminal flow-induced Ca^{2+} response in microperfused collecting ducts of 2-week-old *orpk* mutant mice, which harbor an *IFT88* hypomorphic mutation causing shortened cilia (Liu et al. 2005b). This abrogated response occurred in tubular segments that were not yet dilated, suggesting defective mechanotransduction might mark tubules sensitized to cystogenesis. In addition, magnitude of the Ca^{2+} response was observed to be twofold higher at P14 than at P7 in wild-type mice, an increase that was not observed in *orpk* mice. This suggests that the cilia-mediated, flow-induced intracellular Ca^{2+} response matures at P14. Interestingly, this time point coincides with the developmental window that influences renal cystic disease severity (Piontek et al. 2007).

Absence of PC1 or PC2 function in renal cells also obliterated flow-mediated intracellular Ca^{2+} increase (Nauli et al. 2003). This led to the proposal that ciliary PC1 senses flow and activates the Ca^{2+} channel activity of ciliary PC2, causing Ca^{2+} influx into the cilium, which, in turn, results in increased intracellular Ca^{2+}. FPC, TRPV4, and NEK8 also complex with PC2 at the primary cilium, and cells deficient in these proteins have shown impaired mechanotransduction (Kottgen et al. 2008; Manning et al. 2013; Wang et al. 2007). Yet, the variable development of renal tubular cysts among various *Pkhd1* mutant mouse models, with some displaying absence of cysts (Gallagher et al. 2008; Moser et al. 2005) and others presence of cysts (Garcia-Gonzalez et al. 2007; Woollard et al. 2007), the presence of glomerular but not tubular cysts in *Nek8*-null mice, which die at birth (Manning et al. 2013), contrasting with the *Nek*$^{jck/jck}$ mutant adults, which show a phenotype resembling human ADPKD (Smith et al. 2006), along with the lack of renal cyst formation in *Trpv4*-deficient mice and zebrafish, has questioned the contribution of impaired mechanotransduction to renal cystogenesis (Kottgen et al. 2008). In *Pkhd1*-mutant PCK rats, which model ARPKD and develop renal cysts, TRPV4 mechanosensory function was reduced in cyst-lining epithelial cells and TRPV4 activation attenuated the renal cystic disease (Zaika et al. 2013), demonstrating that impaired fluid flow-induced intracellular Ca^{2+} generation can modulate renal cystic disease, even if it may not be sufficient to initiate renal cystogenesis. Since mutations in *PKHD1* cause ARPKD in humans and rats, but show variable effects in mice, these varying influences of FPC on renal cystogenesis across different species may reflect differential regulation of fluid flow-induced Ca^{2+} responses in mouse, rat, and human kidneys. Likewise, the requirement of TRPV4 in fluid flow-induced Ca^{2+} generation may be cell line or tubule segment specific as illustrated by the lack of involvement of TRPV4 in fluid flow-mediated, PC2- and calcium-dependent induction of *endothelin-1* expression in IMCD cells (Pandit et al. 2015).

Dysfunctional flow sensing or mechanotransduction may contribute to renal cystogenesis via aberrant calcium-mediated or calcium-independent mechanisms. Collecting duct cells derived from the *orpk* mouse mutant exhibit blunted primary cilia, increased apical calcium entry in response to fluid flow (Siroky et al. 2006) and to epidermal growth factor (EGF) treatment, and an augmented PC2- and TRPV4-dependent proliferative response to EGF (Zhang et al. 2013). PC1 undergoes mechano-regulated cleavages within its C-terminal tail (CTT), generating fragments (CTT, P30, P15) that translocate to the nucleus, interact with various transcription transactivators (CHOP, TCF, STAT3, STAT6), and modulate proliferative, apoptotic, or cytokine-stimulated pathways. Cleavage of the CTT is stimulated by loss of mechanostimulation, as in MDCK cultures under static (no-flow) conditions (Low et al. 2006) or in wild-type mouse kidneys subjected to unilateral obstruction of the ureter, and by loss of mechanosensory function, as in *Kif3a*-null kidneys which lack primary cilia (Chauvet et al. 2004). PC2 influences cleavage and subsequent stabilization of PC1 CTT via a calcium-independent mechanism. Membrane-localized (non-cleaved) CTT binds and sequesters STAT6 and its coactivator P100 to prevent their nuclear activity (Low et al. 2006), or maintains canonical (JAK-activated, SOCS inhibitable) signaling via STAT1 and STAT3, thought to modulate cytokine-mediated STAT signaling (Talbot et al. 2011, 2014). Cleavage and nuclear translocation of CTT can be anti-cystogenic, by binding to CHOP, TCF, or β-catenin to prevent interaction with p300, thus inhibiting apoptosis or proliferative Wnt-signaling (Lal et al. 2008; Merrick et al. 2012), or can exacerbate cystic disease via Src-mediated or augmented cytokine-mediated STAT3 activation (Talbot et al. 2011, 2014). Observations of CTT fragments and nuclear staining with antibodies directed to the C-terminus of PC1 in human ADPKD kidneys (Talbot et al. 2011), along with the abilities of leflunomide and pyrimethamine, inhibitors of STAT6 and STAT3, respectively, to ameliorate cystic disease in PKD mouse models (Olsan et al. 2011; Takakura et al. 2011), support the relevance of these CTT-mediated functions in PKD.

For more than a decade, Ca^{2+} influx into the cilium could only be speculated until strategies to visualize Ca^{2+} in the primary cilium were pioneered. Genetically encoded Ca^{2+} indicators were fused to a ciliary targeting sequence of ciliary localized proteins (Delling et al. 2013; Jin et al. 2014a; Su et al. 2013). Jin et al. (2014a) showed that upon mechanical bending of a primary cilium, Ca^{2+} increased in the cilium, which was followed by increased Ca^{2+} in the cytosol, at last providing experimental evidence that primary cilia convert fluid flow shear stress into ciliary Ca^{2+} signals (Fig. 11.3a). Additionally, *Pkd2* siRNA knockdown impeded the increased ciliary Ca^{2+} response, supporting the role of PC2 in mediating flow-induced Ca^{2+} response. This sequence of events has been supported in vivo in the nonmotile cilia of zebrafish Kupffer's vesicle, which is orthologous to the mammalian node. In the node, leftward fluid flow generated by nodal cilia elevates cytosolic calcium in cells on the left side of the node, triggering asymmetric gene expression which is required for left–right asymmetry. In response to leftward fluid flow, Ca^{2+} was shown to increase in nonmotile cilia and, subsequently, in the cytoplasm, supporting that primary cilia transduce mechanical forces into ciliary Ca $^{2+}$ (Yuan et al. 2015). *pkd2* and its putative partner *pkd1-like1* were identified as the molecules required for mechanotransduction in Kupffer's vesicle.

However, by patch clamping the cilium, DeCaen et al. (2013) showed that ciliary ion currents were generated by the gene products of *Pkd1L1* and *Pkd2L1*, while PC1 and PC2 activities were undetectable (DeCaen et al. 2013). In addition, laser disruption of the ciliary tip increased ciliary Ca^{2+}, but caused only small increases in Ca^{2+} at the ciliary base, which did not significantly alter cytosolic Ca^{2+} levels (Delling et al. 2013) (Fig. 11.3b). By developing a transgenic mouse expressing a

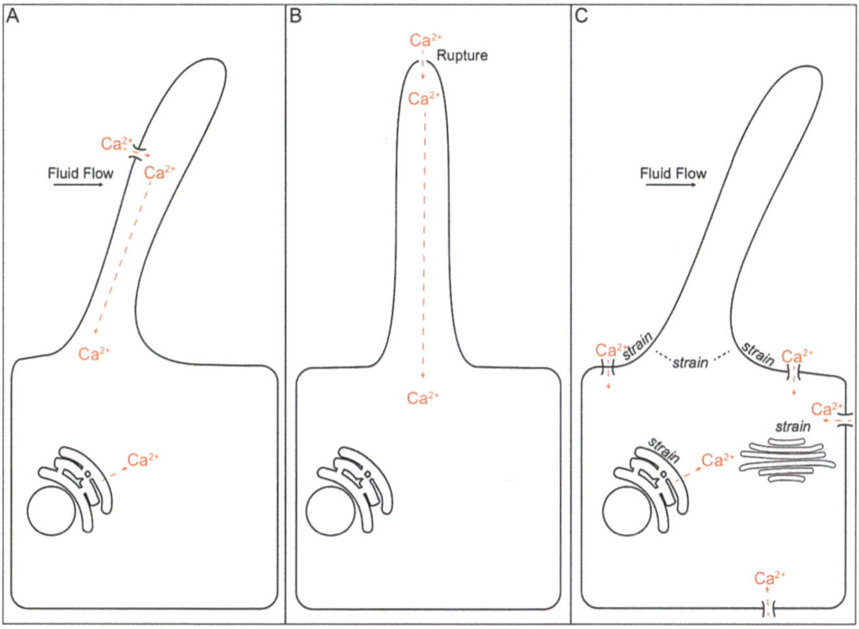

Fig. 11.3 Models of fluid flow shear stress-induced intracellular Ca^{2+} generation. (**a**) Fluid flow induces ciliary Ca^{2+} influx, which releases intracellular Ca^{2+} stores (Jin et al. 2014a; Yuan et al. 2015). (**b**) Ciliary Ca^{2+} influx induced by ciliary membrane rupture increases Ca^{2+} at ciliary base but does not release intracellular Ca^{2+} stores (Delling et al. 2013). (**c**) Bending of primary cilium induced by fluid flow causes strain at base of cilium, which induces Ca^{2+} influx via apical and basolateral membranes and, subsequently, release of intracellular Ca^{+2} stores (Liu et al. 2003; Rydholm et al. 2010; Khayyeri et al. 2015)

ratiometric genetically encoded Ca^{2+} indicator in primary cilia and using swept-field confocal microscopy capturing up to 500–1000 frames/s to image cells and tissues derived from this mouse, fluid flow was not observed to induce Ca^{2+} changes in primary cilia of cultured kidney epithelial cells nor in the renal thick ascending limb of loop of Henle, in crown cells of the embryonic node, or in kinocilia of inner ear hair cells (Delling et al. 2016). In incidences when ciliary Ca^{2+} increased, Ca^{2+} was observed first to increase in the cytoplasm and subsequently to diffuse into the cilium. At supraphysiological flow velocities, ciliary tips were observed to rupture, which was followed by extracellular Ca^{2+} entry. Collectively, these results challenge the notions that primary cilia and the polycystins transduce mechanical strain into ciliary Ca^{2+} and, moreover, that ciliary Ca^{2+} regulates intracellular Ca^{2+}.

Mathematical modeling of ciliary deflection and cellular response has provided additional perspectives (Fig. 11.3c). By integrating data from measurements of shear stress-induced intracellular Ca^{2+} generation in microperfused and split-open rabbit cortical collecting ducts with theoretical modeling, Liu et al. (2003) proposed that activation of increased intracellular Ca^{2+} initiates at the base of the cilium where the cilium transfers mechanical forces to the cytoskeleton (Liu et al. 2003).

Consistent with the ciliary base having an important role in mechanotransduction, another mathematical model suggested that ciliary deflection causes strain to build up at the base of the primary cilium and in the surrounding cell membrane (Rydholm et al. 2010). This strain buildup occurred approximately 30 s following induction of fluid flow coinciding with the 30 s time delay observed to generate a flow-induced intracellular Ca^{2+} response in MDCK cells (Praetorius and Spring 2001). In addition to strain accumulating at the ciliary base, results of another computational study showed that deflection of primary cilia causes strain to accumulate also at the nucleus and Golgi apparatus (Khayyeri et al. 2015). Thus, analysis of transcriptional or posttranslational changes upon fluid flow may merit investigation.

Taken together, the data suggest that the intermediary steps between ciliary deflection and increased intracellular Ca^{2+} require further study. Differences in cell lines, imaging techniques, use of cell monolayers versus perfused or split-open tubules, and species may contribute to the discrepancies. The debate on the role of the primary cilium as a mechanotransducer of ciliary Ca^{2+} continues (Tran and Lechtreck 2016). Super-resolution imaging and additional in vivo studies may be important in defining the mechanosensory role of primary cilia.

11.5 Primary Cilium is a Signaling Organelle

Multiple signaling pathways are disrupted in ADPKD (reviewed in Song et al. 2009; Torres and Harris 2006), and accumulating evidence suggests that a number of these are mediated by primary cilia. We discuss Ca^{2+}/cAMP, LKB1/AMPK, and Wnt signaling, which play important roles in PKD. In addition, we review Hedgehog (Hh) signaling, whose regulation at the primary cilium is the most characterized, and thus, ciliary mechanisms learned from this pathway may enable ideas that can be applied to other signaling molecules relevant to PKD.

11.5.1 Ca^{2+}/cAMP

Ca^{2+} levels are fivefold higher in the cilium than in the cytosol demonstrating that the cilium is a separate Ca^{2+} compartment (Delling et al. 2013). Additionally, studies by Choi et al. (2011) indicate that PC2 is integral to the formation of a complex that regulates cAMP and Ca^{2+} at the primary cilium. The PC2 C-terminus interacts with A-kinase anchoring protein 150 (AKAP150), which creates a scaffold for binding of protein kinase A (PKA); adenylyl cyclases 5 and 6 (AC5/6), which synthesize cAMP; and phosphodiesterase 4C (PDE4C), which catabolizes cAMP (Choi et al. 2011). Loss of PC2 in *Pkd2−/−* derived renal epithelial cells resulted in the absence of ciliary AC5/6 and PDE4C and elevated intracellular cAMP levels. Expression of wild-type PC2, but not of a PC2 mutant lacking Ca^{2+} channel activity, decreased cAMP levels, suggesting that the PC2 Ca^{2+} channel activity specifically regulates cAMP levels. Using proximity labeling, Mick et al. (2015) also detected the ciliary presence of PKA and AC5/6 in IMCD3 cells (Mick et al. 2015). Collectively, these data suggest that cilia are distinct Ca^{2+}/cAMP signaling organelles. The influence of ciliary Ca^{2+} in modulating ADPKD has been demonstrated genetically in mice (Jin et al. 2014b). The voltage-dependent L-type calcium channel, CaV1.2, localizes to primary cilia, but is absent from cilia in *Pkd1−/−* and *Pkd2−/−* derived renal epithelial cells. Deletion of *CaV1.2* in *Pkd1* heterozygous mice caused the growth of few but extremely large renal cysts, while cysts did not occur in single *Pkd1* heterozygotes.

Although mechanisms linking ciliary Ca^{2+}/cAMP to cytosolic pools require further investigation, targeting cAMP formation in preclinical models of ADPKD has proven extremely effective (Gattone et al. 2003; Torres et al. 2004; Wang et al. 2005, 2008). Intracellular cAMP formation can be attenuated by inhibiting arginine vasopressin receptor 2 (AVPR2), which stimulates AC6, or activating the somatostatin receptor, which inhibits AC6. Administration of Tolvaptan, an inhibitor of AVPR2, together with pasireotide, an activator of somatostatin receptors in the $Pkd1^{RC/RC}$ mouse, resulted in lower renal cystic and fibrotic volumes than treatment with either Tolvaptan or pasireotide alone, demonstrating that targeting both receptors simultaneously is more effective than targeting one alone (Hopp et al. 2015). Knockout of *Pde1a*, *Pde1c*, or *Pde3a*, which catabolize cAMP, in a $Pkd2^{-/WS25}$ mouse model exacerbated renal cystogenesis, suggesting that activating these enzymes may also be beneficial (Ye et al. 2015).

The Tolvaptan Efficacy and Safety in Management of Autosomal Dominant Polycystic Kidney Disease and its Outcomes (TEMPO) 3:4 study was a 3-year phase III clinical trial, in which early-stage ADPKD patients anticipated to have rapid renal cyst growth were randomly assigned to either a Tolvaptan or a placebo treatment group (Torres et al. 2011). ADPKD patients administered Tolvaptan showed reduced kidney volume and reduced symptoms of kidney malfunction, including flank pain, hypertension, and albuminuria, compared to patients administered a placebo (Torres et al. 2012). Tolvaptan has been approved for ADPKD patients in Japan, Canada, and Europe. In the USA, the Food and Drug Administration considered the side effects, such as polyuria, excessive thirst, and risk of liver injury, to outweigh the benefits. However, an open-label extension study (TEMPO 4:4) is ongoing (https://clinicaltrials.gov/ct2/show/record/NCT01214421).

11.5.2 LKB1/AMPK

The mammalian target of rapamycin (mTOR) pathway is aberrantly activated in cyst-lining epithelial cells of human ADPKD tissue (Shillingford et al. 2006), and administration of mTOR inhibitor, rapamycin, has attenuated renal cystogenesis in most PKD rodent models tested (Shillingford et al. 2006; Tao et al. 2005; Wahl et al. 2006; reviewed in Ibraghimov-Beskrovnaya and Natoli 2011). mTOR signaling is essential for cell metabolism and growth and is mediated by two multi-protein complexes, mTOR complex 1 (mTORC1) and mTORC2. mTORC1 phosphorylates and activates p706S kinase (S6K) and inhibits 4E-BP1 to stimulate protein synthesis, enabling cell growth. While nutrient and amino acid availability stimulates mTORC1, energy stress and the "energy sensor" molecule, AMP-activated protein kinase (AMPK) inhibits mTORC1 in a ciliary-dependent manner (Boehlke et al. 2010).

Upon fluid flow, the ciliary-localized liver kinase B1 (LKB1) tumor suppressor kinase phosphorylates AMPK, which accumulates at the ciliary base and suppresses mTORC1 and, in turn, cell growth (Boehlke et al. 2010). *Kif3a* cko mice were shown to have larger collecting duct cells, demonstrating the ciliary role in regulating mTOR signaling and cell growth. Further, MDCK cells deficient for Kif3a or Ift88 showed larger cell size only when subjected to fluid flow, demonstrating the importance of fluid shear stress as a stimulus. In LKB1-deficient cells, fluid flow also resulted in larger cells, and while P-AMPK levels of whole cell lysates were not altered, P-AMPK specifically at the basal body was reduced. *Pkd2* kd MDCK cells subjected to fluid flow did not show larger cell size nor elevated levels of P-S6K, suggesting that fluid flow stimulation of LKB1/AMPK signaling is independent of flow-mediated intracellular Ca^{2+} generation.

In addition to flow-mediated suppression of mTOR signaling, other regulators of mTOR signaling have been identified. Lack of ubiquitination of the hepatocyte growth factor (HGF) receptor, c-Met, also activates mTOR signaling in a PKD mouse model, and administration of a c-Met inhibitor attenuated renal cystogenesis in *Pkd1*-null mutant embryos (Qin et al. 2010). Additionally, treatment of MDCK cells with metformin, which activates AMPK, suppressed mTOR signaling and CFTR activity, and administration of metformin to *Pkd1* mouse mutants either prior to or during renal cyst formation attenuated renal cystic disease (Takiar et al. 2011).

Despite the success of mTor inhibitors in attenuating cystic disease in most preclinical models of PKD, clinical trials of sirolimus (rapamycin) and its derivative, everolimus, in ADPKD patients have not shown the same efficacy (Ruggenenti et al. 2016; Serra et al. 2010; Walz et al. 2010). In an 18-month study, no difference in total kidney volume was observed in patients treated with sirolimus compared to patients receiving standard care (Serra et al. 2010). In a 24-month study, ADPKD patients receiving everolimus showed a lower total kidney volume increase, but also a lower parenchymal volume increase, and kidney function was not better than those of patients receiving a placebo (Walz et al. 2010). In another study, ADPKD patients at stages 3b and 4 of chronic kidney disease received sirolimus for 1 year but showed greater decline in glomerular filtration rate (GFR), increased proteinuria, and higher total kidney volume than patients receiving conventional care (Ruggenenti et al. 2016). The study concluded that sirolimus was unsafe and ineffective in ADPKD patients. Another study is ongoing to determine if pulsed administration of sirolimus will be more effective (https://clinicaltrials.gov/ct2/show/NCT02055079). Rapamycin acts mostly on mTORC1, and feedback mechanisms may result in elevated mTOR signaling via mTORC2. A new mTOR inhibitor that inhibits both complexes has shown efficacy in a rat model of PKD (Ravichandran et al. 2015). Alternatively, targeting the pathway upstream of mTORC1/2 at the level of AMPK or cMet might also be more effective. Additionally, administration of lower doses of mTOR inhibitors as part of a combination therapy may also yield more beneficial results. Finally, targeting drug delivery to the kidney may be most effective at reducing toxicity and increasing therapeutic efficacy (Shillingford et al. 2012). Folate receptors are expressed at high levels in the kidney, and thus, folate receptor-mediated endocytosis may be exploited to target compounds to the kidney. Indeed, administration of folate-conjugated rapamycin (FC-rapamycin) to adult mice reduced mTOR signaling in the kidney and not in the spleen. FC-rapamycin also attenuated PKD in *bpk* (*BALB/c polycystic kidneys*) mice, mutant for *Bicaudal C Family RNA-binding protein 1* (*Bicc1*) (Cogswell et al. 2003), which regulates *Pkd2* via microRNA miR-17 (Tran et al. 2010).

11.5.3 Wnt

Wnt signaling is activated upon binding of Wnt ligand to a Frizzled (Fz) receptor. Downstream of this, Wnt signaling partitions into canonical and noncanonical pathways, determined by the presence of co-receptors. While canonical Wnt signaling regulates cell proliferation, differentiation, and cell fate, noncanonical signaling directs tissue organization and morphogenesis and causes transient increases in intracellular Ca^{2+}. For the purpose of this review, we will focus predominantly on the canonical pathway. For a review on noncanonical signaling and ADPKD, see Tran et al. (2014a).

In the canonical pathway, in the absence of ligand, a destruction complex comprised of scaffolding protein, Axin2, adenomatous polyposis coli (APC), casein kinase 1 (CK1), and glycogen synthase kinase 3β (GSK3β) phosphorylates and targets β-catenin for degradation. In the presence of ligand, binding of Wnt to Fz recruits the low-density lipoprotein (LDL)-related protein 6 (LRP6) co-receptor, cytoplasmic Dishevelled (Dvl), Axin2, and GSK3β concomitantly disassembling the destruction complex. Stabilized β-catenin translocates to the nucleus and dimerizes with the T-cell factor/lymphoid enhancer factor (TCF/LEF) transcription factors to activate target genes, such as *Lef1*, *Axin2*, and *c-myc*, promoting cell proliferation (reviewed in Oh and Katsanis 2013).

Two studies show that fluid flow can regulate Wnt signaling. Fluid flow across IMCD cells increased the expression of *inversin*, which is mutated in NPHP2, and downregulated β-catenin (Simons et al. 2005). While experiments in *Xenopus laevis* embryos and mammalian renal epithelial cells suggested that *inversin* antagonizes canonical Wnt signaling while promoting the noncanonical Wnt/Planar Cell Polarity pathway (Simons et al. 2005), kidneys of *inversin*-null mouse mutants showed normal regulation of canonical Wnt signaling (Sugiyama et al. 2011), suggesting regulation of the balance between mammalian canonical and noncanonical Wnt branches may be more intricate in vivo. Fluid flow in the mouse embryonic node has also been shown to regulate Wnt signaling via feedback loops with *Cerl2* (Nakamura et al. 2012). An early event in establishing left–right asymmetry is the decay of *Cerl2* mRNA on the left side of crown cells in the mouse node induced by fluid flow. Wnt3 promotes *Wnt3* expression and further *Cerl2* mRNA degradation, and conversely, Cerl2 promotes Wnt3 degradation. These data suggest that flow-mediated decay of *Cerl2* enables asymmetric Wnt signaling around the node. This contrasts with the effect of fluid flow over IMCD cells, which leads to downregulated β-catenin.

Pathway components, such as β-catenin, APC, and Dishevelled 3, localize to the ciliary axoneme, while P-β-catenin, which is targeted for degradation, is present at the basal body (Corbit et al. 2008; Mick et al. 2015). In *Kif3a−/−* embryos and in Wnt3a ligand-stimulated cells derived from *Kif3a−/−*, *IFT88*$^{orpk/orpk}$, and *Ofd1−/−* mice harboring the Batgal Wnt reporter, reporter activity increased, suggesting that cilia act as a "brake" in canonical Wnt signal transduction (Corbit et al. 2008). In vitro studies suggest that Jouberin (Jbn), encoded by *AHI1*, which is mutated in Joubert Syndrome, facilitates nuclear translocation of β-catenin and that the primary cilium acts to sequester Jouberin and β-catenin away from the nucleus (Lancaster et al. 2011).

Yet, the functionality of primary cilia in mediating Wnt signaling has been unclear. IFT mutant embryos show Hedgehog (Hh) mutant phenotypes and not phenotypes characteristic of misregulated Wnt signaling (reviewed in Eggenschwiler and Anderson 2007). Additionally, although *Kif3a−/−; Batgal* and *Ift88−/−;Batgal* embryos showed increased Wnt reporter activity (Corbit et al. 2008), *Thm1*$^{aln/aln}$; *Batgal* and *Ahi−/−;Batgal* embryos, which have shortened and absent cilia, respectively, showed normal reporter activity (Lancaster et al. 2011; Stottmann et al. 2009). Interestingly, *Dnchc2−/−;Batgal* embryos showed decreased reporter activity in the kidney but increased activity in the midbrain (Lancaster et al. 2011). This varying effect of *Dnchc2* deficiency was demonstrated in vitro. Moderate inhibition of *Dnchc2*, which did not affect cilia structure, dampened Wnt response, while greater inhibition of *Dnchc2*, which shortened cilia, increased Wnt response. Thus, ciliary regulation of Wnt signaling is context dependent; some organs, such as the brain, are more sensitive to dosage of a particular ciliary gene. These data also suggest that deficiency of different ciliary genes differentially influences Wnt signaling output.

The correct balance of Wnt signaling is critical for maintaining renal tubular integrity, and both inappropriate activation and inhibition of the pathway can lead to renal cysts. Renal-specific overexpression of active β-catenin or loss of APC, which negatively regulates β-catenin, caused renal cysts in mice (Qian et al. 2005; Saadi-Kheddouci et al. 2001). Additionally, enhancing canonical Wnt signaling by ablating *Aquaporin 1* (*Aqp1*) in a mouse model of PKD exacerbated renal cystogenesis (Wang et al. 2015). AQP1 interacts with β-catenin, GSK3β, LRP6, and Axin1, and deficiency of AQP1 increased the levels of stabilized β-catenin. Finally, in an *Mks1* knockout mouse model of Meckel Syndrome with shortened cilia, canonical Wnt signaling was upregulated prior to cysts forming in embryonic mutant kidneys (Wheway et al. 2013). Conversely, in *Ahi−/−* mice, which model NPHP, canonical Wnt signaling was reduced at 5 months of age, and renal microcysts and tubular dilations ensued at 12 months of age (Lancaster et al. 2009).

However, misregulated Wnt signaling does not precede cystogenesis in all mouse models of cystic kidney disease. In mice with renal-specific deletion of *Kif3a*, *Ift20*, *Ift140*, or *Tmem67*, upregulation of canonical Wnt signaling was detected after renal cysts were formed, suggesting that other mechanisms underlie initiation of renal cystogenesis (Jonassen et al. 2008, 2012; Leightner et al. 2013; Lin et al. 2003). A direct role for loss of polycystins on Wnt signaling in vivo is also controversial. *Pkd1−/−; TCF-lacZ* embryonic kidneys and *Pkd2/WS25; TCF-lacZ* adult kidneys showed normal reporter activity in cyst-lining epithelia and normal levels of β-catenin in whole kidney lysates, even after cysts had formed (Miller et al. 2011). In direct contrast, canonical Wnt signaling was upregulated in cyst-lining epithelial cells of E17.5 *Pkd1−/−*; *TCF-lacZ* kidneys, and levels of total and active β-catenin of whole kidney lysates were also elevated (Qin et al. 2012). Varying expression levels and patterns by different Wnt reporter lines and the absence of Wnt reporter activity in sites of known canonical Wnt signaling have been reported. Different background strains may also possibly account for the discrepancies observed in reporter activity as well as β-catenin levels.

Recently, PC1 has been shown to bind various Wnt ligands. Binding of WNT9B resulted in increased intracellular Ca^{2+} via Ca^{2+} influx, and WNT9B and WNT3A induced whole-cell currents in WT MEFs but not in *Pkd2−/−* MEFs (Kim et al. 2016b). Combined morpholino knockdowns of dvl2 and pkd1 or of wnt9a and pkd1 in *Xenopus* showed synergistic effects on inducing a pronephric cystic phenotype, indicating these molecules function in the same pathway. These data are the first to identify ligands of PC1 and indicate that misregulated Wnt/Ca^{2+} signaling is causative in ADPKD.

11.5.4 Hedgehog

The first signaling pathway discovered to occur at the primary cilium was the mammalian Hh pathway. The first major discovery was made when ethylnitrosourea (ENU)-mutagenized mouse embryos with Hh signaling phenotypes, such as neural tube patterning defects and polydactyly, were found to have mutations in *Ift* genes (Huangfu et al. 2003). Hh signaling is essential for patterning and development of most vertebrate organs and is also critical for tissue homeostasis (reviewed in Ingham et al. 2011). Hh ligands initiate signaling by binding the transmembrane receptor, Patched (PTCH1), at the cilium. Once bound, PTCH1 diminishes from the cilium, and concomitantly, the transmembrane signal transducer, Smoothened (SMO), is enriched in the cilium and activated (Corbit et al. 2005; Rohatgi et al. 2007). A series of dephosphorylation and phosphorylation events activate full-length Glioblastoma transcription factors, GLI2 and GLI3 (Niewiadomski et al. 2014), which accumulate at the ciliary distal tip (Haycraft et al. 2005). In the absence of ligand, full-length GLI3 protein is processed into GLI3 repressor, which also requires the primary cilium. GLI transcription factors translocate to the nucleus, and the balance of GLI activators (GLIA) to GLI3 repressor determines the level of Hh signaling output (Christensen and Ott 2007; Eggenschwiler and Anderson 2007).

IFT, the BBSome, ciliary Ca^{2+}, and the phospholipid content of the ciliary membrane tune this developmental pathway by ultimately regulating ciliary localization of Hh signaling components. In mice, loss of all IFT proteins misregulates Hh signaling (Nozawa et al. 2013). Loss of most complex B proteins, which mediate anterograde IFT, causes the absence of cilia and inability to transduce the Hh signal (Huangfu and Anderson 2005; Huangfu et al. 2003; Liu et al. 2005a). In contrast, loss of IFT complex A proteins, THM1 and IFT122, which largely mediate retrograde IFT, sequesters proteins in bulb-like structures at the distal tip of shortened cilia and causes inappropriate activation of the Hh pathway due to enhanced GLI2A and GLI3A activities (Qin et al. 2011; Tran et al. 2008). In addition to cilia compartmentalizing the Hh signaling cascade, IFT is integral to Hh transduction by trafficking signaling components within cilia. Unlike most complex B proteins, IFT25 and IFT27 are not essential for ciliogenesis but form a subcomplex that is critical for ciliary import of GLI2 and ciliary export of the BBSome and its cargoes, PTCH1, SMO, and Gpr161, a negative regulator of the pathway (Eguether et al. 2014; Keady et al. 2012; Liew et al. 2014). As a result, loss of IFT25 or IFT27 does not alter cilia structure but disrupts activation of the pathway. Ca^{2+} modulates IFT (Collingridge et al. 2013), and treatment of MEF with Smoothened Agonist (SAG) resulted in elevated ciliary Ca^{2+} after 24 h (Delling et al. 2013). Delling et al. (2013) propose that rather than SAG directly activating Ca^{2+} channels, SAG recruits PC2-L1 to primary cilia to "tune" Ca^{2+} levels for proper function of IFT proteins, such as IFT25, which has a Ca^{2+} binding site. Delling et al. (2013) further propose that other Hh signaling molecules be examined for Ca^{2+} dependence. More recently, the phospholipid content of the ciliary membrane has also been shown to be a critical regulator of the pathway. Phophoinositide phosphatase, *INPP5E*, in which mutations cause ciliopathies (Jacoby et al. 2009), creates a distinct phosphoinositide distribution in the ciliary membrane, whereby phosphoinositide 4-phosphate (PI(4)P) is present in the membrane that ensheathes the axoneme and phosphoinositide 4,5-bisphosphate (PI(4,5) P2) comprises the membrane at the ciliary base (Chavez et al. 2015; Garcia-Gonzalo et al. 2015). Loss of INPP5E caused mislocalization of PI(4,5)P2 to the ciliary membrane enclosing the axoneme, which increased ciliary localization of Hh negative modulators, PI(4,5)P2-binding Tulp3, Gpr161, and IFT140, resulting in inactivation of the pathway.

Like the Wnt pathway, regulation of Hh signaling has been studied mostly in cells that do not undergo shear stress. However, Hh signaling also has a role in chondrocytes, which do undergo mechanical strain. In embryonic chondrocytes, Hh signaling is essential for proliferation and differentiation in the growth plate, while in adult chondrocytes, aberrant activation of Hh signaling leads to degeneration of chondrocytes in osteoarthritis. Chondrocytes subjected to cyclic tensile strain

(CTS) showed increased *Ihh* transcription and increased Hh activity as assessed by levels of *Gli1* transcripts (Thompson et al. 2014). In *Ift88*$^{orpk/orpk}$ cells subjected to CTS, *Ihh* transcription increased but *Gli1* transcription did not, demonstrating that the primary cilium is not required for *Ihh* transcription but for transduction of the pathway. Thus, in chondrocytes, Hh signaling is regulated in a mechanosensitive manner (Shao et al. 2012; Thompson et al. 2014).

Evidence that Hh signaling is increased in renal cystogenesis is emerging. In a genome-wide transcriptome analysis of human ADPKD renal cystic tissue, expression of Hh signaling components, including *Gli2*, was upregulated (Song et al. 2009). Additionally, several mouse models of renal cystic disease show increased Hh signaling. Mutation of the *Glis2* transcription factor, a member of the Kruppel-like C2H2 zinc finger protein subfamily, including the GLI proteins, causes NPHP in humans and mice, and transcriptome analysis of *Glis2−/−* kidneys revealed an upregulation of *Gli1*, a direct target of the Hh pathway (Attanasio et al. 2007). In vitro, Glis2 repressed the Hh pathway and *Glis2* knockdown resulted in transformation of renal epithelial cells to a fibroblast appearance, suggesting that reduced levels of Hh signaling may maintain renal tubular epithelial cells in a differentiated state (Li et al. 2011). Conditional deletion of IFT140, a complex A protein, resulted in cystic kidney disease and increased expression of *Gli* transcripts in cystic kidneys, suggesting increased Hh signaling (Jonassen et al. 2012). Similarly, perinatal deletion of *Thm1*, which also encodes a complex A protein (Tran et al. 2008) and is mutated in patients with ciliopathies (Davis et al. 2011), also caused cystic kidney disease in mice and increased expression of *Gli* transcripts in cystic kidneys (Tran et al. 2014b). Further, *Gli* transcripts were also increased in *jck* and *Pkd1* cko cystic kidneys.

In a study examining the effects of corticosteroid overexposure on kidney development, addition of Hh inhibitor, cyclopamine, reduced hydrocortisone-induced cysts without affecting organ growth, implicating increased Hh signaling in this mechanism of cystogenesis (Chan et al. 2010). Hh inhibitors also prevented cyst formation in cultured embryonic kidneys of *Thm1*, *jck*, and *Pkd1* mutant mice (Tran et al. 2014b). Downregulating Hh signaling in orthologous PKD mouse models in vivo will help determine a functional role for increased Hh signaling in PKD-mediated renal cystogenesis.

11.6 Perspectives

While aberrant regulation of multiple signaling pathways has been implicated in ADPKD, this complexity is not unique to ADPKD but occurs also in cancer and correlates with more advanced disease. This emphasizes the need to understand the molecular mechanisms of early disease and, ultimately, those that initiate disease processes. In ADPKD-derived, *PKD1*-mutant iPS cells, only the absence of ciliary PC1 and PC2, and not abnormalities in cell proliferation, apoptosis, or ciliogenesis, was observed, suggesting that ciliary mislocalization of proteins may be early events (Freedman et al. 2013). Thus, determining the proteins resident in cilia in healthy and disease states may help reveal initiating mechanisms. Advances in proteomics and imaging technologies may facilitate such investigations. Recently, proximity labeling of primary cilia using APEX technology was demonstrated to successfully uncover novel proteins in wild-type and *Ift27−/−* primary cilia (Mick et al. 2015). APEX technology provides remarkable sensitivity to allow temporal "snapshots" of both transient and stable residents of the dynamic ciliary proteome. Performing similar experiments in PC1/PC2-deficient cells, which also have primary cilia, seems feasible. Proximity labeling may also be applied to examining proteins at the ciliary base, the ER, and Golgi apparatus in response to fluid flow and in *PKD1/2* mutant cells to further analyze the cellular events important in PKD biology.

Super-resolution imaging has refined the localization of proteins at the ciliary base. Combining stimulated emission depletion (STED) super-resolution imaging with transmission electron microscopy has provided a nanometer-scale view of the spatial distribution of seven proteins at the transition zone and distal appendices in mammalian cells (Yang et al. 2015). This level of resolution can be applied to proteins uncovered using proteomics technology in the axoneme, ciliary base, ER, and Golgi apparatus. Live imaging of renal proximal tubules of Cilia^{GFP} mice that express ciliary-localized somatostatin receptor 3 (Sstr3)::GFP shows that fluid flow causes cilia to bend and lie almost parallel to the apical cell surface (O'Connor et al. 2013). These experiments can be extended to analyzing cilia behavior at various stages of renal cystogenesis and correlating with fluid flow-induced intracellular Ca $^{2+}$ generation. The achievement of imaging ciliary Ca^{2+} in vivo can also be applied to possibly other organelles and possibly other pathways.

Thus far, targeting cAMP levels by administration of Tolvaptan has proven most effective in treating patients with ADPKD. Still this therapy can be ameliorated to improve efficacy and reduce side effects. Targeting other pathways in combination with Tolvaptan and directing drug delivery to the kidney may potentiate therapeutic benefits. Alternatively, targeting the PC protein defect and ciliary mislocalization has been proposed as a more proximal therapy (Cai et al. 2014; Trudel et al. 2016). Approximately, 30% of mutations are missense mutations for which the protein defect may be potentially targeted. Since ciliary localization of Hh signaling components is regulated by phospholipid content of ciliary membrane, investigations into the role of the ciliary membrane lipid content on PC ciliary localization may also be warranted. Glycosphingolipid metabolism modulates ADPKD in mouse models (Natoli et al. 2010) and is involved in the formation of apical membrane and primary cilia of differentiated MDCK cells (Pescio et al. 2012). Thus, the future of ADPKD research promises exciting discoveries toward finding effective therapies and ultimately a cure.

References

Abdul-Majeed S, Nauli SM (2011) Dopamine receptor type 5 in the primary cilia has dual chemo- and mechano-sensory roles. Hypertension 58:325–331

Abdul-Majeed S, Moloney BC, Nauli SM (2012) Mechanisms regulating cilia growth and cilia function in endothelial cells. Cell Mol Life Sci 69:165–173

Afzelius BA (1976) A human syndrome caused by immotile cilia. Science 193:317–319

Ait-Lounis A, Baas D, Barras E, Benadiba C, Charollais A, Nlend Nlend R, Liegeois D, Meda P, Durand B, Reith W (2007) Novel function of the ciliogenic transcription factor RFX3 in development of the endocrine pancreas. Diabetes 56:950–959

Aldahmesh MA, Li Y, Alhashem A, Anazi S, Alkuraya H, Hashem M, Awaji AA, Sogaty S, Alkharashi A, Alzahrani S et al (2014) IFT27, encoding a small GTPase component of IFT particles, is mutated in a consanguineous family with Bardet–Biedl syndrome. Hum Mol Genet 23:3307–3315

Alten L, Schuster-Gossler K, Beckers A, Groos S, Ulmer B, Hegermann J, Ochs M, Gossler A (2012) Differential regulation of node formation, nodal ciliogenesis and cilia positioning by Noto and Foxj1. Development 139:1276–1284

Ansley SJ, Badano JL, Blacque OE, Hill J, Hoskins BE, Leitch CC, Kim JC, Ross AJ, Eichers ER, Teslovich TM et al (2003) Basal body dysfunction is a likely cause of pleiotropic Bardet–Biedl syndrome. Nature 425:628–633

Attanasio M, Uhlenhaut NH, Sousa VH, O'Toole JF, Otto E, Anlag K, Klugmann C, Treier AC, Helou J, Sayer JA et al (2007) Loss of GLIS2 causes nephronophthisis in humans and mice by increased apoptosis and fibrosis. Nat Genet 39:1018–1024

Avasthi P, Marshall WF (2012) Stages of ciliogenesis and regulation of ciliary length. Differ Res Biol Divers 83:S30–S42

Avasthi P, Marshall W (2013) Ciliary secretion: switching the cellular antenna to 'transmit'. Curr Biol 23:R471–R473

Avasthi P, Marley A, Lin H, Gregori-Puigjane E, Shoichet BK, von Zastrow M, Marshall WF (2012) A chemical screen identifies class a g-protein coupled receptors as regulators of cilia. ACS Chem Biol 7:911–919

Avasthi P, Onishi M, Karpiak J, Yamamoto R, Mackinder L, Jonikas MC, Sale WS, Shoichet B, Pringle JR, Marshall WF (2014) Actin is required for IFT regulation in *Chlamydomonas reinhardtii*. Curr Biol 24:2025–2032

Barr MM, Sternberg PW (1999) A polycystic kidney-disease gene homologue required for male mating behaviour in *C. elegans*. Nature 401:386–389

Beales PL, Bland E, Tobin JL, Bacchelli C, Tuysuz B, Hill J, Rix S, Pearson CG, Kai M, Hartley J et al (2007) IFT80, which encodes a conserved intraflagellar transport protein, is mutated in Jeune asphyxiating thoracic dystrophy. Nat Genet 39:727–729

Berbari NF, Lewis JS, Bishop GA, Askwith CC, Mykytyn K (2008) Bardet–Biedl syndrome proteins are required for the localization of G protein-coupled receptors to primary cilia. Proc Natl Acad Sci USA 105:4242–4246

Berman SA, Wilson NF, Haas NA, Lefebvre PA (2003) A novel MAP kinase regulates flagellar length in Chlamydomonas. Curr Biol 13:1145–1149

Bershteyn M, Atwood SX, Woo WM, Li M, Oro AE (2010) MIM and cortactin antagonism regulates ciliogenesis and hedgehog signaling. Dev Cell 19:270–283

Besschetnova TY, Kolpakova-Hart E, Guan Y, Zhou J, Olsen BR, Shah JV (2010) Identification of signaling pathways regulating primary cilium length and flow-mediated adaptation. Curr Biol 20:182–187

Blacque OE, Li C, Inglis PN, Esmail MA, Ou G, Mah AK, Baillie DL, Scholey JM, Leroux MR (2006) The WD repeat-containing protein IFTA-1 is required for retrograde intraflagellar transport. Mol Biol Cell 17:5053–5062

Bloodgood RA (2012) The future of ciliary and flagellar membrane research. Mol Biol Cell 23:2407–2411

Boehlke C, Kotsis F, Patel V, Braeg S, Voelker H, Bredt S, Beyer T, Janusch H, Hamann C, Godel M et al (2010) Primary cilia regulate mTORC1 activity and cell size through Lkb1. Nat Cell Biol 12:1115–1122

Bonneau D, Raymond F, Kremer C, Klossek JM, Kaplan J, Patte F (1993) Usher syndrome type I associated with bronchiectasis and immotile nasal cilia in two brothers. J Med Genet 30:253–254

Brazelton WJ, Amundsen CD, Silflow CD, Lefebvre PA (2001) The bld1 mutation identifies the Chlamydomonas osm-6 homolog as a gene required for flagellar assembly. Curr Biol 11:1591–1594

Broekhuis JR, Leong WY, Jansen G (2013) Regulation of cilium length and intraflagellar transport. Int Rev Cell Mol Biol 303:101–138

Bujakowska KM, Zhang Q, Siemiatkowska AM, Liu Q, Place E, Falk MJ, Consugar M, Lancelot ME, Antonio A, Lonjou C et al (2015) Mutations in IFT172 cause isolated retinal degeneration and Bardet–Biedl syndrome. Hum Mol Genet 24:230–242

Burghoorn J, Dekkers MP, Rademakers S, de Jong T, Willemsen R, Jansen G (2007) Mutation of the MAP kinase DYF-5 affects docking and undocking of kinesin-2 motors and reduces their speed in the cilia of *Caenorhabditis elegans*. Proc Natl Acad Sci USA 104:7157–7162

Cai Y, Fedeles SV, Dong K, Anyatonwu G, Onoe T, Mitobe M, Gao JD, Okuhara D, Tian X, Gallagher AR et al (2014) Altered trafficking and stability of polycystins underlie polycystic kidney disease. J Clin Invest 124:5129–5144

Calvet JP (2008) Strategies to inhibit cyst formation in ADPKD. Clin J Am Soc Nephrol CJASN 3:1205–1211

Calvet JP, Grantham JJ (2001) The genetics and physiology of polycystic kidney disease. Semin Nephrol 21:107–123

Camner P, Mossberg B, Afzelius BA (1975) Evidence of congenitally nonfunctioning cilia in the tracheobronchial tract in two subjects. Am Rev Respir Dis 112:807–809

Cano DA, Murcia NS, Pazour GJ, Hebrok M (2004) Orpk mouse model of polycystic kidney disease reveals essential role of primary cilia in pancreatic tissue organization. Development 131:3457–3467

Chan SK, Riley PR, Price KL, McElduff F, Winyard PJ, Welham SJ, Woolf AS, Long DA (2010) Corticosteroid-induced kidney dysmorphogenesis is associated with deregulated expression of known cystogenic molecules, as well as *Indian hedgehog*. Am J Physiol Renal Physiol 298: F346–F356

Chapin HC, Rajendran V, Caplan MJ (2010) Polycystin-1 surface localization is stimulated by polycystin-2 and cleavage at the G protein-coupled receptor proteolytic site. Mol Biol Cell 21:4338–4348

Chauvet V, Tian X, Husson H, Grimm DH, Wang T, Hiesberger T, Igarashi P, Bennett AM, Ibraghimov-Beskrovnaya O, Somlo S et al (2004) Mechanical stimuli induce cleavage and nuclear translocation of the polycystin-1 C terminus. J Clin Invest 114:1433–1443

Chavez M, Ena S, Van Sande J, de Kerchove d'Exaerde A, Schurmans S, Schiffmann SN (2015) Modulation of ciliary phosphoinositide content regulates trafficking and sonic hedgehog signaling output. Dev Cell 34:338–350

Chih B, Liu P, Chinn Y, Chalouni C, Komuves LG, Hass PE, Sandoval W, Peterson AS (2012) A ciliopathy complex at the transition zone protects the cilia as a privileged membrane domain. Nat Cell Biol 14:61–72

Choi YH, Suzuki A, Hajarnis S, Ma Z, Chapin HC, Caplan MJ, Pontoglio M, Somlo S, Igarashi P (2011) Polycystin-2 and phosphodiesterase 4C are components of a ciliary A-kinase anchoring protein complex that is disrupted in cystic kidney diseases. Proc Natl Acad Sci USA 108:10679–10684

Christensen ST, Ott CM (2007) Cell signaling. A ciliary signaling switch. Science 317:330–331

Coene KL, Mans DA, Boldt K, Gloeckner CJ, van Reeuwijk J, Bolat E, Roosing S, Letteboer SJ, Peters TA, Cremers FP et al (2011) The ciliopathy-associated protein homologs RPGRIP1 and

RPGRIP1L are linked to cilium integrity through interaction with Nek4 serine/threonine kinase. Hum Mol Genet 20:3592–3605

Cogswell C, Price SJ, Hou X, Guay-Woodford LM, Flaherty L, Bryda EC (2003) Positional cloning of jcpk/bpk locus of the mouse. Mamm Genome 14:242–249

Cole DG, Diener DR, Himelblau AL, Beech PL, Fuster JC, Rosenbaum JL (1998) Chlamydomonas kinesin-II-dependent intraflagellar transport (IFT): IFT particles contain proteins required for ciliary assembly in *Caenorhabditis elegans* sensory neurons. J Cell Biol 141:993–1008

Collingridge P, Brownlee C, Wheeler GL (2013) Compartmentalized calcium signaling in cilia regulates intraflagellar transport. Curr Biol 23:2311–2318

Corbit KC, Aanstad P, Singla V, Norman AR, Stainier DY, Reiter JF (2005) Vertebrate Smoothened functions at the primary cilium. Nature 437:1018–1021

Corbit KC, Shyer AE, Dowdle WE, Gaulden J, Singla V, Chen MH, Chuang PT, Reiter JF (2008) Kif3a constrains beta-catenin-dependent Wnt signalling through dual ciliary and non-ciliary mechanisms. Nat Cell Biol 10:70–76

Craft JM, Harris JA, Hyman S, Kner P, Lechtreck KF (2015) Tubulin transport by IFT is upregulated during ciliary growth by a cilium-autonomous mechanism. J Cell Biol 208:223–237

Davis EE, Zhang Q, Liu Q, Diplas BH, Davey LM, Hartley J, Stoetzel C, Szymanska K, Ramaswami G, Logan CV et al (2011) TTC21B contributes both causal and modifying alleles across the ciliopathy spectrum. Nat Genet 43:189–196

Dawe HR, Smith UM, Cullinane AR, Gerrelli D, Cox P, Badano JL, Blair-Reid S, Sriram N, Katsanis N, Attie-Bitach T et al (2007) The Meckel–Gruber syndrome proteins MKS1 and meckelin interact and are required for primary cilium formation. Hum Mol Genet 16:173–186

Deane JA, Cole DG, Seeley ES, Diener DR, Rosenbaum JL (2001) Localization of intraflagellar transport protein IFT52 identifies basal body transitional fibers as the docking site for IFT particles. Curr Biol 11:1586–1590

DeCaen PG, Delling M, Vien TN, Clapham DE (2013) Direct recording and molecular identification of the calcium channel of primary cilia. Nature 504:315–318

Delling M, DeCaen PG, Doerner JF, Febvay S, Clapham DE (2013) Primary cilia are specialized calcium signalling organelles. Nature 504:311–314

Delling M, Indzhykulian AA, Liu X, Li Y, Xie T, Corey DP, Clapham DE (2016) Primary cilia are not calcium-responsive mechanosensors. Nature 531:656–660

den Hollander AI, Koenekoop RK, Mohamed MD, Arts HH, Boldt K, Towns KV, Sedmak T, Beer M, Nagel-Wolfrum K, McKibbin M et al (2007) Mutations in LCA5, encoding the ciliary protein lebercilin, cause Leber congenital amaurosis. Nat Genet 39:889–895

Dentler W (2013) A role for the membrane in regulating *Chlamydomonas flagellar* length. PLoS One 8:e53366

Eggenschwiler JT, Anderson KV (2007) Cilia and developmental signaling. Annu Rev Cell Dev Biol 23:345–373

Eguether T, San Agustin JT, Keady BT, Jonassen JA, Liang Y, Francis R, Tobita K, Johnson CA, Abdelhamed ZA, Lo CW et al (2014) IFT27 links the BBSome to IFT for maintenance of the ciliary signaling compartment. Dev Cell 31:279–290

Engel BD, Ludington WB, Marshall WF (2009) Intraflagellar transport particle size scales inversely with flagellar length: revisiting the balance-point length control model. J Cell Biol 187:81–89

Esteban MA, Harten SK, Tran MG, Maxwell PH (2006) Formation of primary cilia in the renal epithelium is regulated by the von Hippel–Lindau tumor suppressor protein. J Am Soc Nephrol 17:1801–1806

Fogelgren B, Lin SY, Zuo X, Jaffe KM, Park KM, Reichert RJ, Bell PD, Burdine RD, Lipschutz JH (2011) The exocyst protein Sec10 interacts with Polycystin-2 and knockdown causes PKD-phenotypes. PLoS Genet 7:e1001361

Follit JA, Tuft RA, Fogarty KE, Pazour GJ (2006) The intraflagellar transport protein IFT20 is associated with the Golgi complex and is required for cilia assembly. Mol Biol Cell 17:3781–3792

Franco I, Gulluni F, Campa CC, Costa C, Margaria JP, Ciraolo E, Martini M, Monteyne D, De Luca E, Germena G et al (2014) PI3K class II alpha controls spatially restricted endosomal PtdIns3P and Rab11 activation to promote primary cilium function. Dev Cell 28:647–658

Franco I, Margaria JP, De Santis MC, Ranghino A, Monteyne D, Chiaravalli M, Pema M, Campa CC, Ratto E, Gulluni F, Perez-Morga D, Somlo S, Merlo GR, Boletta A, Hirsch E (2016) Phosphoinositide 3-Kinase-C2α regulates polycystin-2 ciliary entry and protects against kidney cyst formation. J Am Soc Nephrol 27(4):1135–1144. doi:10.1681/ASN.2014100967. Epub 2015 Aug 13

Freedman BS, Lam AQ, Sundsbak JL, Iatrino R, Su X, Koon SJ, Wu M, Daheron L, Harris PC, Zhou J et al (2013) Reduced ciliary polycystin-2 in induced pluripotent stem cells from polycystic kidney disease patients with PKD1 mutations. J Am Soc Nephrol 24:1571–1586

Fujiwara M, Ishihara T, Katsura I (1999) A novel WD40 protein, CHE-2, acts cell-autonomously in the formation of *C. elegans* sensory cilia. Development 126:4839–4848

Gainullin VG, Hopp K, Ward CJ, Hommerding CJ, Harris PC (2015) Polycystin-1 maturation requires polycystin-2 in a dose-dependent manner. J Clin Invest 125:607–620

Gallagher AR, Esquivel EL, Briere TS, Tian X, Mitobe M, Menezes LF, Markowitz GS, Jain D, Onuchic LF, Somlo S (2008) Biliary and pancreatic dysgenesis in mice harboring a mutation in Pkhd1. Am J Pathol 172:417–429

Garcia-Gonzalez MA, Menezes LF, Piontek KB, Kaimori J, Huso DL, Watnick T, Onuchic LF, Guay-Woodford LM, Germino GG (2007) Genetic interaction studies link autosomal dominant and recessive polycystic kidney disease in a common pathway. Hum Mol Genet 16:1940–1950

Garcia-Gonzalo FR, Corbit KC, Sirerol-Piquer MS, Ramaswami G, Otto EA, Noriega TR, Seol AD, Robinson JF, Bennett CL, Josifova DJ et al (2011) A transition zone complex regulates mammalian ciliogenesis and ciliary membrane composition. Nat Genet 43:776–784

Garcia-Gonzalo FR, Phua SC, Roberson EC, Garcia G 3rd, Abedin M, Schurmans S, Inoue T, Reiter JF (2015) Phosphoinositides Regulate ciliary protein trafficking to modulate hedgehog signaling. Dev Cell 34:400–409

Gattone VH 2nd, Wang X, Harris PC, Torres VE (2003) Inhibition of renal cystic disease development and progression by a vasopressin V2 receptor antagonist. Nat Med 9:1323–1326

Geng L, Okuhara D, Yu Z, Tian X, Cai Y, Shibazaki S, Somlo S (2006) Polycystin-2 traffics to cilia independently of polycystin-1 by using an N-terminal RVxP motif. J Cell Sci 119:1383–1395

Guo DF, Rahmouni K (2011) Molecular basis of the obesity associated with Bardet–Biedl syndrome. Trends Endocrinol Metab TEM 22:286–293

Guo DF, Cui H, Zhang Q, Morgan DA, Thedens DR, Nishimura D, Grobe JL, Sheffield VC, Rahmouni K (2016) The BBSome controls energy homeostasis by mediating the transport of the leptin receptor to the plasma membrane. PLoS Genet 12:e1005890

Han SJ, Jang HS, Kim JI, Lipschutz JH, Park KM (2016) Unilateral nephrectomy elongates primary cilia in the remaining kidney via reactive oxygen species. Sci Rep 6:22281

Han YM, Kang GM, Byun K, Ko HW, Kim J, Shin MS, Kim HK, Gil SY, Yu JH, Lee B et al (2014) Leptin-promoted cilia assembly is critical for normal energy balance. J Clin Investig 124:2193–2197

Hartman TR, Liu D, Zilfou JT, Robb V, Morrison T, Watnick T, Henske EP (2009) The tuberous sclerosis proteins regulate formation of the primary cilium via a rapamycin-insensitive and polycystin 1-independent pathway. Hum Mol Genet 18:151–163

Hatayama M, Mikoshiba K, Aruga J (2011) IP3 signaling is required for cilia formation and left–right body axis determination in Xenopus embryos. Biochem Biophys Res Commun 410:520–524

Haycraft CJ, Banizs B, Aydin-Son Y, Zhang Q, Michaud EJ, Yoder BK (2005) Gli2 and Gli3 localize to cilia and require the intraflagellar transport protein polaris for processing and function. PLoS Genet 1:e53

Haycraft CJ, Schafer JC, Zhang Q, Taulman PD, Yoder BK (2003) Identification of CHE-13, a novel intraflagellar transport protein required for cilia formation. Exp Cell Res 284:251–263

He M, Subramanian R, Bangs F, Omelchenko T, Liem KF Jr, Kapoor TM, Anderson KV (2014) The kinesin-4 protein Kif7 regulates mammalian Hedgehog signalling by organizing the cilium tip compartment. Nat Cell Biol 16:663–672

Hearn T, Spalluto C, Phillips VJ, Renforth GL, Copin N, Hanley NA, Wilson DI (2005) Subcellular localization of ALMS1 supports involvement of centrosome and basal body dysfunction in the pathogenesis of obesity, insulin resistance, and type 2 diabetes. Diabetes 54:1581–1587

Heon E, Kim G, Qin S, Garrison JE, Tavares E, Vincent A, Nuangchamnong N, Scott CA, Slusarski DC, Sheffield VC (2016) Mutations in C8ORF37 cause Bardet Biedl syndrome (BBS21). Hum Mol Genet 25(11):2283–2294. Epub 2016 Mar 22

Hilton LK, Gunawardane K, Kim JW, Schwarz MC, Quarmby LM (2013) The kinases LF4 and CNK2 control ciliary length by feedback regulation of assembly and disassembly rates. Curr Biol 23:2208–2214

Hoffmeister H, Babinger K, Gurster S, Cedzich A, Meese C, Schadendorf K, Osten L, de Vries U, Rascle A, Witzgall R (2011) Polycystin-2 takes different routes to the somatic and ciliary plasma membrane. J Cell Biol 192:631–645

Hogan MC, Manganelli L, Woollard JR, Masyuk AI, Masyuk TV, Tammachote R, Huang BQ, Leontovich AA, Beito TG, Madden BJ et al (2009) Characterization of PKD protein-positive exosome-like vesicles. J Am Soc Nephrol 20:278–288

Hong DH, Yue G, Adamian M, Li T (2001) Retinitis pigmentosa GTPase regulator (RPGRr)-interacting protein is stably associated with the photoreceptor ciliary axoneme and anchors RPGR to the connecting cilium. J Biol Chem 276:12091–12099

Hopp K, Hommerding CJ, Wang X, Ye H, Harris PC, Torres VE (2015) Tolvaptan plus pasireotide shows enhanced efficacy in a PKD1 model. J Am Soc Nephrol 26:39–47

Hopp K, Ward CJ, Hommerding CJ, Nasr SH, Tuan HF, Gainullin VG, Rossetti S, Torres VE, Harris PC (2012) Functional polycystin-1 dosage governs autosomal dominant polycystic kidney disease severity. J Clin Invest 122:4257–4273

Hsiao YC, Tuz K, Ferland RJ (2012) Trafficking in and to the primary cilium. Cilia 1:4

Hu J, Wittekind SG, Barr MM (2007) STAM and Hrs down-regulate ciliary TRP receptors. Mol Biol Cell 18:3277–3289

Huangfu D, Anderson KV (2005) Cilia and Hedgehog responsiveness in the mouse. Proc Natl Acad Sci USA 102:11325–11330

Huangfu D, Liu A, Rakeman AS, Murcia NS, Niswander L, Anderson KV (2003) Hedgehog signalling in the mouse requires intraflagellar transport proteins. Nature 426:83–87

Husson H, Moreno S, Smith LA, Smith MM, Russo RJ, Pitstick R, Sergeev M, Ledbetter SR, Bukanov NO, Lane M et al (2016) Reduction of ciliary length through pharmacologic or genetic inhibition of CDK5 attenuates polycystic kidney disease in a model of nephronophthisis. Hum Mol Genet 25(11):2245–2255. Epub 2016 Apr 5

Ibraghimov-Beskrovnaya O, Natoli TA (2011) mTOR signaling in polycystic kidney disease. Trends Mol Med 17:625–633

Ingham PW, Nakano Y, Seger C (2011) Mechanisms and functions of Hedgehog signalling across the metazoa. Nat Rev Genet 12:393–406

Iomini C, Babaev-Khaimov V, Sassaroli M, Piperno G (2001) Protein particles in *Chlamydomonas flagella* undergo a transport cycle consisting of four phases. J Cell Biol 153:13–24

Iomini C, Li L, Esparza JM, Dutcher SK (2009) Retrograde intraflagellar transport mutants identify complex A proteins with multiple genetic interactions in *Chlamydomonas reinhardtii*. Genetics 183:885–896

Ishikawa H, Ide T, Yagi T, Jiang X, Hirono M, Sasaki H, Yanagisawa H, Wemmer KA, Stainier DY, Qin H et al (2014) TTC26/DYF13 is an intraflagellar transport protein required for transport of motility-related proteins into flagella. Elife 3:e01566

Jacoby M, Cox JJ, Gayral S, Hampshire DJ, Ayub M, Blockmans M, Pernot E, Kisseleva MV, Compere P, Schiffmann SN et al (2009) INPP5E mutations cause primary cilium signaling defects, ciliary instability and ciliopathies in human and mouse. Nat Genet 41:1027–1031

Jin X, Mohieldin AM, Muntean BS, Green JA, Shah JV, Mykytyn K, Nauli SM (2014a) Cilioplasm is a cellular compartment for calcium signaling in response to mechanical and chemical stimuli. Cell Mol Life Sci 71:2165–2178

Jin X, Muntean BS, Aal-Aaboda MS, Duan Q, Zhou J, Nauli SM (2014b) L-type calcium channel modulates cystic kidney phenotype. Biochim Biophys Acta 1842(9):1518–1526. doi:10.1016/j.bbadis.2014.06.001. Epub 2014 Jun 9

Jonassen JA, San Agustin J, Follit JA, Pazour GJ (2008) Deletion of IFT20 in the mouse kidney causes misorientation of the mitotic spindle and cystic kidney disease. J Cell Biol 183:377–384

Jonassen JA, SanAgustin J, Baker SP, Pazour GJ (2012) Disruption of IFT complex A causes cystic kidneys without mitotic spindle misorientation. J Am Soc Nephrol 23:641–651

Keady BT, Samtani R, Tobita K, Tsuchya M, San Agustin JT, Follit JA, Jonassen JA, Subramanian R, Lo CW, Pazour GJ (2012) IFT25 links the signal-dependent movement of Hedgehog components to intraflagellar transport. Dev Cell 22:940–951

Keeling J, Tsiokas L, Maskey D (2016) Cellular mechanisms of ciliary length control. Cells 5

Khan SA, Muhammad N, Khan MA, Kamal A, Rehman ZU, Khan S (2016) Genetics of human Bardet–Biedl syndrome, an updates. Clin Genet 90(1):3–15. doi:10.1111/cge.12737. Epub 2016 Feb 9

Khayyeri H, Barreto S, Lacroix D (2015) Primary cilia mechanics affects cell mechanosensation: a computational study. J Theor Biol 379:38–46

Kim JH, Ki SM, Joung JG, Scott E, Heynen-Genel S, Aza-Blanc P, Kwon CH, Kim J, Gleeson JG, Lee JE (2016a) Genome-wide screen identifies novel machineries required for both ciliogenesis and cell cycle arrest upon serum starvation. Biochim Biophys Acta 1863:1307–1318

Kim S, Lee K, Choi JH, Ringstad N, Dynlacht BD (2015) Nek2 activation of Kif24 ensures cilium disassembly during the cell cycle. Nat Commun 6:8087

Kim J, Lee JE, Heynen-Genel S, Suyama E, Ono K, Lee K, Ideker T, Aza-Blanc P, Gleeson JG (2010) Functional genomic screen for modulators of ciliogenesis and cilium length. Nature 464:1048–1051

Kim S, Nie H, Nesin V, Tran U, Outeda P, Bai CX, Keeling J, Maskey D, Watnick T, Wessely O et al (2016b) The polycystin complex mediates Wnt/Ca signalling. Nat Cell Biol 18 (7):752–764. doi:10.1038/ncb3363. Epub 2016 May 23

Kim H, Xu H, Yao Q, Li W, Huang Q, Outeda P, Cebotaru V, Chiaravalli M, Boletta A, Piontek K et al (2014) Ciliary membrane proteins traffic through the Golgi via a Rabep1/GGA1/Arl3-dependent mechanism. Nat Commun 5:5482

Knodler A, Feng S, Zhang J, Zhang X, Das A, Peranen J, Guo W (2010) Coordination of Rab8 and Rab11 in primary ciliogenesis. Proc Natl Acad Sci USA 107:6346–6351

Kottgen M, Buchholz B, Garcia-Gonzalez MA, Kotsis F, Fu X, Doerken M, Boehlke C, Steffl D, Tauber R, Wegierski T et al (2008) TRPP2 and TRPV4 form a polymodal sensory channel complex. J Cell Biol 182:437–447

Kozminski KG, Beech PL, Rosenbaum JL (1995) The Chlamydomonas kinesin-like protein FLA10 is involved in motility associated with the flagellar membrane. J Cell Biol 131:1517–1527

Kozminski KG, Johnson KA, Forscher P, Rosenbaum JL (1993) A motility in the eukaryotic flagellum unrelated to flagellar beating. Proc Natl Acad Sci USA 90:5519–5523

Kurbegovic A, Kim H, Xu H, Yu S, Cruanes J, Maser RL, Boletta A, Trudel M, Qian F (2014) Novel functional complexity of polycystin-1 by GPS cleavage in vivo: role in polycystic kidney disease. Mol Cell Biol 34:3341–3353

Lal M, Song X, Pluznick JL, Di Giovanni V, Merrick DM, Rosenblum ND, Chauvet V, Gottardi CJ, Pei Y, Caplan MJ (2008) Polycystin-1 C-terminal tail associates with beta-catenin and inhibits canonical Wnt signaling. Hum Mol Genet 17:3105–3117

Lancaster MA, Louie CM, Silhavy JL, Sintasath L, Decambre M, Nigam SK, Willert K, Gleeson JG (2009) Impaired Wnt-beta-catenin signaling disrupts adult renal homeostasis and leads to cystic kidney ciliopathy. Nat Med 15:1046–1054

Lancaster MA, Schroth J, Gleeson JG (2011) Subcellular spatial regulation of canonical Wnt signalling at the primary cilium. Nat Cell Biol 13:700–707

Lavagnino M, Gardner K, Sedlak AM, Arnoczky SP (2013) Tendon cell ciliary length as a biomarker of in situ cytoskeletal tensional homeostasis. Muscles Ligaments Tendons J 3:118–121

Lechtreck KF, Johnson EC, Sakai T, Cochran D, Ballif BA, Rush J, Pazour GJ, Ikebe M, Witman GB (2009) The *Chlamydomonas reinhardtii* BBSome is an IFT cargo required for export of specific signaling proteins from flagella. J Cell Biol 187:1117–1132

Leightner AC, Hommerding CJ, Peng Y, Salisbury JL, Gainullin VG, Czarnecki PG, Sussman CR, Harris PC (2013) The Meckel syndrome protein meckelin (TMEM67) is a key regulator of cilia function but is not required for tissue planar polarity. Hum Mol Genet 22:2024–2040

Li B, Rauhauser AA, Dai J, Sakthivel R, Igarashi P, Jetten AM, Attanasio M (2011) Increased hedgehog signaling in postnatal kidney results in aberrant activation of nephron developmental programs. Hum Mol Genet 20(21):4155–4166. doi:10.1093/hmg/ddr339

Liew GM, Ye F, Nager AR, Murphy JP, Lee JS, Aguiar M, Breslow DK, Gygi SP, Nachury MV (2014) The intraflagellar transport protein IFT27 promotes BBSome exit from cilia through the GTPase ARL6/BBS3. Dev Cell 31:265–278

Lin F, Hiesberger T, Cordes K, Sinclair AM, Goldstein LS, Somlo S, Igarashi P (2003) Kidney-specific inactivation of the KIF3A subunit of kinesin-II inhibits renal ciliogenesis and produces polycystic kidney disease. Proc Natl Acad Sci USA 100:5286–5291

Lindstrand A, Davis EE, Carvalho CM, Pehlivan D, Willer JR, Tsai IC, Ramanathan S, Zuppan C, Sabo A, Muzny D et al (2014) Recurrent CNVs and SNVs at the NPHP1 locus contribute pathogenic alleles to Bardet–Biedl syndrome. Am J Hum Genet 94:745–754

Liu W, Murcia NS, Duan Y, Weinbaum S, Yoder BK, Schwiebert E, Satlin LM (2005b) Mechanoregulation of intracellular Ca2+ concentration is attenuated in collecting duct of monocilium-impaired orpk mice. Am J Physiol Renal Physiol 289:F978–F988

Liu A, Wang B, Niswander LA (2005a) Mouse intraflagellar transport proteins regulate both the activator and repressor functions of Gli transcription factors. Development 132:3103–3111

Liu W, Xu S, Woda C, Kim P, Weinbaum S, Satlin LM (2003) Effect of flow and stretch on the [Ca2+]i response of principal and intercalated cells in cortical collecting duct. Am J Physiol Renal Physiol 285:F998–F1012

Louie CM, Gleeson JG (2005) Genetic basis of Joubert syndrome and related disorders of cerebellar development. Hum Mol Genet 14 Spec No. 2:R235–R242

Low SH, Vasanth S, Larson CH, Mukherjee S, Sharma N, Kinter MT, Kane ME, Obara T, Weimbs T (2006) Polycystin-1, STAT6, and P100 function in a pathway that transduces ciliary mechanosensation and is activated in polycystic kidney disease. Dev Cell 10:57–69

Lu Q, Insinna C, Ott C, Stauffer J, Pintado PA, Rahajeng J, Baxa U, Walia V, Cuenca A, Hwang YS et al (2015) Early steps in primary cilium assembly require EHD1/EHD3-dependent ciliary vesicle formation. Nat Cell Biol 17:531

Ludington WB, Wemmer KA, Lechtreck KF, Witman GB, Marshall WF (2013) Avalanche-like behavior in ciliary import. Proc Natl Acad Sci USA 110:3925–3930

Luijten MN, Basten SG, Claessens T, Vernooij M, Scott CL, Janssen R, Easton JA, Kamps MA, Vreeburg M, Broers JL et al (2013) Birt–Hogg–Dube syndrome is a novel ciliopathy. Hum Mol Genet 22:4383–4397

Luo N, Lu J, Sun Y (2012) Evidence of a role of inositol polyphosphate 5-phosphatase INPP5E in cilia formation in zebrafish. Vis Res 75:98–107

Ma M, Tian X, Igarashi P, Pazour GJ, Somlo S (2013) Loss of cilia suppresses cyst growth in genetic models of autosomal dominant polycystic kidney disease. Nat Genet 45:1004–1012
Manning DK, Sergeev M, van Heesbeen RG, Wong MD, Oh JH, Liu Y, Henkelman RM, Drummond I, Shah JV, Beier DR (2013) Loss of the ciliary kinase Nek8 causes left–right asymmetry defects. J Am Soc Nephrol 24:100–112
Marion V, Stoetzel C, Schlicht D, Messaddeq N, Koch M, Flori E, Danse JM, Mandel JL, Dollfus H (2009) Transient ciliogenesis involving Bardet–Biedl syndrome proteins is a fundamental characteristic of adipogenic differentiation. Proc Natl Acad Sci USA 106:1820–1825
Marion V, Stutzmann F, Gerard M, De Melo C, Schaefer E, Claussmann A, Helle S, Delague V, Souied E, Barrey C et al (2012) Exome sequencing identifies mutations in LZTFL1, a BBSome and smoothened trafficking regulator, in a family with Bardet–Biedl syndrome with situs inversus and insertional polydactyly. J Med Genet 49:317–321
Marshall WF, Rosenbaum JL (2001) Intraflagellar transport balances continuous turnover of outer doublet microtubules: implications for flagellar length control. J Cell Biol 155:405–414
Marshall WF, Qin H, Rodrigo Brenni M, Rosenbaum JL (2005) Flagellar length control system: testing a simple model based on intraflagellar transport and turnover. Mol Biol Cell 16:270–278
Masyuk AI, Masyuk TV, LaRusso NF (2008) Cholangiocyte primary cilia in liver health and disease. Dev Dyn 237:2007–2012
Mattera R, Arighi CN, Lodge R, Zerial M, Bonifacino JS (2003) Divalent interaction of the GGAs with the Rabaptin-5-Rabex-5 complex. EMBO J 22:78–88
Mazelova J, Astuto-Gribble L, Inoue H, Tam BM, Schonteich E, Prekeris R, Moritz OL, Randazzo PA, Deretic D (2009) Ciliary targeting motif VxPx directs assembly of a trafficking module through Arf4. EMBO J 28:183–192
Merrick D, Chapin H, Baggs JE, Yu Z, Somlo S, Sun Z, Hogenesch JB, Caplan MJ (2012) The gamma-secretase cleavage product of polycystin-1 regulates TCF and CHOP-mediated transcriptional activation through a p300-dependent mechanism. Dev Cell 22:197–210
Mick DU, Rodrigues RB, Leib RD, Adams CM, Chien AS, Gygi SP, Nachury MV (2015) Proteomics of primary cilia by proximity labeling. Dev Cell 35:497–512
Miller MM, Iglesias DM, Zhang Z, Corsini R, Chu L, Murawski I, Gupta I, Somlo S, Germino GG, Goodyer PR (2011) T-cell factor/beta-catenin activity is suppressed in two different models of autosomal dominant polycystic kidney disease. Kidney Int 80:146–153
Mitchell DR (2007) The evolution of eukaryotic cilia and flagella as motile and sensory organelles. Adv Exp Med Biol 607:130–140
Moser M, Matthiesen S, Kirfel J, Schorle H, Bergmann C, Senderek J, Rudnik-Schoneborn S, Zerres K, Buettner R (2005) A mouse model for cystic biliary dysgenesis in autosomal recessive polycystic kidney disease (ARPKD). Hepatology 41:1113–1121
Mukhopadhyay S, Wen X, Chih B, Nelson CD, Lane WS, Scales SJ, Jackson PK (2010) TULP3 bridges the IFT-A complex and membrane phosphoinositides to promote trafficking of G protein-coupled receptors into primary cilia. Genes Dev 24:2180–2193
Nachury MV, Loktev AV, Zhang Q, Westlake CJ, Peranen J, Merdes A, Slusarski DC, Scheller RH, Bazan JF, Sheffield VC et al (2007) A core complex of BBS proteins cooperates with the GTPase Rab8 to promote ciliary membrane biogenesis. Cell 129:1201–1213
Nakamura T, Saito D, Kawasumi A, Shinohara K, Asai Y, Takaoka K, Dong F, Takamatsu A, Belo JA, Mochizuki A et al (2012) Fluid flow and interlinked feedback loops establish left–right asymmetric decay of Cerl2 mRNA. Nat Commun 3:1322
Natoli TA, Gareski TC, Dackowski WR, Smith L, Bukanov NO, Russo RJ, Husson H, Matthews D, Piepenhagen P, Ibraghimov-Beskrovnaya O (2008) Pkd1 and Nek8 mutations affect cell-cell adhesion and cilia in cysts formed in kidney organ cultures. Am J Physiol Renal Physiol 294:F73–F83
Natoli TA, Smith LA, Rogers KA, Wang B, Komarnitsky S, Budman Y, Belenky A, Bukanov NO, Dackowski WR, Husson H et al (2010) Inhibition of glucosylceramide accumulation results in effective blockade of polycystic kidney disease in mouse models. Nat Med 16:788–792

Nauli SM, Alenghat FJ, Luo Y, Williams E, Vassilev P, Li X, Elia AE, Lu W, Brown EM, Quinn SJ et al (2003) Polycystins 1 and 2 mediate mechanosensation in the primary cilium of kidney cells. Nat Genet 33:129–137

Niewiadomski P, Kong JH, Ahrends R, Ma Y, Humke EW, Khan S, Teruel MN, Novitch BG, Rohatgi R (2014) Gli protein activity is controlled by multisite phosphorylation in vertebrate hedgehog signaling. Cell Rep 6:168–181

Nozawa YI, Lin C, Chuang PT (2013) Hedgehog signaling from the primary cilium to the nucleus: an emerging picture of ciliary localization, trafficking and transduction. Curr Opin Genet Dev 23:429–437

O'Connor AK, Malarkey EB, Berbari NF, Croyle MJ, Haycraft CJ, Bell PD, Hohenstein P, Kesterson RA, Yoder BK (2013) An inducible CiliaGFP mouse model for in vivo visualization and analysis of cilia in live tissue. Cilia 2:8

Oh EC, Katsanis N (2013) Context-dependent regulation of Wnt signaling through the primary cilium. J Am Soc Nephrol 24:10–18

Oishi I, Kawakami Y, Raya A, Callol-Massot C, Izpisua Belmonte JC (2006) Regulation of primary cilia formation and left–right patterning in zebrafish by a noncanonical Wnt signaling mediator, duboraya. Nat Genet 38:1316–1322

Olsan EE, Mukherjee S, Wulkersdorfer B, Shillingford JM, Giovannone AJ, Todorov G, Song X, Pei Y, Weimbs T (2011) Signal transducer and activator of transcription-6 (STAT6) inhibition suppresses renal cyst growth in polycystic kidney disease. Proc Natl Acad Sci USA 108:18067–18072

Otto EA, Loeys B, Khanna H, Hellemans J, Sudbrak R, Fan S, Muerb U, O'Toole JF, Helou J, Attanasio M et al (2005) Nephrocystin-5, a ciliary IQ domain protein, is mutated in Senior–Loken syndrome and interacts with RPGR and calmodulin. Nat Genet 37:282–288

Otto EA, Schermer B, Obara T, O'Toole JF, Hiller KS, Mueller AM, Ruf RG, Hoefele J, Beekmann F, Landau D et al (2003) Mutations in INVS encoding inversin cause nephronophthisis type 2, linking renal cystic disease to the function of primary cilia and left–right axis determination. Nat Genet 34:413–420

Ou Y, Ruan Y, Cheng M, Moser JJ, Rattner JB, van der Hoorn FA (2009) Adenylate cyclase regulates elongation of mammalian primary cilia. Exp Cell Res 315:2802–2817

Pandit MM, Gao Y, van Hoek A, Kohan DE (2015) Osmolar regulation of endothelin-1 production by the inner medullary collecting duct. Life Sci 159:135–139. doi:10.1016/j.lfs.2015.10.037. Epub 2015 Nov 10

Parker JD, Quarmby LM (2003) Chlamydomonas fla mutants reveal a link between deflagellation and intraflagellar transport. BMC Cell Biol 4:11

Pazour GJ, Dickert BL, Vucica Y, Seeley ES, Rosenbaum JL, Witman GB, Cole DG (2000) Chlamydomonas IFT88 and its mouse homologue, polycystic kidney disease gene tg737, are required for assembly of cilia and flagella. J Cell Biol 151:709–718

Pazour GJ, Dickert BL, Witman GB (1999) The DHC1b (DHC2) isoform of cytoplasmic dynein is required for flagellar assembly. J Cell Biol 144:473–481

Pazour GJ, Wilkerson CG, Witman GB (1998) A dynein light chain is essential for the retrograde particle movement of intraflagellar transport (IFT). J Cell Biol 141:979–992

Pedersen LB, Miller MS, Geimer S, Leitch JM, Rosenbaum JL, Cole DG (2005) Chlamydomonas IFT172 is encoded by FLA11, interacts with CrEB1, and regulates IFT at the flagellar tip. Curr Biol 15:262–266

Perkins LA, Hedgecock EM, Thomson JN, Culotti JG (1986) Mutant sensory cilia in the nematode *Caenorhabditis elegans*. Dev Biol 117:456–487

Pescio LG, Favale NO, Marquez MG, Sterin-Speziale NB (2012) Glycosphingolipid synthesis is essential for MDCK cell differentiation. Biochim Biophys Acta 1821:884–894

Piao T, Luo M, Wang L, Guo Y, Li D, Li P, Snell WJ, Pan J (2009) A microtubule depolymerizing kinesin functions during both flagellar disassembly and flagellar assembly in Chlamydomonas. Proc Natl Acad Sci USA 106:4713–4718

Piontek K, Menezes LF, Garcia-Gonzalez MA, Huso DL, Germino GG (2007) A critical developmental switch defines the kinetics of kidney cyst formation after loss of Pkd1. Nat Med 13:1490–1495

Poole CA, Flint MH, Beaumont BW (1985) Analysis of the morphology and function of primary cilia in connective tissues: a cellular cybernetic probe? Cell Motil 5:175–193

Porath B, Gainullin VG, Cornec-Le Gall E, Dillinger EK, Heyer CM, Hopp K, Edwards ME, Madsen CD, Mauritz SR, Banks CJ et al (2016) Mutations in GANAB, encoding the glucosidase IIalpha subunit, cause autosomal-dominant polycystic kidney and liver disease. Am J Hum Genet 98:1193–1207

Porter ME, Bower R, Knott JA, Byrd P, Dentler W (1999) Cytoplasmic dynein heavy chain 1b is required for flagellar assembly in Chlamydomonas. Mol Biol Cell 10:693–712

Praetorius HA, Spring KR (2001) Bending the MDCK cell primary cilium increases intracellular calcium. J Membr Biol 184:71–79

Praetorius HA, Spring KR (2003) Removal of the MDCK cell primary cilium abolishes flow sensing. J Membr Biol 191:69–76

Pugacheva EN, Jablonski SA, Hartman TR, Henske EP, Golemis EA (2007) HEF1-dependent Aurora A activation induces disassembly of the primary cilium. Cell 129:1351–1363

Qian CN, Knol J, Igarashi P, Lin F, Zylstra U, Teh BT, Williams BO (2005) Cystic renal neoplasia following conditional inactivation of apc in mouse renal tubular epithelium. J Biol Chem 280:3938–3945

Qin J, Lin Y, Norman RX, Ko HW, Eggenschwiler JT (2011) Intraflagellar transport protein 122 antagonizes Sonic Hedgehog signaling and controls ciliary localization of pathway components. Proc Natl Acad Sci USA 108:1456–1461

Qin S, Taglienti M, Cai L, Zhou J, Kreidberg JA (2012) c-Met and NF-kappaB-dependent overexpression of Wnt7a and -7b and Pax2 promotes cystogenesis in polycystic kidney disease. J Am Soc Nephrol 23:1309–1318

Qin S, Taglienti M, Nauli SM, Contrino L, Takakura A, Zhou J, Kreidberg JA (2010) Failure to ubiquitinate c-Met leads to hyperactivation of mTOR signaling in a mouse model of autosomal dominant polycystic kidney disease. J Clin Invest 120:3617–3628

Qin H, Wang Z, Diener D, Rosenbaum J (2007) Intraflagellar transport protein 27 is a small G protein involved in cell-cycle control. Curr Biol 17:193–202

Quinlan RJ, Tobin JL, Beales PL (2008) Modeling ciliopathies: primary cilia in development and disease. Curr Top Dev Biol 84:249–310

Ravichandran K, Zafar I, Ozkok A, Edelstein CL (2015) An mTOR kinase inhibitor slows disease progression in a rat model of polycystic kidney disease. Nephrol Dial Transplant 30:45–53

Rohatgi R, Milenkovic L, Scott MP (2007) Patched1 regulates hedgehog signaling at the primary cilium. Science 317:372–376

Romio L, Fry AM, Winyard PJ, Malcolm S, Woolf AS, Feather SA (2004) OFD1 is a centrosomal/basal body protein expressed during mesenchymal-epithelial transition in human nephrogenesis. J Am Soc Nephrol 15:2556–2568

Rosenbaum JL, Child FM (1967) Flagellar regeneration in protozoan flagellates. J Cell Biol 34:345–364

Rosenbaum JL, Witman GB (2002) Intraflagellar transport. Nat Rev Mol Cell Biol 3:813–825

Ruggenenti P, Gentile G, Perico N, Perna A, Barcella L, Trillini M, Cortinovis M, Ferrer Siles CP, Reyes Loaeza JA, Aparicio MC et al (2016) Effect of sirolimus on disease progression in patients with autosomal dominant polycystic kidney disease and CKD stages 3b-4. Clin J Am Soc Nephrol 11(5):785–794. doi:10.2215/CJN.09900915. Epub 2016 Feb 22

Ruiz-Perez VL, Goodship JA (2009) Ellis-van Creveld syndrome and Weyers acrodental dysostosis are caused by cilia-mediated diminished response to hedgehog ligands. Am J Med Genet C Semin Med Genet 151C:341–351

Rydholm S, Zwartz G, Kowalewski JM, Kamali-Zare P, Frisk T, Brismar H (2010) Mechanical properties of primary cilia regulate the response to fluid flow. Am J Physiol Renal Physiol 298: F1096–F1102

Saadi-Kheddouci S, Berrebi D, Romagnolo B, Cluzeaud F, Peuchmaur M, Kahn A, Vandewalle A, Perret C (2001) Early development of polycystic kidney disease in transgenic mice expressing an activated mutant of the beta-catenin gene. Oncogene 20:5972–5981

Sanchez de Diego A, Alonso Guerrero A, Martinez AC, van Wely KH (2014) Dido3-dependent HDAC6 targeting controls cilium size. Nat Commun 5:3500

Saraga-Babic M, Vukojevic K, Bocina I, Drnasin K, Saraga M (2012) Ciliogenesis in normal human kidney development and post-natal life. Pediatr Nephrol 27:55–63

Schaefer E, Stoetzel C, Scheidecker S, Geoffroy V, Prasad MK, Redin C, Missotte I, Lacombe D, Mandel JL, MullerJ et al (2016) Identification of a novel mutation confirms the implication of IFT172 (BBS20) in Bardet–Biedl syndrome. J Hum Genet 61(5):447–450. doi:10.1038/jhg.2015.162. Epub 2016 Jan 14

Scheidecker S, Etard C, Pierce NW, Geoffroy V, Schaefer E, Muller J, Chennen K, Flori E, Pelletier V, Poch O et al (2014) Exome sequencing of Bardet–Biedl syndrome patient identifies a null mutation in the BBSome subunit BBIP1 (BBS18). J Med Genet 51:132–136

Schwarz N, Hardcastle AJ, Cheetham ME (2012) Arl3 and RP2 mediated assembly and traffic of membrane associated cilia proteins. Vis Res 75:2–4

Seo S, Baye LM, Schulz NP, Beck JS, Zhang Q, Slusarski DC, Sheffield VC (2010) BBS6, BBS10, and BBS12 form a complex with CCT/TRiC family chaperonins and mediate BBSome assembly. Proc Natl Acad Sci USA 107:1488–1493

Seo S, Zhang Q, Bugge K, Breslow DK, Searby CC, Nachury MV, Sheffield VC (2011) A novel protein LZTFL1 regulates ciliary trafficking of the BBSome and Smoothened. PLoS Genet 7: e1002358

Serra AL, Poster D, Kistler AD, Krauer F, Raina S, Young J, Rentsch KM, Spanaus KS, Senn O, Kristanto P et al (2010) Sirolimus and kidney growth in autosomal dominant polycystic kidney disease. N Engl J Med 363:820–829

Shao YY, Wang L, Welter JF, Ballock RT (2012) Primary cilia modulate Ihh signal transduction in response to hydrostatic loading of growth plate chondrocytes. Bone 50:79–84

Sharma N, Kosan ZA, Stallworth JE, Berbari NF, Yoder BK (2011) Soluble levels of cytosolic tubulin regulate ciliary length control. Mol Biol Cell 22:806–816

Shillingford JM, Leamon CP, Vlahov IR, Weimbs T (2012) Folate-conjugated rapamycin slows progression of polycystic kidney disease. J Am Soc Nephrol 23:1674–1681

Shillingford JM, Murcia NS, Larson CH, Low SH, Hedgepeth R, Brown N, Flask CA, Novick AC, Goldfarb DA, Kramer-Zucker A et al (2006) The mTOR pathway is regulated by polycystin-1, and its inhibition reverses renal cystogenesis in polycystic kidney disease. Proc Natl Acad Sci USA 103:5466–5471

Signor D, Wedaman KP, Orozco JT, Dwyer ND, Bargmann CI, Rose LS, Scholey JM (1999) Role of a class DHC1b dynein in retrograde transport of IFT motors and IFT raft particles along cilia, but not dendrites, in chemosensory neurons of living *Caenorhabditis elegans*. J Cell Biol 147:519–530

Simons M, Gloy J, Ganner A, Bullerkotte A, Bashkurov M, Kronig C, Schermer B, Benzing T, Cabello OA, Jenny A et al (2005) Inversin, the gene product mutated in nephronophthisis type II, functions as a molecular switch between Wnt signaling pathways. Nat Genet 37:537–543

Siroky BJ, Ferguson WB, Fuson AL, Xie Y, Fintha A, Komlosi P, Yoder BK, Schwiebert EM, Guay-Woodford LM, Bell PD (2006) Loss of primary cilia results in deregulated and unabated apical calcium entry in ARPKD collecting duct cells. Am J Physiol Renal Physiol 290:F1320–F1328

Smith LA, Bukanov NO, Husson H, Russo RJ, Barry TC, Taylor AL, Beier DR, Ibraghimov-Beskrovnaya O (2006) Development of polycystic kidney disease in juvenile cystic kidney mice: insights into pathogenesis, ciliary abnormalities, and common features with human disease. J Am Soc Nephrol 17:2821–2831

Song L, Dentler WL (2001) Flagellar protein dynamics in Chlamydomonas. J Biol Chem 276:29754–29763

Song X, Di Giovanni V, He N, Wang K, Ingram A, Rosenblum ND, Pei Y (2009) Systems biology of autosomal dominant polycystic kidney disease (ADPKD): computational identification of gene expression pathways and integrated regulatory networks. Hum Mol Genet 18:2328–2343

Sorokin S (1962) Centrioles and the formation of rudimentary cilia by fibroblasts and smooth muscle cells. J Cell Biol 15:363–377

Spalluto C, Wilson DI, Hearn T (2012) Nek2 localises to the distal portion of the mother centriole/basal body and is required for timely cilium disassembly at the G2/M transition. Eur J Cell Biol 91:675–686

Stephens RE (1997) Synthesis and turnover of embryonic sea urchin ciliary proteins during selective inhibition of tubulin synthesis and assembly. Mol Biol Cell 8:2187–2198

Stottmann RW, Tran PV, Turbe-Doan A, Beier DR (2009) Ttc21b is required to restrict sonic hedgehog activity in the developing mouse forebrain. Dev Biol 335:166–178

Su X, Driscoll K, Yao G, Raed A, Wu M, Beales PL, Zhou J (2014) Bardet–Biedl syndrome proteins 1 and 3 regulate the ciliary trafficking of polycystic kidney disease 1 protein. Hum Mol Genet 23:5441–5451

Su S, Phua SC, DeRose R, Chiba S, Narita K, Kalugin PN, Katada T, Kontani K, Takeda S, Inoue T (2013) Genetically encoded calcium indicator illuminates calcium dynamics in primary cilia. Nat Methods 10:1105–1107

Su X, Wu M, Yao G, El-Jouni W, Luo C, Tabari A, Zhou J (2015) Regulation of polycystin-1 ciliary trafficking by motifs at its C-terminus and polycystin-2 but not by cleavage at the GPS site. J Cell Sci 128:4063–4073

Sugiyama N, Tsukiyama T, Yamaguchi TP, Yokoyama T (2011) The canonical Wnt signaling pathway is not involved in renal cyst development in the kidneys of inv mutant mice. Kidney Int 79:957–965

Sun Z, Amsterdam A, Pazour GJ, Cole DG, Miller MS, Hopkins N (2004) A genetic screen in zebrafish identifies cilia genes as a principal cause of cystic kidney. Development 131:4085–4093

Takakura A, Nelson EA, Haque N, Humphreys BD, Zandi-Nejad K, Frank DA, Zhou J (2011) Pyrimethamine inhibits adult polycystic kidney disease by modulating STAT signaling pathways. Hum Mol Genet 20:4143–4154

Takiar V, Nishio S, Seo-Mayer P, King JD Jr, Li H, Zhang L, Karihaloo A, Hallows KR, Somlo S, Caplan MJ (2011) Activating AMP-activated protein kinase (AMPK) slows renal cystogenesis. Proc Natl Acad Sci USA 108:2462–2467

Talbot JJ, Shillingford JM, Vasanth S, Doerr N, Mukherjee S, Kinter MT, Watnick T, Weimbs T (2011) Polycystin-1 regulates STAT activity by a dual mechanism. Proc Natl Acad Sci USA 108:7985–7990

Talbot JJ, Song X, Wang X, Rinschen MM, Doerr N, LaRiviere WB, Schermer B, Pei YP, Torres VE, Weimbs T (2014) The cleaved cytoplasmic tail of polycystin-1 regulates Src-dependent STAT3 activation. J Am Soc Nephrol 25:1737–1748

Tam LW, Ranum PT, Lefebvre PA (2013) CDKL5 regulates flagellar length and localizes to the base of the flagella in Chlamydomonas. Mol Biol Cell 24:588–600

Tam LW, Wilson NF, Lefebvre PA (2007) A CDK-related kinase regulates the length and assembly of flagella in Chlamydomonas. J Cell Biol 176:819–829

Tao S, Kakade VR, Woodgett JR, Pandey P, Suderman ED, Rajagopal M, Rao R (2015) Glycogen synthase kinase-3beta promotes cyst expansion in polycystic kidney disease. Kidney Int 87:1164–1175

Tao Y, Kim J, Schrier RW, Edelstein CL (2005) Rapamycin markedly slows disease progression in a rat model of polycystic kidney disease. J Am Soc Nephrol 16:46–51

Thiel C, Kessler K, Giessl A, Dimmler A, Shalev SA, von der Haar S, Zenker M, Zahnleiter D, Stoss H, Beinder E et al (2011) NEK1 mutations cause short-rib polydactyly syndrome type majewski. Am J Hum Genet 88:106–114

Thompson CL, Chapple JP, Knight MM (2014) Primary cilia disassembly down-regulates mechanosensitive hedgehog signalling: a feedback mechanism controlling ADAMTS-5 expression in chondrocytes. Osteoarthr Cartil 22:490–498

Torres VE, Harris PC (2006) Mechanisms of disease: autosomal dominant and recessive polycystic kidney diseases. Nat Clin Pract Nephrol 2:40–55 quiz 55

Torres VE, Chapman AB, Devuyst O, Gansevoort RT, Grantham JJ, Higashihara E, Perrone RD, Krasa HB, Ouyang J, Czerwiec FS (2012) Tolvaptan in patients with autosomal dominant polycystic kidney disease. N Engl J Med 367:2407–2418

Torres VE, Meijer E, Bae KT, Chapman AB, Devuyst O, Gansevoort RT, Grantham JJ, Higashihara E, Perrone RD, Krasa HB et al (2011) Rationale and design of the TEMPO (tolvaptan efficacy and safety in management of autosomal dominant polycystic kidney disease and its outcomes) 3–4 study. Am J Kidney Dis 57:692–699

Torres VE, Wang X, Qian Q, Somlo S, Harris PC, Gattone VH 2nd (2004) Effective treatment of an orthologous model of autosomal dominant polycystic kidney disease. Nat Med 10:363–364

Tran PV, Lechtreck KF (2016) An age of enlightenment for cilia: the FASEB summer research conference on the "Biology of cilia and flagella". Dev Biol 409:319–328

Tran PV, Sharma M, Li X, Calvet JP (2014a) Developmental signaling: does it bridge the gap between cilia dysfunction and renal cystogenesis? Birth Defects Res C Embryo Today 102:159–173

Tran PV, Talbott GC, Turbe-Doan A, Jacobs DT, Schonfeld MP, Silva LM, Chatterjee A, Prysak M, Allard BA, Beier DR (2014b) Downregulating Hedgehog signaling reduces renal cystogenic potential of mouse models. J Am Soc Nephrol 25(10):2201–2212. doi:10.1681/ASN.2013070735. Epub 2014 Apr 3

Tran PV, Haycraft CJ, Besschetnova TY, Turbe-Doan A, Stottmann RW, Herron BJ, Chesebro AL, Qiu H, Scherz PJ, Shah JV et al (2008) THM1 negatively modulates mouse sonic hedgehog signal transduction and affects retrograde intraflagellar transport in cilia. Nat Genet 40:403–410

Tran U, Zakin L, Schweickert A, Agrawal R, Doger R, Blum M, De Robertis EM, Wessely O (2010) The RNA-binding protein bicaudal C regulates polycystin 2 in the kidney by antagonizing miR-17 activity. Development 137:1107–1116

Trudel M, Yao Q, Qian F (2016) The role of G-protein-coupled receptor proteolysis site cleavage of polycystin-1 in renal physiology and polycystic kidney disease. Cells 5

Upadhyay VS, Muntean BS, Kathem SH, Hwang JJ, Aboualaiwi WA, Nauli SM (2014) Roles of dopamine receptor on chemosensory and mechanosensory primary cilia in renal epithelial cells. Front Physiol 5:72

van der Vaart A, Rademakers S, Jansen G (2015) DLK-1/p38 MAP kinase signaling controls cilium length by regulating RAB-5 mediated endocytosis in *Caenorhabditis elegans*. PLoS Genet 11:e1005733

Verghese E, Ricardo SD, Weidenfeld R, Zhuang J, Hill PA, Langham RG, Deane JA (2009) Renal primary cilia lengthen after acute tubular necrosis. J Am Soc Nephrol 20:2147–2153

Verghese E, Weidenfeld R, Bertram JF, Ricardo SD, Deane JA (2008) Renal cilia display length alterations following tubular injury and are present early in epithelial repair. Nephrol Dial Transplant 23:834–841

Wahl PR, Serra AL, Le Hir M, Molle KD, Hall MN, Wuthrich RP (2006) Inhibition of mTOR with sirolimus slows disease progression in Han:SPRD rats with autosomal dominant polycystic kidney disease (ADPKD). Nephrol Dial Transplant 21:598–604

Walczak-Sztulpa J, Eggenschwiler J, Osborn D, Brown DA, Emma F, Klingenberg C, Hennekam RC, Torre G, Garshasbi M, Tzschach A et al (2010) Cranioectodermal Dysplasia, Sensenbrenner syndrome, is a ciliopathy caused by mutations in the IFT122 gene. Am J Hum Genet 86:949–956

Wallace DP (2011) Cyclic AMP-mediated cyst expansion. Biochim Biophys Acta 1812:1291–1300

Walz G, Budde K, Mannaa M, Nurnberger J, Wanner C, Sommerer C, Kunzendorf U, Banas B, Horl WH, Obermuller N et al (2010) Everolimus in patients with autosomal dominant polycystic kidney disease. N Engl J Med 363:830–840
Wang X, Gattone V 2nd, Harris PC, Torres VE (2005) Effectiveness of vasopressin V2 receptor antagonists OPC-31260 and OPC-41061 on polycystic kidney disease development in the PCK rat. J Am Soc Nephrol 16:846–851
Wang W, Li F, Sun Y, Lei L, Zhou H, Lei T, Xia Y, Verkman AS, Yang B (2015) Aquaporin-1 retards renal cyst development in polycystic kidney disease by inhibition of Wnt signaling. FASEB J 29:1551–1563
Wang J, Silva M, Haas LA, Morsci NS, Nguyen KC, Hall DH, Barr MM (2014) *C. elegans* ciliated sensory neurons release extracellular vesicles that function in animal communication. Curr Biol 24:519–525
Wang X, Wu Y, Ward CJ, Harris PC, Torres VE (2008) Vasopressin directly regulates cyst growth in polycystic kidney disease. J Am Soc Nephrol 19:102–108
Wang S, Zhang J, Nauli SM, Li X, Starremans PG, Luo Y, Roberts KA, Zhou J (2007) Fibrocystin/polyductin, found in the same protein complex with polycystin-2, regulates calcium responses in kidney epithelia. Mol Cell Biol 27:3241–3252
Ward HH, Brown-Glaberman U, Wang J, Morita Y, Alper SL, Bedrick EJ, Gattone VH 2nd, Deretic D, Wandinger-Ness A (2011) A conserved signal and GTPase complex are required for the ciliary transport of polycystin-1. Mol Biol Cell 22:3289–3305
Wei W, Hackmann K, Xu H, Germino G, Qian F (2007) Characterization of cis-autoproteolysis of polycystin-1, the product of human polycystic kidney disease 1 gene. J Biol Chem 282:21729–21737
Weimbs T (2007) Polycystic kidney disease and renal injury repair: common pathways, fluid flow, and the function of polycystin-1. Am J Physiol Renal Physiol 293:F1423–F1432
Westlake CJ, Baye LM, Nachury MV, Wright KJ, Ervin KE, Phu L, Chalouni C, Beck JS, Kirkpatrick DS, Slusarski DC et al (2011) Primary cilia membrane assembly is initiated by Rab11 and transport protein particle II (TRAPPII) complex-dependent trafficking of Rabin8 to the centrosome. Proc Natl Acad Sci USA 108:2759–2764
Wheway G, Abdelhamed Z, Natarajan S, Toomes C, Inglehearn C, Johnson CA (2013) Aberrant Wnt signalling and cellular over-proliferation in a novel mouse model of Meckel–Gruber syndrome. Dev Biol 377:55–66
White MC, Quarmby LM (2008) The NIMA-family kinase, Nek1 affects the stability of centrosomes and ciliogenesis. BMC Cell Biol 9:29
Williams CL, Li C, Kida K, Inglis PN, Mohan S, Semenec L, Bialas NJ, Stupay RM, Chen N, Blacque OE et al (2011) MKS and NPHP modules cooperate to establish basal body/transition zone membrane associations and ciliary gate function during ciliogenesis. J Cell Biol 192:1023–1041
Wloga D, Camba A, Rogowski K, Manning G, Jerka-Dziadosz M, Gaertig J (2006) Members of the NIMA-related kinase family promote disassembly of cilia by multiple mechanisms. Mol Biol Cell 17:2799–2810
Wood CR, Rosenbaum JL (2015) Ciliary ectosomes: transmissions from the cell's antenna. Trends Cell Biol 25:276–285
Wood CR, Huang K, Diener DR, Rosenbaum JL (2013) The cilium secretes bioactive ectosomes. Curr Biol 23:906–911
Woollard JR, Punyashtiti R, Richardson S, Masyuk TV, Whelan S, Huang BQ, Lager DJ, vanDeursen J, Torres VE, Gattone VH et al (2007) A mouse model of autosomal recessive polycystic kidney disease with biliary duct and proximal tubule dilatation. Kidney Int 72:328–336
Wright KJ, Baye LM, Olivier-Mason A, Mukhopadhyay S, Sang L, Kwong M, Wang W, Pretorius PR, Sheffield VC, Sengupta P et al (2011) An ARL3-UNC119-RP2 GTPase cycle targets myristoylated NPHP3 to the primary cilium. Genes Dev 25:2347–2360

Xu Q, Zhang Y, Wei Q, Huang Y, Li Y, Ling K, Hu J (2015) BBS4 and BBS5 show functional redundancy in the BBSome to regulate the degradative sorting of ciliary sensory receptors. Sci Rep 5:11855

Yamaguchi T, Hempson SJ, Reif GA, Hedge AM, Wallace DP (2006) Calcium restores a normal proliferation phenotype in human polycystic kidney disease epithelial cells. J Am Soc Nephrol 17:178–187

Yamaguchi T, Nagao S, Kasahara M, Takahashi H, Grantham JJ (1997) Renal accumulation and excretion of cyclic adenosine monophosphate in a murine model of slowly progressive polycystic kidney disease. Am J Kidney Dis 30:703–709

Yamaguchi T, Wallace DP, Magenheimer BS, Hempson SJ, Grantham JJ, Calvet JP (2004) Calcium restriction allows cAMP activation of the B-Raf/ERK pathway, switching cells to a cAMP-dependent growth-stimulated phenotype. J Biol Chem 279:40419–40430

Yang TT, Su J, Wang WJ, Craige B, Witman GB, Tsou MF, Liao JC (2015) Superresolution pattern recognition reveals the architectural map of the ciliary transition zone. Sci Rep 5:14096

Ye H, Wang X, Sussman CR, Hopp K, Irazabal MV, Bakeberg JL, LaRiviere WB, Manganiello VC, Vorhees CV, Zhao H et al (2015) Modulation of polycystic kidney disease severity by phosphodiesterase 1 and 3 subfamilies. J Am Soc Nephrol 27(5):1312–1320. doi:10.1681/ASN.2015010057. Epub 2015 Sep 15

Ying G, Avasthi P, Irwin M, Gerstner CD, Frederick JM, Lucero MT, Baehr W (2014) Centrin 2 is required for mouse olfactory ciliary trafficking and development of ependymal cilia planar polarity. J Neurosci 34:6377–6388

Yoder BK, Hou X, Guay-Woodford LM (2002) The polycystic kidney disease proteins, polycystin-1, polycystin-2, polaris, and cystin, are co-localized in renal cilia. J Am Soc Nephrol 13:2508–2516

Young RW (1971) The renewal of rod and cone outer segments in the rhesus monkey. J Cell Biol 49:303–318

Young RW, Bok D (1969) Participation of the retinal pigment epithelium in the rod outer segment renewal process. J Cell Biol 42:392–403

Yuan S, Zhao L, Brueckner M, Sun Z (2015) Intraciliary calcium oscillations initiate vertebrate left–right asymmetry. Curr Biol 25:556–567

Zaika O, Mamenko M, Berrout J, Boukelmoune N, O'Neil RG, Pochynyuk O (2013) TRPV4 dysfunction promotes renal cystogenesis in autosomal recessive polycystic kidney disease. J Am Soc Nephrol 24:604–616

Zhang ZR, Chu WF, Song B, Gooz M, Zhang JN, Yu CJ, Jiang S, Baldys A, Gooz P, Steele S et al (2013) TRPP2 and TRPV4 form an EGF-activated calcium permeable channel at the apical membrane of renal collecting duct cells. PLoS One 8:e73424

Zhang Q, Nishimura D, Seo S, Vogel T, Morgan DA, Searby C, Bugge K, Stone EM, Rahmouni K, Sheffield VC (2011) Bardet–Biedl syndrome 3 (Bbs3) knockout mouse model reveals common BBS-associated phenotypes and Bbs3 unique phenotypes. Proc Natl Acad Sci USA 108:20678–20683

Zhang Q, Yu D, Seo S, Stone EM, Sheffield VC (2012) Intrinsic protein–protein interaction-mediated and chaperonin-assisted sequential assembly of stable Bardet–Biedl syndrome protein complex, the BBSome. J Biol Chem 287:20625–20635

Zhou J (2009) Polycystins and primary cilia: primers for cell cycle progression. Annu Rev Physiol 71:83–113

Zimmermann KW (1898) Beitrage zur Kenntniss einiger Drusen und Epithelien. Arch Mikrosk Anat 52:552–706

Zuo X, Fogelgren B, Lipschutz JH (2011) The small GTPase Cdc42 is necessary for primary ciliogenesis in renal tubular epithelial cells. J Biol Chem 286:22469–22477

Zuo X, Guo W, Lipschutz JH (2009) The exocyst protein Sec10 is necessary for primary ciliogenesis and cystogenesis in vitro. Mol Biol Cell 20:2522–2529

Chapter 12
Inflammation and Fibrosis in Polycystic Kidney Disease

Cheng Jack Song, Kurt A. Zimmerman, Scott J. Henke, and Bradley K. Yoder

Abstract Polycystic kidney disease (PKD) is a commonly inherited disorder characterized by cyst formation and fibrosis (Wilson, N Engl J Med 350:151–164, 2004) and is caused by mutations in cilia or cilia-related proteins, such as polycystin 1 or 2 (Oh and Katsanis, Development 139:443–448, 2012; Kotsis et al., Nephrol Dial Transplant 28:518–526, 2013). A major pathological feature of PKD is the development of interstitial inflammation and fibrosis with an associated accumulation of inflammatory cells (Grantham, N Engl J Med 359:1477–1485, 2008; Zeier et al., Kidney Int 42:1259–1265, 1992; Ibrahim, Sci World J 7:1757–1767, 2007). It is unclear whether inflammation is a driving force for cyst formation or a consequence of the pathology (Ta et al., Nephrology 18:317–330, 2013) as in some murine models cysts are present prior to the increase in inflammatory cells (Phillips et al., Kidney Blood Press Res 30:129–144, 2007; Takahashi et al., J Am Soc Nephrol JASN 1:980–989, 1991), while in other models the increase in inflammatory cells is present prior to or coincident with cyst initiation (Cowley et al., Kidney Int 43:522–534, 1993, Kidney Int 60:2087–2096, 2001). Additional support for inflammation as an important contributor to cystic kidney disease is the increased expression of many pro-inflammatory cytokines in murine models and human patients with cystic kidney disease (Karihaloo et al., J Am Soc Nephrol JASN 22:1809–1814, 2011; Swenson-Fields et al., Kidney Int, 2013; Li et al., Nat Med 14:863–868, 2008a). Based on these data, an emerging model in the field is that disruption of primary cilia on tubule epithelial cells leads to abnormal cytokine cross talk between the epithelium and the inflammatory cells contributing to cyst growth and fibrosis (Ta et al., Nephrology 18:317–330, 2013). These cytokines are produced by interstitial fibroblasts, inflammatory cells, and tubule epithelial cells and activate multiple pathways including the JAK-STAT and NF-κB signaling (Qin et al., J Am Soc Nephrol JASN 23:1309–1318, 2012; Park et al., Am J Nephrol 32:169–178, 2010; Bhunia et al., Cell 109:157–168, 2002). Indeed, inflammatory cells are

C.J. Song • K.A. Zimmerman • S.J. Henke • B.K. Yoder (✉)
Department of Cell, Developmental and Integrative Biology, University of Alabama at Birmingham, Birmingham, AL, USA
e-mail: byoder@uab.edu

R.K. Miller (ed.), *Kidney Development and Disease*, Results and Problems in Cell Differentiation 60, DOI 10.1007/978-3-319-51436-9_12

responsible for producing several of the pro-fibrotic growth factors observed in PKD patients with fibrosis (Nakamura et al., Am J Nephrol 20:32–36, 2000; Wilson et al., J Cell Physiol 150:360–369, 1992; Song et al., Hum Mol Genet 18:2328–2343, 2009; Schieren et al., Nephrol Dial Transplant 21:1816–1824, 2006). These growth factors trigger epithelial cell proliferation and myofibroblast activation that stimulate the production of extracellular matrix (ECM) genes including collagen types 1 and 3 and fibronectin, leading to reduced glomerular function with approximately 50% of ADPKD patients progressing to end-stage renal disease (ESRD). Therefore, treatments designed to reduce inflammation and slow the rate of fibrosis are becoming important targets that hold promise to improve patient life span and quality of life. In fact, recent studies in several PKD mouse models indicate that depletion of macrophages reduces cyst severity. In this chapter, we review the potential mechanisms of interstitial inflammation in PKD with a focus on ADPKD and discuss the role of interstitial inflammation in progression to fibrosis and ESRD.

12.1 Introduction

Polycystic kidney diseases (PKD) are a group of genetically inheritable disorders that are characterized by the formation of cysts in the kidney and other organs (Wilson 2004). Two forms of autosomal PKD exist, depending on their underlying genetic mutation. Autosomal dominant PKD (ADPKD) affects approximately 1 in 1000 people and is caused by a mutation in either *PKD1* or *PKD2*, which encode the polycystin-1 (PC1) or polycystin-2 (PC2) protein, respectively (Wilson 2004). Autosomal recessive PKD (ARPKD) affects 1 in 20,000 individuals and is caused by a mutation in *PKHD1*, which encodes the protein fibrocystin/polyductin (Wilson 2004). For both forms of the disease, the affected proteins localize to the primary cilia although both proteins are also found in regions outside of the primary cilium (Wilson 2004; Oh and Katsanis 2012). While the function of primary cilia in the kidney remains uncertain, it has been proposed that they are involved in mechanosensation to detect fluid flow through the tubule lumen, in regulation of cell proliferation and oriented cell divisions, as well as in cell-to-cell and cell-to-matrix signaling (Yoder 2007; Yoder et al. 2002; Zimmerman and Yoder 2015; Kotsis et al. 2013; Basten and Giles 2013). In ADPKD, cyst growth is associated with changes in epithelial cell proliferation and differentiation, fluid secretion, and basement membrane abnormalities (Wilson 2004). Expansion of cysts leads to obstruction and compression of surrounding nephrons that can significantly reduce kidney function (Grantham et al. 2011). During late stages of disease progression, cyst formation is accompanied by ECM deposition and fibrosis, further reducing glomerular filtration and eventually leading to ESRD (Wilson 2004; Wilson and Goilav 2007; Wilson and Falkenstein 1995). Importantly, disruption of primary cilia or the polycystins using a variety of mouse models causes cyst formation in the kidney and liver along with associated fibrosis in both organs (Yoder et al. 1995; Ma et al. 2013). Recently, in vivo studies using conditional cilia and polycystin mutant mice showed that the rate of cyst formation depends on the timing of gene deletion. Disruption of cilia or polycystin proteins prior to postnatal day 12–14 caused rapid cyst formation. In contrast, deletion of these proteins after postnatal day 14 caused slow focal cyst formation (Davenport et al. 2007; Piontek et al. 2007; Lin et al. 2003). However, the protracted rate of cyst formation in the adult-induced mutants can be greatly exacerbated by injury (Sharma et al. 2013; Takakura et al. 2009; Patel et al. 2008). Expression profiling data indicated there is an increased response to injury with marked elevation of innate immune response genes in cilia or polycystin mutant samples, leading to the idea that inflammatory cells are involved in PKD progression (Mrug et al. 2008). This is further supported by data showing reduced cystic pathology and improved renal function in mouse PKD models in which macrophages were depleted (Karihaloo et al. 2011; Swenson-Fields et al. 2013). The involvement of other inflammatory cell types in PKD is less understood. Inflammatory cells can serve as potent producers of pro-inflammatory, pro-mitotic, and pro-fibrotic cytokines; therefore, understanding the involvement of inflammatory cells in cyst formation and fibrosis and their relationship with signaling pathways regulated by the primary cilia located on the epithelium will provide innovative insights into the mechanism of disease progression. More important, this understanding will open new targets for therapeutic intervention to slow the progression of cystic kidney disorders in humans.

12.2 Inflammation in the Pathogenesis of PKD

12.2.1 Immune Cells in PKD

Accumulation of inflammatory cells in the renal interstitial space is becoming appreciated as a hallmark of human and animal models of PKD (Fig. 12.1a, b) (Ta et al. 2013; Grantham 2008). One of the most studied inflammatory cells is the macrophage. Macrophages are involved in innate immunity and in tissue development, repair, and homeostasis (Ginhoux et al. 2015; Wynn et al. 2013). In kidney, they express surface markers including F4/80, CD11b, and Ly6c in mice and CD14 and CD16 in humans, which are used to identify different subtypes of macrophage (Nelson et al. 2012). An interesting feature of macrophages is their ability to become polarized and express either pro-inflammatory or anti-inflammatory cytokines in response to signals found in the tissue microenvironment. This paradigm is supported by data showing two distinct responses to treatment with cytokines in vitro (Gordon and Taylor 2005; Mantovani et al. 2002). Macrophages stimulated with LPS or interferon-gamma demonstrate a pro-inflammatory Th1-like phenotype with potent antimicrobial and antitumor activity. This macrophage is referred

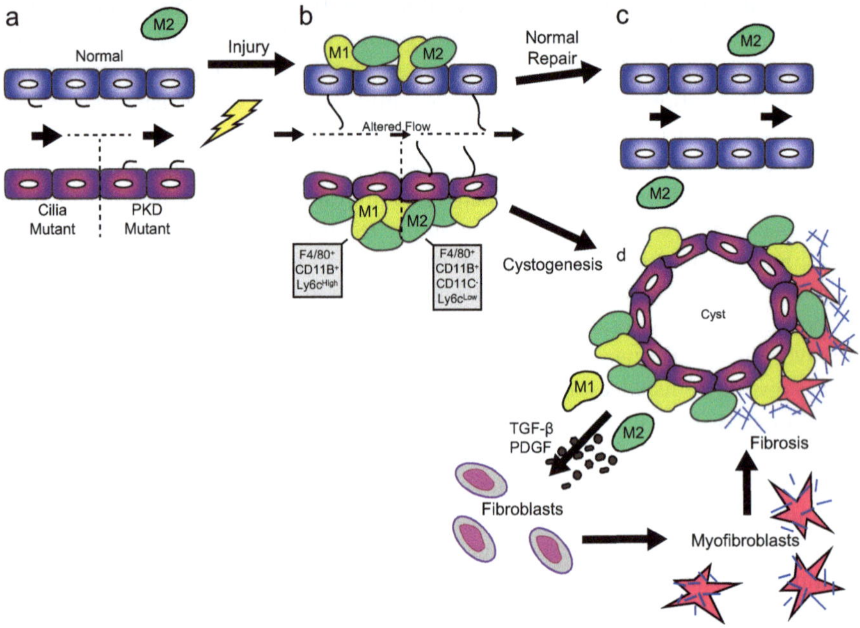

Fig. 12.1 Proposed model focused on the involvement of macrophages following renal injury under normal conditions and in the presence of ciliary or polycystin dysfunction. (**a**) During steady state, primary cilia protrude into the lumen of renal tubules. Macrophages (M2 like) are present in the interstitial space surrounding the renal tubules. (**b**) Following injury, there is increased accumulation of macrophage populations (M1-like) or (M2-like) in the interstitial space of the kidney. Furthermore, cilia length initially increases following injury followed by cilia regression during stages of rapid epithelial cell proliferation. (**c**) During the repair phase in the normal mouse kidney, the injured tubules proliferate, downregulate inflammatory cell accumulation, reform primary cilia, and fully repair the injured epithelium. (**d**) However, in the cilia or PKD mutant mouse kidney, there is an increased accumulation of macrophage populations (M1 and M2) which fail to properly resolve following injury. The persistent increase in macrophage number leads to enhanced secretion of pro-inflammatory and pro-fibrotic cytokines which causes transition of fibroblasts to a myofibroblast phenotype. These myofibroblasts produce large amounts of extracellular matrix proteins leading to deposition of collagen into the extracellular matrix and fibrosis in the PKD kidney

to as the classically activated or M1-like macrophage. M1 macrophages are characterized by expression of iNOS, TNF-α, and IL-1β amongst others and have a pro-inflammatory function (Sica and Mantovani 2012; Benoit et al. 2008). On the other hand, naïve macrophages treated with interleukin-4 or 13 produce an anti-inflammatory response and are referred to as alternatively activated or M2-like macrophages (Adams and Hamilton 1984). These M2 macrophages express arginase 1 and IL-10 and have anti-inflammatory and wound healing functions (Sica and Mantovani 2012; Benoit et al. 2008). However, several in vivo studies demonstrated that macrophages cannot be clearly defined into either M1- or M2-like states, but have a wide degree of plasticity depending on their environment (Helm et al. 2014; Kratochvill et al. 2015; Stables et al. 2011; Xue et al. 2014; Ginhoux and Jung 2014; Lavin et al. 2014). It is now believed that macrophages which are present in vivo display a range of phenotypes that fall somewhere in between the M1 and M2 spectrum and that they are able to rapidly switch these phenotypes based on external cues. Intriguingly, evidence suggests that M1 macrophages can transition to an M2-like polarization following phagocytosis of necrotic or apoptotic cells (Huen and Cantley 2015; Fadok et al. 1998) or when cultured with renal epithelium demonstrating the importance of the tissue microenvironment on macrophage polarization (Swenson-Fields et al. 2013).

The origin of macrophages in specific tissues is an area of intense debate. The old model regarding macrophage origin suggests that all macrophages present in tissue can be maintained from a constant supply of bone marrow-derived circulating monocytes (van Furth and Cohn 1968; Katz et al. 1979). However, the new model of macrophage origin has demonstrated that there are two distinct populations of macrophages that can be distinguished by their progenitors, developmental history, and renewal mechanisms (Schulz et al. 2012). One subtype of macrophage is derived from bone marrow progenitor cells and is referred to as monocytes or infiltrating macrophages (in mice F4/80^{lo} and Cd11b^{hi}) once they reach their destination organ (Schulz et al. 2012; Merad et al. 2002; Ginhoux et al. 2010). In contrast, tissue-resident macrophages (in mice F4/80^{hi} and Cd11b^{lo}) develop in the yolk sac, enter tissues early during development, and originate from embryonic progenitors (Schulz et al. 2012). The yolk sac-derived macrophages persist in some adult tissues (e.g., epidermis, liver) independent of bone marrow-derived hematopoietic stem cells (HSCs) (Schulz et al. 2012; Ginhoux et al. 2010). They can self-proliferate in response to growth factors such as CSF-1 and other proliferative cytokines (Ginhoux and Jung 2014). In normal kidney, about half of the resident macrophages are derived from the yolk sac and maintained through self-renewal while the remainders are derived from the hematopoietic lineage (Schulz et al. 2012). Resident renal macrophages have important developmental roles influencing ureteric bud branching, producing cytokines, and monitoring surrounding cells for abnormal cell death (Rae et al. 2007; Pollard 2009). The second distinct population of renal macrophage, the infiltrating macrophage, rapidly accumulates in the kidney in response to injury (e.g., following renal ischemia–reperfusion injury; IRI), produces large amounts of pro-inflammatory cytokines, and is associated with worsened kidney injury and fibrosis. Bone marrow-derived infiltrating macrophages can be further classified based on expression of the surface marker Ly6c (Ly6c^{hi}, Ly6c^{int}, Ly6c^{lo}) (Clements et al. 2015; Lin et al. 2009). Following IR injury, Ly6c^{hi} infiltrating macrophages present in the kidney and produce pro-inflammatory cytokines (Clements et al. 2015). Several days after infiltration, these Ly6c^{hi} macrophages downregulate expression of Ly6c (Ly6c^{lo}) associated with kidney repair and production of extracellular matrix genes (Lin et al. 2009). Despite functional similarities to M1 and M2 macrophages, gene expression analysis comparing Ly6c^{hi}, Ly6c^{int}, and Ly6c^{lo} macrophages to M1- and M2-like macrophages indicates that these populations do not fit into the canonical M1/M2 classification (Clements et al. 2015).

Both infiltrating and resident macrophage numbers increase in PKD. While data indicate that macrophages impact the rate of cyst formation based on depletion studies in mice, it is not clear whether inflammation is important for cyst initiation or is simply a consequence of the pathology (Ta et al. 2013). In the Lewis polycystic kidney (*lpk*) and *pck* rats, data indicate cysts are present prior to the increase in inflammatory cells (Phillips et al. 2007; Takahashi et al. 1991); however, in Han: SPRD rats inflammatory cells are present prior to or coincident with cyst initiation (Cowley et al. 1993, 2001). Furthermore, using a heterozygous ADPKD mouse model exposed to IRI, two groups demonstrated an increased number of neutrophils, macrophages, and pro-inflammatory cytokines prior to cyst formation (Bastos et al. 2009; Prasad et al. 2009). The functional importance of macrophages in cystic disease progression was demonstrated by depletion of macrophages using liposomal clodronate (LC), a phagocytic poison that leads to macrophage death. Treatment of mice with LC improved cystic indices and renal function compared to controls, suggesting that macrophages do have a role in promoting cyst formation (Karihaloo et al. 2011; Swenson-Fields et al. 2013).

The overlap between resident and infiltrating macrophage populations in mice and humans remains uncertain as many of the macrophage surface markers are not shared between species. In human PKD, CD68-positive macrophages, which resemble M2-like or resident macrophages in mice, are present in the renal interstitium of ADPKD patients at both early and advanced stages of kidney disease (Zeier et al. 1992). More recently, HAM56 (a marker for human macrophages)-positive macrophages were shown to be distributed throughout the interstitium of kidneys from ARPKD and ADPKD patients including sites adjacent to cysts. However, the ratio of M2-like macrophages out of total macrophages was similar between control and PKD kidneys (Swenson-Fields et al. 2013), suggesting that the markers used in this study were not specific for human M2-like macrophages or that the ratio of M2-like macrophages was unaltered in the cystic kidneys. Interestingly, epithelial cells derived from control and cystic human kidneys produced soluble factors which induced polarization of macrophages to an M2-like phenotype in vitro (Swenson-Fields et al. 2013). However, marker characterization studies of the M2-like macrophages found within ADPKD kidneys have not revealed the origins of these macrophages (yolk sac or bone marrow derived). The M2-like macrophage response in mice was first noted in a rapidly progressive model of ARPKD, the Cys1-cpk mouse (Mrug et al. 2008); however, when these macrophages enter the kidney remains controversial. In contrast, CD-68 (ED-1)-positive macrophages have been detected in heterozygous Han:SPRD-Cy rats when cystic dilatations were initiating, suggesting a possible causal rather than a responsive role for macrophages in cyst development (Cowley et al. 2001). Subsequent studies using the $PKD1^{fl/fl}$; Pkhd1-Cre mouse also demonstrated the presence of alternatively activated macrophages based on expression of surface markers (F4/80+, CD45+, CD11c-) and low levels of Ly6C ($Ly6C^{lo}$) (Karihaloo et al. 2011).

The involvement of other immune cells such as T lymphocytes, mast cells, and neutrophils has been reported in human PKD patients and rodent models (Zeier et al. 1992; Ibrahim 2007; Takahashi et al. 1991; Vogler et al. 1999; Kaspareit-

Rittinghausen et al. 1989; McPherson et al. 2004; Bernhardt et al. 2007). Currently, there is significant debate about the presence of dendritic cells in the kidney. Several groups have shown that dendritic cells are present in the kidney based on their surface expression of CD11b, F4/80, and CD11c (Li et al. 2008b; Hochheiser et al. 2013; Hull et al. 2015). However, CD11c has also been reported to be expressed by tissue-resident macrophages, raising the possibility that resident macrophages and dendritic cells may be overlapping populations (Cao et al. 2015). In addition to macrophages, the number of CD4-positive T cells is increased in the interstitial space of ADPKD patients and in animal models of PKD (Zeier et al. 1992; Takahashi et al. 1991; Vogler et al. 1999; Kaspareit-Rittinghausen et al. 1989). Furthermore, it is believed that mast cells may be involved in PKD progression through production of pro-inflammatory molecules including chymase and AngII (McPherson et al. 2004; Nadel 1991; Dell'Italia and Husain 2002). Finally, an elevated number of neutrophils have been reported in human and canine cystic kidney disease models (Bernhardt et al. 2007; O'Leary et al. 1999). However, conclusions about the contribution of neutrophils to disease pathogenesis are limited as the markers used for neutrophils in those studies have subsequently been shown to be non-neutrophil specific. While a wide variety of immune cells are increased in cystic models, their importance in cyst formation remains to be determined. Future studies should be directed at better defining these populations and their respective roles in cyst formation and fibrosis.

12.2.2 Dysregulation of Chemokines and Cytokines in PKD

Chemokines are important for immune cell infiltration, activation, and polarization and in regulating immune cell behavior (Fig. 12.2b, c). Therefore, determining the involvement of chemokines and cytokines in mouse models of PKD will provide insight into their role in modulating immune cells and inflammation associated with cystic disease.

One of the best studied chemokines is MCP-1 (Ccl2), which binds to its cognate receptor, CCR2. CCR2 is a member of the CC chemokine family of G-protein coupled receptors and is typically expressed on the surface of monocytes or T cells (Li et al. 2008b; Bruhl et al. 2004). CCR2 mediates cell infiltration in renal inflammatory states such as diabetic nephropathy, glomerulonephritis, and ischemia–reperfusion injury. More important, MCP-1 is markedly elevated in the cyst fluid of ADPKD patients, and increased levels of MCP-1 are associated with worsened renal function and cystic disease outcome (Zheng et al. 2003). The increased expression of MCP-1 in mouse and rat models of PKD appears to parallel that observed in humans. In Han:SPRD rats, MCP-1 mRNA was elevated in homozygous mutant rats compared to wild-type controls (Cowley et al. 2001). Interestingly, heterozygous Han:SPRD cy/+ males displayed higher MCP-1 levels compared to females, which correlates with observations in humans that males often experience higher MCP-1 concentration in urine and serum than females

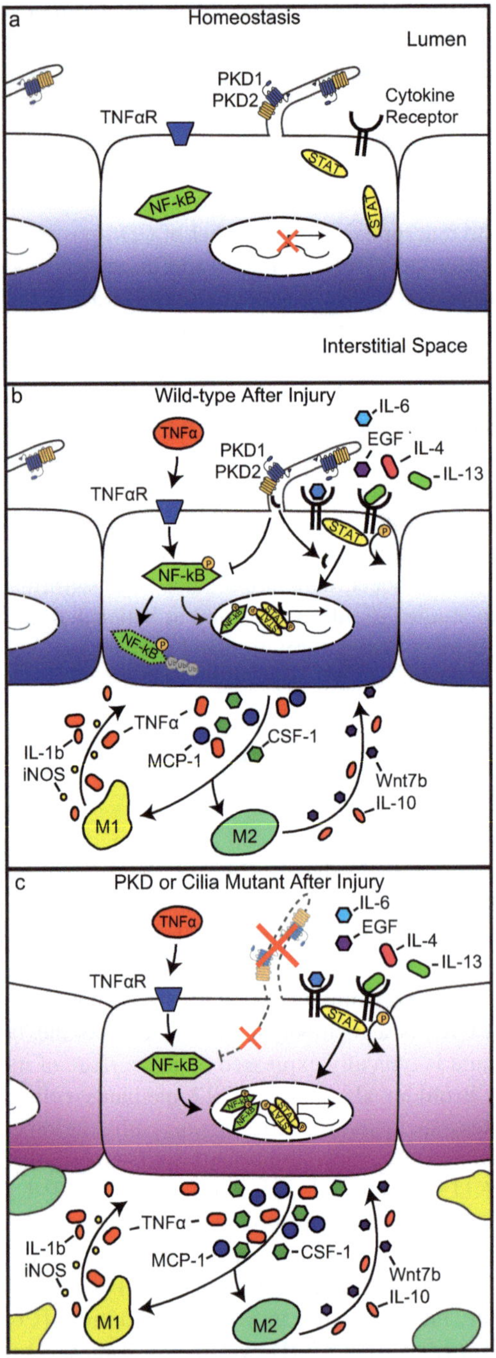

Fig. 12.2 Inflammatory signaling pathways activated under normal or injured conditions in control, cilia, or PKD mutant epithelial cells. (**a**) During homeostasis, NF-κB and JAK-STAT signaling pathways are not activated. (**b**) In injured wild-type epithelial cells, PC1 undergoes C-terminal tail (CTT) proteolytic cleavage. The CTT of PC1 translocates into the nucleus where it

(Cowley et al. 2001). In vitro models suggest that MCP-1 is produced by tubule epithelial cells as *PKD1-/-* tubule cells have significantly higher expression of MCP-1 compared to control cells. However, in a mouse adriamycin nephropathy model, data suggest that renal F4/80+ CD11c+ mononuclear phagocytes in kidney are also a major source of MCP-1 (Cao et al. 2015). Based on the increased MCP-1 expression across multiple cystic models and human patients, MCP-1 has been proposed as a biomarker for PKD. However, there is no evidence demonstrating a direct causative role for MCP-1 in disease progression as inhibition of MCP-1 signaling using Bindarit did not affect cyst formation (Zoja et al. 2015). Nevertheless, Bindarit treatment did reduce the renal accumulation of ED-1-positive inflammatory cells, suggesting that CCR2-mediated inflammatory cell infiltrates are not a critical component of PKD.

In addition to MCP-1, expression of several other chemokines/cytokines is increased in cystic kidney disease models. This includes osteopontin, IL-1beta, CCL6, CX3CL1, colony-stimulating factor 3 (CSF3), CXCL1, CSF1, CCL28, IL33, IL8, IL6, and IL17D and CXCL8 that function in the recruitment of myeloid-derived cells (macrophages, mast cells, neutrophils) into the cystic kidney. The importance of these chemokines in disease progression still needs to be thoroughly evaluated.

In contrast to the above cytokines, TNF-α appears to have a significant role in cyst formation (Fig. 12.2b, c). In human ADPKD cyst fluid, levels of TNF-α were dramatically increased compared to controls (Li et al. 2008a; Gardner et al. 1991). Using the *cpk* mouse model, Nakamura et al. demonstrated that cystic mice had increased expression of TNF-α mRNA compared to wild-type controls (Nakamura et al. 1993). Furthermore, TNF-α expression increased with age and cyst severity, suggesting that TNF-α accumulation is associated with cyst progression (Li et al. 2008a; Nakamura et al. 1993). Addition of TNF-α to cultured *PKD2+/-* and wild-type embryonic kidney explants induced cystogenesis that could be blocked by the TNF-α inhibitor etanercept (Li et al. 2008a). Inhibition of TNF-α-converting enzyme (TACE) led to a significant reduction in cystic index, serum creatinine, and BUN levels in the *bpk* mouse model of PKD (Dell et al. 2001). Taken together, these data implicate TNF-α as a key contributor to cyst formation.

Fig. 12.2 (continued) serves as a co-activator of the STAT signaling pathway. The NF-κB pathway is also activated through ligands, such as TNF-α, binding to their cognate receptors on the cell surface. Collectively, activation of the NF-κB and JAK-STAT signaling pathways induces transient activation of pro-inflammatory genes such as MCP-1, TNF-α, and CSF-1. These cytokines rapidly recruit macrophages to the site of injury leading to production of pro-inflammatory (IL-1β, TNF-α, iNOS) and anti-inflammatory cytokines (IL-10, Wnt7b) that feed back to the injured epithelium leading to tissue repair. (**c**) In injured PKD or cilia mutant epithelial cells, there is persistently increased NF-κB and JAK-STAT pathway activation. The persistent activation of NF-κB and JAK-STAT signaling leads to enhanced production of pro-inflammatory, pro-fibrotic, and chemoattractant cytokines and increased macrophage accumulation. Accumulated macrophages produce cystogenic cytokines such as TNF-α or Wnt7b that may be responsible for driving cyst formation

12.2.3 Inflammatory Pathways in PKD

Recent studies in PKD disease models demonstrate that disruption of ciliary proteins specifically in the tubule epithelium causes cyst formation that is associated with an altered inflammatory response (Ta et al. 2013). Importantly, immune cells do not possess cilia, raising the possibility that cilia on tubule epithelium have a role in regulating inflammatory responses and that dysregulation of signals between the epithelium and inflammatory cells contributes to cyst development (Fig. 12.2) (Qin et al. 2012; Dinsmore and Reiter 2016; Wann and Knight 2012; Wann et al. 2014; Low et al. 2006). The signaling pathway involved and the potential role for either PC1 or PC2 in the inflammatory response are unknown although mice heterozygous for PC1 or PC2 show enhanced inflammation following renal injury compared to their wild-type littermates (Bastos et al. 2009; Prasad et al. 2009). The global gene analysis of human *PKD1* kidneys found that many inflammatory pathways are activated in PKD including the JAK-STAT pathway and the NF-κB signaling pathway (Song et al. 2009). NF-κB is an important inflammatory pathway activated in response to a variety of cytokines including TNF-α, IL-1β, IL6, and MCP-1 (Pahl 1999). Following ligand binding, phosphorylation of the NF-κB inhibitor, IκB, results in its degradation and subsequent activation and nuclear translocation of NF-κB. Data from several studies support a role for NF-κB in cyst formation (Qin et al. 2012). *PKD1* knockout epithelial cells have increased NF-κB-induced luciferase reporter activity and enhanced phosphorylation of the NF-κB subunit P65 compared to control cells (Qin et al. 2012). Furthermore, Park et al. demonstrated there is an increase of NF-κB protein and that it is phosphorylated in epithelial cells from *PKD2* knockout mice, but not in the control mice (Park et al. 2010). Collectively, these data indicate that upregulation of NF-κB pathway activity occurs in animal models of PKD. Wnt7b may function downstream of NF-κB as studies by Qin et al. showed that increased NF-κB signaling results in overexpression of Wnt7b, which is also present in *PKD1* mutant mice. Importantly, inhibition of NF-κb or Wnt7b results in a significant decrease in cyst severity in the *PKD1* mutant mouse and in organ culture models (Qin et al. 2012).

The JAK-STAT pathway is activated during inflammation and is believed to contribute to cyst formation as demonstrated by increased STAT3 activity in cyst-lining epithelial cells of PKD mutant animals (Fig. 12.2c) (Weimbs and Talbot 2013). In vitro and In vivo studies have shown that PC1 and PC2 are required for JAK1/2 activation and that the c-terminal tail of PC1 regulates STAT3 activation through the src tyrosine kinase (Talbot et al. 2014; Bhunia et al. 2002). The JAK-STAT pathway is also regulated by the suppressor of cytokine signaling (SOCS-1) which limits the inflammatory activity of macrophages. SOCS-1 knockout leads to the development of renal cysts in mice, but it is unknown whether this effect is mediated by inflammation or other contributions of the JAK-STAT signaling pathway (Metcalf et al. 2002).

12.3 Fibrosis in PKD

12.3.1 Extracellular Matrix in PKD

The hallmark of tubule interstitial fibrosis is the accumulation of extracellular matrix (ECM) proteins such as collagen I and III (Zeisberg and Neilson 2010). Similar to many renal disorders, abnormalities in ECM accumulation were also found in kidneys from PKD patients and from orthologous animal models of PKD (Wilson et al. 1992; Wilson and Falkenstein 1995; Wilson and Sherwood 1991; Geng et al. 1999; Wilson and Burrow 1999; Guay-Woodford 2003; Wilson 2008). While ECM accumulation was often described in end-stage kidney disease (ESRD), abnormal ECM composition and increased basement membrane thickness were also observed in early-stage, pre-dialysis ADPKD patients (Wilson et al. 1992). This suggests that tissue remodeling and ECM deposition may be an essential component of cyst progression rather than simply being the result of cyst formation and inflammation. The ADPKD-associated changes include abnormalities in ECM production, composition and turnover, and differences in content of collagen types I, III, and basement membrane collagen type IV, fibronectin, and heparin sulfate proteoglycan (HSPG) (Wilson et al. 1992). These abnormalities were most commonly noted in regions surrounding cysts and interstitial fibroblasts (Norman 2011). While many factors may regulate the activity of pro-fibrotic pathways, recent studies in animal models indicate that polycystins may control production of ECM more directly and that inadequate regulation of this process is responsible for PKD-associated vascular abnormalities (Liu et al. 2014). Importantly, upregulation of the major pro-fibrotic growth factor, transforming growth factor β (TGF-β), was observed in *PKD1* haploinsufficient mice and loss of *PKD1* leads to increased responsiveness of the cystic epithelium and fibroblasts to TGF-β (Wilson et al. 1992; Song et al. 2009; Hassane et al. 2010). These observations may explain enhanced Smad2/3 nuclear localization observed in PKD patients and animal models (Hassane et al. 2010). Similarly, loss of polycystins is associated with notochord collagen overexpression in zebrafish that is thought to contribute to their body axis curvature defects (Mangos et al. 2010). Together, these data suggest that the PKD-associated changes in ECM composition may be a direct consequence of polycystin mutation.

12.3.2 Dysregulation of ECM Turnover in PKD

Development of fibrosis in PKD and other fibrotic models is dependent on both the amount of ECM that is produced and the extent of matrix turnover which occurs during disease progression (Norman 2011; Catania et al. 2007). Among the many factors that influence ECM turnover, matrix metalloproteinases (MMPs) and tissue inhibitors of metalloproteinases (TIMPs) are perhaps the most extensively studied in PKD (Nakamura et al. 2000; Hassane et al. 2010; Catania et al. 2007; Schaefer et al. 1996; Obermuller et al. 2001; Rankin et al. 1996). Despite the clear evidence that MMPs are dysregulated in PKD, the importance and cause of these changes remains controversial. Schaefer and colleagues showed that MMP2 was present in proximal tubule cells from normal rat kidney but significantly decreased in the cystic Han:SPRD rat (Schaefer et al. 1996). This group also demonstrated that TIMP-1 and TIMP-2 mRNA levels were increased in the mutant rat (Schaefer et al. 1996). Paradoxically, MMP14 mRNA was increased in cyst-lining epithelia and distal tubules in the Han:SPRD rat (Obermuller et al. 2001). Inhibition of MMP activity with batimastat reduced cyst number and kidney weight suggesting a detrimental role for MMPs in cyst progression (Obermuller et al. 2001). Likewise, conditioned media from kidney tubule cells isolated from the *cpk* mouse contained elevated amounts of MMP-2, MMP-3, MMP-9, and TIMP1 (Rankin et al. 1996). The increased expression of MMPs in PKD models was recently confirmed by Hassane et al. who showed increased expression of MMP-2 using a *PKD1* mouse model of cyst formation (Hassane et al. 2010). In human ADPKD, patient serum contains increased levels of multiple MMPs, TIMPs, and plasminogen activator inhibitors (PAI) in both pre-dialysis and end-stage ADPKD kidneys (Nakamura et al. 2000). Data demonstrating an important role of the plasminogen activators and inhibitors in PKD are limited, although increased expression of PAI-1 has been reported in both mouse and human PKD samples (Norman 2011; Eddy 2009). It is believed that PAI are pro-fibrotic due to their ability to recruit macrophage and myofibroblasts to the tubulointerstitial area rather than their inhibitory effects on the serine proteases tPA and uPA (Norman 2011; Eddy 2009). Together, these data suggest that MMP expression is dysregulated in PKD and may lead to cyst progression by promoting ECM turnover and allowing expansion of cystic epithelial cells.

The functional consequences of changes in ECM-degrading enzymes in PKD are not fully understood. However, they likely include changes in cellular function and immune regulation. One mechanism of immune regulation may occur through breakdown of collagen I into a PGP fragment. PGP is a key regulator of inflammatory neutrophil accumulation, a process that is central to the pathogenesis of chronic obstructive pulmonary disease (Snelgrove et al. 2010). Since one of two MMPs that participate in generation of PGP is increased in ADPKD (MMP-9), PGP or other ECM collagen fragments may exert important immunoregulatory functions in PKD kidneys.

The interactions of cells with ECM or its breakdown products are mediated by matrix receptors including integrins and syndecans (Geiger et al. 2001). Following binding of ECM to these receptors, focal adhesion complexes form allowing for intracellular signaling to occur through a variety of pathways (Ehrhardt et al. 2002; Wozniak et al. 2004). To date, it has been demonstrated that several ECM receptor components are increased in PKD patients including various integrins, syndecan-4, as well as the integrin-associated focal adhesion kinase (Geng et al. 1999; Wilson and Burrow 1999; Zeltner et al. 2008; Wallace et al. 2014; Joly et al. 2003). Mitigation of renal cyst formation in response to disruption of these pathways (e.g., by genetic targeting of integrin in mice) points to functional relevance of the cell–ECM interactions in the pathogenesis of PKD (Lee et al. 2015). Of note, a hypomorphic mutation in the laminin alpha5 gene was sufficient to drive renal cyst formation (Shannon et al. 2006).

12.3.3 Fibrotic Cells in PKD

Myofibroblasts are specialized cells that can exert large contractile forces, mediate wound healing, and substantially contribute to the expansion of ECM and development of renal tubulointerstitial fibrosis (Fig. 12.1d) (Qi et al. 2006). A hallmark of myofibroblasts is expression of cytoskeletal alpha smooth muscle actin (αSMA) which is incorporated into stress fibers. While αSMA expressing interstitial cells were found in focal areas during early-stage ADPKD, they were found in abundance in end-stage ADPKD kidneys (Hassane et al. 2010). αSMA-positive cells were also found in kidneys from *PKD1*-deficient mice (Hassane et al. 2010) and end-stage PKD kidneys from Han:SPRD-Cy rats (Song et al. 2009; Schieren et al. 2006). Cells expressing high amounts of αSMA are potent producers of ECM proteins including type I and III collagen. Production of excess ECM proteins leads to interstitial fibrosis, tubule and nephron loss, a decline in glomerular filtration rate, and eventual end-stage renal disease. Although the origin of renal PKD myofibroblasts remains unknown, it appears that these cells can differentiate from several different precursors, including resident interstitial fibroblasts, or pericytes (Kuo et al. 1997). Such transformation may be regulated by TGF-β or perhaps by PKD-related alterations in calcium flux (Follonier Castella et al. 2010). Targeting of αSMA-positive cells for anti-fibrotic therapy to help reduce scarring and retain renal function is an attractive idea, although, to date, no such treatment exists.

The origin of myofibroblasts in the kidney of PKD mouse models remains unknown; however, in a UUO model of renal fibrosis, it was shown that a majority of myofibroblasts in the kidney were derived through proliferation of resident fibroblasts although bone marrow-derived fibrocytes also contributed (~35%) to the myofibroblast population (LeBleu et al. 2013). Using fate mapping studies, this group concluded the vascular pericytes likely do not contribute to renal fibrosis. In contrast, Humphreys and colleagues determined that resident pericytes were major contributors to renal myofibroblasts following UUO-induced injury (Humphreys et al. 2010). While the exact reasoning for the discrepancy is unknown, both groups showed that epithelial-to-mesenchymal transition (EMT) was a minor contributor to the SMA+ myofibroblast population following UUO-induced renal injury. Interestingly, EMT may be enhanced in PKD through factors such as TGF-β, platelet-derived growth factor (PDGF), FGF-1, and Insulin-like growth factor (IGF) I and II, all of which are upregulated in cystic disease (Norman 2011; Kuo et al. 1997). Togawa et al. showed that expression of E-cadherin and beta-catenin, components important for epithelial adhesion, was significantly attenuated in tubules from PCK rats (Togawa et al. 2011). Furthermore, cystic epithelial cells found in fibrotic regions showed increased mesenchymal markers including vimentin and fibronectin (Togawa et al. 2011). Gene expression analysis from human ADPKD samples showed significant alterations in the expression of genes related to smooth muscle which led the authors to speculate on a possible EMT in these patients (Schieren et al. 2006). However, a recent study concluded that EMT provides only a minor contribution to the SMA-positive population in PKD kidneys using a *PKD1* mouse model (Hassane et al. 2010). Despite evidence that EMT plays an important role in the formation of myofibroblasts in the kidney (Zeisberg and Neilson 2010), the extent of EMT and its importance for the pathogenesis of cysts and fibrosis in PKD patients is unknown.

The involvement of macrophages in promoting accumulation and degradation of extracellular matrix proteins is well established. Macrophages secrete growth factors and cytokines that induce myofibroblast activation and extracellular matrix production leading to fibrosis (Fig. 12.1d) (Vernon et al. 2010). The most well-defined macrophage-secreted growth factor is TGF-β, which drives extracellular matrix production and fibrosis through binding to the TGF-β receptor present on myofibroblasts (Wynn and Barron 2010). In addition to TGF-β, macrophages are major producers of PDGF and drive extracellular matrix production and fibrosis in a bleomycin model of idiopathic pulmonary fibrosis (Bonner 2004). Macrophages also produce other pro-fibrotic growth factors including galectin-3, insulin-like growth factor-I, IL-4, and IL-13, factors that are potent inducers of extracellular matrix production and fibrosis (Wynn and Barron 2010). Specifically, galectin-3 is critical for activation of renal fibroblasts and knockout of galectin-3 reduced fibrosis following unilateral ureteric obstruction (UUO) (Henderson et al. 2008). Despite extensive studies demonstrating that macrophages promote fibrosis, other data indicate that macrophages produce several matrix-degrading proteases including MMP1, MMP2, MMP8, MMP9, and MMP13 and TIMPs (Wynn and Barron 2010). For example, overexpression of MM9 in alveolar macrophages significantly reduced fibrotic lesions and hydroxyproline content compared to controls (Cabrera et al. 2007). Furthermore, adoptive transfer of bone marrow-derived macrophages into leukocyte-depleted mice significantly attenuated fibrosis following UUO-induced injury (Nishida et al. 2005). Collectively, these data suggest that different macrophage subtypes are responsible for promoting or regressing fibrosis through production of matrix promoting or degrading factors although this concept has not been extensively studied in PKD models.

12.3.4 Fibrotic Cytokines and Signaling Pathways

A number of intracellular signaling pathways are activated in cyst-lining epithelial cells in ADPKD, many of which are also activated in fibrosis (Norman 2011). TGF-β, a major fibrogenic cytokine, is highly expressed in cystic epithelia in human, rat, and mouse models of PKD. Studies in rodent models suggest that increased TGF-β expression and signaling correlate with later stages of the disease and that TGF-β is important in cyst progression and fibrosis rather than cyst initiation (Hassane et al. 2010). Activin A is a cytokine that belongs to the TGF-β family of growth factors. During development, activin A is produced by the ureteric bud (UB) and negatively regulates its branching; furthermore, activin A is also involved in kidney regeneration following injury, suggesting an important role for this cytokine during kidney formation/regeneration (Maeshima et al. 2003; Ritvos et al. 1995). More important, inhibition of activin A signaling slows the progression of PKD (Leonhard et al. 2016). Additionally, activin A expression is increased in a *PKD1* mouse, model raising the possibility that altered activin A expression has a role in renal cystic disease (Hassane et al. 2010). Its expression is also associated with conditions wherein loss of cilia or polycystins causes rapid cyst formation (e.g., after injury or in juvenile ages). These studies indicate that activin A is a critical cytokine in cyst progression and ESRD since it is involved in epithelial regeneration and is a member of the TGF-β family of pro-fibrotic growth factors capable of inducing ECM gene expression in renal cells (Yamashita et al. 2004).

In addition to TGF-β, several other pro-fibrotic growth factors were associated with ECM production in PKD including epithelial growth factor (EGF), fibroblast growth factor-1 (FGF-1), and hepatocyte growth factor (HGF). Levels of these pro-fibrotic cytokines increased in PKD over time and their highest levels were often found in end-stage or near-end-stage PKD kidneys (Kuo et al. 1997; Du and Wilson 1995). Furthermore, overexpression of the bone morphogenic protein receptor (ALK-3) or knockout of Bmp-7 triggers cyst formation in mice (Hu et al. 2003; Jena et al. 1997). Importantly, in complementary studies, BMP-7 treatment attenuated renal cystic disease progression in a *jck* mouse model of PKD (Leonhard et al. 2016). Similarly, EGF is another important regulator of cystic epithelial cell proliferation and inhibition of this pathway is being targeted for therapeutic development (Du and Wilson 1995). It could be presumed that targeting multiple pro-fibrotic pathways may be an effective means of intervention for patients progressing toward ESRD. Overall, upregulation of multiple pro-fibrotic growth factor components suggests that these cytokines work in conjunction to increase extracellular matrix during PKD progression. However, the specific pathways involved in fibrosis associated with ADPKD remain to be defined.

12.3.5 Anti-Fibrotic Therapies in PKD

Slowing cyst expansion and the development of fibrosis is key to prolonging patient life span and improving palliative care. Although progress has been made (Boor et al. 2007; Fine and Norman 2008; Vilayur and Harris 2009; Albaqumi et al. 2008; Grgic et al. 2009), development of effective anti-fibrotic strategies in PKD patients is very limited. Therefore, it is imperative that we better understand the mechanisms underlying the initiation and progression of fibrosis. Identifying key molecular mechanisms that are impacted during the development of cysts and how they ultimately contribute to ESRD should provide novel targets for anti-fibrotic therapies.

References

Adams DO, Hamilton TA (1984) The cell biology of macrophage activation. Annu Rev Immunol 2:283–318

Albaqumi M, Srivastava S, Li Z, Zhdnova O, Wulff H, Itani O, Wallace DP, Skolnik EY (2008) KCa3.1 potassium channels are critical for cAMP-dependent chloride secretion and cyst growth in autosomal-dominant polycystic kidney disease. Kidney Int 74:740–749

Basten SG, Giles RH (2013) Functional aspects of primary cilia in signaling, cell cycle and tumorigenesis. Cilia 2:6

Bastos AP, Piontek K, Silva AM, Martini D, Menezes LF, Fonseca JM, Fonseca II, Germino GG, Onuchic LF (2009) Pkd1 haploinsufficiency increases renal damage and induces microcyst formation following ischemia/reperfusion. J Am Soc Nephrol JASN 20:2389–2402

Benoit M, Desnues B, Mege JL (2008) Macrophage polarization in bacterial infections. J Immunol 181:3733–3739

Bernhardt WM, Wiesener MS, Weidemann A, Schmitt R, Weichert W, Lechler P, Campean V, Ong AC, Willam C, Gretz N, Eckardt KU (2007) Involvement of hypoxia-inducible transcription factors in polycystic kidney disease. Am J Pathol 170:830–842

Bhunia AK, Piontek K, Boletta A, Liu L, Qian F, Xu PN, Germino FJ, Germino GG (2002) PKD1 induces p21(waf1) and regulation of the cell cycle via direct activation of the JAK-STAT signaling pathway in a process requiring PKD2. Cell 109:157–168

Bonner JC (2004) Regulation of PDGF and its receptors in fibrotic diseases. Cytokine Growth Factor Rev 15:255–273

Boor P, Sebekova K, Ostendorf T, Floege J (2007) Treatment targets in renal fibrosis. Nephrol Dial Transplant 22:3391–3407

Bruhl H, Cihak J, Schneider MA, Plachy J, Rupp T, Wenzel I, Shakarami M, Milz S, Ellwart JW, Stangassinger M, Schlondorff D, Mack M (2004) Dual role of CCR2 during initiation and progression of collagen-induced arthritis: evidence for regulatory activity of CCR2+ T cells. J Immunol 172:890–898

Cabrera S, Gaxiola M, Arreola JL, Ramirez R, Jara P, D'Armiento J, Richards T, Selman M, Pardo A (2007) Overexpression of MMP9 in macrophages attenuates pulmonary fibrosis induced by bleomycin. Int J Biochem Cell Biol 39:2324–2338

Cao Q, Wang Y, Wang XM, Lu J, Lee VW, Ye Q, Nguyen H, Zheng G, Zhao Y, Alexander SI, Harris DC (2015) Renal F4/80+ CD11c+ mononuclear phagocytes display phenotypic and functional characteristics of macrophages in health and in adriamycin nephropathy. J Am Soc Nephrol JASN 26:349–363

Catania JM, Chen G, Parrish AR (2007) Role of matrix metalloproteinases in renal pathophysiologies. Am J Physiol Renal Physiol 292:F905–F911

Clements M, Gershenovich M, Chaber C, Campos-Rivera J, Du P, Zhang M, Ledbetter S, Zuk A (2015) Differential Ly6C expression after renal ischemia-reperfusion identifies unique macrophage populations. J Am Soc Nephrol JASN 27:159–170

Cowley BD Jr, Gudapaty S, Kraybill AL, Barash BD, Harding MA, Calvet JP, Gattone VH 2nd (1993) Autosomal-dominant polycystic kidney disease in the rat. Kidney Int 43:522–534

Cowley BD Jr, Ricardo SD, Nagao S, Diamond JR (2001) Increased renal expression of monocyte chemoattractant protein-1 and osteopontin in ADPKD in rats. Kidney Int 60:2087–2096

Davenport JR, Watts AJ, Roper VC, Croyle MJ, van Groen T, Wyss JM, Nagy TR, Kesterson RA, Yoder BK (2007) Disruption of intraflagellar transport in adult mice leads to obesity and slow-onset cystic kidney disease. Curr Biol 17:1586–1594

Dell KM, Nemo R, Sweeney WE Jr, Levin JI, Frost P, Avner ED (2001) A novel inhibitor of tumor necrosis factor-alpha converting enzyme ameliorates polycystic kidney disease. Kidney Int 60:1240–1248

Dell'Italia LJ, Husain A (2002) Dissecting the role of chymase in angiotensin II formation and heart and blood vessel diseases. Curr Opin Cardiol 17:374–379

Dinsmore C, Reiter JF (2016) Endothelial primary cilia inhibit atherosclerosis. EMBO Rep 17:156–166

Du J, Wilson PD (1995) Abnormal polarization of EGF receptors and autocrine stimulation of cyst epithelial growth in human ADPKD. Am J Phys 269:C487–C495

Eddy AA (2009) Serine proteases, inhibitors and receptors in renal fibrosis. Thromb Haemost 101:656–664

Ehrhardt A, Ehrhardt GR, Guo X, Schrader JW (2002) Ras and relatives—job sharing and networking keep an old family together. Exp Hematol 30:1089–1106

Fadok VA, Bratton DL, Konowal A, Freed PW, Westcott JY, Henson PM (1998) Macrophages that have ingested apoptotic cells in vitro inhibit proinflammatory cytokine production through autocrine/paracrine mechanisms involving TGF-beta, PGE2, and PAF. J Clin Invest 101:890–898

Fine LG, Norman JT (2008) Chronic hypoxia as a mechanism of progression of chronic kidney diseases: from hypothesis to novel therapeutics. Kidney Int 74:867–872

Follonier Castella L, Gabbiani G, McCulloch CA, Hinz B (2010) Regulation of myofibroblast activities: calcium pulls some strings behind the scene. Exp Cell Res 316:2390–2401

Gardner KD Jr, Burnside JS, Elzinga LW, Locksley RM (1991) Cytokines in fluids from polycystic kidneys. Kidney Int 39:718–724

Geiger B, Bershadsky A, Pankov R, Yamada KM (2001) Transmembrane crosstalk between the extracellular matrix--cytoskeleton crosstalk. Nat Rev Mol Cell Biol 2:793–805

Ginhoux F, Jung S (2014) Monocytes and macrophages: developmental pathways and tissue homeostasis. Nat Rev Immunol 14:392–404

Ginhoux F, Greter M, Leboeuf M, Nandi S, See P, Gokhan S, Mehler MF, Conway SJ, Ng LG, Stanley ER, Samokhvalov IM, Merad M (2010) Fate mapping analysis reveals that adult microglia derive from primitive macrophages. Science 330:841–845

Ginhoux F, Schultze JL, Murray PJ, Ochando J, Biswas SK (2015) New insights into the multidimensional concept of macrophage ontogeny, activation and function. Nat Immunol 17:34–40

Gordon S, Taylor PR (2005) Monocyte and macrophage heterogeneity. Nat Rev Immunol 5:953–964

Grantham JJ (2008) Clinical practice. Autosomal dominant polycystic kidney disease. N Engl J Med 359:1477–1485

Grantham JJ, Mulamalla S, Swenson-Fields KI (2011) Why kidneys fail in autosomal dominant polycystic kidney disease. Nat Rev Nephrol 7:556–566

Grgic I, Kiss E, Kaistha BP, Busch C, Kloss M, Sautter J, Muller A, Kaistha A, Schmidt C, Raman G, Wulff H, Strutz F, Grone HJ, Kohler R, Hoyer J (2009) Renal fibrosis is attenuated by targeted disruption of KCa3.1 potassium channels. Proc Natl Acad Sci USA 106:14518–14523

Guay-Woodford LM (2003) Murine models of polycystic kidney disease: molecular and therapeutic insights. Am J Physiol Renal Physiol 285:F1034–F1049

Hassane S, Leonhard WN, van der Wal A, Hawinkels LJ, Lantinga-van Leeuwen IS, ten Dijke P, Breuning MH, de Heer E, Peters DJ (2010) Elevated TGFbeta-Smad signalling in experimental Pkd1 models and human patients with polycystic kidney disease. J Pathol 222:21–31

Helm O, Held-Feindt J, Grage-Griebenow E, Reiling N, Ungefroren H, Vogel I, Kruger U, Becker T, Ebsen M, Rocken C, Kabelitz D, Schafer H, Sebens S (2014) Tumor-associated macrophages exhibit pro- and anti-inflammatory properties by which they impact on pancreatic tumorigenesis. Int J Cancer 135:843–861

Henderson NC, Mackinnon AC, Farnworth SL, Kipari T, Haslett C, Iredale JP, Liu FT, Hughes J, Sethi T (2008) Galectin-3 expression and secretion links macrophages to the promotion of renal fibrosis. Am J Pathol 172:288–298

Hochheiser K, Heuser C, Krause TA, Teteris S, Ilias A, Weisheit C, Hoss F, Tittel AP, Knolle PA, Panzer U, Engel DR, Tharaux PL, Kurts C (2013) Exclusive CX3CR1 dependence of kidney DCs impacts glomerulonephritis progression. J Clin Invest 123:4242–4254

Hu MC, Piscione TD, Rosenblum ND (2003) Elevated SMAD1/beta-catenin molecular complexes and renal medullary cystic dysplasia in ALK3 transgenic mice. Development 130:2753–2766
Huen SC, Cantley LG (2015) Macrophage-mediated injury and repair after ischemic kidney injury. Pediatr Nephrol 30:199–209
Hull TD, Kamal AI, Boddu R, Bolisetty S, Guo L, Tisher CC, Rangarajan S, Chen B, Curtis LM, George JF, Agarwal A (2015) Heme oxygenase-1 regulates myeloid cell trafficking in AKI. J Am Soc Nephrol JASN 26:2139–2151
Humphreys BD, Lin SL, Kobayashi A, Hudson TE, Nowlin BT, Bonventre JV, Valerius MT, McMahon AP, Duffield JS (2010) Fate tracing reveals the pericyte and not epithelial origin of myofibroblasts in kidney fibrosis. Am J Pathol 176:85–97
Ibrahim S (2007) Increased apoptosis and proliferative capacity are early events in cyst formation in autosomal-dominant, polycystic kidney disease. Sci World J 7:1757–1767
Jena N, Martin-Seisdedos C, McCue P, Croce CM (1997) BMP7 null mutation in mice: developmental defects in skeleton, kidney, and eye. Exp Cell Res 230:28–37
Joly D, Morel V, Hummel A, Ruello A, Nusbaum P, Patey N, Noel LH, Rousselle P, Knebelmann B (2003) Beta4 integrin and laminin 5 are aberrantly expressed in polycystic kidney disease: role in increased cell adhesion and migration. Am J Pathol 163:1791–1800
Karihaloo A, Koraishy F, Huen SC, Lee Y, Merrick D, Caplan MJ, Somlo S, Cantley LG (2011) Macrophages promote cyst growth in polycystic kidney disease. J Am Soc Nephrol JASN 22:1809–1814
Kaspareit-Rittinghausen J, Rapp K, Deerberg F, Wcislo A, Messow C (1989) Hereditary polycystic kidney disease associated with osteorenal syndrome in rats. Vet Pathol 26:195–201
Katz SI, Tamaki K, Sachs DH (1979) Epidermal Langerhans cells are derived from cells originating in bone marrow. Nature 282:324–326
Kotsis F, Boehlke C, Kuehn EW (2013) The ciliary flow sensor and polycystic kidney disease. Nephrol Dial Transplant 28:518–526
Kratochvill F, Neale G, Haverkamp JM, Van de Velde LA, Smith AM, Kawauchi D, McEvoy J, Roussel MF, Dyer MA, Qualls JE, Murray PJ (2015) TNF counterbalances the emergence of M2 tumor macrophages. Cell Rep 12:1902–1914
Kuo NT, Norman JT, Wilson PD (1997) Acidic FGF regulation of hyperproliferation of fibroblasts in human autosomal dominant polycystic kidney disease. Biochem Mol Med 61:178–191
Lavin Y, Winter D, Blecher-Gonen R, David E, Keren-Shaul H, Merad M, Jung S, Amit I (2014) Tissue-resident macrophage enhancer landscapes are shaped by the local microenvironment. Cell 159:1312–1326
LeBleu VS, Taduri G, O'Connell J, Teng Y, Cooke VG, Woda C, Sugimoto H, Kalluri R (2013) Origin and function of myofibroblasts in kidney fibrosis. Nat Med 19:1047–1053
Lee K, Boctor S, Barisoni LM, Gusella GL (2015) Inactivation of integrin-beta1 prevents the development of polycystic kidney disease after the loss of polycystin-1. J Am Soc Nephrol JASN 26:888–895
Leonhard WN, Kunnen SJ, Plugge AJ, Pasternack A, Jianu SB, Veraar K, El Bouazzaoui F, Hoogaars WM, Ten Dijke P, Breuning MH, De Heer E, Ritvos O, Peters DJ (2016) Inhibition of activin signaling slows progression of polycystic kidney disease. J Am Soc Nephrol JASN 27:3589–3599
Li L, Huang L, Sung SS, Vergis AL, Rosin DL, Rose CE Jr, Lobo PI, Okusa MD (2008b) The chemokine receptors CCR2 and CX3CR1 mediate monocyte/macrophage trafficking in kidney ischemia-reperfusion injury. Kidney Int 74:1526–1537
Li X, Magenheimer BS, Xia S, Johnson T, Wallace DP, Calvet JP, Li R (2008a) A tumor necrosis factor-alpha-mediated pathway promoting autosomal dominant polycystic kidney disease. Nat Med 14:863–868
Lin SL, Castano AP, Nowlin BT, Lupher ML Jr, Duffield JS (2009) Bone marrow Ly6C^{high} monocytes are selectively recruited to injured kidney and differentiate into functionally distinct populations. J Immunol 183:6733–6743

Lin F, Hiesberger T, Cordes K, Sinclair AM, Goldstein LS, Somlo S, Igarashi P (2003) Kidney-specific inactivation of the KIF3A subunit of kinesin-II inhibits renal ciliogenesis and produces polycystic kidney disease. Proc Natl Acad Sci USA 100:5286–5291

Liu D, Wang CJ, Judge DP, Halushka MK, Ni J, Habashi JP, Moslehi J, Bedja D, Gabrielson KL, Xu H, Qian F, Huso D, Dietz HC, Germino GG, Watnick T (2014) A Pkd1-Fbn1 genetic interaction implicates TGF-beta signaling in the pathogenesis of vascular complications in autosomal dominant polycystic kidney disease. J Am Soc Nephrol JASN 25:81–91

Low SH, Vasanth S, Larson CH, Mukherjee S, Sharma N, Kinter MT, Kane ME, Obara T, Weimbs T (2006) Polycystin-1, STAT6, and P100 function in a pathway that transduces ciliary mechanosensation and is activated in polycystic kidney disease. Dev Cell 10:57–69

Ma M, Tian X, Igarashi P, Pazour GJ, Somlo S (2013) Loss of cilia suppresses cyst growth in genetic models of autosomal dominant polycystic kidney disease. Nat Genet 45:1004–1012

Maeshima A, Yamashita S, Maeshima K, Kojima I, Nojima Y (2003) Activin a produced by ureteric bud is a differentiation factor for metanephric mesenchyme. J Am Soc Nephrol JASN 14:1523–1534

Mangos S, Lam PY, Zhao A, Liu Y, Mudumana S, Vasilyev A, Liu A, Drummond IA (2010) The ADPKD genes pkd1a/b and pkd2 regulate extracellular matrix formation. Dis Model Mech 3:354–365

Mantovani A, Sozzani S, Locati M, Allavena P, Sica A (2002) Macrophage polarization: tumor-associated macrophages as a paradigm for polarized M2 mononuclear phagocytes. Trends Immunol 23:549–555

McPherson EA, Luo Z, Brown RA, LeBard LS, Corless CC, Speth RC, Bagby SP (2004) Chymase-like angiotensin II-generating activity in end-stage human autosomal dominant polycystic kidney disease. J Am Soc Nephrol JASN 15:493–500

Merad M, Manz MG, Karsunky H, Wagers A, Peters W, Charo I, Weissman IL, Cyster JG, Engleman EG (2002) Langerhans cells renew in the skin throughout life under steady-state conditions. Nat Immunol 3:1135–1141

Metcalf D, Mifsud S, Di Rago L, Nicola NA, Hilton DJ, Alexander WS (2002) Polycystic kidneys and chronic inflammatory lesions are the delayed consequences of loss of the suppressor of cytokine signaling-1 (SOCS-1). Proc Natl Acad Sci USA 99:943–948

Mrug M, Zhou J, Woo Y, Cui X, Szalai AJ, Novak J, Churchill GA, Guay-Woodford LM (2008) Overexpression of innate immune response genes in a model of recessive polycystic kidney disease. Kidney Int 73:63–76

Nadel JA (1991) Biology of mast cell tryptase and chymase. Ann N Y Acad Sci 629:319–331

Nakamura T, Ebihara I, Fukui M, Osada S, Tomino Y, Masaki T, Goto K, Furuichi Y, Koide H (1993) Increased endothelin and endothelin receptor mRNA expression in polycystic kidneys of cpk mice. J Am Soc Nephrol JASN 4:1064–1072

Nakamura T, Ushiyama C, Suzuki S, Ebihara I, Shimada N, Koide H (2000) Elevation of serum levels of metalloproteinase-1, tissue inhibitor of metalloproteinase-1 and type IV collagen, and plasma levels of metalloproteinase-9 in polycystic kidney disease. Am J Nephrol 20:32–36

Nelson PJ, Rees AJ, Griffin MD, Hughes J, Kurts C, Duffield J (2012) The renal mononuclear phagocytic system. J Am Soc Nephrol JASN 23:194–203

Nishida M, Okumura Y, Fujimoto S, Shiraishi I, Itoi T, Hamaoka K (2005) Adoptive transfer of macrophages ameliorates renal fibrosis in mice. Biochem Biophys Res Commun 332:11–16

Norman J (2011) Fibrosis and progression of autosomal dominant polycystic kidney disease (ADPKD). Biochim Biophys Acta 1812:1327–1336

O'Leary CA, Mackay BM, Malik R, Edmondston JE, Robinson WF, Huxtable CR (1999) Polycystic kidney disease in bull terriers: an autosomal dominant inherited disorder. Aust Vet J 77:361–366

Obermuller N, Morente N, Kranzlin B, Gretz N, Witzgall R (2001) A possible role for metalloproteinases in renal cyst development. Am J Physiol Renal Physiol 280:F540–F550

Oh EC, Katsanis N (2012) Cilia in vertebrate development and disease. Development 139:443–448

Pahl HL (1999) Activators and target genes of Rel/NF-kappaB transcription factors. Oncogene 18:6853–6866

Park EY, Seo MJ, Park JH (2010) Effects of specific genes activating RAGE on polycystic kidney disease. Am J Nephrol 32:169–178

Patel V, Li L, Cobo-Stark P, Shao X, Somlo S, Lin F, Igarashi P (2008) Acute kidney injury and aberrant planar cell polarity induce cyst formation in mice lacking renal cilia. Hum Mol Genet 17:1578–1590

Phillips JK, Hopwood D, Loxley RA, Ghatora K, Coombes JD, Tan YS, Harrison JL, McKitrick DJ, Holobotvskyy V, Arnolda LF, Rangan GK (2007) Temporal relationship between renal cyst development, hypertension and cardiac hypertrophy in a new rat model of autosomal recessive polycystic kidney disease. Kidney Blood Press Res 30:129–144

Piontek K, Menezes LF, Garcia-Gonzalez MA, Huso DL, Germino GG (2007) A critical developmental switch defines the kinetics of kidney cyst formation after loss of Pkd1. Nat Med 13:1490–1495

Pollard JW (2009) Trophic macrophages in development and disease. Nat Rev Immunol 9:259–270

Prasad S, McDaid JP, Tam FW, Haylor JL, Ong AC (2009) Pkd2 dosage influences cellular repair responses following ischemia-reperfusion injury. Am J Pathol 175:1493–1503

Qi W, Chen X, Poronnik P, Pollock CA (2006) The renal cortical fibroblast in renal tubulointerstitial fibrosis. Int J Biochem Cell Biol 38:1–5

Qin S, Taglienti M, Cai L, Zhou J, Kreidberg JA (2012) c-Met and NF-kappaB-dependent overexpression of Wnt7a and -7b and Pax2 promotes cystogenesis in polycystic kidney disease. J Am Soc Nephrol JASN 23:1309–1318

Rae F, Woods K, Sasmono T, Campanale N, Taylor D, Ovchinnikov DA, Grimmond SM, Hume DA, Ricardo SD, Little MH (2007) Characterisation and trophic functions of murine embryonic macrophages based upon the use of a Csf1r-EGFP transgene reporter. Dev Biol 308:232–246

Rankin CA, Suzuki K, Itoh Y, Ziemer DM, Grantham JJ, Calvet JP, Nagase H (1996) Matrix metalloproteinases and TIMPS in cultured C57BL/6J-cpk kidney tubules. Kidney Int 50:835–844

Ritvos O, Tuuri T, Eramaa M, Sainio K, Hilden K, Saxen L, Gilbert SF (1995) Activin disrupts epithelial branching morphogenesis in developing glandular organs of the mouse. Mech Dev 50:229–245

Schaefer L, Han X, Gretz N, Hafner C, Meier K, Matzkies F, Schaefer RM (1996) Tubular gelatinase A (MMP-2) and its tissue inhibitors in polycystic kidney disease in the Han:SPRD rat. Kidney Int 49:75–81

Schieren G, Rumberger B, Klein M, Kreutz C, Wilpert J, Geyer M, Faller D, Timmer J, Quack I, Rump LC, Walz G, Donauer J (2006) Gene profiling of polycystic kidneys. Nephrol Dial Transplant 21:1816–1824

Schulz C, Gomez Perdiguero E, Chorro L, Szabo-Rogers H, Cagnard N, Kierdorf K, Prinz M, Wu B, Jacobsen SE, Pollard JW, Frampton J, Liu KJ, Geissmann F (2012) A lineage of myeloid cells independent of Myb and hematopoietic stem cells. Science 336:86–90

Shannon MB, Patton BL, Harvey SJ, Miner JH (2006) A hypomorphic mutation in the mouse laminin alpha5 gene causes polycystic kidney disease. J Am Soc Nephrol JASN 17:1913–1922

Sharma N, Malarkey EB, Berbari NF, O'Connor AK, Vanden Heuvel GB, Mrug M, Yoder BK (2013) Proximal tubule proliferation is insufficient to induce rapid cyst formation after cilia disruption. J Am Soc Nephrol JASN 24:456–464

Sica A, Mantovani A (2012) Macrophage plasticity and polarization: in vivo veritas. J Clin Invest 122:787–795

Snelgrove RJ, Jackson PL, Hardison MT, Noerager BD, Kinloch A, Gaggar A, Shastry S, Rowe SM, Shim YM, Hussell T, Blalock JE (2010) A critical role for LTA4H in limiting chronic pulmonary neutrophilic inflammation. Science 330:90–94

Song X, Di Giovanni V, He N, Wang K, Ingram A, Rosenblum ND, Pei Y (2009) Systems biology of autosomal dominant polycystic kidney disease (ADPKD): computational identification of gene expression pathways and integrated regulatory networks. Hum Mol Genet 18:2328–2343

Stables MJ, Shah S, Camon EB, Lovering RC, Newson J, Bystrom J, Farrow S, Gilroy DW (2011) Transcriptomic analyses of murine resolution-phase macrophages. Blood 118:e192–e208

Swenson-Fields KI, Vivian CJ, Salah SM, Peda JD, Davis BM, van Rooijen N, Wallace DP, Fields TA (2013) Macrophages promote polycystic kidney disease progression. Kidney Int 83:855–864

Ta MH, Harris DC, Rangan GK (2013) Role of interstitial inflammation in the pathogenesis of polycystic kidney disease. Nephrology 18:317–330

Takahashi H, Calvet JP, Dittemore-Hoover D, Yoshida K, Grantham JJ, Gattone VH 2nd (1991) A hereditary model of slowly progressive polycystic kidney disease in the mouse. J Am Soc Nephrol JASN 1:980–989

Takakura A, Contrino L, Zhou X, Bonventre JV, Sun Y, Humphreys BD, Zhou J (2009) Renal injury is a third hit promoting rapid development of adult polycystic kidney disease. Hum Mol Genet 18:2523–2531

Talbot JJ, Song X, Wang X, Rinschen MM, Doerr N, LaRiviere WB, Schermer B, Pei YP, Torres VE, Weimbs T (2014) The cleaved cytoplasmic tail of polycystin-1 regulates Src-dependent STAT3 activation. J Am Soc Nephrol JASN 25:1737–1748

Togawa H, Nakanishi K, Mukaiyama H, Hama T, Shima Y, Sako M, Miyajima M, Nozu K, Nishii K, Nagao S, Takahashi H, Iijima K, Yoshikawa N (2011) Epithelial-to-mesenchymal transition in cyst lining epithelial cells in an orthologous PCK rat model of autosomal-recessive polycystic kidney disease. Am J Physiol Renal Physiol 300:F511–F520

van Furth R, Cohn ZA (1968) The origin and kinetics of mononuclear phagocytes. J Exp Med 128:415–435

Vernon MA, Mylonas KJ, Hughes J (2010) Macrophages and renal fibrosis. Semin Nephrol 30:302–317

Vilayur E, Harris DC (2009) Emerging therapies for chronic kidney disease: what is their role? Nat Rev Nephrol 5:375–383

Vogler C, Homan S, Pung A, Thorpe C, Barker J, Birkenmeier EH, Upadhya P (1999) Clinical and pathologic findings in two new allelic murine models of polycystic kidney disease. J Am Soc Nephrol JASN 10:2534–2539

Wallace DP, White C, Savinkova L, Nivens E, Reif GA, Pinto CS, Raman A, Parnell SC, Conway SJ, Fields TA (2014) Periostin promotes renal cyst growth and interstitial fibrosis in polycystic kidney disease. Kidney Int 85:845–854

Wann AK, Knight MM (2012) Primary cilia elongation in response to interleukin-1 mediates the inflammatory response. Cell Mol Life Sci 69:2967–2977

Wann AK, Chapple JP, Knight MM (2014) The primary cilium influences interleukin-1beta-induced NFkappaB signalling by regulating IKK activity. Cell Signal 26:1735–1742

Weimbs T, Talbot JJ (2013) STAT3 signaling in polycystic kidney disease. Drug Discov Today Dis Mech 10:e113–e118

Wilson PD (2004) Polycystic kidney disease. N Engl J Med 350:151–164

Wilson PD (2008) Mouse models of polycystic kidney disease. Curr Top Dev Biol 84:311–350

Wilson PD, Burrow CR (1999) Cystic diseases of the kidney: role of adhesion molecules in normal and abnormal tubulogenesis. Exp Nephrol 7:114–124

Wilson PD, Falkenstein D (1995) The pathology of human renal cystic disease. Curr Top Pathol 88:1–50

Wilson PD, Goilav B (2007) Cystic disease of the kidney. Annu Rev Pathol 2:341–368

Wilson PD, Sherwood AC (1991) Tubulocystic epithelium. Kidney Int 39:450–463

Wilson PD, Geng L, Li X, Burrow CR (1999) The PKD1 gene product, "polycystin-1," is a tyrosine-phosphorylated protein that colocalizes with alpha2beta1-integrin in focal clusters in adherent renal epithelia. Lab Invest J Tech Methods Pathol 79:1311–1323

Wilson PD, Hreniuk D, Gabow PA (1992) Abnormal extracellular matrix and excessive growth of human adult polycystic kidney disease epithelia. J Cell Physiol 150:360–369
Wozniak MA, Modzelewska K, Kwong L, Keely PJ (2004) Focal adhesion regulation of cell behavior. Biochim Biophys Acta 1692:103–119
Wynn TA, Barron L (2010) Macrophages: master regulators of inflammation and fibrosis. Semin Liver Dis 30:245–257
Wynn TA, Chawla A, Pollard JW (2013) Macrophage biology in development, homeostasis and disease. Nature 496:445–455
Xue J, Schmidt SV, Sander J, Draffehn A, Krebs W, Quester I, De Nardo D, Gohel TD, Emde M, Schmidleithner L, Ganesan H, Nino-Castro A, Mallmann MR, Labzin L, Theis H, Kraut M, Beyer M, Latz E, Freeman TC, Ulas T, Schultze JL (2014) Transcriptome-based network analysis reveals a spectrum model of human macrophage activation. Immunity 40:274–288
Yamashita S, Maeshima A, Kojima I, Nojima Y (2004) Activin A is a potent activator of renal interstitial fibroblasts. J Am Soc Nephrol JASN 15:91–101
Yoder BK (2007) Role of primary cilia in the pathogenesis of polycystic kidney disease. J Am Soc Nephrol JASN 18:1381–1388
Yoder BK, Hou X, Guay-Woodford LM (2002) The polycystic kidney disease proteins, polycystin-1, polycystin-2, polaris, and cystin, are co-localized in renal cilia. J Am Soc Nephrol JASN 13:2508–2516
Yoder BK, Richards WG, Sweeney WE, Wilkinson JE, Avener ED, Woychik RP (1995) Insertional mutagenesis and molecular analysis of a new gene associated with polycystic kidney disease. Proc Assoc Am Physicians 107:314–323
Zeier M, Fehrenbach P, Geberth S, Mohring K, Waldherr R, Ritz E (1992) Renal histology in polycystic kidney disease with incipient and advanced renal failure. Kidney Int 42:1259–1265
Zeisberg M, Neilson EG (2010) Mechanisms of tubulointerstitial fibrosis. J Am Soc Nephrol JASN 21:1819–1834
Zeltner R, Hilgers KF, Schmieder RE, Porst M, Schulze BD, Hartner A (2008) A promoter polymorphism of the alpha 8 integrin gene and the progression of autosomal-dominant polycystic kidney disease. Nephron Clin Pract 108:c169–c175
Zheng D, Wolfe M, Cowley BD Jr, Wallace DP, Yamaguchi T, Grantham JJ (2003) Urinary excretion of monocyte chemoattractant protein-1 in autosomal dominant polycystic kidney disease. J Am Soc Nephrol JASN 14:2588–2595
Zimmerman K, Yoder BK (2015) SnapShot: sensing and signaling by cilia. Cell 161(692–692):e691
Zoja C, Corna D, Locatelli M, Rottoli D, Pezzotta A, Morigi M, Zanchi C, Buelli S, Guglielmotti A, Perico N, Remuzzi A, Remuzzi G (2015) Effects of MCP-1 inhibition by bindarit therapy in a rat model of polycystic kidney disease. Nephron 129:52–61

Chapter 13
Epithelial to Mesenchymal Transition (EMT) and Endothelial to Mesenchymal Transition (EndMT): Role and Implications in Kidney Fibrosis

Ana S. Cruz-Solbes and Keith Youker

Abstract Tubulointerstitial injury is one of the hallmarks of renal disease. In particular, interstitial fibrosis has a prominent role in the development and progression of kidney injury. Collagen-producing fibroblasts are responsible for the ECM deposition. However, the origin of those activated fibroblasts is not clear. This chapter will discuss in detail the concept of epithelial to mesenchymal transition (EMT) and endothelial to mesenchymal transition (EndMT) in the context of fibrosis and kidney disease. In short, EMT and EndMT involve a change in cell shape, loss of polarity and increased motility associated with increased collagen production. Thus, providing a new source of fibroblasts. However, many controversies exist regarding the existence of EMT and EndMT in kidney disease, as well as its burden and role in disease development. The aim of this chapter is to provide an overview of the concepts and profibrotic pathways and to present the evidence that has been published in favor and against EMT and EndMT.

Chronic renal disease is associated with marked tubulointerstitial injury. This is independent of the primary process or etiology. The tubulointerstitial injury is usually in the form of tubular dilation or interstitial fibrosis (Liu 2004; Nath 1992). The focus of this chapter is a controversial contributor to renal fibrosis: epithelial to mesenchymal transition (EMT).

Elizabeth Hay is one of the first women to receive a medical degree from John Hopkins Hospital. She described an epithelial–mesenchymal shift using chick primitive streak formation in the late 1960s. Her life's work was devoted to collagen production, extracellular matrix, and EMT. Her discoveries would later be recognized as a new source of fibroblasts.

A.S. Cruz-Solbes, M.D. (✉) • K. Youker, Ph.D.
Houston Methodist Hospital, Houston, TX, USA
e-mail: sofiacruzs@gmail.com

R.K. Miller (ed.), *Kidney Development and Disease*, Results and Problems in Cell Differentiation 60, DOI 10.1007/978-3-319-51436-9_13

Many novel findings have been published since Dr. Hay initially described EMT. The most relevant one is the actual definition of EMT; initially, EMT stood for epithelial to mesenchymal transformation, and now the T stands for transition. Although this might seem like a small detail, the modification from transformation to transition grants the phenomenon more progressive and reversible characteristics than the definitiveness of transformation. In addition, the existence and relevance of a reverse transition (mesenchymal to epithelial) has also been introduced to the scientific community.

EMT is a controversial subject, especially in kidney disease. Some scientists dedicate their professional life to it, while some others are skeptical about its presence and role of myofibroblasts transitioning from epithelial and endothelial cell types in fibrotic diseases. Even more so when the lack of consistency in the results is taken into account.

This chapter discusses several issues regarding EMT and endothelial to mesenchymal transition (EndMT). It includes the basics of EMT, EndMT, and MET; the types of EMT; identification markers and pathways; its implications with fibrosis in end-stage disease; EMT in kidney disease; EndMT in kidney disease; and controversies associated with this topic.

13.1 Epithelial to Mesenchymal Transition—The Basics

EMT is a phenomenon where inherent epithelial markers are lost and mesenchymal characteristics are gained. Simplistically speaking, an epithelial cell transitions into a mesenchymal cell, losing all the characteristics inherent to epithelial cells and gaining mesenchymal features in the process. The end result is a motile, collagen-producing cell from epithelial origin. This has been identified in both physiologic and pathologic scenarios.

The main function of epithelial cells is to provide structure and prevent permeability; therefore, its most prominent characteristic is tight junctions and polarity. These cells are adhered to the basal membranes and to each other; this serves as an anchor to the tissue and, therefore, lacks motility. Epithelial cells can be identified through immunohistochemistry by identification of E-cadherin, one of the molecules responsible for adhesion among cells. Besides providing stability, epithelial cell's specific functions vary based on localization and histologic characteristics; nonetheless, its roles can be grouped into secretion, selective absorption, protection (barrier function), and transcellular transport.

Histologically, the epithelium is categorized based on the shape of the cells, number of layers, and the presence or absence of cilia and keratin (Fig. 13.1). Keratin, a tough resistant protein, is important for certain epithelial tissues since it prevents water absorption through the tissue. This is most important in the outermost layer of the epithelial cells on the body's surface. Histological characteristics of each epithelial tissue are based on both function and anatomical location of the tissue. For example, the esophagus is lined with nonkeratinized squamous multilayered epithelial cells since its main role is to protect against abrasions while at the same time absorbing water unlike the kidney tubules which are lined with simple

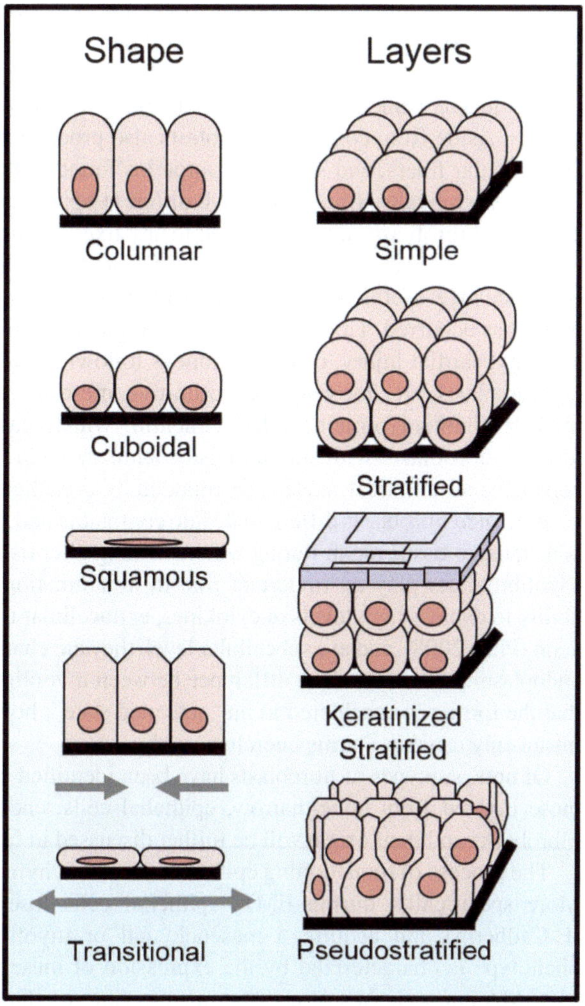

Fig. 13.1 Epithelial tissue is divided based on the shape and number of layers. Regarding shape it can be divided into columnar, cuboidal, squamous, or transitional (seen only in urinary tract). Regarding layers it can be divided into simple, stratified, keratinized stratified, and pseudostratified. Therefore, epithelial tissue can be described based on both characteristics (i.e., simple squamous, cuboidal stratified, columnar kertinized stratified)

cuboidal epithelium whose principal function is to secrete or absorb electrolytes, water, and other substances like urea.

Specialized epithelial tissue may undergo changes due to its environment, one example is transitional epithelium. Transitional epithelium has cells that can change from squamous/flat to cuboidal based on the amount of tension on the epithelium. This tissue is only found in the urinary bladder since it allows the organ to expand based on physical needs.

A specialized type of squamous simple epithelium lines the cavities specifically of blood vessels and serves as a semipermeable barrier that allows materials to cross either by diffusion or filtration.

On the other hand, the prototype for mesenchymal cells is fibroblasts whose function is to deposit collagen as a repair mechanism often secondary to an insult.

Its role is to maintain structural integrity of connective tissues, and it achieves its goal by secreting precursors of extracellular matrix. These cells are capable of producing collagen as a consequence of several mechanisms, including but not limited to mechanical stress, chemical injury, cytokines, and inflammatory substances. Aside from collagen, fibroblasts also produce glycosaminoglycans, elastic and reticular fibers, and some glycoproteins found in the extracellular matrix.

Morphologically, fibroblasts have different appearances based on their location and state. Fibroblasts are not adhered to any fixed membrane nor are they restricted by polarizing attachment and do not form monolayers. These are mobile cells. This is particularly useful considering that they are needed in remote tissues after an injury has occurred. This could be, for example, the skin for wound repair, the heart after myocardial injury, or in the kidneys following injury.

Fibroblasts are derived from primitive mesenchyme and express the intermediate filament protein called vimentin, which has been classically used to identify fibroblasts with immunohistochemistry techniques. Histologically, these cells have an elliptical nucleus surrounded by branched cytoplasm.

Activated fibroblasts differentiate into contractile and secretory myofibroblasts that contribute to tissue repair during wound healing after tissue injury (Hinz et al. 2007). Myofibroblasts play an important role in inflammation primarily because of their ability to express high levels of cytokines, extracellular matrix, and α-smooth muscle actin (Phan 2008), and at a subcellular level, they are characterized by abundant rough endoplasmic reticulum. The difference between myofibroblasts and myofibrocytes is that the former is considered as an "activated state"; however, many authors tend to mistakenly use these terms interchangeably.

Of note, subtypes of fibroblasts have been identified based on their origin such as those derived from bone-marrow, epithelial cells, endothelial cells and pericytes. Fibroblasts and their origin will be further discussed in future sections of this chapter.

The process of transitioning epithelial to mesenchymal cells takes multiple steps. More specifically, during EMT, epithelial cells lose their specific cell marker (E-Cadherin) and acquire a mesenchymal or myofibroblastic phenotype. This phenotype is characterized by the expression of mesenchymal cell products such as αSMA (smooth muscle actin), vimentin (intermediate filament), and type 1 collagen (Piera-Velazquez et al. 2011). The completion of an EMT is signaled by the degradation of underlying basement membrane and the formation of a mesenchymal cell that can now migrate away from the epithelial layer in which it originated (Kalluri and Weinberg 2009).

EMT has been described in many contexts such as in embryologic development, endocardial cushion formation, metastatic properties in breast cancer, and fibrotic response in interstitial pulmonary fibrosis. Its role is probably more extensive than what is previously thought.

EMT is about disaggregating epithelial units and reshaping them into fibroblasts with motility properties and collagen production abilities (Kalluri and Neilson 2003). As it was stated by Masayuki Iwano et al., EMT of terminally differentiated epithelium, in its purest sense, produces a tissue fibroblast (Iwano et al. 2002). Thus, by providing a new source of fibroblasts, EMT becomes a possible contributor to fibrosis in end-stage organ disease and, as such, may become a new potential anti-fibrotic target.

13.2 Endothelial to Mesenchymal Transition—What Is It?

Endothelial to mesenchymal transition (EndMT) is a subtype of EMT where specific epithelial cells known as endothelial cells transition into a mesenchymal cell type. As previously described, endothelial cells are a single layer of cells lining vessels which serve as a permeable membrane. EndMT was first discovered in the context of heart development around 1975 (Markwald et al. 1975), describing its existence and relevance in the formation of valves and septum during embryogenesis. EndMT has also been described in several publications as a contributor to the development and progression of pulmonary (Coll-Bonfill et al. 2015), cardiac (Zeisberg et al. 2007), and renal fibrosis (Zeisberg et al. 2008; Li et al. 2009).

Compared to the number of publications regarding EMT mechanisms and pathways, EndMT is not as thoroughly studied. As of today, it is assumed that EMT and EndMT share similar molecular mechanisms (Piera-Velazquez et al. 2011) as well as similar triggers and inhibitors. Therefore, the mechanisms for identifying EndMT are very similar to EMT, with a few exceptions to take into account. For example, dual staining (E-cadherin and αSMA, vimentin, or type 1 collagen) is one of the most utilized methods to detect EMT/EndMT; however, the marker used to identify the epithelium is E-cadherin, while the marker used to identify endothelium is VE-Cadherin (where V stands for vascular).

13.3 Mesenchymal to Epithelial Transition—New Concept, Same Phenomenon?

Mesenchymal to epithelial transition (MET) is a process where mesenchymal cells transition into epithelial cells. MET is the opposite of EMT. It has been postulated as a possible contributor to recovery from fibrotic lesions. Little is known about MET as compared to EMT.

One hypothesis is that EMT and MET are the opposite ends of the same plasticity spectrum. Where, depending on the stimuli, an adult cell may lose its epithelial characteristics and become a collagen-producing fibroblast (EMT) or a mesenchymal cell may gain epithelial traits (MET). A balance between these two is necessary to maintain tissue integrity and/or to repair injury.

MET occurs during embryogenesis and early development, specifically during nephrogenesis (Davies 1996), somitogensis (Nakaya et al. 2004), cardiogenesis (Nakajima et al. 2000), and hepatogenesis (Li et al. 2011). In nephrogenesis, MET occurs at the ureteric bud in order to form the renal epithelium, which will ultimately form the nephron.

On the other hand, MET has also been used to describe oncogenesis. The current theory is that EMT allows for the migration from the primary tumor into the bloodstream, and MET is responsible for anchoring the resulting circulating atypical cells to produce a secondary tumor (Yang and Weinberg 2008).

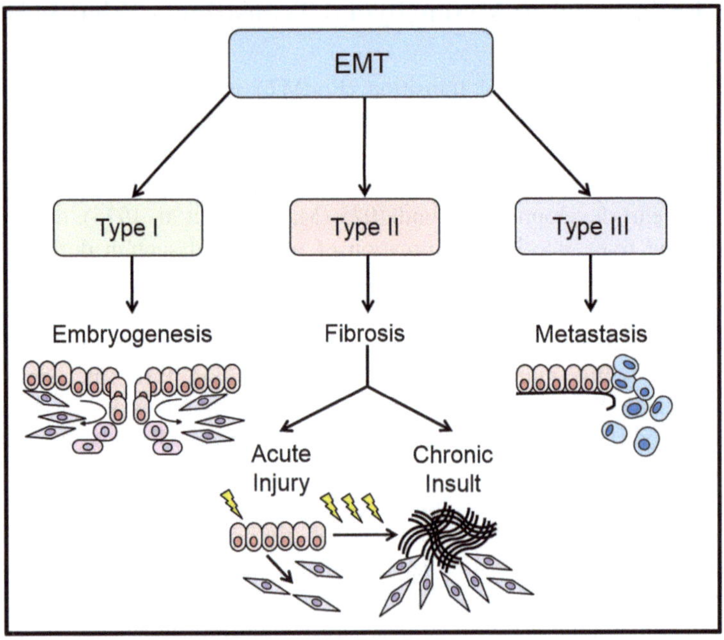

Fig. 13.2 Three types of EMT have been described. Type 1 is associated with embryogenesis, such as development of cardiac cushion. Type 2 is associated with fibrosis, induced either by acute injury such as ischemic event or chronic insult such as persistent inflammatory state. Type 3 is associated with metastasis in neoplastic masses as seen in breast cancer. *EMT* epithelial to mesenchymal transition

MET is poorly understood, but it has increasingly become an area of interest, mainly in the setting of embryology and cancer dissemination. Further studies are warranted to correctly pinpoint its role and possible induction or repression, primarily in the fibrotic setting. It will be interesting to see if MET plays an important role in end-stage organ fibrosis.

13.4 Types of EMT

EMT can be subdivided into three subtypes (type 1, type 2, and type 3) based on the biological context in which they occur. This classification system was created in Polonia in 2007 and subsequently revised and agreed upon in an EMT meeting in 2008 (Kalluri and Weinberg 2009) (Fig. 13.2 and Table 13.1).

EMT type 1 comprises embryonic and developmental EMT programs (Kovacic et al. 2012). Historically, it was the first EMT described and is currently the most studied and understood. This type of EMT is associated with implantation and organ development. Its purpose is to generate diverse cell types that share common mesenchymal characteristics (Kalluri and Weinberg 2009; Zeisberg and Neilson

Table 13.1 Types of EMT

<table>
<tr><th></th><th>Characteristic</th><th>Fibrosis</th><th>Invasive phenotype</th><th>Extra</th><th>Common for all types</th></tr>
<tr><td>Type 1 EMT</td><td>Implantation, embryo formation, and organ development</td><td>No</td><td>No</td><td>Mesenchymal cells have potential to subsequently undergo MET to generate secondary epithelia</td><td rowspan="3">Transcription factors
• SNAI1
• SLUG
• ZEB1
• TWIST
• Goosecoid
• FOXC2
Mesenchymal markers
• FSP1
• αSMA
• Collagen 1
• DDR2
• Vimentin
• Desmin
• Cytokeratin
Epithelial markers
• E-Cadherin</td></tr>
<tr><td>Type 2 EMT</td><td>Wound healing, tissue regeneration, and organ fibrosis</td><td>Yes</td><td>No</td><td>Acute versus chronic</td></tr>
<tr><td>Type 3 EMT</td><td>Neoplastic cells invade and metastasize</td><td>No</td><td>Yes</td><td></td></tr>
</table>

2009; Kovacic et al. 2012; Lee and Nelson 2012). For example, the trophoectoderm cells, which are precursors of the cytotrophoblast, undergo an EMT to facilitate invasion of the endometrium and subsequent anchoring of the placenta, enabling its function in nutrient and gas exchange (Aplin et al. 1998; Bischof et al. 2006). This type of EMT is not associated with fibrosis nor invasive phenotypes (Zeisberg and Neilson 2009; Kovacic et al. 2012). It was previously thought that EMT occurred only in this context and cells lost their plasticity once definite adult cell types were formed.

Type 2 EMT occurs secondary to inflammation (Kalluri and Weinberg 2009) and is associated with wound healing, tissue regeneration, and organ fibrosis (Kalluri and Weinberg 2009; Lee and Nelson 2012; Lopez-Novoa and Nieto 2009; Wynn 2008). After injury occurs, inflammatory cytokines, mechanical stress, or chemical/hormonal substances stimulate EMT with the purpose of creating a new pool of collagen-producing fibroblasts to repair the insult. However, if the cytokine bath occurs secondary to persistent injury, the absolute pool of fibroblasts generated increases dramatically (Kalluri and Neilson 2003; Zeisberg and Neilson 2009; Neilson 2007; Hay 2005). These types of EMT are not associated with embryologic development or metastatic phenotype. Type 2 EMT can be further subdivided based on duration of inflammatory signals.

The type 2 EMT phenomenon that occurs acutely has an initiating signal, and once the injury is resolved, EMT stops. In the acute setting, it is a physiologic response to an inflammatory event. For example, in the setting of cardiac ischemia, an increased pool of collagen-producing fibroblasts may heal the injury more efficiently and potentially avoid rupture of the ventricle free wall. EMT provides a new source of fibroblasts to deposit scar tissue to avoid worsening injury. Once the acute inflammatory state is resolved, the EMT-activating trigger decreases and is eventually deactivated. Clinically, the event is usually self-limiting. Therefore, the fibrosis produced is enough for wound healing or injury repair but not enough to accumulate and interfere with organ function.

Chronic EMT 2, on the other hand, responds to ongoing inflammation, eventually leading to destruction of organ architecture and function (Kalluri and Weinberg 2009). In the setting of chronic disease, inflammatory cytokines and detrimental signaling are constant, which means that epithelial cells will continue to transition into collagen-producing myofibroblasts for as long as the insult is present. As a consequence, epithelial cell number will decrease and the activated fibroblast population will persistently increase, as well as the collagen production and deposition; this translates into increased fibrosis and decreased functional viable tissue. This chapter mainly focuses on this subtype of EMT, specifically in regard to kidney disease.

Theoretically, these subtypes of type 2 EMT are easily differentiated. The definitions are pretty straightforward. However, in practice, it is not clear when an acute injury becomes a chronic one or when a wound healing response stops and a pathological fibrotic process begins. Questions have been raised in regard to acute events perpetuating chronic-type responses; such is the case, for example, in acute kidney injury, where it is not certain if the event is self-limiting or contributes to irreversible organ damage.

Type 3 EMT aids in metastasis of malignant cells (Kalluri and Weinberg 2009; Zeisberg and Neilson 2009; Kovacic et al. 2012; Lee and Nelson 2012; Thiery 2002). It happens in cells that have previously undergone neoplastic genetic and epigenetic changes (Kalluri and Weinberg 2009) and grants the cells the ability to migrate. These cells are now capable of migrating away from the primary tumor site through the bloodstream and potentially initiating a secondary tumor at a remote location (Lee and Nelson 2012; Thiery 2002; Miyazawa et al. 2000). These types of EMT are not associated with embryologic development or fibrotic deposition.

These classifications of EMT are based on physiologic context. In vitro data provide information for all kinds of EMT by elucidating mechanisms, pathways, and possible identification markers. In vivo studies, on the other hand, can be specifically designed to focus on any subtype of EMT. Several mechanisms and pathways have been postulated as initiators and perpetuators of EMT. The pathways of EMT activation have been primarily studied in EMT type 1, and an assumption is made that EMT types 2 and 3 share common pathways. This assumption was based primarily on in vivo animal data and in vitro data, showing consistency in pathways and mechanisms of EMT, regardless of the physiologic subtype.

13.5 EMT—Individual Cell Transitioning Process

The transition of an epithelial or endothelial cell into a mesenchymal cell was discussed earlier in a broad manner. This section aims to explain the actual transitioning process in more detail, highlighting the relevant markers that can be

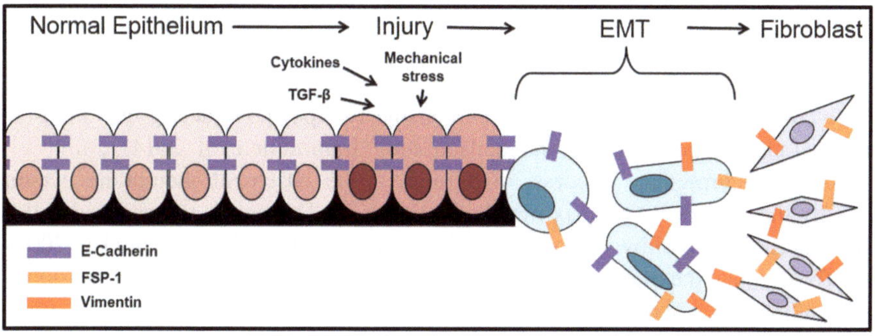

Fig. 13.3 A diagram of the process of EMT is displayed. Epithelial cells with epithelial markers (such as E-Cadherin—*Purple*) are static and adhered to the basal membrane. These cells are subjected to injury (i.e., cytokines, TGF-B, or mechanical stress). Due to this, epithelial cells start transitioning into mesenchymal cells (motile cells which have mesenchymal markers such as FSP-1—*Yellow* and Vimentin—*Orange*). Epithelial cells lose inherent markers (E-cadherin) and eventually only mesenchymal markers are present, which represent the end of EMT. Mesenchymal cells (i.e., fibroblasts) are the final products. EMT cells are considered positive if they have both epithelial and mesenchymal markers in a single point in time

used to identify a cell undergoing EMT or EndMT. A simplistic diagram demonstrating a single transitioning cell is shown in Fig. 13.3.

By co-incubating human tubular epithelial cells (renal) with the cytokine TGFβ1, Yang et al. were able to propose the following sequence of events in EMT (Yang and Liu 2001):

1. Loss of epithelial cell adhesion initiated by E-cadherin decrease
2. De novo αSMA expression and actin reorganization
3. Disruption of tubular basement membrane mainly secondary to MMP2
4. Enhanced cell migration and invasion

The epithelial cell needs to lose its characteristic markers and identifiers that aid in its specific function prior to transitioning. The first and pivotal step in epithelial to mesenchymal cell transition is the downregulation of E-cadherin (Lamouille et al. 2014). E-cadherin is a transmembrane protein that is expressed in all epithelial cells and functions to anchor epithelial cells to its neighboring epithelial cells, providing stability. Significant decreased E-cadherin is considered a prototypical EMT identifier and, as such, has been seen in embryonic development, tissue fibrosis, and cancer metastasis (Zeisberg and Neilson 2009; Hay and Zuk 1995). In the case of transitioning endothelial cells, the pivotal step is decreased VE-Cadherin, whose function is the same as E-cadherin (Nakajima et al. 2000; Zeisberg et al. 2002). For research purposes, E-cadherin expression is used as a marker for epithelial cell presence, and downregulation is synonym of EMT initiation.

Once E-cadherin is downregulated, cell-to-cell adhesion is weakened (Kalluri and Neilson 2003; Zeisberg and Neilson 2009; Kovacic et al. 2012; Lamouille et al. 2014; Strutz et al. 2002; Ikenouchi et al. 2003; Yilmaz and Christofori 2009; Huber et al. 2005). The epithelium loses cell to cell junctions, desmosomes, and, eventually, apical-basal polarity (Kalluri and Neilson 2003; Lamouille et al. 2014).

Next, αSMA expression is induced and actin reorganization occurs (Saito 2013). As a consequence, the transitioning cell changes shape and structure into a mesenchymal phenotype (Zeisberg and Neilson 2009; Lamouille et al. 2014).

In the intermediate process, disruption of the basement membrane occurs, mostly due to MMP2 (matrix metallopeptidase 2). The newly transitioned cell is now motile which allows for invasive phenotypical potential (Yang and Liu 2001).

To complete the transition, the new mesenchymal cells activate the expression of additional genes and proteins: αSMA, collagen 1, collagen III, vimentin, fibronectin, and fibroblast-specific protein 1 (FSP-1) (Kovacic et al. 2012; Strutz et al. 2002; Lu et al. 2001; Rastaldi et al. 2002). The end product of EMT is a collagen-producing, FSP-1, αSMA, and vimentin-expressing myofibroblast lacking any evidence of epithelial characteristics or genotype.

For research purposes, this detailed process is relevant in the identification of EMT. A cell is catalogued as positive for EMT if it contains markers for both epithelial cells and mesenchymal cells, such as CD31 or E-Cadherin and FSP-1, αSMA, or vimentin. A potential pitfall for this method is that only cells that are actively in the transitioning process will be identified since cells whose transition has terminated or is close to initiation will not be positive for both markers. Also, FSP-1, αSMA, and vimentin are not 100% specific to mesenchymal cells, which might produce some false-positive dual stained cells.

13.6 TGFβ-dependent EMT Pathways

Several activation or triggering mechanisms have been implicated in the EMT and EndMT pathways. It is broadly divided into the TGFβ-dependent and TGF-β-independent pathways. In this chapter, the focus will be on the TGFβ-dependent pathway. Primarily because it is the best described trigger for EMT and EndMT and secondarily because of its pro-fibrotic properties as inducer and perpetuator of resident fibroblasts independent of EMT activation. The key players in the TGF-β-dependent pathway will be described and, afterward, their interaction.

Transforming growth factor β (TGFβ) is considered to be the prototypical cytokine for induction of EMT (Lee et al. 2011; Sanford et al. 1997; Xu et al. 2009). TGFβ is a cytokine belonging to the TGFβ superfamily and is a potent stimulator of most inflammatory pathways as well as being involved in some anti-inflammatory pathways. TGFβ signal is initiated when the ligand (mainly TGFβ1) binds to its cognate TGFβ type II receptor (TβRII). This leads to the phosphorylation of TβRI (also known as ALK5) (Lebrin et al. 2004). Once phosphorylated, activated TβRI phosphorylates canonical SMAD2/3, which in turn translocate to the nucleus and control gene expression. TGFβ has been implicated as a master switch of fibrosis in many tissues (Lopez-Novoa and Nieto 2009; Zeisberg et al. 2007) for both EMT-mediated fibrosis and non-EMT fibrosis, and this is a common pathway for both. Of note, TGFβ can also bind to activin receptor-like kinase 1 (ALK1) exclusively in endothelial cells, which induces SMAD1/5 to potentiate angiogenic programs (Xavier et al. 2015). Fibrosis will be the end result of this pathway, either through EMT-induced myofibroblast proliferation, TGFβ activation of native fibroblasts, or a combination of both.

Several genes and transcription factors have been associated with that first step of E-cadherin downregulation; most transcription factors involved are part of the SNAI1 family, ZEB1, SIP1/ZEB2, and bHLH factors (Saito 2013; Peinado et al. 2007; Zhou et al. 2004; Wu et al. 2009a, b; Du et al. 2010). For research purposes, these transcription factors, genes, and protein expression are used as surrogates for EMT or EndMT activity.

SNAI1 is considered a key organizer of EMT (Kovacic et al. 2012; Ikenouchi et al. 2003) and one of the most potent stimulators of E-cadherin downregulation (Kovacic et al. 2012; Ikenouchi et al. 2003). Through their carboxyl-terminal zinc finger domains (Xu et al. 2009; Peinado et al. 2007), SNAI1 and SLUG repress epithelial genes by binding to E-box DNA sequences in the proximal promoter region of the E-cadherin gene (Lin et al. 2010; Peinado et al. 2004; Herranz et al. 2008; Dong et al. 2013; von Gise and Pu 2012; Vincent et al. 2009). This binding of E-box sequences in the proximal promoter region of E-cadherin by SNAI1/SLUG additionally recruits inflammatory cytokines and represses gene expression. SNAI1 is dependent on SMAD 2/3 activation (Lee et al. 2011; Peinado et al. 2003; Thuault et al. 2006; Lo et al. 2007). SNAI1 regulates various aspects of the EMT such as RhoA and vimentin upregulation, stimulating cytoskeletal remodeling and activating expression of matrix metalloproteases, as well as providing protection from cell death (von Gise and Pu 2012; Zeisberg and Neilson 2009). SNAI1 has a more potent E-cadherin downregulation than SLUG (Bolos et al. 2003; von Gise and Pu 2012). SNAI1 is quickly degraded in cytoplasm with a half-life of only 25 min in human cell lines (Zhou et al. 2004; Yook et al. 2006; Dominguez et al. 2003). Thus, SNAI1 persistence in nuclear space increases results in triggering the EMT cascade.

TWIST is a member of the bHLH-family of transcription factors. It is upregulated during mesoderm development, neural tube formation, and tissue fibrosis and tumor metastasis (Yu et al. 2008; Chen and Behringer 1995; Kida et al. 2007; Yang et al. 2004). The gene TWIST has been described as a participant of EMT in all three types of EMT and is considered an independent pathway from SNAI1 and SLUG in repressing E-cadherin and upregulates fibronectin and N-cadherin (Yang et al. 2004, 2007).

In regard to the EMT pathway, TGFβ activates and upregulates expression of the transcription factors, SNAI1 and SNAI2 (aka SLUG), and directly downregulates the expression of E-cadherin (Cano et al. 2000; Lamouille et al. 2014; Lee et al. 2011; Zeisberg et al. 2007; Peinado et al. 2003; Thuault et al. 2006; Lo et al. 2007) (Fig. 13.4). SNAI1 is phosphorylated by GSK3B to facilitate its nucleocytoplasmic transport by exposing a nuclear export sequence (Zeisberg and Neilson 2009; Zhou et al. 2004; Yook et al. 2006; Boutet et al. 2006; Yu et al. 2008; Dominguez et al. 2003). Also, TGFβ represses the expression of Id (inhibitor of differentiation protein), which inhibits EMT (Kondo et al. 2004). This Id repression is required for further E-cadherin downregulation.

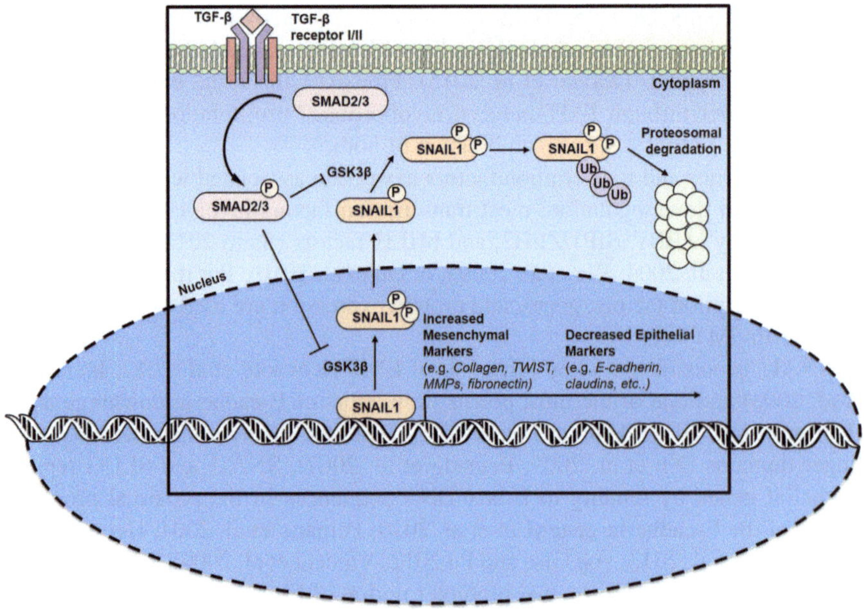

Fig. 13.4 The EMT pathway starts secondary to injury; this diagram only represents the SMAD-dependent TGF-B pathway. TGF-B binds to TGF-B Receptor II extracytoplasmic domain which phosphorylates TGF-B Receptor I cytoplasmic domain, which, in turn, phosphorylates SMAD2/3. SMAD2/3 inhibits GSKB signaling in nucleus and cytoplasm. GSKB has a dual phosphorylation role: first it phosphorylates SNAI1 for translocation from nucleus to cytoplasm and second it phosphorylates SNAI1 to promote its proteosomal degradation through ubiquitination. SNAI1 has a short cytoplasmic half-life, but it is rather stable intranuclear. Therefore, by inhibiting GSKB signaling, SNAI1 maintains its intranuclear capacity and promotes the loss of epithelial markers (i.e., E-cadherin, claudins) and increased mesenchymal markers (i.e., collagen, TWIST, MMPS, fibronectin). *TGF-B* tumor growth factor B; *GSKB* glycogen synthase kinase 3 beta; *MMPS* matrix metalloproteinases

13.7 Organ Fibrosis and EMT/EndMT

Fibrosis is the hallmark of any end-stage organ disease, and, as such, it is a topic of great interest in both the research and clinical community. A total of 45% of deaths can be attributed to fibrotic disease of some kind. Fibrosis is a key component in the progression of chronic kidney disease, and in 2014, an estimated 13% of the general population has been diagnosed with some degree of chronic kidney disease (Rockey et al. 2015). Overall, fibrosis is the consequence of collagen deposition in viable functional tissue. Interstitial collagen deposition interferes with healthy parenchymal architecture and functionality. Damage secondary to fibrosis production depends on the spatial location of the collagen deposition, the extent of it, and the tissue's intrinsic function.

For example, collagen deposition in the alveolar space of the lungs disrupts normal parenchymal architecture and causes a significant decrease in the area of oxygen diffusion, impairing respiration. At the same time, fibrosis decreases lung compliance (ability to stretch and expand) and reservoir capacity. In the heart, for example, accumulation of interstitial fibrosis has several consequences. First, it interferes with electrical conduction of cardiomyocytes by interrupting the architecture which decreases efficient contractility. Independent of electric conduction aberrance, fibrotic deposition decreases contractility by interfering with compliance and effective contractile tissue. This, in turn, decreases stroke volume and cardiac output and increases the odds of intracardiac thrombi formation. Within the kidney when looking at, interstitial fibrosis in renal tubules and glomeruli interferes with absorption into the nephron and reabsorption of substrates from the nephron. Impaired renal function is associated with acid–base disturbances and electrolyte imbalances. If collagen I deposition occurs in the perivascular space, renal ischemia can occur. Renal fibrosis, characterized by tubulointerstitial fibrosis, tubular atrophy, and glomerulosclerosis, disrupts the normal architecture of the kidney and ultimately leads to end-stage renal disease (He et al. 2013).

The hallmark of fibrosis is the activation and accumulation of fibroblasts or myofibroblasts and the deposition of abundant extracellular matrix (Piera-Velazquez et al. 2011). Although it was previously thought that all collagen-producing fibroblasts were thought to originate solely from resident fibroblasts, recent data have shown that various types of other cells have the ability to be a source of collagen-producing fibroblasts. Pericytes, bone marrow, endothelial, and epithelial cells have all been postulated as possible sources of activated fibroblasts. This will be discussed throughout this chapter, specifically in the context of renal disease.

EMT contributes to fibrogenesis as a repair mechanism in response to an inflammation. EMT has been identified in human tissue in the context of renal, pulmonary, and liver fibrosis (Kalluri and Neilson 2003). Additionally, EMT is found to occur in most animal models of organ fibrosis to different extents. EMT is thought to be a double-injury mechanism: increasing the amount of viable collagen-producing myofibroblasts and interfering with parenchymal architecture through epithelial cell loss (Lee and Nelson 2012). For instance, in 2011, Deng et al. identified biliary epithelial cells forming small- and medium-sized bile ducts in human patients with biliary atresia transition to myofibroblasts, which may account for prominent bile ductular proliferation and directly contribute to fibrogenesis in biliary atresia (Deng et al. 2011).

In the heart, Zeisberg et al. reported that up to 35% of fibroblasts originated from endothelial cells using lineage tracing techniques in an aortic-banding pressure-overload mouse model (Zeisberg et al. 2007). A few years later, in 2010, Hashimoto et al. published similar findings in a bleomycin-pulmonary hypertension mice model, where up to 16% of lung fibroblasts were derived from endothelial cells (Hashimoto et al. 2010). Specific to fibrotic kidney disease, findings will be discussed in the next sections with more detail.

EMT correlates with higher fibrosis percentage in organ failure animal models and decreased functionality surrogates. In the heart, lung, and kidney, EMT presence has been correlated with worse functional markers as well as overall poor outcomes. This opens up the possibility of blocking or decreasing EMT and possibly having a positive impact on the disease progression.

13.8 EMT in Fibrotic Kidney Disease

End-stage renal dysfunction is characterized by progressive loss of kidney function and increased deposition of extracellular matrix leading to generalized fibrosis (Liu 2004). Progressive chronic kidney disease is the consequence of destructive fibrosis (He et al. 2013). Thus, the scientific community is interested in fibrotic pathways and mechanisms, in the hope of potentially halting or inhibiting them.

Several theories exist on the origin of collagen-producing myofibroblasts in kidney disease. Although the overall agreement is that the majority of them arise from resident fibroblast population, other possible sources have been postulated. In this section, tubular epithelial-derived myofibroblasts will be discussed.

A NCBI database search yields over 400 publication results under "EMT Kidney Fibrosis." Questions regarding its existence and role in vivo have been raised mainly due to inconsistencies in the result.

EMT has been identified, mainly using transgenic epithelial lineage tracing technology, in tubular epithelial cells in vitro and in several animal models in vivo (Zeisberg et al. 2003; Liu 2010; Yang and Liu 2001; Burns et al. 2006; Holian et al. 2008; Lan 2003; Liu et al. 2000; Aresu et al. 2007; Shimizu et al. 2006; Bedi et al. 2008; Djamali et al. 2005). Efforts have been made to try to elucidate mechanisms and key players in the transition process. Some strategies to decrease or halt EMT in kidney fibrosis have also been implemented in research settings. Many of the relevant publications will be described in this section.

In 1995, through subtractive and differential hybridization of murine cells for transcript comparison, fibroblast-specific protein 1 (FSP-1) was associated with cells of mesenchymal origin or has markers of fibroblasts (Strutz et al. 1995). However, it was noted that in fibrotic kidneys secondary to persistent inflammation, many fibroblasts could be identified in interstitial sites of collagen deposition and also in tubular epithelium adjacent to the inflammatory process. This led to the hypothesis that fibroblasts may arise from local conversion of epithelium.

A few years later, a subtotal nephrectomy (right subcapsular nephrectomy and infarction of 2/3 of the left kidney) rat model was implemented to investigate the potential role of tubular epithelial cells in progressive fibrosis in the remnant kidney (Ng et al. 1998). Phenotypic evidence of tubular epithelial-myofibroblast transdifferentiation was supported by ultrastructural studies identifying actin microfilaments and dense bodies within tubular epithelial cells. More so, they noted that histologically, tubular epithelial cells lost apical-basal polarity and tight junctions, became elongated, detached from the basal membrane and neighboring cells, and even appeared to migrate into peritubular interstitium through the damaged basement membrane. It should be noted though that these changes were seen in $<15\%$ of the tubules.

In 2001, through the incubation of human tubular epithelial cells with TGFβ1, Yang et al. saw that it induced de novo expression of αSMA, loss of epithelial marker (E-cadherin), transformation of myofibroblastic morphology, and production of interstitial matrix (Yang and Liu 2001). These findings were repeated in vivo using a unilateral ureteral obstruction mouse model.

Iwano et al. published their landmark study in 2002, where they aimed to identify the sources of interstitial kidney fibroblasts using bone marrow chimeras and transgenic reporter mice in a unilateral ureteral obstruction nephropathy in order to perform lineage tracing (Iwano et al. 2002). They concluded that a substantial number of organ fibroblasts appear from sources other than the epithelium. However, the percentage of transitioning cells was not reported. It did, however, highlight the potential plasticity of differentiated cells in adult tissue under pathologic conditions.

Several authors have documented EMT of tubular epithelial cells of humans (Nightingale et al. 2004; Yang and Liu 2001) and animals (Li et al. 2004) in vitro. In these studies, EMT was mainly identified by de novo expression of a mesenchymal marker (αSMA or FSP-1) and loss of epithelial marker (E-Cadherin).

Based on the assumption that EMT of tubular cells contributed to the renal accumulation of matrix proteins associated with diabetic nephropathy, a study was done to examine the role of prosclerotic growth factor (CTGF) in EMT induction (Burns et al. 2006). They noted that CTGF partially induces EMT in vitro; this was not blocked by neutralizing anti-TGFβ antibodies, suggesting that this phenomenon is independent of TGFβ action. In vivo, a decrease of E-cadherin in tubules of diabetic rats was noted, as well as an increase in CTGF and αSMA.

A few years later, Zhang et al. investigated the role of plasminogen in renal fibrosis using both in vivo (unilateral ureteral obstruction mice model) and in vitro (Zhang et al. 2007). They noted a significant decrease of EMT, identified by loss of E-cadherin and acquisition of αSMA, in mice missing plasminogen (50 versus 76% in normal mice). These in vivo findings were also supported by in vitro tubular epithelial cell transition into fibronectin-producing, fibroblast-like cells, which was blocked by knocking down protease-activated receptor-1 (PAR-1) using silencing RNA.

For the first time ever, Rastaldi et al. aimed to identify EMT in human renal biopsies obtained from patients with different degrees of renal fibrosis and different etiologies (Rastaldi et al. 2002). After analyzing 133 biopsies, they concluded that human tubular epithelial cells can transition to extracellular matrix protein-producing mesenchymal cells. They observed loss of epithelial antigens in 8–10% of tubular cross-sections. In 2005, Nishitani et al. identified some EMT cells after looking at 142 samples from human patients with IgA nephropathy (Nishitani et al. 2005). In 2006, Hertig et al. studied biopsies obtained 3 months after renal transplant from 56 patients and reported that EMT was virtually absent in implantation biopsies, whereas 41% of the grafts were EMT positive in the absence of advanced chronic allograft nephropathy (Hertig et al. 2006). These results suggest that EMT is present in human renal disease.

Interestingly, Lovisa et al. proposed a new concept in 2015 based on findings on three different mouse models: unilateral ureteral obstruction nephropathy, nephrotoxic serum-induced nephritis, and folic acid-induced nephropathy (Lovisa et al. 2015). They proposed a partial EMT program for tubular epithelial cells, where they remain associated with their basement membrane but express markers of both epithelial and mesenchymal cells. This caused an arrest in the G2 phase of the cell cycle and lower expression of solute and solvent transporters. Also, they noted that transgenic expression of TWIST1 or SNAI1 was sufficient to promote TGF-β1-induced G2 arrest of tubular epithelial cells, limiting the cells' potential for repair and regeneration.

Around that same time, Grande et al. showed that the reactivation of SNAI1 in mouse renal epithelial cells is required for the development of fibrosis in the kidney (Grande et al. 2015). They concluded that SNAI1 reactivation induced partial EMT that relayed signals to interstitium to promote myofibroblast differentiation, fibrogenesis, and to sustain inflammation, without directly contributing to the myofibroblast pool.

The concept of partial EMT is an intriguing one: tubular epithelial cells are activated secondary to an inflammatory response and, via SNAI1, contribute to fibrosis production and express mesenchymal markers, without becoming motile or damaging epithelial structure.

13.9 EndMT in Fibrotic Kidney Disease

In the context of kidney disease and EndMT, several papers have been published trying to identify the existence and, if it exists, quantify its implications in fibrosis with the intention to halt the process in different nephropathy animal models and human patients (Table 13.2). However, data have not been consistent, especially in the percentage of fibroblasts derived from endothelial cells in a development disease. The varying contributions of endothelial cells from study to study make it difficult to interpret the significance of EndMT in fibrosis.

Zeisberg et al. conducted a landmark experiment in 2008 to confirm the contribution of EndMT in renal fibrosis in three different animal models: unilateral ureteral obstruction nephropathy, streptozotocin-induced diabetic nephropathy, and a surrogate for Alport syndrome (Zeisberg et al. 2008). They concluded that approximately 30–50% of fibroblasts co-expressed the endothelial marker CD31 and mesenchymal markers FSP-1 and αSMA. Using dual transgenic mice, they confirmed this finding with endothelial lineage tracing.

Approximately, a year later, Li et al. concluded that EndMT occurs and contributes to the early development of diabetic renal interstitial fibrosis. They did so by looking at a streptozotocin-induced diabetic nephropathy model using endothelial-lineage traceable mice to identify if EndMT occurred in diabetic renal interstitial fibrosis (Li et al. 2009). They concluded that about 10–23.5% of renal interstitial myofibroblasts in the diabetic nephropathy model were of endothelial origin compared to 0.2% in control mice. They complemented their findings with in vitro studies using primary culture of fluorescently labeled adult mouse kidney cells and they identified through confocal microscopy and real-time PCR de novo expression of αSMA and loss of expression of VE-Cadherin and CD31 secondary to induction by TGFβ1.

Table 13.2 Summary of current literature of EndMT in kidney disease

Authors	Model	EndMT cells	Extra
Zeisberg et al. (2008)	Animal; Unilateral ureteral obstruction nephropathy, Streptozotocin-induced diabetic nephropathy, Surrogate for Alport syndrome	30–50%	
Li et al. (2009)	Animal; Streptozotocin-induced diabetic nephropathy	10–23.5%	
LeBleu et al. (2013)	Animal; Unilateral ureteral obstruction nephropathy	10%	35% bone marrow-derived 5% EMT
Deng et al. (2016)	Humans; Biopsies from primary and secondary nephropathies Animal; Unilateral ureteral obstruction nephropathy	Presence	

A few years later, LeBleu et al. engineered a meticulous study to identify myofibroblast source and function in kidney disease induced by unilateral ureteral obstruction using multiple genetically modified mice (LeBleu et al. 2013). They concluded that the total pool of myofibroblasts is split: 50% arising from resident fibroblasts via proliferation and 50% did not. The 50% that did not arise from resident fibroblasts was further subdivided into 35% via bone marrow differentiation, 10% via EndMT, and 5% via EMT. They also concluded that vascular pericytes likely do not contribute to the pool of myofibroblasts.

In early 2016, Deng et al. demonstrated that in human patients with primary and secondary nephropathy, αSMA, as a surrogate for mesenchymal marker, was expressed in endothelium (Deng et al. 2016). Renal biopsies from adult patients with primary nephropathy (e.g., IgA nephropathy) and secondary nephropathy (e.g., diabetic or lupus nephritis) were analyzed for EndMT presence per double labeling of CD31 and αSMA. Sequential sections of abnormal microvasculature were reconstructed using 3D models and analyzed; the percentage of αSMA+ CD31+ peritubular capillary numbers was substantially increased compared to controls. This suggests that EndMT contributes to interstitial fibrosis in kidney disease in both primary and secondary nephropathies. They also followed this method of analysis in unilateral ureteral obstruction mouse model and found the same to be true which they further confirmed in an in vitro human umbilical vein endothelial cell culture stimulated by TGFβ1.

Based on recent publications like those described previously, some attempts to halt or ameliorate EndMT in kidney disease have been published in the last few years. This research was conducted under the assumption that EndMT is a definite contributor to fibrosis in renal disease.

In 2013, Curci et al. investigated the role of complement in the induction of EndMT in an ischemic/reperfusion injury swine model (Curci et al. 2014). By inhibiting complement activity within the model, peritubular capillary density was preserved and significant reduction in renal fibrosis was observed, as well as very few dual stained cells were identified, contrary to those seen without complement inhibition. Similar findings documenting EndMT's contribution to fibrosis were reported by Basile et al. in an ischemic/reperfusion injury rat model (Basile et al. 2011).

Kanasaki et al. observed that the DPP4 (dipeptidyl peptidase-4) inhibitor currently in the market as an antidiabetic drug under the name linagliptin ameliorates kidney fibrosis in diabetic mice without altering blood glucose levels by inhibiting EndMT and restoring the microRNA *mir-29* (Kanasaki et al. 2014) both in vivo and in vitro.

Xavier et al. addressed the question whether endothelial TGFβ signals contribute to renal fibrosis by using mice with partial ablation of TβRII in renal fibrosis secondary to folic acid nephropathy and unilateral ureteral obstruction (Xavier et al. 2015). The data demonstrated that restricting TGFβ signaling in vascular endothelial cells is enough to reduce EndMT, decrease TGFβ-induced canonical SMAD2 signaling, improve angiogenesis, and reduce fibrosis. They concluded that endothelial TGFβ and EndMT are implicated in regulating angiogenic and fibrotic response to injury in two nephropathy models.

EndMT of tubular epithelial cells has been successfully associated with collagen-producing myofibroblasts in vitro. In vivo evidence of the phenomenon has been shown by some authors; however, the real burden of this phenomenon on the progression of the disease or prognosis of this has not been conclusive. The inconsistency within the results is due partially to the fact that EndMT lacks a gold standard for identification.

13.10 Other Sources of Myofibroblasts

Some authors have dismissed the implication of EMT and its role in fibrotic kidney disease altogether, while others have concluded that although EMT or EndMT participates in myofibroblast pool production, other sources suggest that EndMT plays a more important role in fibrosis. For example, as mentioned previously, LeBleu et al. demonstrated that although some myofibroblasts arise from EMT and EndMT (5% and 10%, respectively), approximately 35% arise from bone marrow. Other sources of myofibroblasts that have been identified will be discussed.

In 2007, Broekema et al. aimed to identify the contribution of bone marrow-derived cells to renal interstitial myofibroblast population in relation to fibrotic changes after ischemia/reperfusion injury by unilateral engraftment in the tubular epithelium in rats (Broekema et al. 2007). An average of 32% of myofibroblasts was derived from the bone marrow which were considered functional per procollagen I production capacity. Interestingly, increased TGFβ and decreased BMP-7 were observed in interstitial myofibroblasts but not in bone marrow-derived myofibroblasts.

In 2008, Lin et al. used a transgenic reporter mouse model to examine the origins of collagen 1A1 producing cells in the kidney in fibrotic renal disease (Lin et al. 2008). They concluded that either vascular injury or vascular factors are the most likely triggers for pericyte migration and differentiation into myofibroblast. Based on these results, they suggested that the scientific community should refocus fibrosis research to injury of the vasculature rather than injury to the epithelium.

Humphreys et al. in 2010 assessed the contribution of renal epithelial cells to myofibroblasts using genetically modified Flox-stop reporter mice in a ureteral obstructive nephropathy model (Humphreys et al. 2010) both in vivo and in vitro. Although renal epithelial cells acquired mesenchymal markers in vitro, they found no evidence that epithelial cells migrate outside of the tubular basement membrane and differentiate into interstitial myofibroblasts in vivo. Interestingly, they noted that pericyte cells, and not epithelial cells, may account for a large majority of myofibroblasts.

13.11 Controversies—Discussion

The concept, role, and burden of EMT and EndMT are controversial. Results have been met with skepticism by some while others believe that the data are significant enough to warrant attention. Some of the controversies revolving around this subject will be discussed here: the concept of adult cell plasticity, presence of EMT/EndMT in human samples, role and relevance of EMT/EndMT in fibrotic disease, risk of inhibition of EMT/EndMT, existence of MET in fibrotic disease, and inconsistencies in EMT/EndMT in kidney disease.

According to previous physiologic paradigms, cells derive from an embryonic lineage and commit to a cell line. Fully mature, developed cells have limited plasticity and transitioning cells break these concepts of adult cell plasticity. Therefore, resistance has been offered regarding its existence in the context of metastatic and fibrotic disease. These skepticisms have decreased significantly, though, partially thanks to in vitro studies in the area of EMT but also secondary to the extraordinary results of stem cell research. EMT and EndMT are currently accepted as a definite process in vitro but only as a potential process in vivo.

The concept that EMT/EndMT occurs in human tissue and not only in vitro is something that has not been fully accepted by the scientific community. For obvious reasons, lineage tracing methods are not a possibility in human samples, and the fact that EMT/EndMT is thought to be an early, time-dependent process complicates the identification of it. Therefore, the sensitivity of the current techniques is very poor. Current evidence of EMT/EndMT in human samples is based on dual staining techniques; this means the only cells quantified as positive are in the process of transitioning, while fully transformed collagen-producing fibrocytes that have already lost all epithelial or endothelial markers are excluded from results.

Furthermore, mesenchymal markers lack specificity since FSP-1 and αSMA have also been identified in inflammatory and endothelial cells. It is a very ambitious goal to scientifically prove that EMT or EndMT significantly contributes to fibrosis in end-stage organ dysfunction without explicit lineage tracing methods. However, current information obtained from human samples is not negligible; quantification of EMT or EndMT positive cells and its correlation with degree of fibrosis, epithelial contribution, and clinical outcomes can shed some light on this topic. However, animal in vivo data is needed to comprehend mechanism since results obtained from human samples are merely descriptive.

Another area of controversy is the role and relevance of EMT and EndMT in the progression of fibrotic disease. Assuming EMT or EndMT exists in vivo, its actual contribution has been questioned in more than one occasion. In various animal models of different organ fibrosis, the percentage of collagen-producing myofibroblasts varies dramatically from approximately 5 to 50%. Also, the potency of these myofibroblasts originating either from epithelial cells or endothelial cells has not been approached. Furthermore, TGFβ is an EMT/EndMT trigger but also triggers fibrosis independent of EMT/EndMT. As these pathways are deeply interconnected, it is difficult to elucidate between the fibrotic changes seen secondary to TGFβ (independent of EMT/EndMT pathway) and actual EMT/EndMT contribution. EMT or EndMT's contribution might be overestimated when correlating the number of dual stained cells with fibrosis percentage due to the attribution of TGFβ's pro-fibrotic consequences.

EMT and EndMT occur in the setting of inflammation as a natural process of wound healing. Thus, inhibiting the pathway as an anti-fibrotic mechanism may have other detrimental effects. If we manage to block EMT or EndMT as an anti-fibrotic strategy in chronic fibrotic disease (such as chronic kidney disease or chronic heart failure), what would happen to the tissue if it is subjected to an acute immediate injury? Would the collagen production of resident fibroblasts be sufficient to avoid dysfunction in the long term? Or will the fibrotic response be insufficient because of a decreased myofibroblast pool? Early blockade of TGFβ using anti-TGFβ antibodies has decreased survival in a myocardial infarction mouse model (Ikeuchi et al. 2004). However, a TGFβ blockade would decrease EMT/EndMT as well as any pro-fibrotic TGFβ-induced cascade. Further studies will be warranted to identify an anti-fibrotic target therapy within the EMT/EndMT pathway.

MET has been identified in embryology and, with some controversy, in oncology setting. Little is known about MET within fibrotic tissue and its signaling mechanism is not clearly defined. As of now, the same set of genes associated with EMT have been postulated for MET (SNAI1, SLUG, TWIST). This is based on the premise that EMT and MET are ends of the spectrum of a same transitional pathway and their participation in the injury–repair balance. Although this hypothesis makes sense theoretically, few data has been published regarding this matter.

EMT and EndMT have been identified in vitro, in vivo using animal models, and in human kidney samples. Authors have inhibited either EMT or EndMT and reported decreased fibrosis and increased renal function. However, other authors have also published myofibroblasts originating from other cell types and rejecting the notion of EMT/EndMT or accepting the phenomenon but doubting its relevance and role in end-stage fibrotic renal disease. The inconsistency in results and conflicting data has opened the floor for debate. For instance, while EMT and EndoMT can be triggered in vitro under different circumstances, EndoMT quantification using in vivo animal models ranged from 50% (Iwano et al. 2002) to 10% (LeBleu et al. 2013). Even using the same animal model and lineage tracing techniques, the variability among published results is significant. Rather than answering the question of EMT/EndMT's contribution to kidney fibrotic disease, it opens the topic for discussion and interpretation. The variation in results could be explained by the fact that EMT/EndMT is an ongoing time-dependent process that, depending on the time between analysis and onset of injury, might have not occurred yet or the transition might be completed by then. Another possible explanation could be the lack of specificity in the markers currently used for EMT/EndMT identification. Last but not least, partial EMT might also explain the variability in all the previous publications where they observed the same phenomenon at different stages. These are speculations based on assumptions, and further studies with more developed methods are necessary to truly identify the contribution of EMT/EndMT to renal fibrotic disease.

13.12 Conclusion

EMT is the transition of epithelial cells into mesenchymal cells and this can occur in the context of development, injury, or metastasis. Epithelial cells lose inherent markers and characteristics such as attachment to the basal membrane and neighboring cells while gaining mesenchymal cell characteristics and markers like motility and collagen-producing capacity. Endothelial to Mesenchymal Transition (EndMT) is a subtype of EMT where the transitioning cell is an epithelial cell, a specialized vascular epithelial, into collagen-producing mesenchymal cells.

Three types of EMT have been described based on their pathophysiologic context: type 1 EMT has been described in embryogenesis and development, type 2 EMT in the setting of inflammation and fibrosis, both acute and chronic conditions, and type 3 EMT has been associated with increased metastatic characteristics in cancer.

On the other hand, mesenchymal to epithelial transition (MET), mechanistically the opposite of EMT, has been clearly identified in embryologic development and partially implicated in the development of a secondary tumor site in the oncologic setting. Further studies are warranted to elucidate a role of MET within fibrotic disease, either as a promoter of anti-fibrotic characteristics or as a faulty mechanism within the injury–recovery spectrum. Considering the extent of fibrotic disease in end-stage organ failure and its clinical outcome implications, the scientific community is in need of a new anti-fibrotic pathway to focus on. Current therapies focus on preventing the symptoms and progressions, but few actually aim to reverse the current fibrosis. Therefore, MET offers a potential new opportunity of therapeutic targets and interventions.

EMT has been identified as a contributor of end-stage organ fibrosis in the kidney, heart, and liver. By providing a new source of collagen-producing myofibroblasts, the potential of fibrosis deposition is increased. Thus, EMT provides a new source of activated myofibroblasts and a new mechanism of initiating and developing fibrosis. This has the potential to become a novel therapeutic target.

The primary focus of this chapter is the contribution of EMT and EndMT to kidney fibrotic disease, and many relevant research proposals have been described as evidence of the current state of this topic.

In regard to EMT and EndMT in kidney disease, several manuscripts have been published on this matter. EMT and EndMT have been extensively and repeatedly proven in vitro. However, conflicting data exists for in vivo models. EMT was proposed as an important contributor to renal fibrosis since 1995, and conclusive data has not been reported yet, even with access to lineage tracing transgenic animal models. Results range from attributing up to 50% of collagen-producing myofibroblasts to endothelial origin, to 10% from endothelial cells and 5% to epithelial cells to not finding any fibroblasts derived from epithelial origin. Authors have also identified collagen-producing myofibroblasts originating from bone marrow and pericytes, as opposed to those from epithelial or endothelial origin. EMT in human samples of kidney disease has also been described, primarily in biopsy specimens of nephropathies and posttransplant grafts. EMT and EndMT have also been inhibited both in vitro and in vivo using animal models with a subsequent decrease in fibrosis and increase in kidney function.

A relatively new concept of partial EMT has been proposed and may open the table for discussion on the role and contribution of EMT to renal fibrosis in end-stage disease.

EMT and EndMT in fibrotic disease development and progression are controversial. Results have been met with skepticism mainly due to the concept of adult cell plasticity, identification of EMT/EndMT in human tissue, the actual relevance and contribution of EMT/EndMT to organ fibrosis, risk of inhibiting EMT/EndMT pathway, the concept of MET in fibrotic disease and injury–repair balance, and finally, and most importantly, the inconsistencies in the literature regarding EMT and EndMT in renal fibrotic disease.

Fibrosis is the hallmark of most chronic diseases, which leads to loss of parenchymal architecture and pathologic dysfunction. EMT and EndMT provide a new source of collagen-producing myofibroblasts and, as such, a potential therapeutic anti-fibrotic target. Further studies are warranted and as technology improves, mechanisms and pathway implicated in this process might become clearer. EMT Even though there is controversy regarding EMT, one thing is sure, it represents a new opportunity: an opportunity for discussion and an opportunity for discovery; a pathway that proposes a new mechanism for collagen deposition; and, as such, a new therapeutic option to prevent, delay, and treat fibrosis.

References

Aplin JD, Haigh T, Vicovac L, Church HJ, Jones CJ (1998) Anchorage in the developing placenta: an overlooked determinant of pregnancy outcome? Hum Fertil (Camb) 1:75–79

Aresu L, Rastaldi MP, Scanziani E, Baily J, Radaelli E, Pregel P, Valenza F (2007) Epithelial-mesenchymal transition (EMT) of renal tubular cells in canine glomerulonephritis. Virchows Arch 451:937–942

Basile DP, Friedrich JL, Spahic J, Knipe N, Mang H, Leonard EC, Changizi-Ashtiyani S, Bacallao RL, Molitoris BA, Sutton TA (2011) Impaired endothelial proliferation and mesenchymal transition contribute to vascular rarefaction following acute kidney injury. Am J Physiol Renal Physiol 300:F721–F733

Bedi S, Vidyasagar A, Djamali A (2008) Epithelial-to-mesenchymal transition and chronic allograft tubulointerstitial fibrosis. Transplant Rev (Orlando) 22:1–5

Bischof P, Aplin JD, Bentin-Ley U, Brannstrom M, Casslen B, Castrillo JL, Classen-Linke I, Critchley HO, Devoto L, D'Hooghe T, Horcajadas JA, Groothuis P, Ivell R, Pongrantz I, Macklon NS, Sharkey A, Vicovac L, White JO, Winterhager E, Von Wolff M, Simon C, Stavreus-Evers A (2006) Implantation of the human embryo: research lines and models. From the implantation research network 'Fruitful'. Gynecol Obstet Investig 62:206–216

Bolos V, Peinado H, Perez-Moreno MA, Fraga MF, Esteller M, Cano A (2003) The transcription factor Slug represses E-cadherin expression and induces epithelial to mesenchymal transitions: a comparison with Snail and E47 repressors. J Cell Sci 116:499–511

Boutet A, De Frutos CA, Maxwell PH, Mayol MJ, Romero J, Nieto MA (2006) Snail activation disrupts tissue homeostasis and induces fibrosis in the adult kidney. EMBO J 25:5603–5613

Broekema M, Harmsen MC, Van Luyn MJ, Koerts JA, Petersen AH, Van Kooten TG, Van Goor H, Navis G, Popa ER (2007) Bone marrow-derived myofibroblasts contribute to the renal interstitial myofibroblast population and produce procollagen I after ischemia/reperfusion in rats. J Am Soc Nephrol 18:165–175

Burns WC, Twigg SM, Forbes JM, Pete J, Tikellis C, Thallas-Bonke V, Thomas MC, Cooper ME, Kantharidis P (2006) Connective tissue growth factor plays an important role in advanced glycation end product-induced tubular epithelial-to-mesenchymal transition: implications for diabetic renal disease. J Am Soc Nephrol 17:2484–2494

Cano A, Perez-Moreno MA, Rodrigo I, Locascio A, Blanco MJ, Del Barrio MG, Portillo F, Nieto MA (2000) The transcription factor snail controls epithelial-mesenchymal transitions by repressing E-cadherin expression. Nat Cell Biol 2:76–83

Chen ZF, Behringer RR (1995) Twist is required in head mesenchyme for cranial neural tube morphogenesis. Genes Dev 9:686–699

Coll-Bonfill N, Musri MM, Ivo V, Barbera JA, Tura-Ceide O (2015) Transdifferentiation of endothelial cells to smooth muscle cells play an important role in vascular remodelling. Am J Stem Cells 4:13–21

Curci C, Castellano G, Stasi A, Divella C, Loverre A, Gigante M, Simone S, Cariello M, Montinaro V, Lucarelli G, Ditonno P, Battaglia M, Crovace A, Staffieri F, Oortwijn B, Van Amersfoort E, Gesualdo L, Grandaliano G (2014) Endothelial-to-mesenchymal transition and renal fibrosis in ischaemia/reperfusion injury are mediated by complement anaphylatoxins and Akt pathway. Nephrol Dial Transplant 29:799–808

Davies JA (1996) Mesenchyme to epithelium transition during development of the mammalian kidney tubule. Acta Anat (Basel) 156:187–201

Deng Y, Guo Y, Liu P, Zeng R, Ning Y, Pei G, Li Y, Chen M, Guo S, Li X, Han M, Xu G (2016) Blocking protein phosphatase 2A signaling prevents endothelial-to-mesenchymal transition and renal fibrosis: a peptide-based drug therapy. Sci Rep 6:19821

Deng YH, Pu CL, Li YC, Zhu J, Xiang C, Zhang MM, Guo CB (2011) Analysis of biliary epithelial-mesenchymal transition in portal tract fibrogenesis in biliary atresia. Dig Dis Sci 56:731–740

Djamali A, Reese S, Yracheta J, Oberley T, Hullett D, Becker B (2005) Epithelial-to-mesenchymal transition and oxidative stress in chronic allograft nephropathy. Am J Transplant 5: 500–509

Dominguez D, Montserrat-Sentis B, Virgos-Soler A, Guaita S, Grueso J, Porta M, Puig I, Baulida J, Franci C, Garcia De Herreros A (2003) Phosphorylation regulates the subcellular location and activity of the snail transcriptional repressor. Mol Cell Biol 23:5078–5089

Dong C, Wu Y, Wang Y, Wang C, Kang T, Rychahou PG, Chi YI, Evers BM, Zhou BP (2013) Interaction with Suv39H1 is critical for Snail-mediated E-cadherin repression in breast cancer. Oncogene 32:1351–1362

Du C, Zhang C, Hassan S, Biswas MH, Balaji KC (2010) Protein kinase D1 suppresses epithelial-to-mesenchymal transition through phosphorylation of snail. Cancer Res 70:7810–7819

Grande MT, Sanchez-Laorden B, Lopez-Blau C, De Frutos CA, Boutet A, Arevalo M, Rowe RG, Weiss SJ, Lopez-Novoa JM, Nieto MA (2015) Snail1-induced partial epithelial-to-mesenchymal transition drives renal fibrosis in mice and can be targeted to reverse established disease. Nat Med 21:989–997

Hashimoto N, Phan SH, Imaizumi K, Matsuo M, Nakashima H, Kawabe T, Shimokata K, Hasegawa Y (2010) Endothelial-mesenchymal transition in bleomycin-induced pulmonary fibrosis. Am J Respir Cell Mol Biol 43:161–172

Hay ED (2005) The mesenchymal cell, its role in the embryo, and the remarkable signaling mechanisms that create it. Dev Dyn 233:706–720

Hay ED, Zuk A (1995) Transformations between epithelium and mesenchyme: normal, pathological, and experimentally induced. Am J Kidney Dis 26:678–690

He J, Xu Y, Koya D, Kanasaki K (2013) Role of the endothelial-to-mesenchymal transition in renal fibrosis of chronic kidney disease. Clin Exp Nephrol 17:488–497

Herranz N, Pasini D, Diaz VM, Franci C, Gutierrez A, Dave N, Escriva M, Hernandez-Munoz I, Di Croce L, Helin K, Garcia De Herreros A, Peiro S (2008) Polycomb complex 2 is required for E-cadherin repression by the Snail1 transcription factor. Mol Cell Biol 28:4772–4781

Hertig A, Verine J, Mougenot B, Jouanneau C, Ouali N, Sebe P, Glotz D, Ancel PY, Rondeau E, Xu-Dubois YC (2006) Risk factors for early epithelial to mesenchymal transition in renal grafts. Am J Transplant 6:2937–2946

Hinz B, Phan SH, Thannickal VJ, Galli A, Bochaton-Piallat ML, Gabbiani G (2007) The myofibroblast: one function, multiple origins. Am J Pathol 170:1807–1816

Holian J, Qi W, Kelly DJ, Zhang Y, Mreich E, Pollock CA, Chen XM (2008) Role of Kruppel-like factor 6 in transforming growth factor-beta1-induced epithelial-mesenchymal transition of proximal tubule cells. Am J Physiol Renal Physiol 295:F1388–F1396

Huber MA, Kraut N, Beug H (2005) Molecular requirements for epithelial-mesenchymal transition during tumor progression. Curr Opin Cell Biol 17:548–558

Humphreys BD, Lin SL, Kobayashi A, Hudson TE, Nowlin BT, Bonventre JV, Valerius MT, Mcmahon AP, Duffield JS (2010) Fate tracing reveals the pericyte and not epithelial origin of myofibroblasts in kidney fibrosis. Am J Pathol 176:85–97

Ikenouchi J, Matsuda M, Furuse M, Tsukita S (2003) Regulation of tight junctions during the epithelium-mesenchyme transition: direct repression of the gene expression of claudins/occludin by Snail. J Cell Sci 116:1959–1967

Ikeuchi M, Tsutsui H, Shiomi T, Matsusaka H, Matsushima S, Wen J, Kubota T, Takeshita A (2004) Inhibition of TGF-beta signaling exacerbates early cardiac dysfunction but prevents late remodeling after infarction. Cardiovasc Res 64:526–535

Iwano M, Plieth D, Danoff TM, Xue C, Okada H, Neilson EG (2002) Evidence that fibroblasts derive from epithelium during tissue fibrosis. J Clin Invest 110:341–350

Kalluri R, Neilson EG (2003) Epithelial-mesenchymal transition and its implications for fibrosis. J Clin Invest 112:1776–1784

Kalluri R, Weinberg RA (2009) The basics of epithelial-mesenchymal transition. J Clin Invest 119:1420–1428

Kanasaki K, Shi S, Kanasaki M, He J, Nagai T, Nakamura Y, Ishigaki Y, Kitada M, Srivastava SP, Koya D (2014) Linagliptin-mediated DPP-4 inhibition ameliorates kidney fibrosis in streptozotocin-induced diabetic mice by inhibiting endothelial-to-mesenchymal transition in a therapeutic regimen. Diabetes 63:2120–2131

Kida Y, Asahina K, Teraoka H, Gitelman I, Sato T (2007) Twist relates to tubular epithelial-mesenchymal transition and interstitial fibrogenesis in the obstructed kidney. J Histochem Cytochem 55:661–673

Kondo M, Cubillo E, Tobiume K, Shirakihara T, Fukuda N, Suzuki H, Shimizu K, Takehara K, Cano A, Saitoh M, Miyazono K (2004) A role for Id in the regulation of TGF-beta-induced epithelial-mesenchymal transdifferentiation. Cell Death Differ 11:1092–1101

Kovacic JC, Mercader N, Torres M, Boehm M, Fuster V (2012) Epithelial-to-mesenchymal and endothelial-to-mesenchymal transition: from cardiovascular development to disease. Circulation 125:1795–1808

Lamouille S, Xu J, Derynck R (2014) Molecular mechanisms of epithelial-mesenchymal transition. Nat Rev Mol Cell Biol 15:178–196

Lan HY (2003) Tubular epithelial-myofibroblast transdifferentiation mechanisms in proximal tubule cells. Curr Opin Nephrol Hypertens 12:25–29

Lebleu VS, Taduri G, O'Connell J, Teng Y, Cooke VG, Woda C, Sugimoto H, Kalluri R (2013) Origin and function of myofibroblasts in kidney fibrosis. Nat Med 19:1047–1053

Lebrin F, Goumans MJ, Jonker L, Carvalho RL, Valdimarsdottir G, Thorikay M, Mummery C, Arthur HM, Ten Dijke P (2004) Endoglin promotes endothelial cell proliferation and TGF-beta/ALK1 signal transduction. EMBO J 23:4018–4028

Lee K, Nelson CM (2012) New insights into the regulation of epithelial-mesenchymal transition and tissue fibrosis. Int Rev Cell Mol Biol 294:171–221

Lee K, Gjorevski N, Boghaert E, Radisky DC, Nelson CM (2011) Snail1, Snail2, and E47 promote mammary epithelial branching morphogenesis. EMBO J 30:2662–2674

Li J, Qu X, Bertram JF (2009) Endothelial-myofibroblast transition contributes to the early development of diabetic renal interstitial fibrosis in streptozotocin-induced diabetic mice. Am J Pathol 175:1380–1388

Li JH, Wang W, Huang XR, Oldfield M, Schmidt AM, Cooper ME, Lan HY (2004) Advanced glycation end products induce tubular epithelial-myofibroblast transition through the RAGE-ERK1/2 MAP kinase signaling pathway. Am J Pathol 164:1389–1397

Li B, Zheng YW, Sano Y, Taniguchi H (2011) Evidence for mesenchymal-epithelial transition associated with mouse hepatic stem cell differentiation. PLoS One 6:e17092

Lin SL, Kisseleva T, Brenner DA, Duffield JS (2008) Pericytes and perivascular fibroblasts are the primary source of collagen-producing cells in obstructive fibrosis of the kidney. Am J Pathol 173:1617–1627

Lin T, Ponn A, Hu X, Law BK, Lu J (2010) Requirement of the histone demethylase LSD1 in Snai1-mediated transcriptional repression during epithelial-mesenchymal transition. Oncogene 29:4896–4904

Liu Y (2004) Epithelial to mesenchymal transition in renal fibrogenesis: pathologic significance, molecular mechanism, and therapeutic intervention. J Am Soc Nephrol 15:1–12

Liu Y (2010) New insights into epithelial-mesenchymal transition in kidney fibrosis. J Am Soc Nephrol 21:212–222

Liu Y, Rajur K, Tolbert E, Dworkin LD (2000) Endogenous hepatocyte growth factor ameliorates chronic renal injury by activating matrix degradation pathways. Kidney Int 58:2028–2043

Lo HW, Hsu SC, Xia W, Cao X, Shih JY, Wei Y, Abbruzzese JL, Hortobagyi GN, Hung MC (2007) Epidermal growth factor receptor cooperates with signal transducer and activator of transcription 3 to induce epithelial-mesenchymal transition in cancer cells via up-regulation of TWIST gene expression. Cancer Res 67:9066–9076

Lopez-Novoa JM, Nieto MA (2009) Inflammation and EMT: an alliance towards organ fibrosis and cancer progression. EMBO Mol Med 1:303–314

Lovisa S, Lebleu VS, Tampe B, Sugimoto H, Vadnagara K, Carstens JL, Wu CC, Hagos Y, Burckhardt BC, Pentcheva-Hoang T, Nischal H, Allison JP, Zeisberg M, Kalluri R (2015) Epithelial-to-mesenchymal transition induces cell cycle arrest and parenchymal damage in renal fibrosis. Nat Med 21:998–1009

Lu J, Landerholm TE, Wei JS, Dong XR, Wu SP, Liu X, Nagata K, Inagaki M, Majesky MW (2001) Coronary smooth muscle differentiation from proepicardial cells requires rhoA-mediated actin reorganization and p160 rho-kinase activity. Dev Biol 240:404–418
Markwald RR, Fitzharris TP, Smith WN (1975) Sturctural analysis of endocardial cytodifferentiation. Dev Biol 42:160–180
Miyazawa K, Iwaya K, Kuroda M, Harada M, Serizawa H, Koyanagi Y, Sato Y, Mizokami Y, Matsuoka T, Mukai K (2000) Nuclear accumulation of beta-catenin in intestinal-type gastric carcinoma: correlation with early tumor invasion. Virchows Arch 437:508–513
Nakajima Y, Yamagishi T, Hokari S, Nakamura H (2000) Mechanisms involved in valvuloseptal endocardial cushion formation in early cardiogenesis: roles of transforming growth factor (TGF)-beta and bone morphogenetic protein (BMP). Anat Rec 258:119–127
Nakaya Y, Kuroda S, Katagiri YT, Kaibuchi K, Takahashi Y (2004) Mesenchymal-epithelial transition during somitic segmentation is regulated by differential roles of Cdc42 and Rac1. Dev Cell 7:425–438
Nath KA (1992) Tubulointerstitial changes as a major determinant in the progression of renal damage. Am J Kidney Dis 20:1–17
Neilson EG (2007) Plasticity, nuclear diapause, and a requiem for the terminal differentiation of epithelia. J Am Soc Nephrol 18:1995–1998
Ng YY, Huang TP, Yang WC, Chen ZP, Yang AH, Mu W, Nikolic-Paterson DJ, Atkins RC, Lan HY (1998) Tubular epithelial-myofibroblast transdifferentiation in progressive tubulointerstitial fibrosis in 5/6 nephrectomized rats. Kidney Int 54:864–876
Nightingale J, Patel S, Suzuki N, Buxton R, Takagi KI, Suzuki J, Sumi Y, Imaizumi A, Mason RM, Zhang Z (2004) Oncostatin M, a cytokine released by activated mononuclear cells, induces epithelial cell-myofibroblast transdifferentiation via Jak/Stat pathway activation. J Am Soc Nephrol 15:21–32
Nishitani Y, Iwano M, Yamaguchi Y, Harada K, Nakatani K, Akai Y, Nishino T, Shiiki H, Kanauchi M, Saito Y, Neilson EG (2005) Fibroblast-specific protein 1 is a specific prognostic marker for renal survival in patients with IgAN. Kidney Int 68:1078–1085
Peinado H, Ballestar E, Esteller M, Cano A (2004) Snail mediates E-cadherin repression by the recruitment of the Sin3A/histone deacetylase 1 (HDAC1)/HDAC2 complex. Mol Cell Biol 24:306–319
Peinado H, Olmeda D, Cano A (2007) Snail, Zeb and bHLH factors in tumour progression: an alliance against the epithelial phenotype? Nat Rev Cancer 7:415–428
Peinado H, Quintanilla M, Cano A (2003) Transforming growth factor beta-1 induces snail transcription factor in epithelial cell lines: mechanisms for epithelial mesenchymal transitions. J Biol Chem 278:21113–21123
Phan SH (2008) Biology of fibroblasts and myofibroblasts. Proc Am Thorac Soc 5:334–337
Piera-Velazquez S, Li Z, Jimenez SA (2011) Role of endothelial-mesenchymal transition (EndoMT) in the pathogenesis of fibrotic disorders. Am J Pathol 179:1074–1080
Rastaldi MP, Ferrario F, Giardino L, Dell'Antonio G, Grillo C, Grillo P, Strutz F, Muller GA, Colasanti G, D'Amico G (2002) Epithelial-mesenchymal transition of tubular epithelial cells in human renal biopsies. Kidney Int 62:137–146
Rockey DC, Bell PD, Hill JA (2015) Fibrosis—a common pathway to organ injury and failure. N Engl J Med 373:96
Saito A (2013) EMT and EndMT: regulated in similar ways? J Biochem 153:493–495
Sanford LP, Ormsby I, Gittenberger-De Groot AC, Sariola H, Friedman R, Boivin GP, Cardell EL, Doetschman T (1997) TGFbeta2 knockout mice have multiple developmental defects that are non-overlapping with other TGFbeta knockout phenotypes. Development 124:2659–2670
Shimizu M, Kondo S, Urushihara M, Takamatsu M, Kanemoto K, Nagata M, Kagami S (2006) Role of integrin-linked kinase in epithelial-mesenchymal transition in crescent formation of experimental glomerulonephritis. Nephrol Dial Transplant 21:2380–2390
Strutz F, Okada H, Lo CW, Danoff T, Carone RL, Tomaszewski JE, Neilson EG (1995) Identification and characterization of a fibroblast marker: FSP-1. J Cell Biol 130:393–405
Strutz F, Zeisberg M, Ziyadeh FN, Yang CQ, Kalluri R, Muller GA, Neilson EG (2002) Role of basic fibroblast growth factor-2 in epithelial-mesenchymal transformation. Kidney Int 61: 1714–1728

Thiery JP (2002) Epithelial-mesenchymal transitions in tumour progression. Nat Rev Cancer 2: 442–454

Thuault S, Valcourt U, Petersen M, Manfioletti G, Heldin CH, Moustakas A (2006) Transforming growth factor-beta employs HMGA2 to elicit epithelial-mesenchymal transition. J Cell Biol 174:175–183

Vincent T, Neve EP, Johnson JR, Kukalev A, Rojo F, Albanell J, Pietras K, Virtanen I, Philipson L, Leopold PL, Crystal RG, De Herreros AG, Moustakas A, Pettersson RF, Fuxe J (2009) A SNAIL1-SMAD3/4 transcriptional repressor complex promotes TGF-beta mediated epithelial-mesenchymal transition. Nat Cell Biol 11:943–950

Von Gise A, Pu WT (2012) Endocardial and epicardial epithelial to mesenchymal transitions in heart development and disease. Circ Res 110:1628–1645

Wu Y, Deng J, Rychahou PG, Qiu S, Evers BM, Zhou BP (2009a) Stabilization of snail by NF-kappaB is required for inflammation-induced cell migration and invasion. Cancer Cell 15: 416–428

Wu Y, Evers BM, Zhou BP (2009b) Small C-terminal domain phosphatase enhances snail activity through dephosphorylation. J Biol Chem 284:640–648

Wynn TA (2008) Cellular and molecular mechanisms of fibrosis. J Pathol 214:199–210

Xavier S, Vasko R, Matsumoto K, Zullo JA, Chen R, Maizel J, Chander PN, Goligorsky MS (2015) Curtailing endothelial TGF-beta signaling is sufficient to reduce endothelial-mesenchymal transition and fibrosis in CKD. J Am Soc Nephrol 26:817–829

Xu J, Lamouille S, Derynck R (2009) TGF-beta-induced epithelial to mesenchymal transition. Cell Res 19:156–172

Yang J, Liu Y (2001) Dissection of key events in tubular epithelial to myofibroblast transition and its implications in renal interstitial fibrosis. Am J Pathol 159:1465–1475

Yang J, Weinberg RA (2008) Epithelial-mesenchymal transition: at the crossroads of development and tumor metastasis. Dev Cell 14:818–829

Yang J, Mani SA, Donaher JL, Ramaswamy S, Itzykson RA, Come C, Savagner P, Gitelman I, Richardson A, Weinberg RA (2004) Twist, a master regulator of morphogenesis, plays an essential role in tumor metastasis. Cell 117:927–939

Yang Z, Zhang X, Gang H, Li X, Li Z, Wang T, Han J, Luo T, Wen F, Wu X (2007) Up-regulation of gastric cancer cell invasion by Twist is accompanied by N-cadherin and fibronectin expression. Biochem Biophys Res Commun 358:925–930

Yilmaz M, Christofori G (2009) EMT, the cytoskeleton, and cancer cell invasion. Cancer Metastasis Rev 28:15–33

Yook JI, Li XY, Ota I, Hu C, Kim HS, Kim NH, Cha SY, Ryu JK, Choi YJ, Kim J, Fearon ER, Weiss SJ (2006) A Wnt-Axin2-GSK3beta cascade regulates Snail1 activity in breast cancer cells. Nat Cell Biol 8:1398–1406

Yu W, Kamara H, Svoboda KK (2008) The role of twist during palate development. Dev Dyn 237: 2716–2725

Zeisberg M, Neilson EG (2009) Biomarkers for epithelial-mesenchymal transitions. J Clin Invest 119:1429–1437

Zeisberg M, Hanai J, Sugimoto H, Mammoto T, Charytan D, Strutz F, Kalluri R (2003) BMP-7 counteracts TGF-beta1-induced epithelial-to-mesenchymal transition and reverses chronic renal injury. Nat Med 9:964–968

Zeisberg M, Maeshima Y, Mosterman B, Kalluri R (2002) Renal fibrosis. Extracellular matrix microenvironment regulates migratory behavior of activated tubular epithelial cells. Am J Pathol 160:2001–2008

Zeisberg EM, Potenta SE, Sugimoto H, Zeisberg M, Kalluri R (2008) Fibroblasts in kidney fibrosis emerge via endothelial-to-mesenchymal transition. J Am Soc Nephrol 19:2282–2287

Zeisberg EM, Tarnavski O, Zeisberg M, Dorfman AL, Mcmullen JR, Gustafsson E, Chandraker A, Yuan X, Pu WT, Roberts AB, Neilson EG, Sayegh MH, Izumo S, Kalluri R (2007) Endothelial-to-mesenchymal transition contributes to cardiac fibrosis. Nat Med 13:952–961

Zhang G, Kernan KA, Collins SJ, Cai X, Lopez-Guisa JM, Degen JL, Shvil Y, Eddy AA (2007) Plasmin(ogen) promotes renal interstitial fibrosis by promoting epithelial-to-mesenchymal transition: role of plasmin-activated signals. J Am Soc Nephrol 18:846–859

Zhou BP, Deng J, Xia W, Xu J, Li YM, Gunduz M, Hung MC (2004) Dual regulation of Snail by GSK-3beta-mediated phosphorylation in control of epithelial-mesenchymal transition. Nat Cell Biol 6:931–940

MIX
Papier aus verantwortungsvollen Quellen
Paper from responsible sources
FSC® C105338

If you have any concerns about our products,
you can contact us on
ProductSafety@springernature.com

In case Publisher is established outside the EU,
the EU authorized representative is:
Springer Nature Customer Service Center GmbH
Europaplatz 3, 69115 Heidelberg, Germany

Printed by Libri Plureos GmbH
in Hamburg, Germany